New Engineering Mathematics

Volume II

New Engineering Mathematics

Volume II

A Chandra Babu
C R Seshan

Alpha Science International Ltd.
Oxford, U.K.

A Chandra Babu
C R Seshan
Department of Mathematics
The American College
Madurai, India

Alpha Science International Ltd.
7200 The Quorum, Oxford Business Park North
Garsington Road, Oxford OX4 2JZ, U.K.

Printed from the camera-ready copy provided by the Authors

ISBN 1-84265-292-3

Printed in India

PREFACE

We have great pleasure in bringing out **New Engineering Mathematics** in three volumes.

Each Chapter begins with an introduction, the relevant definitions and concepts. A detailed discussion of the theory, inclusion of a large number of model problems and exercises with separate short and long answer questions are special features of these volumes.

Volume I contains 292 short answer questions and 726 long answer questions, of which 231 are solved as examples. Volume II contains 467 short answer questions and 832 long answer questions, of which 295 are solved as examples, to provide enough practice and to generate confidence in the subject. Volume III contains 206 short answer questions and 619 long answer questions, of which 184 are solved as examples, to provide enough practice and initiate further study.

A close reading of the theory portions is sufficient to answer all the questions in Part-A of the exercises. Further a student who works out the examples given under each topic will be able to solve all the problems in the exercises. The example problems are chosen with great care so that they cover necessary information and techniques for problem solving.

We hope this book will be useful to students of all levels. Critical evaluation and suggestions for improvement from students and teachers will be thankfully acknowledged and adapted in future editions.

We wish to express our sincere thanks to the post graduate students of American College Mr. Hasan Mohammed K. and Mr. Elango Valan S., who took great care in typing the manuscript. We are also thankful to Narosa Publishing House, who have brought out this book in a very short span of time. We have great pleasure to dedicate this book to our parents and teachers.

A Chandra Babu
C R Seshan

CONTENTS

VOLUME II

VOLUME III

CHAPTER 1

MULTIPLE INTEGRALS

1.0 INTRODUCTION

Integration can be defined as the reverse process of differentiation or as the limit of a sum. Let the function f(x) be defined on a closed interval [a, b] and is piecewise continuous there. Then the Riemann integral $\int_a^b f(x)dx$ is defined as

$$\int_a^b f(x)dx = \lim_{\substack{n\to\infty \\ \text{Max}\,\lambda_i \to 0}} \sum_{i=1}^{n} f(s_i)(x_i - x_{i-1}),$$ where $a = x_0 < x_1 < x_2 < \ldots < x_n = b$ is a partition of [a, b] into closed subintervals, $s_i \in [x_{i-1}, x_i]$ and $\lambda_i = |x_i - x_{i-1}|$.

The extensions of Riemann integral to two and three dimensions are called double and triple integrals or in general multiple integrals. In this chapter, we discuss the various methods of evaluating the double integrals $\iint_R f(x, y)\, dx\, dy$ and the triple integrals $\iiint_V f(x, y, z)\, dx\, dy\, dz$, when f is a function, continuous inside and on the boundary of the region R or V. One can also think of the multiple integral in the n-dimensional space $\mathbb{R}^n$, denoted by $\iint_V \cdots \int f(x_1, x_2, \cdots, x_n)\, dx_1\, dx_2 \cdots dx_n$

1.1 DOUBLE INTEGRALS

Definition 1.1.1 Let f(x, y) be a continuous function in the closed and bounded region R in the two dimensional space $\mathbb{R}^2$. Let the region R be subdivided in any manner into n subregions $\Delta R_1, \Delta R_2, \ldots, \Delta R_n$ of areas $\Delta A_1, \Delta A_2, \ldots, \Delta A_n$. Let (ς_i, η_i) be any point in the ith subregion ΔR_i. Then the **double integral** of f(x, y) over the region R is defined as

$$\lim_{\substack{n\to\infty \\ \text{Max}|\Delta A_i| \to 0}} \sum_{i=1}^{n} f(\varsigma_i, \eta_i)\Delta A_i$$ and is denoted by $\iint_R f(x, y)\, dA$ or $\iint_R f(x, y)\, dx\, dy$

1.1.1 PROPERTIES OF DOUBLE INTEGRALS

1. If f(x, y) and g(x, y) are continuous function in the region R and a and b are real numbers, then $\iint_R [af(x,y) \pm bg(x,y)]\,dx\,dy = a\iint_R f(x,y)dx\,dy + b\iint_R g(x,y)\,dx\,dy$

2. If $f(x, y) \geq 0$ for all $(x, y) \in R$, then $\iint_R f(x,y)\,dx\,dy \geq 0$

3. If f(x, y) is continuous in the region R $\left|\iint_R f(x,y)\,dx\,dy\right| \leq \iint_R |f(x,y)|\,dx\,dy$

4. If $m \leq f(x, y) \leq M$ for all $(x, y) \in R$ and A is the area of the region R, then $mA \leq \iint_R f(x,y)dxdy \leq MA$

5. If f(x, y) and g(x, y) are any two continuous functions in the region R and $f(x, y) \leq g(x, y)$ for all $(x, y) \in R$, then $\iint_R f(x,y)\,dx\,dy \leq \iint_R g(x,y)\,dx\,dy$

1.1.2 APPLICATIONS OF DOUBLE INTEGRALS

1. Area of a region.

If f(x, y) = 1 then $\iint_R dx\,dy = \lim_{\substack{n\to\infty \\ \text{Max}|\Delta A_i|\to 0}} \sum_{i=1}^{n} \Delta A_i = A$, the area of the region R.

2. Volume under a surface.

If z = f(x, y) is a surface in $\mathbb{R}^3$ and z is non-negative over the region R then $\iint_R f(x,y)\,dx\,dy = \iint_R z\,dx\,dy$ gives the volume of the region below the surface and above the region R in the xy-plane, and bounded by the cylinder $0 \leq z \leq f(x,y), (x,y) \in \partial R$, the boundary of the region R.

3. Average.

$\frac{1}{A}\iint_R f(x,y)\,dx\,dy$ gives the average value of f(x, y) over R, where A is the area of the region R.

4. Centre of gravity.

Let f(x, y) = ρ(x, y) be the density function (mass per unit area) of a distribution of mass in the region R in the xy-plane. Then

(i) $M = \iint_R f(x,y)\,dx\,dy$ gives the **total mass** of R.

(ii) $\bar{x} = \frac{1}{M}\iint_R x.f(x,y)\,dx\,dy,\ \bar{y} = \frac{1}{M}\iint_R y.f(x,y)\,dx\,dy$ gives the coordinates of the **centre of gravity** $(\bar{x}, \bar{y})$ of the mass M in R.

5. Moment of Inertia.

$I_x = \iint_R y^2.f(x,y)\,dx\,dy$ and $I_y = \iint_R x^2.f(x,y)\,dx\,dy$ gives the **moment of inertia** of the mass M in R about the x and y axes respectively. $I_0 = I_x + I_y$ is the moment of inertia of the mass about the origin.

1.1.3 EVALUATION OF DOUBLE INTEGRALS

Let f(x, y) be a function defined and continuous in a region R in the xy-plane. To evaluate the double integral $I = \iint_R f(x,y)\,dx\,dy$, (1)

one has to perform two successive integrations. We consider the following three cases.

Case 1. All the limits are constants.

$R = \{ (x, y) / a \le x \le b, c \le y \le d \}$

R is a rectangular region with sides parallel to the coordinate axes. For any fixed $x \in [a, b]$, consider the integral $\int_c^d f(x,y)\,dy$. The value of this integral is a function of x and hence can be integrated w.r.t x and we get $\int_a^b \left[\int_c^d f(x,y)\,dy \right] dx$ (2).

Here, we first integrate along the vertical strip PQ (Fig. 1.1) and then slide it from AD to BC parallel to the x-axis. Similarly we can define $\int_c^d \left[\int_a^b f(x,y)\,dx \right] dy$ (3)

i.e., we first integrate along the horizontal strip P′Q′ and then slide it from AB to DC, parallel to the y-axis. The integrals (2) and (3) are called **iterated integrals.** Both the integrals give the same value. Thus for double integrals over a rectangular region with sides parallel to the coordinate axes or equivalently, for double integrals in which both pairs of limits are constants, it does not matter whether we first integrate w.r.t x and then w.r.t y or vice versa.

Case 2. Limits for x are constants.

$R = \{(x, y) / \varphi(x) \le y \le \chi(x), a \le x \le b\}$ where $\varphi(x)$ and $\chi(x)$ are continuous functions such that $\varphi(x) \le \chi(x)$ for all $x \in [a, b]$. (Fig. 1.2) Then

$$I = \int_{x=a}^{b} \left[\int_{y=\varphi(x)}^{\chi(x)} f(x,y)\,dy \right] dx \qquad (4)$$

While evaluating the inner integral in (4), x is treated as a constant. The iterated integral in the right hand side of (4) is also written as $\int_a^b \int_{\varphi(x)}^{\chi(x)} f(x,y)\,dy\,dx$ or $\int_a^b dx \int_{\varphi(x)}^{\chi(x)} f(x,y)\,dy$.

Case 3. Limits for y are constants.

R = {(x, y) / φ(y) ≤ x ≤ χ(y), c ≤ y ≤ d} where φ(y) and χ(y) are continuous functions such that φ(y) ≤ χ(y) for all y ∈ [c, d]. (Fig. 1.3) Then

$$I = \int_{y=c}^{d} \left[\int_{x=\varphi(y)}^{\chi(y)} f(x,y)\,dx \right] dy \qquad (5)$$

While evaluating the inner integral in (5), y is treated as a constant. The iterated integral in the right hand side of (5) can also be written as $\int_{c}^{d}\int_{\varphi(y)}^{\chi(y)} f(x,y)\,dx\,dy$ or $\int_{c}^{d} dy \int_{\varphi(y)}^{\chi(y)} f(x,y)\,dx$.

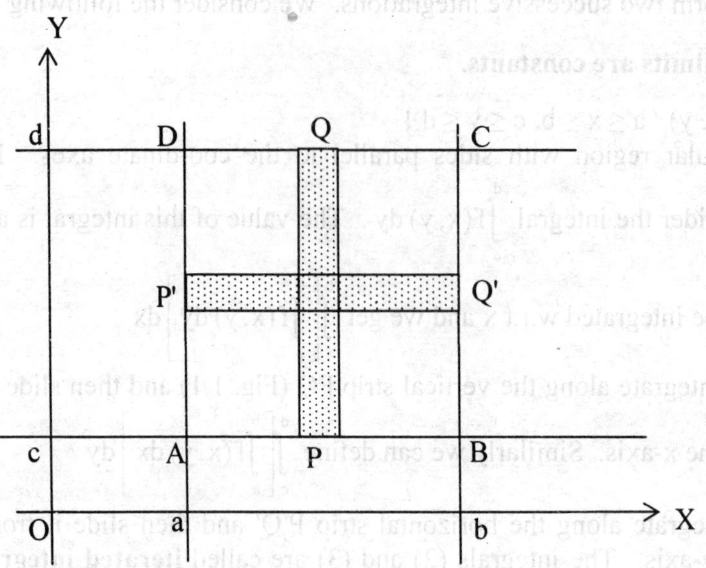

Fig. 1.1

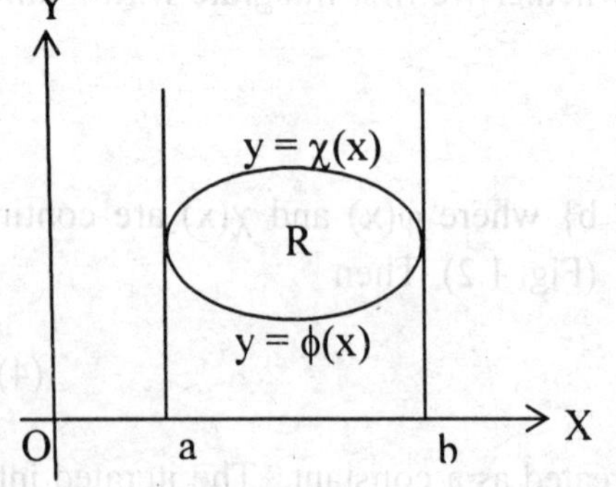

Fig. 1.2

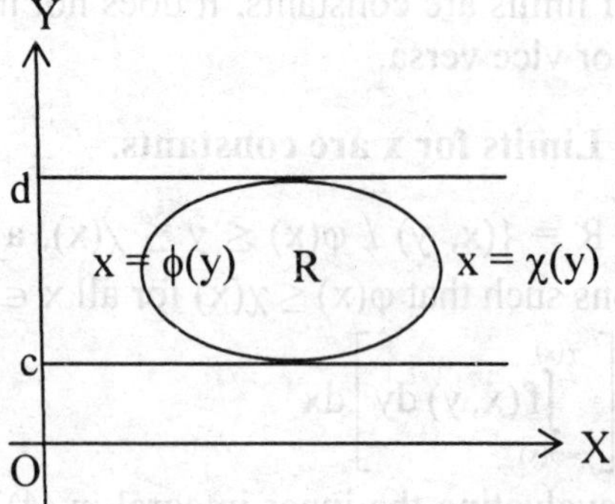

Fig. 1.3

If the region of integration R is such that it can not be written in one of the forms discussed above, then we divide R into a finite number of subregions $R_1, R_2, \ldots, R_m$ such that each of these subregions can be represented in one of the above forms and we get the double integral over R by adding the integrals over these subregions.

i.e., $$\iint_R f(x,y)\,dx\,dy = \sum_{i=1}^{m}\left[\iint_{R_i} f(x,y)\,dx\,dy\right] \qquad (6)$$

For example, in Fig 1.4, $R = R_1 \cup R_2$ is a partition of R and we have

$$\iint_R f(x,y)\,dx\,dy = \iint_{R_1} f(x,y)\,dx\,dy + \iint_{R_2} f(x,y)\,dx\,dy$$

$$= \int_a^c\left[\int_{\varphi_1(x)}^{\chi(x)} f(x,y)\,dy\right]dx + \int_c^b\left[\int_{\varphi_2(x)}^{\chi(x)} f(x,y)\,dy\right]dx$$

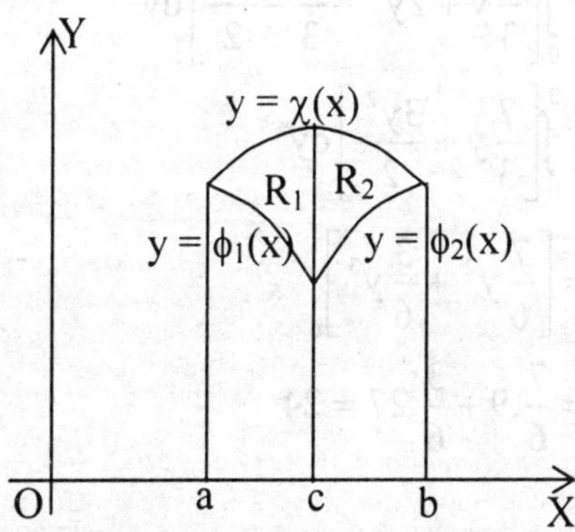

Figure 1.4

1.1.4 DOUBLE INTEGRATION IN POLAR COORDINATES

(i) To evaluate the double integral $\int_{\theta_1}^{\theta_2}\int_{r_1}^{r_2} f(r,\theta)\,dr\,d\theta$ where r_1 and r_2 are functions of θ and θ_1, θ_2 are constants, we first integrate w.r.t r between the limits $r = r_1$ and $r = r_2$ keeping θ fixed and the resulting expression is integrated w.r.t θ from θ_1 to θ_2. The order of integration is changed when θ_1 and θ_2 are functions of r and r_1, r_2 are constants.

(ii) The double integral in Cartesian coordinates $I = \iint_R f(x,y)\,dx\,dy$, can be transformed into a double integral in polar coordinates by substituting $x = r\cos\theta$, $y = r\sin\theta$ and $dx\,dy = r\,dr\,d\theta$. Thus $I = \iint_R f(r\cos\theta, r\sin\theta)\,r\,dr\,d\theta$

Note: If A is the area of the region R enclosed by curves whose equations are in polar coordinates, then $A = \iint_R r\,dr\,d\theta$.

Example 1.1.1 Evaluate the double integral $\int_0^3 \int_1^2 xy(x+y)\,dx\,dy$

Solution:
Since all the limits are constants, order of integration is immaterial.

$$\int_0^3 \int_1^2 xy(x+y)\,dx\,dy = \int_0^3 \left[\int_1^2 (x^2y + xy^2)\,dx \right] dy$$

$$= \int_0^3 \left[\frac{x^3y}{3} + \frac{x^2y^2}{2} \right]_1^2 dy$$

$$= \int_0^3 \left[\frac{8}{3}y + 2y^2 - \frac{y}{3} - \frac{y^2}{2} \right] dy$$

$$= \int_0^3 \left[\frac{7}{3}y + \frac{3y^2}{2} \right] dy$$

$$= \left[\frac{7}{6}y^2 + \frac{3}{6}y^3 \right]_0^3$$

$$= \frac{7}{6}.9 + \frac{3}{6}.27 = 24$$

Example 1.1.2 Evaluate $\int_0^{\pi/2} \int_0^{\infty} \frac{r\,dr\,d\theta}{(r^2+a^2)^2}$

Solution:
Since all the limits are constants, the order of integration is immaterial.

$$I = \int_0^{\pi/2} \left[\int_0^{\infty} \frac{r}{(r^2+a^2)^2} dr \right] d\theta$$

$$= \int_0^{\pi/2} \left[\frac{-1}{2} \frac{1}{r^2+a^2} \right]_0^{\infty} d\theta$$

$$= \int_0^{\pi/2} \frac{1}{2a^2} d\theta$$

$$= \left[\frac{1}{2a^2}\theta. \right]_0^{\pi/2} = \frac{\pi}{4a^2}$$

Example 1.1.3 Evaluate $\int_0^a \int_0^{\sqrt{a^2-x^2}} y^3 dy\, dx$.

Solution:

Since the inner limits are functions of x, first integrate w.r.t. y.

$$I = \int_0^a \left[\int_0^{\sqrt{a^2-x^2}} y^3 dy \right] dx$$

$$= \int_0^a \left[\frac{y^4}{4} \right]_0^{\sqrt{a^2-x^2}} dx$$

$$= \int_0^a \frac{1}{4}\left(a^2 - x^2\right)^2 dx$$

$$= \frac{1}{4}\int_0^a \left(a^4 - 2a^2x^2 + x^4\right) dx$$

$$= \frac{1}{4}\left[a^4x - 2a^2\frac{x^3}{3} + \frac{x^5}{5} \right]_0^a$$

$$= \frac{1}{4}\left[a^5 - \frac{2}{3}a^5 + \frac{a^5}{5} \right] = \frac{2a^5}{15}$$

Example 1.1.4 Evaluate $\int_0^{2\pi} \int_{a\sin\theta}^{a} r\, dr\, d\theta$.

Solution:

Since the inner limits are functions of θ, first integrate w.r.t r.

$$I = \int_0^{2\pi} \left[\int_{a\sin\theta}^{a} r\, dr \right] d\theta$$

$$= \int_0^{2\pi} \left[\frac{r^2}{2} \right]_{a\sin\theta}^{a} d\theta$$

$$= \int_0^{2\pi} \left(\frac{a^2}{2} - \frac{a^2}{2}\sin^2\theta \right) d\theta$$

$$= \frac{a^2}{2}\int_0^{2\pi} \left(1 - \sin^2\theta\right) d\theta$$

$$= \frac{a^2}{2}\int_0^{2\pi} \frac{1+\cos 2\theta}{2} d\theta \qquad \left(\text{Since } 1 - \sin^2\theta = \cos^2\theta = \frac{1+\cos 2\theta}{2}\right)$$

$$= \frac{a^2}{2}\left[\frac{\theta}{2} + \frac{\sin 2\theta}{4}\right]_0^{2\pi}$$

$$= \frac{a^2}{2}[\pi] = \frac{\pi a^2}{2}$$

Example 1.1.5 Evaluate $\iint xy\,dx\,dy$ taken over the positive quadrant bounded by $\frac{x}{a} + \frac{y}{b} = 1$.

Solution:

The region of integration is OAB (Fig 1.5) bounded by the lines $y = 0$, $x = 0$ and $\frac{x}{a} + \frac{y}{b} = 1$

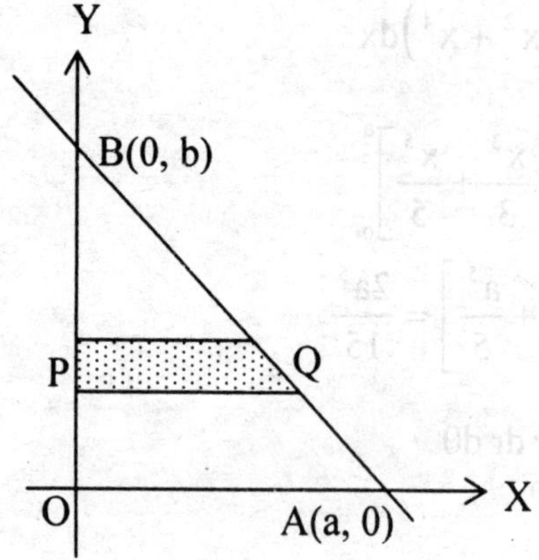

Fig 1.5

First integrate w.r.t x along the strip PQ, keeping y constant. On PQ, the limits for x are $x = 0$ and $x = a(1 - y/b)$. Then integrate w.r.t y and the limits for y are $y = 0$ and $y = b$.

$$\text{Thus } \iint xy\,dx\,dy = \int_0^b \left[\int_0^{a(1-y/b)} xy\,dx\right] dy$$

$$= \int_0^b \left[\frac{x^2 y}{2}\right]_0^{a(1-y/b)} dy$$

$$= \int_0^b \frac{a^2}{2}(1 - y/b)^2 . y\,dy$$

$$= \frac{a^2}{2.b^2}\int_0^b y(b-y)^2\,dy$$

$$= \frac{a^2}{2.b^2}\int_0^b \left(b^2y - 2by^2 + y^3\right)dy$$

$$= \frac{a^2}{2b^2}\left(b^2\frac{y^2}{2} - 2b\frac{y^3}{3} + \frac{y^4}{4}\right)_0^b$$

$$= \frac{a^2}{2b^2}\left(\frac{b^4}{2} - \frac{2}{3}b^4 + \frac{1}{4}b^4\right)_0^b = \frac{a^2b^2}{24}$$

Example 1.1.6 Evaluate $\iint_R y\,dx\,dy$ where the region of integration R is the region in the first quadrant bounded by the ellipse $\frac{x^2}{a^2}+\frac{y^2}{b^2}=1$.

Solution:

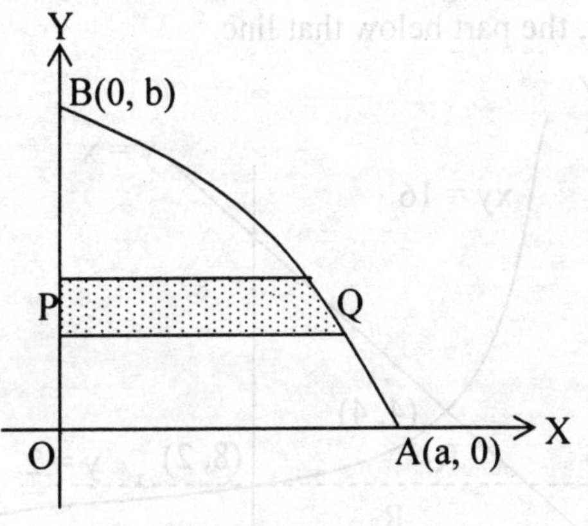

Fig. 1.6

The region of integration is OAB (Fig. 1.6). First integrate w.r.t x, along the strip PQ, keeping y constant. On PQ, the limits for x are x = 0 and $x = \frac{a}{b}\sqrt{b^2 - y^2}$. Then integrate w.r.t y and the limits for y are y = 0 and y = b. Thus,

$$\iint_R y\,dx\,dy = \int_0^b \left[\int_0^{\frac{a}{b}\sqrt{b^2-y^2}} y\,dx\right] dy$$

$$= \int_0^b [y.x]_0^{\frac{a}{b}\sqrt{b^2-y^2}}\,dy$$

$$= \int_0^b \left(y.\tfrac{a}{b}\sqrt{b^2 - y^2}\right)dy$$

$$= \frac{a}{(-2)b}\int_0^b \sqrt{b^2 - y^2}.(-2y)\,dy$$

$$= \frac{-a}{2b}\left[\frac{(b^2 - y^2)^{3/2}}{3/2}\right]_0^b = \frac{a}{2b}.\frac{2}{3}.b^3 = \frac{ab^2}{3}.$$

Example 1.1.7 Evaluate $\iint_R x^2\,dx\,dy$ where R is the region in the first quadrant bounded by the hyperbola xy = 16 and the lines y = x, y = 0 and x = 8.

Solution:

The hyperbola xy = 16 and the line y = x intersect at (4, 4). The hyperbola and the line x = 8 intersect at (8, 2). (Fig. 1.7)

Divide the region R into two subregions R_1 and R_2 where R_1 denotes the part of R lying above the line y = 2 and R_2, the part below that line.

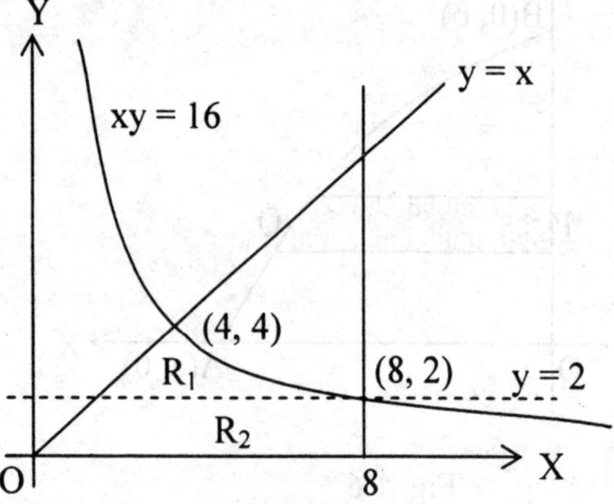

Fig. 1.7

$$\iint_R x^2\,dx\,dy = \iint_{R_1} x^2\,dx\,dy + \iint_{R_2} x^2\,dx\,dy$$

$$= \int_2^4\left[\int_y^{16/y} x^2 dx\right]dy + \int_0^2\left[\int_y^8 x^2 dx\right]dy$$

$$= \int_2^4\left[\frac{x^3}{3}\right]_y^{16/y} dy + \int_0^2\left[\frac{x^3}{3}\right]_y^8 dy$$

$$= \frac{1}{3}\int_2^4 \left(\frac{16^3}{y^3} - y^3\right) dy + \frac{1}{3}\int_0^2 \left(8^3 - y^3\right) dy$$

$$= \frac{1}{3}\left(16^3 \frac{y^{-2}}{-2} - \frac{y^4}{4}\right)_2^4 + \frac{1}{3}\left(8^3.y - \frac{y^4}{4}\right)_0^2$$

$$= \frac{1}{3}(324) + \frac{1}{3}(1020) = 108 + 340 = 448$$

Example 1.1.8 Evaluate $\iint_R \sqrt{x^2 + y^2}\, dx\, dy$ where R is the region in the xy-plane bounded by $x^2 + y^2 = 4$ and $x^2 + y^2 = 9$.

Solution:

Let $I = \iint_R \sqrt{x^2 + y^2}\, dx\, dy$

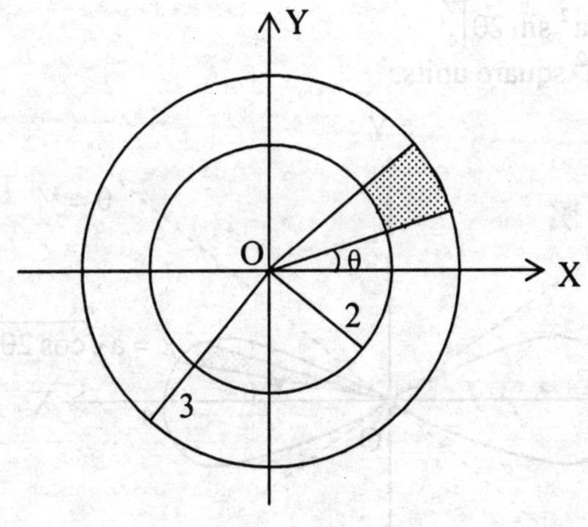

Fig. 1.8

We shall evaluate this by transforming it to polar coordinates by substituting $x = r\cos\theta$, $y = r\sin\theta$ and $dx\, dy = r\, dr\, d\theta$. Then $\sqrt{x^2 + y^2} = r$

$$I = \iint_R r.r.dr.d\theta = \int_0^{2\pi}\left[\int_{r=2}^{3} r^2 dr\right] d\theta$$

$$= \int_0^{2\pi}\left[\frac{r^3}{3}\right]_2^3 d\theta$$

$$= \frac{19}{3}\int_0^{2\pi} d\theta = \frac{19}{3} 2\pi = \frac{38\pi}{3}$$

Example 1.1.9 Find the area of the region bounded by the lemniscate $r^2 = a^2\cos 2\theta$.

Solution:

Area of a region R in the xy-plane is given by $I = \iint_R dxdy$

In polar coordinates, $I = \iint_R r dr d\theta = 4 \times$ Area in the first quadrant (Fig. 1.9)

$$= 4.\int_{\theta=0}^{\pi/4}\left[\int_{r=0}^{a\sqrt{\cos 2\theta}} r dr\right] d\theta$$

$$= 4.\int_{0}^{\pi/4}\left[\frac{r^2}{2}\right]^{a\sqrt{\cos 2\theta}} d\theta$$

$$= 2.\int_{0}^{\pi/4} a^2 \cos 2\theta \, d\theta$$

$$= \left[a^2 \sin 2\theta\right]_0^{\pi/4}$$

$= a^2$ square units.

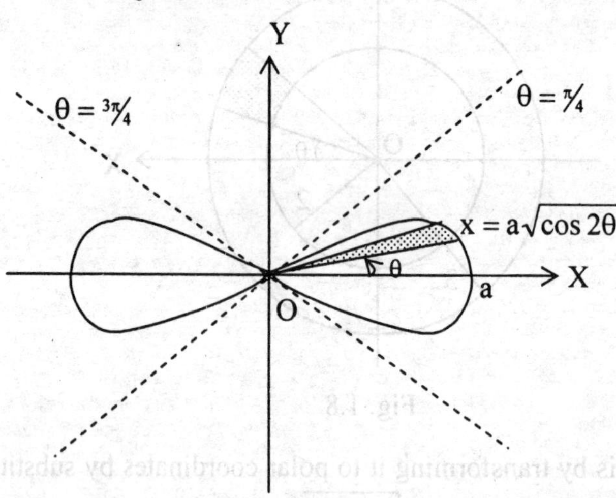

Fig. 1.9

Example 1.1.10 Find the area bounded by the parabolas $y^2 = 4 - x$ and $y^2 = 4 - 4x$.

Solution:

If R is the region bounded by the parabolas, then $A = \iint_R dx\,dy$

The parabolas meet the x-axis at x = 1 and x = 4. Both the parabolas meet the y-axis at the points (0, 2) and (0, -2). The region R is symmetric about the x-axis. (Fig 1.10)

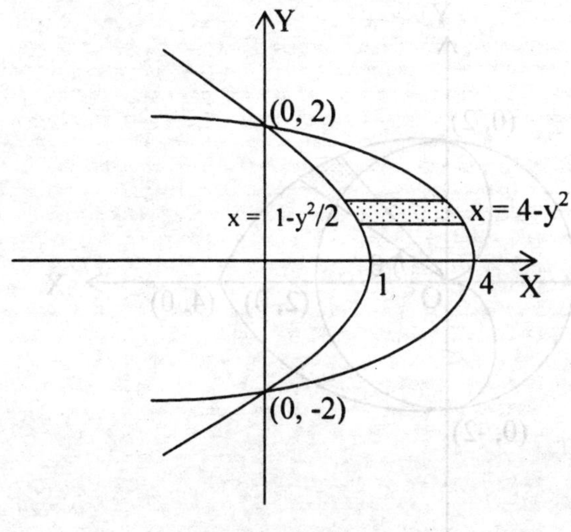

Fig. 1.10

$$\therefore \qquad A = 2\int_0^2 \left[\int_{1-y^2/4}^{4-y^2} dx\right] dy$$

$$= 2\int_0^2 \left[\left(4 - y^2\right) - \left(1 - \tfrac{y^2}{4}\right)\right] dy$$

$$= 2\int_0^2 \left(3 - 3\tfrac{y^2}{4}\right) dy$$

$$= 6\left[y - \tfrac{y^3}{12}\right]_0^2 = 6\left[2 - \tfrac{8}{12}\right] = 8 \text{ square units.}$$

Example 1.1.11 Find the area outside the circle $r = 2$ and inside the cardioid $r = 2(1 + \cos\theta)$.

Solution:

Let R be the region outside the circle and inside the cardioid. Then R is symmetric about the x-axis (Fig. 1.11)

$\therefore$ The required area $A = 2\iint_R r\,dr\,d\theta$

$$= 2\int_0^{\pi/2} \left[\int_2^{2(1+\cos\theta)} r.dr\right] d\theta$$

$$= 2\int_0^{\pi/2} \left[\frac{r^2}{2}\right]_2^{2(1+\cos\theta)} d\theta$$

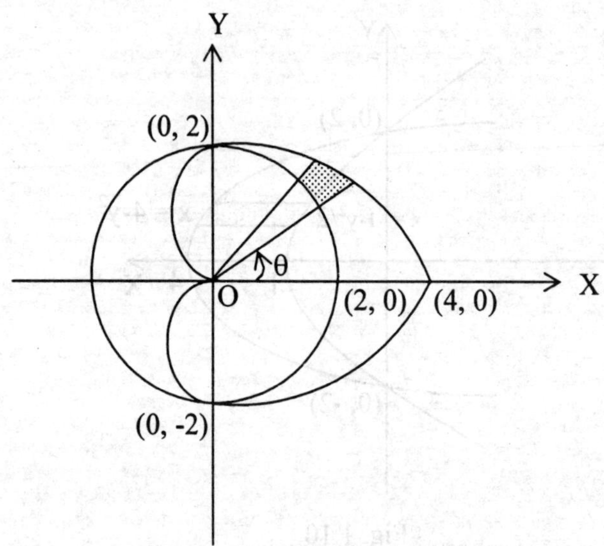

Fig. 1.11

$$= 2\int_0^{\pi/2}\left[4(1+\cos\theta)^2 - 4\right]d\theta$$

$$= 8\int_0^{\pi/2}\left[2\cos\theta + \cos^2\theta\right]d\theta$$

$$= 8\int_0^{\pi/2}\left[2\cos\theta + \frac{1+\cos 2\theta}{2}\right]d\theta$$

$$= 8\left[2\sin\theta + \theta/2 + \frac{\sin 2\theta}{4}\right]_0^{\pi/2} = 8\left[2+\frac{\pi}{4}\right] = 16 + 2\pi \text{ sq. units.}$$

Example 1.1.12 Find the volume of the ellipsoid $\frac{x^2}{a^2}+\frac{y^2}{b^2}+\frac{z^2}{c^2}=1.$

Solution:

Let R be the region in the xy-plane obtained by the projection of the surface $z = c\sqrt{1-\frac{x^2}{a^2}-\frac{y^2}{b^2}}$ in the first octant (Fig 1.12). Then the volume of the ellipsoid is

$$V = 8\iint_R z\,dx\,dy = 8.\int_0^a\left[\int_0^{b\sqrt{1-\frac{x^2}{a^2}}} c\sqrt{1-\frac{x^2}{a^2}-\frac{y^2}{b^2}}\,dy\right]dx$$

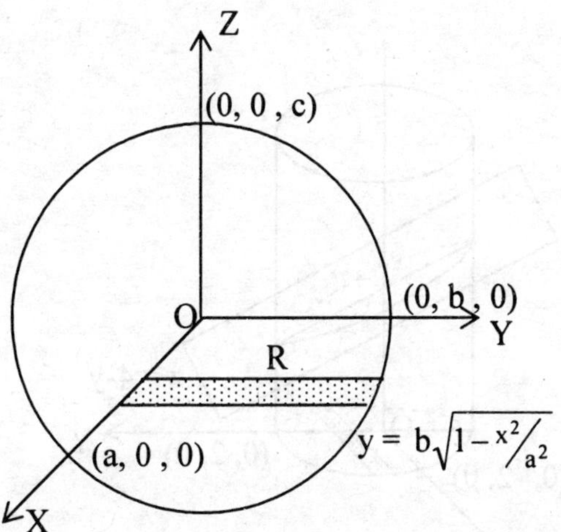

Fig. 1.12

$$= 8.\int_0^a \left[\int_0^{bt} c\sqrt{t^2 - \frac{y^2}{b^2}}\, dy \right] dx \text{, where } t = \sqrt{1 - \frac{x^2}{a^2}}. \quad \text{(keeping x constant)}$$

Put $y = bt\sin\theta$. Then we have,

$$V = 8c \int_0^a \left[\int_0^{\pi/2} t\cos\theta . bt\cos\theta \, d\theta \right] dx$$

$$= 8bc \int_0^a \left[\int_0^{\pi/2} t^2 \cos^2\theta \, d\theta \right] dx$$

$$= 8bc \int_0^a \left[\frac{t^2}{2} \int_0^{\pi/2} (1 + \cos 2\theta)\, d\theta \right] dx$$

$$= 8bc \int_0^a \frac{t^2}{2} \frac{\pi}{2} dx$$

$$= 2bc\pi \int_0^a \left(1 - \frac{x^2}{a^2} \right) dx$$

$$= 2bc\pi \left(x - \frac{x^3}{3a^2} \right)_0^a = 2bc\pi \left(a - \frac{a}{3} \right) = \frac{4}{3}\pi abc \text{ cubic units.}$$

Example 1.1.13 Find the volume bounded by the cylinder $x^2 + y^2 = 4$ and the planes $y + z = 4$ and $z = 0$.

Solution:

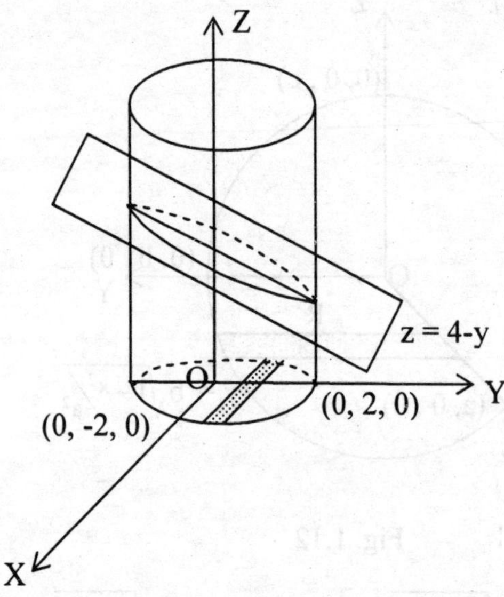

Fig. 1.13

The required volume lies below the surface z = 4-y and is bounded by the cylinder $x^2 + y^2 = 4$ and plane z = 0. Let R be the region in the xy-plane bounded by the circle $x^2 + y^2 = 4$, z = 0.

Then $V = \iint_R z\,dx\,dy$

$$= \int_{-2}^{2}\left[\int_{-\sqrt{4-y^2}}^{\sqrt{4-y^2}}(4-y)dx\right]dy$$

$$= \int_{-2}^{2}\Big[(4-y)x\Big]_{-\sqrt{4-y^2}}^{\sqrt{4-y^2}}dy$$

$$= \int_{-2}^{2}2(4-y)\sqrt{4-y^2}\,dy$$

$$= 16\int_{0}^{2}\sqrt{4-y^2}\,dy \qquad \text{(Second term vanishes being an odd function of y)}$$

$$= 16\int_{0}^{\pi/2}2\cos\theta\, 2\cos\theta\, d\theta \qquad \text{(By putting } y = 2\sin\theta\text{)}$$

$$= 64\int_{0}^{\pi/2}\cos^2\theta\, d\theta$$

$$= 32\int_0^{\pi/2}(1+\cos 2\theta)\,d\theta$$

$$= 32\left[\theta+\frac{\sin 2\theta}{2}\right]_0^{\pi/2} = 16\,\pi \text{ cubic units.}$$

Example 1.1.14 Find the volume bounded above by the paraboloid $x^2 + 4y^2 = z$, below by the plane $z = 0$, and laterally by the cylinders $y^2 = x$ and $x^2 = y$.

Solution:

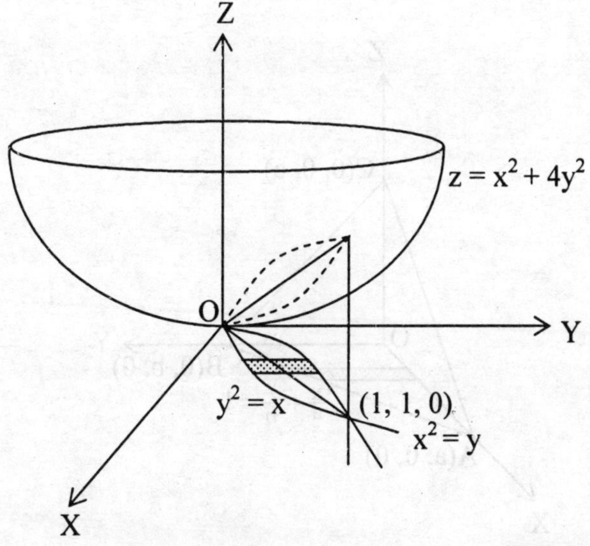

Fig. 1.14

Let R be the region bounded by the parabolas $y^2 = x$ and $x^2 = y$ in the xy-plane (Fig 1.14). The parabolas intersect at (1, 1, 0). The required volume is obtained by integrating the surface $z = x^2 + 4y^2$ over the region R.

i.e., $V = \iint\limits_R z\,dx\,dy$

$$= \int_0^1\left[\int_{x^2}^{\sqrt{x}}(x^2+4y^2)\,dy\right]dx$$

$$= \int_0^1\left[x^2y+4\frac{y^3}{3}\right]_{x^2}^{\sqrt{x}}dx$$

$$= \int_0^1\left[x^{5/2}+\frac{4}{3}x^{3/2}-x^4-\frac{4}{3}x^6\right]dx$$

$$= \left[\frac{x^{7/2}}{7/2} + \frac{4}{3}\frac{x^{5/2}}{5/2} - \frac{x^5}{5} - \frac{4}{3}\frac{x^7}{7} \right]_0^1$$

$$= \frac{2}{7} + \frac{8}{15} - \frac{1}{5} - \frac{4}{21} = \frac{3}{7} \text{ cubic units.}$$

Example 1.1.15 Find the volume of the tetrahedron bounded by the plane $\frac{x}{a}+\frac{y}{b}+\frac{z}{c}=1$ and the coordinate planes.

Solution:

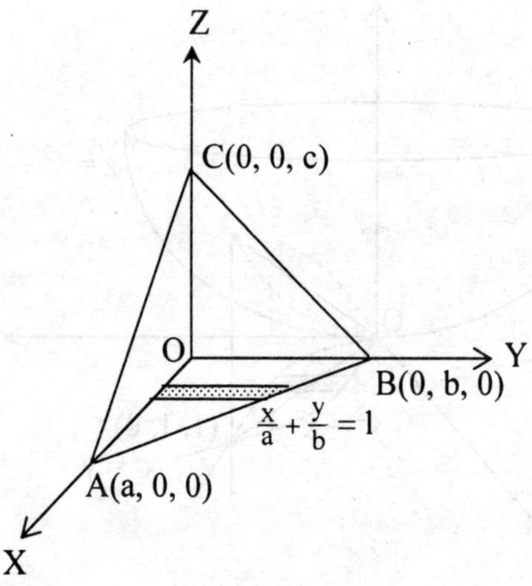

Fig. 1.15

Let the plane $\frac{x}{a}+\frac{y}{b}+\frac{z}{c}=1$ meet the axes at A, B, C. (Fig 1.15) Let R be the region represented by the triangles area OAB in the xy-plane. The volume of the tetrahedron is obtained by integrating the surface $z = c\left(1-\frac{x}{a}-\frac{y}{b}\right)$ over the region R.

$$\therefore \quad V = \iint_R z\,dx\,dy$$

$$= \int_0^a \left[\int_0^{b(1-x/a)} c\left(1-\frac{x}{a}-\frac{y}{b}\right) dy \right] dx$$

$$= c\int_0^a \left[y - \frac{xy}{a} - \frac{y^2}{2b} \right]_0^{b(1-x/a)} dx$$

$$= c\int_0^a \left[b(1 - x/a) - \frac{x}{a} b(1 - x/a) - \frac{b^2}{2b}(1 - x/a)^2 \right] dx$$

$$= cb\int_0^a \left[(1 - x/a)^2 - \frac{1}{2}(1 - x/a)^2 \right] dx$$

$$= \frac{cb}{2}\int_0^a (1 - x/a)^2 dx$$

$$= \frac{cb}{2}\left[\frac{(1 - x/a)^3}{3(-1/a)} \right]_0^a = \frac{abc}{6}$$

EXERCISE 1.1

PART - A

1. Define double integral.
2. State the formula in double integrals to find the area of a region in xy-plane in Cartesian coordinates.
3. State the formula in double integrals to find the area of a region in polar coordinates.
4. State the formula in double integrals to find the volume below a given surface and above a region R in the xy-plane.
5. Evaluate $\int_0^a \int_0^\theta r dr d\theta$.
6. Find the limits for $\iint_R x dx dy$, where R is the region bounded by the lines x = 0, y = 0, x + y = 1.
7. Evaluate $\int_0^a \int_0^b (x^2 + y^2) dx dy$.
8. Evaluate $\int_1^2 \int_1^x xy^2 dy dx$.
9. Evaluate $\int_0^1 \int_{\sqrt{y}}^{2-y} x^2 dx dy$.
10. Evaluate $\int_0^\pi \int_0^{a(1+\cos\theta)} r^2 \sin\theta \, dr d\theta$.
11. Evaluate $\int_{-\pi/2}^{\pi/2} \int_0^{2\cos\theta} r^2 dr d\theta$.

12. Evaluate $\int_0^{\pi}\int_0^{\cos\theta} r\sin\theta\, dr d\theta$.
13. Evaluate $\int_0^{1}\int_x^{\sqrt{x}} (x^2+y^2)\, dxdy$.
14. Evaluate $\int_0^{4}\int_0^{x^2} e^{y/x}\, dydx$.
15. Evaluate $\int_0^{1}\int_{x^2}^{x} xy^2 dydx$.
16. Evaluate $\iint_R y\, dx\, dy$ when R is the region between the parabola $x^2 = y$ and the line $x + y = 2$.
17. Find the limits for $\iint_R dx\, dy$ where R is the region between $y = 2x$ and $y = x^2$ lying to the left of $x = 1$.
18. Find $\iint_R dx\, dy$ where R is the region bounded by $y = 2x$ and $y = x^2$ and lying to the left of $x = 1$.
19. Evaluate $\iint_R x.dx\, dy$ where R is the region bounded by $y = x^2$ and $y = x^3$.
20. Evaluate $\iint_R x^2.dx\, dy$ where R is the region bounded by $y = x$ and $y = 2x$ and $x = 2$.
21. Evaluate $\iint_R y.dx\, dy$, where R is the region bounded by $y = 0$, $y^2 = 4x$ and $y^2 = 5-x$ and lying above the x-axis.
22. Find the area bounded by $y = x^2$, $y = 4 - x^2$.
23. Find the area bounded by $y^2 = 4 - 2x$, $x \geq 0$, $y \geq 0$.
24. Find the area bounded by $x = y^2$, $x + y = 2$.
25. Find the area lying between the parabola $y = 4x-x^2$ and the line $y = x$.

PART - B

26. Evaluate $\iint_R xy\, dx\, dy$ where R is the positive quadrant of the circle $x^2 + y^2 = a^2$.
27. Evaluate $\iint_R (1+x+y)\, dx\, dy$, where R is the region bounded by the lines $y = -x$, $x = \sqrt{y}$, $y = 0$ and $y = 2$.
28. Find $\iint_R (x+y+a)\, dx\, dy$, where R is the region bounded by the circle $x^2 + y^2 = a^2$.

29. Evaluate $\iint_R (x^2 + y^2)\,dx\,dy$, R being the region bounded by $y = x^2$, $x = 2$ and $y = 1$.
30. Evaluate $\iint_R x^2y^2\,dx\,dy$, where R is the circular disc $x^2 + y^2 \le 1$.
31. Evaluate $\iint_R e^{y/x}\,dx\,dy$, where R is the region bounded by $y = x$, $y = 0$ and $x = 1$.
32. Evaluate $\iint_R xy\,dy\,dx$, where R is the region bounded by the curves $x = y^2$, $x = 2 - y$, $y = 0$ and $y = 1$.
33. Find the area bounded by the parabolas $y^2 = 4 - x$ and $y^2 = 4 - 4x$.
34. Find the area bounded by $x^2 = 4y$ and $8y = x^2 + 16$.
35. Find the area inside the circle $r = 4 \sin\theta$ and outside the lemniscate $\rho^2 = 8\cos 2\theta$.
36. Find the area outside $r = 4$ and inside $r = 8\cos\theta$.
37. Find the area inside $r = 2(1 - \cos\theta)$.
38. Evaluate $\iint r^2 \sin\theta\,dr\,d\theta$ over the cardioid $r = a(1 + \cos\theta)$.
39. Evaluate $\iint r^2 \cos\theta\,dr\,d\theta$ over one loop of the lemniscate $r^2 = a^2 \cos 2\theta$.
40. Evaluate $\iint r^2 \sin^2\theta\,dr\,d\theta$ over the area of the circle $r = a\cos\theta$.
41. Find the volume in the first octant between the planes $z = 0$ and $z = x + y + 2$, and inside the cylinder $x^2 + y^2 = 16$.
42. Find the volume cut from $9x^2 + 4y^2 + 36z = 36$ by the plane $z = 0$.
43. Find the volume in the first octant bounded by $x^2 + z = 9$, $3x + 4y = 24$, $x = 0$, $y = 0$ and $z = 0$.
44. Find the volume in the first octant bounded by $xy = 4z$, $y = x$ and $x = 4$.
45. Find the volume common to the cylinders $x^2 + y^2 = 16$ and $x^2 + z^2 = 16$.
46. Find the volume of the solid which is below the plane $z = 2x + 3$ and above the xy-plane and bounded by $y^2 = x$, $x = 0$ and $x = 2$.
47. Find the volume of the solid which is bounded by the cylinder $x^2 + y^2 = 1$ and the planes $y + z = 1$ and $z = 0$.
48. Find the volume of the solid which is bounded by the paraboloid $z = 9 - x^2 - 4y^2$ and the coordinate planes $x \ge 0$, $y \ge 0$, $z \ge 0$.
49. Find the volume of the solid which is bounded by the surfaces $z = 0$, $3z = x^2 + y^2$ and the cylinder $x^2 + y^2 = 9$.
50. Find the volume of the solid which is in the first octant bounded by the cylinders $x^2 + y^2 = a^2$ and $y^2 + z^2 = a^2$.
51. Find the volume common to the cylinders $x^2 + y^2 = a^2$ and $x^2 + z^2 = a^2$.
52. Find the volume bounded by the xy-plane, the cylinder $x^2 + y^2 = 1$ and the plane $x + y + z = 3$
53. Find the volume bounded by the xy-plane, the paraboloid $2z = x^2 + y^2$ and the cylinder $x^2 + y^2 = 4$.

1.2 CHANGE OF ORDER OF INTEGRATION

In a double integral, if the limits of integration are constants, then the order of integration is not important. One can integrate w.r.t x and then w.r.t y or vice versa. But in a double integral with variable limits, the change of order of integration changes the limits of integration. Sometimes it is convenient to evaluate a double integral by changing the order of integration. In such cases the limits of integration are suitably modified. It may be necessary to split the given region of integration into subregions.

Example 1.2.1 By changing the order of integration evaluate $\int_0^a \int_0^{2\sqrt{ax}} x^2 dxdy$.

Solution:

The region of integration is bounded by $y = 0$, $y = 2\sqrt{ax}$, $x = 0$ and $x = a$. This is the region in the first quadrant bounded by the parabola $y^2 = 4ax$ and the line $x = a$ (Fig 1.16). In the given integral, we first integrate w.r.t y between $y = 0$ and $y = 2\sqrt{ax}$, along the strip PQ and the slide this vertical strip PQ from $x = 0$ to $x = a$. To change the order of integration, first integrate w.r.t x between $x = \frac{y^2}{4a}$ and $x = a$ along the strip P′Q′ and then slide this horizontal strip P′Q′ from $y = 0$ to $y = 2a$.

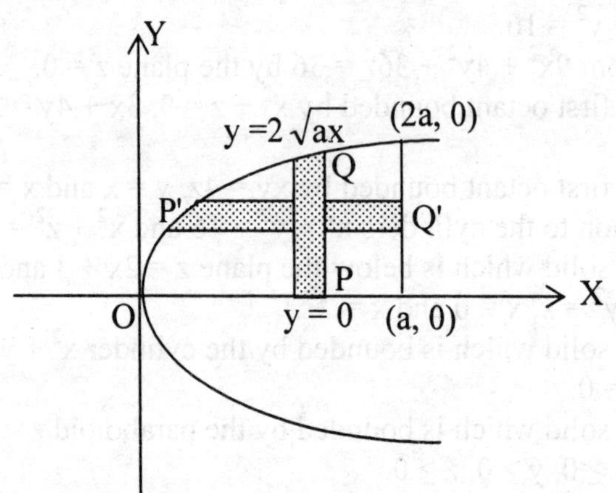

Fig. 1.16

$$\therefore \quad I = \int_{y=0}^{2a} \left[\int_{x=y^2/4a}^{a} x^2 dx \right] dy$$

$$= \int_{y=0}^{2a} \left[\frac{x^3}{3} \right]_{y^2/4a}^{a} dy$$

$$= \int_{y=0}^{2a} \left[\frac{a^3}{3} - \frac{y^6}{3.64.a^3} \right] dy$$

$$= \left[\frac{a^3}{3} y - \frac{y^7}{3.64.a^3.7} \right]_0^{2a}$$

$$= \frac{2a^4}{3} - \frac{(2a)^7}{3.64.a^3.7} = \frac{2a^4}{3} - \frac{2a^4}{21} = \frac{4}{7} a^4$$

Example 1.2.2 Change the order of integration of the double integral $\int_0^a \int_{x^2/a}^{2a-x} xy\, dy\, dx$ and hence evaluate it.

Solution:

The region of integration is bounded by $y = \frac{x^2}{a}$, $y = 2a-x$, $x = 0$ and $x = a$. This region R is in the first quadrant, bounded by the parabola $x^2 = ay$, the line $x + y = 2a$ and the y-axis (Fig 1.17).

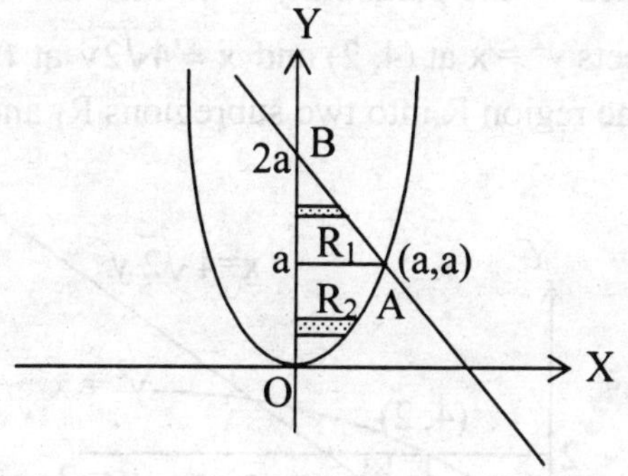

Fig. 1.17

To change the order of integration, divide the region R into two subregions R_1 and R_2. R_1 from $y = 0$ to $y = a$, and R_2 from $y = a$ to $y = 2a$.

$$\therefore \quad I = \iint_{R_1} xy\, dx\, dy + \iint_{R_2} xy\, dx\, dy$$

$$= \int_0^a \left[\int_0^{\sqrt{ay}} xy\, dx \right] dy + \int_a^{2a} \left[\int_0^{2a-y} xy\, dx \right] dy$$

$$= \int_0^a \left[y.\frac{x^2}{2} \right]_0^{\sqrt{ay}} dy + \int_a^{2a} \left[y.\frac{x^2}{2} \right]_0^{2a-y} dy$$

$$= \int_0^a \frac{ay^2}{2}\,dy + \int_a^{2a} \frac{y}{2}(2a-y)^2\,dy$$

$$= \int_0^a \frac{ay^2}{2}\,dy + \int_a^{2a} \left(2a^2y - 2ay^2 + \tfrac{1}{2}y^3\right)dy$$

$$= \frac{a^4}{6} + \left[a^2y^2 - \frac{2ay^3}{3} + \frac{y^4}{8}\right]_a^{2a}$$

$$= \frac{a^4}{6} + 4a^4 - \frac{16}{3}a^4 + 2a^4 - a^4 + \frac{2}{3}a^4 - \frac{a^4}{8}$$

$$= \frac{3a^4}{8}$$

Example 1.2.3 Change the order of integration of $\int_0^2 \int_{y^2}^{4y\sqrt{2}} f(x,y)\,dy\,dx$

Solution:

The region of integration R is bounded by $x = y^2$, $x = 4\sqrt{2}y$, $y = 0$ and $y = 2$. This region is in the first quadrant bounded by the parabola $y^2 = x$, line $x = 4\sqrt{2}y$ and the line $y = 2$. (Fig 1.18) The line $y = 2$ meets $y^2 = x$ at $(4, 2)$ and $x = 4\sqrt{2}y$ at $(8\sqrt{2}, 2)$. To change the order of integration, divide the region R into two subregions R_1 and R_2.

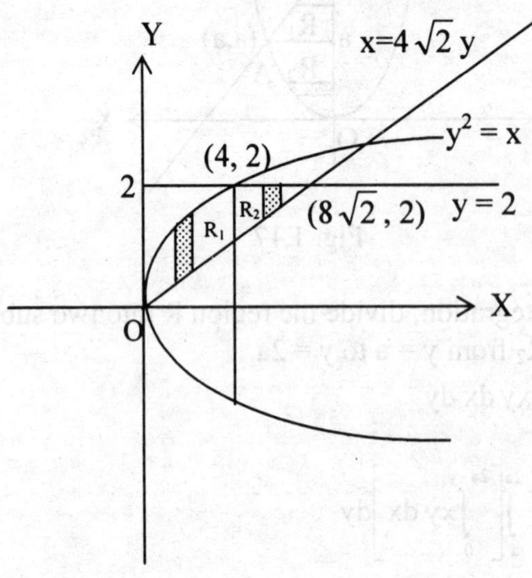

Fig. 1.18

In each of these regions, first integrate along vertical strips and slide these strips from $x = 0$ to $x = 4$ in R_1 and from $x = 4$ to $x = 4\sqrt{2}$ in R_2.

$$\therefore \quad I = \iint_{R_1} f(x,y)\,dy\,dx + \iint_{R_2} f(x,y)\,dy\,dx$$

$$= \int_{x=0}^{4}\left[\int_{y=\frac{x}{4\sqrt{2}}}^{\sqrt{x}} f(x,y)\,dy\right]dx + \int_{x=4}^{8\sqrt{2}}\left[\int_{y=\frac{x}{4\sqrt{2}}}^{2} f(x,y)\,dy\right]dx$$

$$= \int_{0}^{4}\int_{\frac{x}{4\sqrt{2}}}^{\sqrt{x}} f(x,y)\,dydx + \int_{4}^{8\sqrt{2}}\int_{\frac{x}{4\sqrt{2}}}^{2} f(x,y)\,dydx$$

Example 1.2.4 Evaluate $\int_0^\infty\int_x^\infty \frac{e^{-y}}{y}\,dy\,dx$

Solution:

The region of integration is bounded by the lines $y = x$ and $x = 0$. (Fig 1.19) The evaluation of the double integral will be easier if we first integrate w.r.t x. Hence change the order of integration.

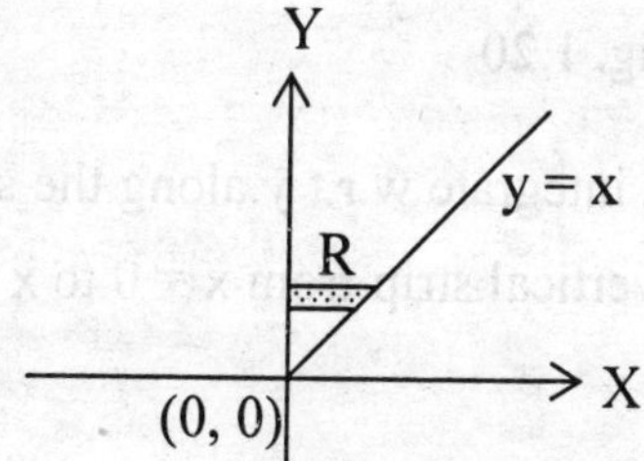

Fig.1.19

$$\therefore \quad I = \int_{y=0}^{\infty}\left[\int_{x=0}^{y} \frac{e^{-y}}{y}\,dx\right]dy$$

$$= \int_0^\infty \left[\frac{e^{-y}}{y}.x\right]_0^y dy$$

$$= \int_0^\infty e^{-y}\,dy$$

$$= \left[\frac{e^{-y}}{-1}\right]_0^\infty$$

$$= 1$$

Example 1.2.5 Change the order of integration and then evaluate $\int_0^a \int_{a-\sqrt{a^2-y^2}}^{a+\sqrt{a^2-y^2}} dy\, dx.$

Solution:

The region of integration is bounded by $x = a \pm \sqrt{a^2 - y^2}$ and the lines y = 0 and y = a. i.e., the circle $x^2 + y^2 - 2ax = 0$ and the x-axis (Fig. 1.20).

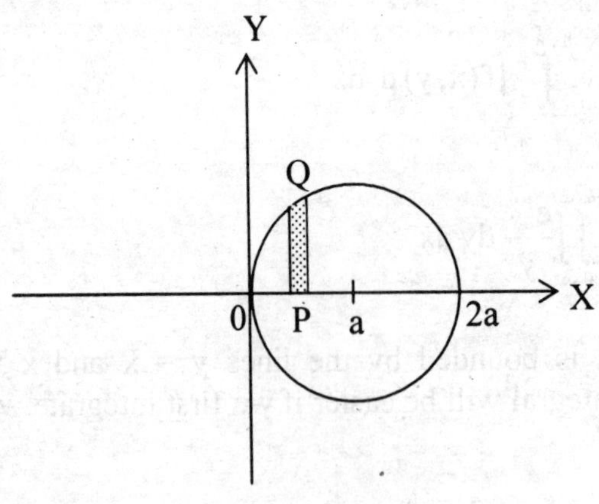

Fig. 1.20

To change the order of integration, first integrate w.r.t y along the strip PQ between y = 0 and $y = \sqrt{2ax - x^2}$ and then slide this vertical strip from x = 0 to x = 2a.

$$\therefore \quad I = \int_0^{2a} \left[\int_0^{\sqrt{2ax-x^2}} dy \right] dx$$

$$= \int_0^{2a} \sqrt{2ax - x^2}\, dx$$

$$= \int_0^{2a} \sqrt{a^2 - (x-a)^2}\, dx$$

$$= \left[\frac{1}{2}.x.\sqrt{a^2 - (x-a)^2} + \frac{a^2}{2} \sin^{-1} \frac{x-a}{a} \right]_0^{2a}$$

$$= \frac{a^2}{2} \left(\frac{\pi}{2} + \frac{\pi}{2} \right)$$

$$= \frac{\pi a^2}{2}$$

EXERCISE 1.2

PART-A

Change the order of integration in the following integrals (Ex: 1-10)

1. $\int_{1}^{2}\int_{3}^{4} f(x,y)\,dy\,dx$

2. $\int_{0}^{2}\int_{y^2}^{4\sqrt{2y}} f(x,y)\,dy\,dx$

3. $\int_{-1}^{-2}\int_{-x}^{2-x^2} f(x,y)\,dy\,dx$

4. $\int_{0}^{b} dx \int_{a^2/2}^{x} f(x,y)\,dy$

5. $\int_{0}^{a}\int_{0}^{\sqrt{a^2-x^2}} f(x,y)\,dx\,dy$

6. $\int_{-1}^{1}\int_{0}^{\sqrt{1-x^2}} f(x,y)\,dy\,dx$

7. $\int_{0}^{1}\int_{y}^{y^{2/3}} dx\,dy$

8. $\int_{0}^{1}\int_{3y}^{3} e^{x^2}\,dx\,dy$

9. $\int_{0}^{1}\int_{0}^{x^2} xe^{y}\,dy\,dx$

10. $\int_{0}^{1}\int_{x^2}^{2-x} xy\,dx\,dy$

PART-B

Change the order of integration in the following integrals and then evaluate them. (Ex: 11-21)

11. $\int_{0}^{a}\int_{y}^{a} \frac{x}{x^2+y^2}\,dx\,dy$

12. $\int_{0}^{2}\int_{0}^{1-x^2} (x+y)\,dx\,dy$

13. $\int_{0}^{b}\int_{y}^{b} \frac{x}{\sqrt{x^2+y^2}}\,dx\,dy$

14. $\int_{0}^{a}\int_{\sqrt{ax-x^2}}^{\sqrt{ax}} y\,dy\,dx$

15. $\int_{0}^{a}\int_{y^2/a}^{2a-y} xy\,dx\,dy$

16. $\int_{0}^{a}\int_{y^2/a}^{y} \frac{y\,dx\,dy}{(a-x)\sqrt{ax-y^2}}$

17. $\int_{0}^{3}\int_{1}^{\sqrt{4-y}} (x+y)\,dx\,dy$

18. $\int_{0}^{1}\int_{x}^{\sqrt{2-x^2}} \frac{x\,dy\,dx}{\sqrt{x^2+y^2}}$

19. $\int_{0}^{1}\int_{x^2}^{2-x} xy\,dy\,dx$

20. $\int_{0}^{4a}\int_{x^2/4a}^{2\sqrt{ax}} dy\,dx$

21. $\int_{0}^{\infty}\int_{0}^{x} x.e^{-x^2/y}\,dy\,dx$

1.3 TRIPLE INTEGRALS

Triple integrals are defined in a manner similar to a double integral.

Definition 1.3.1 Let f(x, y, z) be a function defined and continuous in a closed and bounded region R in $\mathbb{R}^3$. Divide R into n sub regions $\Delta R_1, \Delta R_2, \ldots, \Delta R_n$ of volumes $\Delta V_1, \Delta V_2, \ldots, \Delta V_n$. Let $(\varsigma_i, \eta_i, \delta_i)$ be any point in the ith subregion ΔR_i. Then the **triple integral** of f(x, y, z) over the region R is defined as $\lim\limits_{\substack{n\to\infty \\ \text{Max}|\Delta V_i|\to 0}} \sum_{i=1}^{n} f(\varsigma_i, \eta_i, \delta_i).\Delta V_i$ and is denoted by

$$\iiint_R f(x, y, z)\, dV \text{ or } \iiint_R f(x, y, z)\, dx\, dy\, dz.$$

The triple integral can be computed by expressing it as the iterated integral

$$\int_{x=a}^{b}\left[\int_{y=\varphi(x)}^{\chi(x)}\left[\int_{z=g(x,y)}^{h(x,y)} f(x,y,z)dz\right]dy\right]dx$$

The order of integration can be changed by suitably modifying the limits of integration and if necessary by subdividing the region R into a finite number of subregions, in which case the triple integral over R is obtained by summing the triple integrals over these subregions. The properties of triple integrals are similar to those of double integrals.

1.3.1 APPLICATIONS OF TRIPLE INTEGRALS

1. **Volume of a region**

   If f(x, y, z) = 1, then the triple integral over R gives the volume V of the region R.

   i.e., $V = \iiint_R dx\, dy\, dz.$

2. **Centre of gravity**

   Let f(x, y, z) = ρ(x, y, z) be the density of a mass in the region R.

   Then $M = \iiint_R f(x, y, z)\, dx\, dy\, dz$ gives the total mass of the solid in R

   If $\bar{x} = \frac{1}{M}\iiint_R x.f(x, y, z)\, dx\, dy\, dz$

   $\bar{y} = \frac{1}{M}\iiint_R y.f(x, y, z)\, dx\, dy\, dz$

   $\bar{z} = \frac{1}{M}\iiint_R z.f(x, y, z)\, dx\, dy\, dz$, where f(x, y, z) = ρ(x, y, z), then $(\bar{x}, \bar{y}, \bar{z})$ is the **centre of mass** or **centre of gravity** of the solid of mass M in the region R.

3. Moment of inertia

If $f(x, y, z) = \rho(x, y, z)$ is the density function of a mass in the region R, then

$$I_x = \iiint_R (y^2 + z^2).f(x, y, z)\, dx\, dy\, dz$$

$$I_y = \iiint_R (x^2 + z^2).f(x, y, z)\, dx\, dy\, dz$$

$I_z = \iiint_R (x^2 + y^2).f(x, y, z)\, dx\, dy\, dz$ are the **moments of inertia** of the mass about the x-axis, y-axis and z-axis respectively.

Example 1.3.1 Evaluate $\int_0^a \int_0^x \int_0^{x+y} e^{x+y+z} dz\, dy\, dx$

Solution:

$$\int_0^a \int_0^x \int_0^{x+y} e^{x+y+z} dz\, dy\, dx = \int_0^a \left[\int_0^x \left[\int_0^{x+y} e^{x+y+z} dz \right] dy \right] dx$$

$$= \int_0^a \left[\int_0^x \left[e^{x+y+z} \right]_0^{x+y} dy \right] dx$$

$$= \int_0^a \left[\int_0^x \left[e^{x+y+x+y} - e^{x+y} \right] dy \right] dx$$

$$= \int_0^a \left[e^{2(x+y)}.\frac{1}{2} - e^{x+y} \right]_0^x dx$$

$$= \int_0^a \left[\frac{1}{2}e^{4x} - \frac{1}{2}e^{2x} - e^{2x} + e^x \right] dx$$

$$= \int_0^a \left[\frac{1}{2}e^{4x} - \frac{3}{2}e^{2x} + e^x \right] dx$$

$$= \int_0^a \left[\frac{e^{4x}}{8} - \frac{3e^{2x}}{4} + e^x \right]_0^a dx$$

$$= \frac{e^{4a}}{8} - \frac{3e^{2a}}{4} + e^a - \frac{1}{8} + \frac{3}{4} - 1$$

$$= \frac{e^{4a}}{8} - \frac{3e^{2a}}{4} + e^a - \frac{3}{8}$$

Example 1.3.2 Evaluate $\int_0^{\pi/2} \int_0^{a \sin\theta} \int_0^{\frac{a^2 - r^2}{a}} r\, dz\, dr\, d\theta$

Solution:

$$\int_0^{\pi/2}\int_0^{a\sin\theta}\int_0^{\frac{a^2-r^2}{a}} r\,dz\,dr\,d\theta = \int_0^{\pi/2}\left[\int_0^{a\sin\theta}\left[\int_0^{\frac{a^2-r^2}{a}} r\,dz\right]dr\right]d\theta$$

$$= \int_0^{\pi/2}\left[\int_0^{a\sin\theta}\left[rz\right]_0^{\frac{a^2-r^2}{a}} dr\right]d\theta$$

$$= \int_0^{\pi/2}\left[\int_0^{a\sin\theta} r.\left(\frac{a^2-r^2}{a}\right)dr\right]d\theta$$

$$= \int_0^{\pi/2}\left[a.\frac{r^2}{2}-\frac{1}{a}.\frac{r^4}{4}\right]_0^{a\sin\theta} d\theta$$

$$= \int_0^{\pi/2}\left[\frac{a^3}{2}\sin^2\theta-\frac{a^3}{4}.\sin^4\theta\right]d\theta$$

$$= \frac{a^3}{2}.\frac{1}{2}.\frac{\pi}{2}-\frac{a^3}{4}.\frac{3}{4}.\frac{1}{2}.\frac{\pi}{2}$$

$$= \frac{5a^3\pi}{64}$$

Example 1.3.3 Compute $\iiint_R z\,dx\,dy\,dz$ over the region R in the first octant bounded by the planes $y = 0$, $z = 0$, $x + y = 2$, $2y + x = 6$ and the cylinder $y^2 + z^2 = 4$.

Solution:

The limits for z are $z = 0$ and $z = \sqrt{4-y^2}$ (Fig 1.21)
The limits for x are $x = 2 - y$ and $x = 6 - 2y$, obtained from the two given planes.
Then the limits for y are $y = 0$ and $y = 2$.
Hence the given integral is

$$V = \int_{y=0}^{2}\int_{x=2-y}^{6-2y}\int_{z=0}^{\sqrt{4-y^2}} z\,dz\,dx\,dy$$

$$= \int_{y=0}^{2}\int_{x=2-y}^{6-2y}\left[\frac{z^2}{2}\right]_0^{\sqrt{4-y^2}} dx\,dy.$$

$$= \int_{y=0}^{2}\int_{x=2-y}^{6-2y}\frac{1}{2}\left(4-y^2\right)dx\,dy.$$

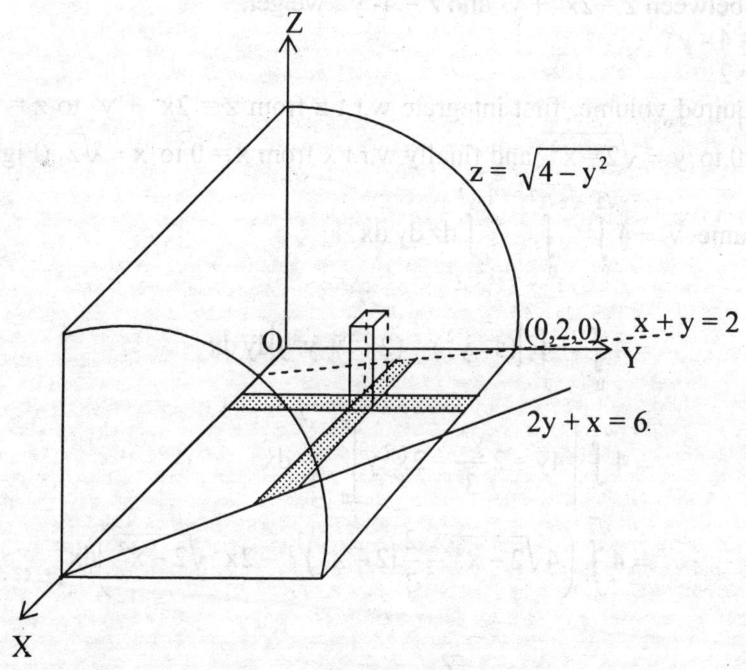

Fig. 1.21

$$= \int_{y=0}^{2} \left[\frac{1}{2}(4-y^2)x\right]_{2-y}^{6-2y} dy$$

$$= \int_{y=0}^{2} \frac{1}{2}(4-y^2)[(6-2y)-(2-y)]dy$$

$$= \int_{y=0}^{2} \frac{1}{2}(4-y^2)(4-y)\,dy$$

$$= \frac{1}{2}\int_{y=0}^{2} \left(16-4y-4y^2+y^3\right) dy$$

$$= \frac{1}{2}\left(16-4\frac{y^2}{2}-4\frac{y^3}{3}+\frac{y^4}{4}\right)_0^2$$

$$= \frac{26}{3}$$

Example 1.3.4 Find the volume bounded by the paraboloid $z = 2x^2 + y^2$ and the cylinder $z = 4 - y^2$.

Solution:

By eliminating z between $z = 2x^2 + y^2$ and $z = 4 - y^2$, we get,

$$2x^2 + y^2 = 4 - y^2$$

i.e., $x^2 + y^2 = 2.$

To obtain the required volume, first integrate w.r.t z from $z = 2x^2 + y^2$ to $z = 4-y^2$, then w.r.t. y from y = 0 to $y = \sqrt{2-x^2}$ and finally w.r.t x from x = 0 to $x = \sqrt{2}$ (Fig. 1.22)

$$\text{The required volume } V = 4\int_0^{\sqrt{2}}\int_0^{\sqrt{2-x^2}}\int_{2x^2+y^2}^{4-y^2} dz\,dy\,dx$$

$$= 4\int_0^{\sqrt{2}}\int_0^{\sqrt{2-x^2}} \left[(4-y^2)-(2x^2+y^2)\right]dy\,dx$$

$$= 4\int_0^{\sqrt{2}} \left[4y - 2\frac{y^3}{3} - 2x^2y\right]_0^{\sqrt{2-x^2}} dx$$

$$= 4\int_0^{\sqrt{2}} \left(4\sqrt{2-x^2} - \frac{2}{3}(2-x^2)^{3/2} - 2x^2\sqrt{2-x^2}\right)dx$$

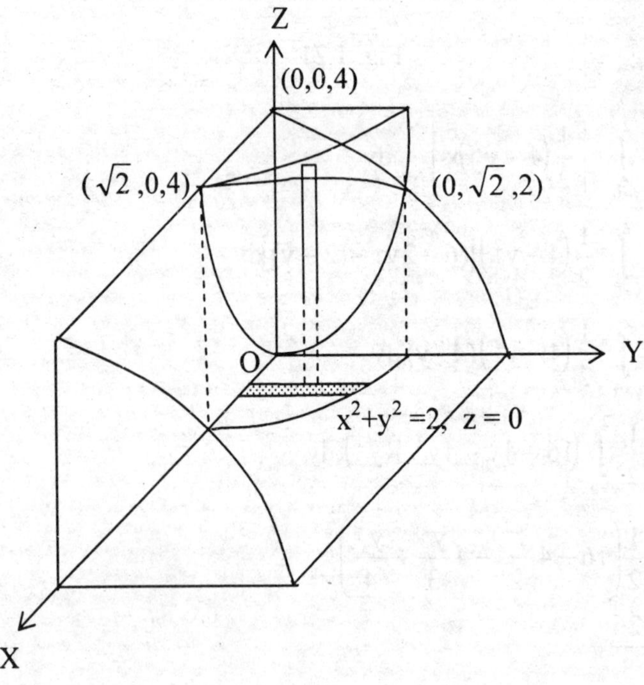

Fig. 1.22

$$= 4\int_0^{\sqrt{2}} \left(2\sqrt{2-x^2}\,(2-x^2) - \frac{2}{3}(2-x^2)^{3/2}\right)dx$$

$$= 4\int_0^{\sqrt{2}} \frac{4}{3}(2-x^2)^{3/2}\,dx$$

$$= \frac{16}{3}\int_0^{\pi/2} 2\sqrt{2}..\cos^3\theta.\sqrt{2}\cos\theta.d\theta \qquad \text{(Substituting } x = \sqrt{2}\sin\theta,\ dx = \sqrt{2}\cos\theta\, d\theta\text{)}$$

$$= \frac{16}{3}.4\int_0^{\pi/2} \cos^4\theta..d\theta$$

$$= \frac{64}{3}.\frac{3}{4}.\frac{1}{2}.\frac{\pi}{2}$$

$= 4\pi$ cubic units.

Example 1.3.5 Find the volume bounded by the parabolic cylinder $z = 4-x^2$ and the planes $x = 0$, $y = 0$, $y = 6$ and $z = 0$.

Solution:

To obtain the required volume first integrate w.r.t z from $z = 0$ to $z = 4 - x^2$, then integrate w.r.t y from $y = 0$ to $y = 6$ and finally w.r.t x from $x = 0$ to $x = 2$ (Fig 1.23).

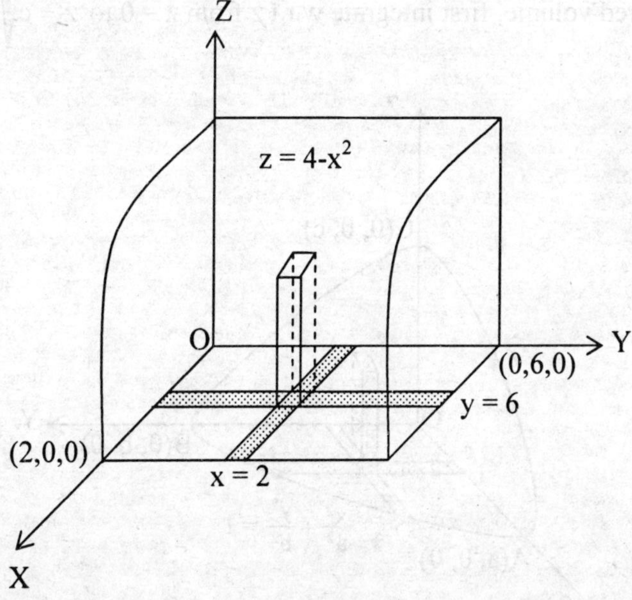

Fig.1.23

Hence the required volume is

$$V = \int_{x=0}^{2} \int_{y=0}^{6} \int_{z=0}^{4-x^2} dz\, dy\, dx$$

$$= \int_{x=0}^{2} \int_{y=0}^{6} (4 - x^2)\, dy\, dx$$

$$= \int_{0}^{2} \left[(4 - x^2)\, y\right]_0^6 dx$$

$$= \int_{0}^{2} (4 - x^2)\, 6\, dx$$

$$= \left[24x - 6.\frac{x^3}{3}\right]_0^2$$

= 48 -16 = 32 cubic units.

Example 1.3.6 Find the volume of the ellipsoid $\frac{x^2}{a^2} + \frac{y^2}{b^2} + \frac{z^2}{c^2} = 1.$

Solution:

The cross section of the ellipsoid by the xy-plane is the ellipse $\frac{x^2}{a^2} + \frac{y^2}{b^2} = 1.$

To obtain the required volume, first integrate w.r.t z from z = 0 to $z = c\sqrt{1 - \frac{x^2}{a^2} - \frac{y^2}{b^2}}$,

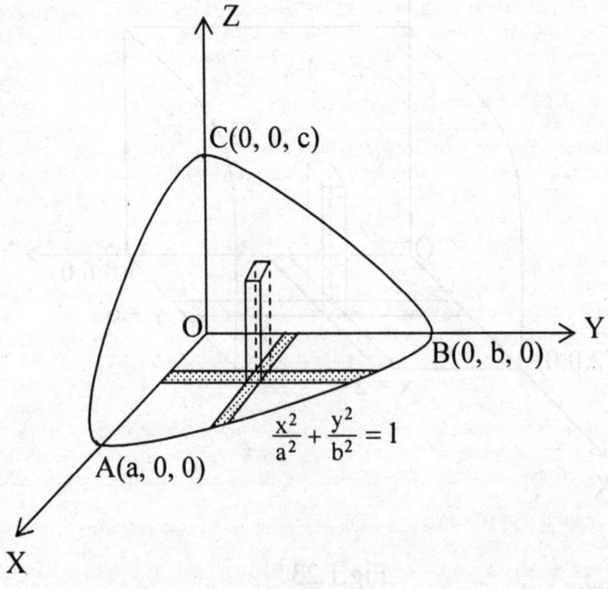

Fig. 1.24

then w.r.t y from y = 0 to $y = b\sqrt{1 - x^2/a^2}$ and finally w.r.t x from x = 0 to x = a (Fig 1.24)

$$\therefore \quad V = 8\int_0^a \int_{y=0}^{b\sqrt{1-\frac{x^2}{a^2}}} \int_{z=0}^{c\sqrt{1-\frac{x^2}{a^2}-\frac{y^2}{b^2}}} dz\, dy\, dx$$

$$= 8\int_0^a \int_0^{b\sqrt{1-\frac{x^2}{a^2}}} c\sqrt{1-\frac{x^2}{a^2}-\frac{y^2}{b^2}}\, dy\, dx$$

$$= 8\int_0^a \int_0^{\rho} \frac{c}{b}\sqrt{\rho^2 - y^2}\, dy\, dx, \text{ where } \rho = b\sqrt{1-\frac{x^2}{a^2}}$$

$$= \frac{8c}{b}\int_0^a \left[\frac{y}{2}\sqrt{\rho^2 - y^2} + \frac{\rho^2}{2}\sin^{-1}\left(y/\rho\right)\right]_0^{\rho} dx$$

$$= \frac{8c}{b}\int_0^a \frac{\rho^2}{2}\frac{\pi}{2}\, dx$$

$$= 2\pi bc \int_0^a \left(1 - x^2/a^2\right) dx$$

$$= 2\pi bc \left[x - \frac{x^3}{3a^2}\right]_0^a$$

$$= 2\pi bc\left(a - \frac{a}{3}\right)$$

$$= \frac{4}{3}\pi abc \text{ cubic units.}$$

Example 1.3.7 Find the volume of the tetrahedron bounded by the coordinate planes and the plane $\frac{x}{a} + \frac{y}{b} + \frac{z}{c} = 1$.

Solution:

To obtain the required volume, first integrate w.r.t z from z = 0 to $z = c\left(1 - \frac{x}{a} - \frac{y}{b}\right)$, then w.r.t y from y = 0 to $y = b\left(1 - \frac{x}{a}\right)$ and finally w.r.t x from x = 0 to x = a (Fig 1.25)

$$\therefore \quad V = \int_0^a \int_{y=0}^{b\left(1-\frac{x}{a}\right)} \int_{z=0}^{c\left(1-\frac{x}{a}-\frac{y}{b}\right)} dz\, dy\, dx$$

$$= \int_0^a \int_0^{b\left(1-\frac{x}{a}\right)} c\left(1-\frac{x}{a}-\frac{y}{b}\right) dy\, dx.$$

$$= \int_0^a \int_0^{\rho} \frac{c}{b}(\rho - y)\, dy\, dx. \quad \text{where } \rho = b\left(1-\frac{x}{a}\right)$$

$$= \frac{c}{b}\int_0^a \left(\rho y - \frac{y^2}{2}\right)_0^{\rho} dx$$

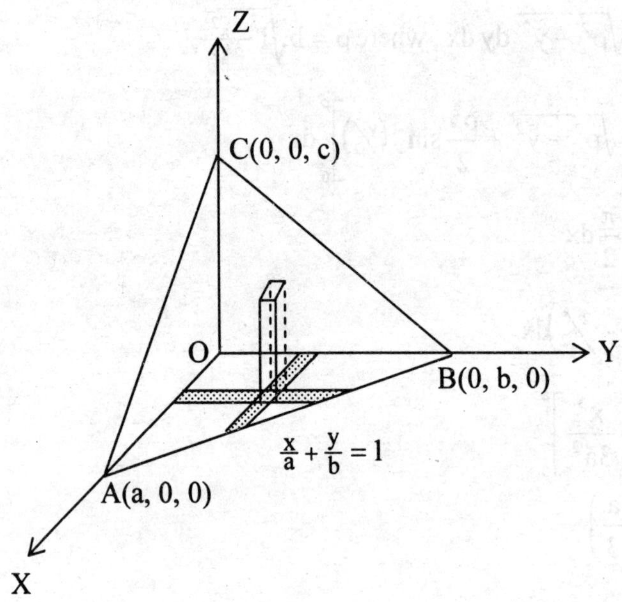

Fig 1.25

$$= \frac{c}{b}\int_0^a \frac{\rho^2}{2} dx$$

$$= \frac{c}{2b}\int_0^a \frac{b^2}{a^2}(a-x)^2 dx$$

$$= \frac{bc}{2a^2}\left[\frac{(a-x)^3}{-3}\right]_0^a$$

$$= \frac{bc}{2a^2}\cdot\frac{a^3}{3}$$

$$= \frac{abc}{6}$$

EXERCISE 1.3

PART-A

1. Define triple integral.
2. State any application of triple integral.
3. State the formula for finding the volume of a region using triple integrals.

Evaluate the following triple integrals (Ex: 4 -18)

4. $\int_0^a \int_0^b \int_0^c dx\,dy\,dz$

5. $\int_0^1 \int_0^2 \int_0^3 x\,dx\,dy\,dz$

6. $\int_0^1 \int_{y^2}^1 \int_0^{1-x} x\,dx\,dy\,dz$

7. $\int_0^1 \int_0^{1-x} \int_0^{2-x} xyz\,dz\,dy\,dx$

8. $\int_0^{\pi/2} \int_0^1 \int_0^2 zr^2 \sin\theta\,dz\,dr\,d\theta$

9. $\int_0^1 \int_{x^2}^x \int_0^{xy} dz\,dy\,dx$

10. $\int_0^{\pi} \int_0^{\pi/4} \int_0^{\sec\varphi} \sin 2\varphi\,dr\,d\varphi\,d\theta$

11. $\int_{x=0}^{2} \int_{y=0}^{6} \int_{z=0}^{4-x^2} dz\,dy\,dx$

12. $\int_0^4 \int_0^x \int_0^{\sqrt{x+y}} z\,dx\,dy\,dz$

13. $\int_0^1 \int_0^{1-x} \int_0^{x+y} e^z\,dx\,dy\,dz$

14. $\int_0^{2a} \int_0^x \int_y^x xyz\,dx\,dy\,dz$

15. $\int_0^{\pi/2} \int_0^{2a\cos\theta} \int_0^{r^2/2a} r\,d\theta\,dr\,dz$

16. $\int_0^a \int_0^x \int_0^y xyz.dz\,dy\,dx$

17. $\int_0^a \int_0^b \int_0^c (x+y+z).dx\,dy\,dz$

18. $\int_0^1 dx \int_0^2 dy \int_1^2 x^2yz\,dz$

PART-B

19. Evaluate $\iiint (x+y+z)\,dx\,dy\,dz$ over the tetrahedron bounded by the planes $x = 0$, $y = 0$, $z = 0$ and $x + y + z = 1$.
20. Evaluate $\iiint xyz\,dx\,dy\,dz$ over the positive octant of the sphere $x^2 + y^2 + z^2 = a^2$.
21. Evaluate $\iiint x^2yz\,dx\,dy\,dz$ over the tetrahedron bounded by the planes $x = 0$, $y = 0$, $z = 0$ and $\frac{x}{a} + \frac{y}{b} + \frac{z}{c} = 1$.

22. Evaluate $\iiint (x+y+z+1)^2 \, dx\,dy\,dz$ taken over the region defined by $x \geq 0, y \geq 0, z \geq 0, x+y+z \geq 1$.
23. Evaluate $\iiint_R (x^2+y^2)\, dx\,dy\,dz$ where R is the region bounded by the surface $x^2+y^2=2z$ and the plane $z=2$.
24. Evaluate $\iiint (x+y+z)\, dx\,dy\,dz$ over the region bounded by the surface $x^2+y^2=a^2, z=0, z=h$.
25. Evaluate $\iiint_R (x^2+y^2+z^2)\, dx\,dy\,dz$ where R is the region bounded by the planes $x+y+z=a, x=0, y=0$ and $z=0$.
26. Evaluate $\iiint_R xyz(x^2+y^2+z^2)\, dx\,dy\,dz$ where R is the positive octant of the sphere $x^2+y^2+z^2=a^2$.
27. Find the volume of the cylinder $x^2+y^2=a^2$ above the xy-plane cut by the plane $x+y+z=a$.
28. Find the volume bounded by the paraboloid $x^2+y^2=az$, the cylinder $x^2+y^2=2ay$ and $z=0$.
29. Find the volume cut off from the paraboloid $z=x^2+y^2$ by the plane $z=2x$.
30. Find the volume lying above the xy-plane inside the cylinder $x^2+y^2=4$ and under the cone $x^2+y^2=z^2$.
31. Find the volume enclosed by the two cylinders $x^2+y^2=2ax$ and $z^2=2ax$.
32. Find the volume included between the elliptic paraboloid $x^2+y^2=2z$, the cylinder $x^2+y^2=a^2$ and the plane $z=0$.

1.4 CHANGE OF VARIABLES

By changing the variables using a suitable transformation, a given integral can be transformed into a simpler integral involving the new variables. Thus in certain cases, the evaluation of a double or triple integral becomes easier, when we change the given variables into a new set of variables. Considering the region of integration and the integrand a suitable transformation is chosen for the change of variables.

1.4.1 CHANGE OF VARIABLES IN DOUBLE INTEGRALS

Let f(x, y) be continuous over a region R in the xy-plane. Let the variables x, y be changed to new variables u, v by the transformation $x = \varphi(u, v)$, $y = \chi(u, v)$ where the functions $\varphi(u, v)$ and $\chi(u, v)$ are continuous and have continuous first order partial derivatives in the region R′ in the uv-plane, which corresponds to the region R in the xy-plane. Then $\iint_R f(x,y)dx\,dy = \iint_{R'} F(u,v)\,|J|\,du\,dv$ where $F(u, v) = f(\varphi(u, v), \chi(u, v))$

and $J = \frac{\partial(x,y)}{\partial(u,v)} = \begin{vmatrix} \frac{\partial x}{\partial u} & \frac{\partial x}{\partial v} \\ \frac{\partial y}{\partial u} & \frac{\partial y}{\partial v} \end{vmatrix} \neq 0$ is the **functional determinant** or **Jacobian of transformation** from (x, y) to (u, v) coordinates.

For example, to change Cartesian coordinates (x, y) to polar coordinates (r, θ), we have $x = r\cos\theta$, $y = r\sin\theta$. Then,

$$J = \begin{vmatrix} \frac{\partial x}{\partial r} & \frac{\partial x}{\partial \theta} \\ \frac{\partial y}{\partial r} & \frac{\partial y}{\partial \theta} \end{vmatrix} = \begin{vmatrix} \cos\theta & -r\sin\theta \\ \sin\theta & r\cos\theta \end{vmatrix} = r$$

$$\therefore \iint_R f(x,y)\,dx\,dy = \iint_{R'} f(r\cos\theta, r\sin\theta)\,r\,dr\,d\theta,$$

where R′ is the region in the rθ-plane corresponding to R in the xy-plane

1.4.2 CHANGE OF VARIABLES IN TRIPLE INTEGRALS

Let f(x, y, z) be continuous over a region R in $\mathbb{R}^3$. Let x, y, z be transformed into u, v, w by the transformation,

$x = \alpha(u, v, w)$, $y = \beta(u, v, w)$, $z = \gamma(u, v, w)$.

Assume that the functions α(u, v, w), β(u, v, w), γ(u, v, w) are continuous and have continuous partial derivatives in the region R′ corresponding to the region R, under the given transformation.

Then $\iiint_R f(x,y,z)\,dx\,dy\,dz = \iiint_{R'} F(u,v,w)\,|J|\,du\,dv\,dw$,

where $F(u, v, w) = f(\alpha(u, v, w), \beta(u, v, w), \gamma(u, v, w))$

$$\text{and } J = \frac{\partial(x,y,z)}{\partial(u,v,w)} = \begin{vmatrix} \frac{\partial x}{\partial u} & \frac{\partial x}{\partial v} & \frac{\partial x}{\partial w} \\ \frac{\partial y}{\partial u} & \frac{\partial y}{\partial v} & \frac{\partial y}{\partial w} \\ \frac{\partial z}{\partial u} & \frac{\partial z}{\partial v} & \frac{\partial z}{\partial w} \end{vmatrix} \neq 0$$

is the **functional determinant** or **Jacobian of transformation** from (x, y, z) to (u, v, w) coordinates.

SPECIAL CASES

1. **Transformation from Cartesian coordinates (x, y, z) to cylindrical coordinates (r, θ, z)**

   The transformation equations are $x = r\cos\theta$, $y = r\sin\theta$, $z = z$ (Fig. 1.26)

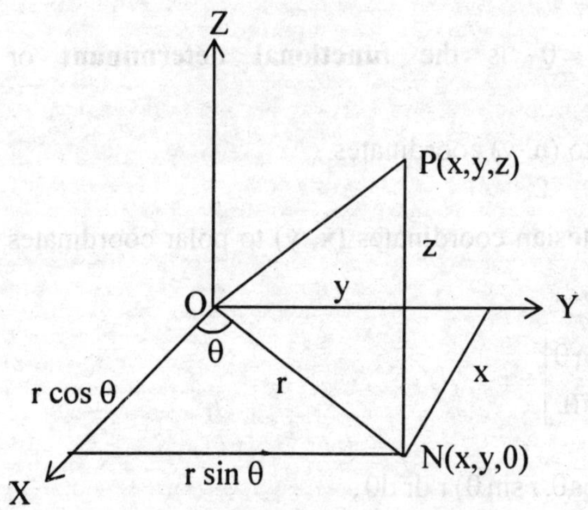

Fig. 1.26

$$J = \begin{vmatrix} \frac{\partial x}{\partial r} & \frac{\partial x}{\partial \theta} & \frac{\partial x}{\partial z} \\ \frac{\partial y}{\partial r} & \frac{\partial y}{\partial \theta} & \frac{\partial y}{\partial z} \\ \frac{\partial z}{\partial r} & \frac{\partial z}{\partial \theta} & \frac{\partial z}{\partial z} \end{vmatrix} = \begin{vmatrix} \cos\theta & -r\sin\theta & 0 \\ \sin\theta & r\cos\theta & 0 \\ 0 & 0 & 1 \end{vmatrix} = r$$

$$\therefore \quad \iiint_R f(x, y, z)\,dx\,dy\,dz = \iiint_{R'} f(r\cos\theta, r\sin\theta, z)\, r\,dr\,d\theta\,dz$$

2. Transformation from Cartesian coordinates (x, y, z) to spherical polar coordinates (r, θ, φ)

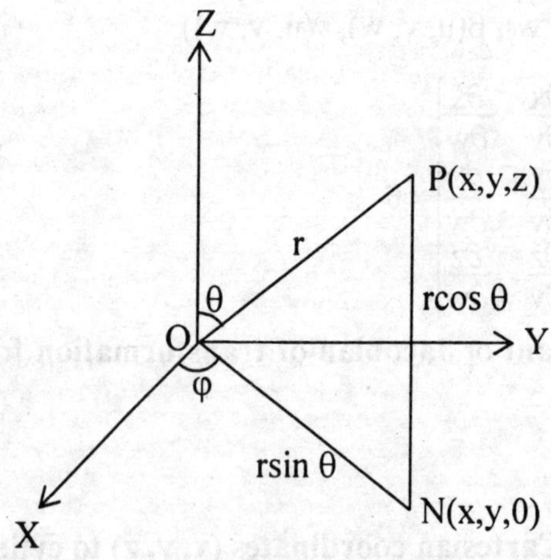

Fig. 1.27

The transformation equations are x = r sin θ cos φ, y = r sin θ sin φ, z = r cos θ (Fig 1.27)

$$J = \begin{vmatrix} \frac{\partial x}{\partial r} & \frac{\partial x}{\partial \theta} & \frac{\partial x}{\partial \varphi} \\ \frac{\partial y}{\partial r} & \frac{\partial y}{\partial \theta} & \frac{\partial y}{\partial \varphi} \\ \frac{\partial z}{\partial r} & \frac{\partial z}{\partial \theta} & \frac{\partial z}{\partial \varphi} \end{vmatrix} = \begin{vmatrix} \sin\theta\cos\varphi & r\cos\theta\cos\varphi & -r\sin\theta\sin\varphi \\ \sin\theta\sin\varphi & r\cos\theta\sin\varphi & r\sin\theta\cos\varphi \\ \cos\theta & -r\sin\theta & 0 \end{vmatrix} = r^2\sin\theta$$

$$\therefore \iiint_R f(x,y,z)\,dx\,dy\,dz = \iiint_{R'} f(r\sin\theta\cos\varphi, r\sin\theta\sin\varphi, r\cos\theta)\, r^2\sin\theta\, dr\, d\theta\, d\varphi$$

Example1.4.1 By transforming into polar coordinates evaluate the double integral $\iint_R e^{-(x^2+y^2)}dy\,dx$, where R is the region bounded by the circle $x^2 + y^2 = a^2$.

Solution:

Substituting x = r cos θ, y = r sin θ, we get dx dy = r dr dθ and $e^{-(x^2+y^2)} = e^{-r^2}$

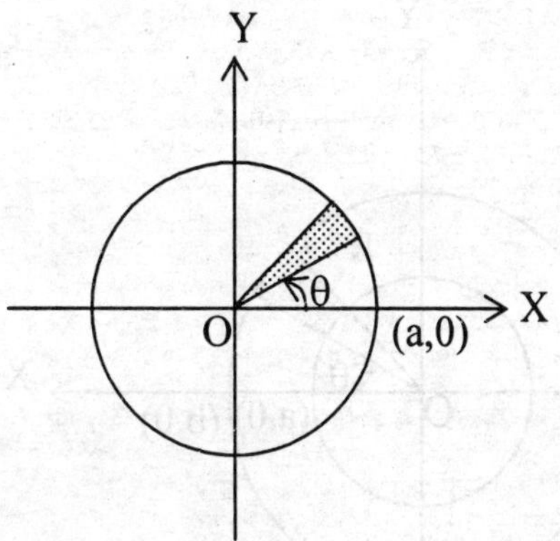

Fig. 1.28

The limits of integration are $0 \le r \le a$ and $0 \le \theta \le 2\pi$ (Fig 1.28)

$$\therefore \text{ The given integral } I = \int_{\theta=0}^{2\pi}\int_{r=0}^{a} e^{-r^2}\, r\, dr\, d\theta$$

$$= \int_{\theta=0}^{2\pi}\left[\left(\frac{-1}{2}\right)e^{-r^2}\right]_0^a d\theta$$

$$= \frac{-1}{2}\int_{\theta=0}^{2\pi}\left[e^{-a^2} - 1\right]d\theta$$

$$= \frac{-1}{2}\left[e^{-a^2} - 1\right]2\pi$$

$$= \pi\left[1 - e^{-a^2}\right]$$

Example 1.4.2 Evaluate $\iint_R \frac{xy}{\sqrt{x^2+y^2}}\,dx\,dy$, after transforming to polar coordinates, where R is the region between the circles $x^2 + y^2 = a^2$ and $x^2 + y^2 = b^2$ in the first quadrant $(a < b)$.

Solution:

Put $x = r\cos\theta, \quad y = r\sin\theta$

Then $dxdy = r\,dr\,d\theta$, and

$$\frac{xy}{\sqrt{x^2+y^2}} = \frac{r\cos\theta\, r\sin\theta}{r}$$

$$= r\cos\theta\sin\theta$$

The limits of integration are $a \le r \le b$ and $0 \le \theta \le \pi/2$.

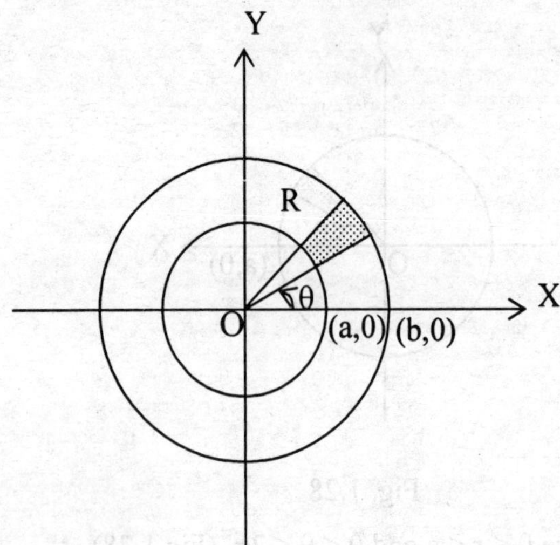

Fig. 1.29

$$\therefore \quad I = \int_0^{\pi/2}\int_a^b r\cos\theta\sin\theta\, r\,dr\,d\theta$$

$$= \int_0^{\pi/2}\left[\frac{r^3}{3}\right]_a^b \cos\theta\sin\theta\,d\theta$$

$$= \frac{1}{3}(b^3 - a^3)\int_0^{\pi/2}\cos\theta\sin\theta\,d\theta$$

$$= \frac{1}{3}(b^3 - a^3)\left[\frac{\sin^2\theta}{2}\right]_0^{\pi/2}$$

$$= \frac{1}{6}(b^3 - a^3)$$

Example 1.4.3 Evaluate $\iint \frac{xy(x+y)^2}{x^2+y^2} dx\,dy$ taken over the sector in the first quadrant bounded by the straight lines y = 0, y = x and the circle $x^2 + y^2 = 1$.

Solution:

Transform the integral into polar coordinates by substituting

$x = r\cos\theta$, $y = r\sin\theta$. Then $dx\,dy = r\,dr\,d\theta$, and

$$\frac{xy(x+y)^2}{x^2+y^2} = \frac{r\cos\theta\, r\sin\theta(r\cos\theta + r\sin\theta)^2}{r^2} = r^2\cos\theta\sin\theta(1+\sin 2\theta)$$

The limits of integration are $o \le \theta \le \pi/4$ and $0 \le r \le 1$(Fig. 1.30)

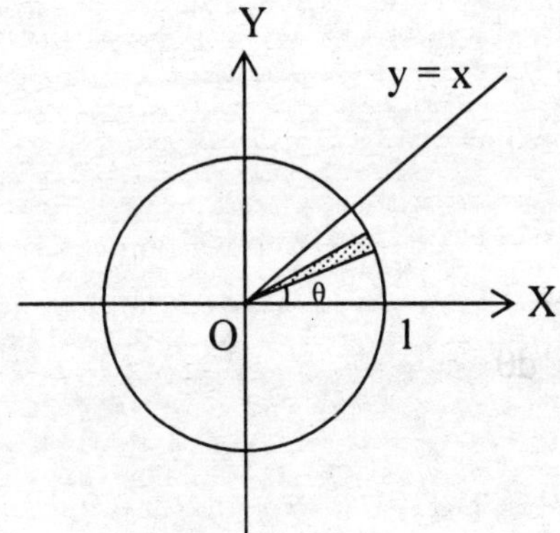

Fig. 1.30

$$I = \int_0^{\pi/4}\int_0^1 r^2\cos\theta\sin\theta(1+\sin 2\theta)\, r dr\, d\theta$$

$$= \int_0^{\pi/4}\int_0^1 r^3 \sin\theta\cos\theta\,(1+\sin 2\theta)\,dr\,d\theta$$

$$= \int_0^{\pi/4} \frac{1}{4}\left(\sin\theta\cos\theta + \sin\theta\cos\theta\sin 2\theta\right)d\theta$$

$$= \frac{1}{4}\int_0^{\pi/4}\sin\theta\cos\theta\,d\theta + \frac{1}{8}\int_0^{\pi/4}\sin^2 2\theta\,d\theta$$

$$= \frac{1}{4}\left[\frac{\sin^2\theta}{2}\right]_0^{\pi/4} + \frac{1}{16}\int_0^{\pi/2}\sin^2 t\,dt \text{ (Substituting } 2\theta = t)$$

$$= \frac{1}{16} + \frac{1}{16}\cdot\frac{1}{2}\cdot\frac{\pi}{2}$$

$$= \frac{4+\pi}{64}$$

Example 1.4.4 Find the volume within the cylinder r = 4cos θ bounded by the sphere $r^2 + z^2 = 16$ and above the plane z = 0.

Solution:

We use cylindrical coordinates.

Volume in cylindrical coordinates is given by $\iiint_R r\,dr\,d\theta\,dz$

To evaluate this integral over R, first integrate w.r.t z from z = 0 to $z = \sqrt{16 - r^2}$, then integrate w.r.t r from r = 0 to r = 4cos θ and finally integrate w.r.t θ from θ = 0 to θ = π

$$\therefore\ V = \int_0^{\pi}\int_0^{4\cos\theta}\int_0^{\sqrt{16-r^2}} r\,dz\,dr\,d\theta$$

$$= \int_0^{\pi}\int_0^{4\cos\theta} r\sqrt{16 - r^2}\ dr\,d\theta$$

$$= \int_0^{\pi}\left[\left(-\frac{1}{2}\right)\frac{(16-r^2)^{3/2}}{3/2}\right]_0^{4\cos\theta} d\theta$$

$$= \frac{-64}{3}\int_0^{\pi}(\sin^3\theta - 1)d\theta$$

$$= \frac{-64}{3}\int_0^{\pi}\left(\frac{3}{4}\sin\theta - \frac{1}{4}\sin 3\theta - 1\right)d\theta$$

$$= \frac{-64}{3}\left[-\frac{3}{4}\cos\theta + \frac{1}{12}\cos 3\theta - \theta\right]_0^{\pi}$$

$$= \frac{-64}{3}\left[\frac{3}{4} - \frac{1}{12} - \pi + \frac{3}{4} - \frac{1}{12}\right]$$

$$= \frac{-64}{3}\left[\frac{4}{3} - \pi\right]$$

$$= \frac{64}{9}(3\pi - 4) \text{ cubic units}$$

Example 1.4.5 Find the volume of the region above the xy-plane bounded by the paraboloid $z = x^2 + y^2$ and the cylinder $x^2 + y^2 = a^2$.

Solution:

The required volume $= 4\iiint\limits_R dx\,dy\,dz$ where R is the region in the first octant.

Transforming the variables to cylindrical coordinates by the transformations $x = r\cos\theta$, $y = r\sin\theta$, $z = z$, we have $dx\,dy\,dz = r dr\,d\theta\,dz$

The equation $z = x^2 + y^2$ of the parabolid becomes $z = r^2$ and the equation $x^2 + y^2 = a^2$ of the cyclider becomes $r^2 = a^2$ i.e., $r = a$. The limits for the region of integration R are $z = 0$ to r^2, $r = 0$ to a and $\theta = o$ to $\pi/2$

$$\therefore I = 4\int_0^{\pi/2}\int_0^a\int_0^{r^2} r\,dz\,dr\,d\theta$$

$$= 4\int_0^{\pi/2}\int_0^a [r.z]_0^{r^2}\,dr\,d\theta$$

$$= 4\int_0^{\pi/2}\int_0^a r^3 dr\,d\theta$$

$$= 4\int_0^{\pi/2}\left[\frac{r^4}{4}\right]_0^a d\theta$$

$$= 4\int_0^{\pi/2}\frac{a^4}{4}d\theta$$

$$= a^4.\frac{\pi}{2} = \frac{\pi a^4}{2}$$

Example 1.4.6 Find the volume of the region bounded above by the sphere $x^2 + y^2 + z^2 = a^2$ and below by the cone $x^2 + y^2 = z^2\tan^2\alpha$ where $0 \le \alpha \le \pi$.

Solution:

We transform the variables x, y, z to spherical polar coordinates r, θ, φ.

$x = r\sin\theta\cos\varphi$, $y = r\sin\theta\sin\varphi$, $z = r\cos\theta$. Then $dx\,dy\,dz = r^2\sin\theta\,dr\,d\theta\,d\varphi$.

Equation of the sphere becomes,

$r^2\sin^2\theta\cos^2\varphi + r^2\sin^2\theta\sin^2\varphi + r^2\cos^2\theta = a^2$

i.e., $r = a$.

Equation of the cone becomes,

$r^2\sin^2\theta\cos^2\varphi + r^2\sin^2\theta\sin^2\varphi = r^2\cos^2\theta.\tan^2\alpha$

i.e., $r^2\sin^2\theta = r^2\cos^2\theta\tan^2\alpha$

i.e., $\tan^2\theta = \tan^2\alpha$

i.e., $\theta = \alpha$

$$\text{Required volume} = 4\int_{\varphi=0}^{\pi/2}\int_{\theta=0}^{\alpha}\int_{r=0}^{a} r^2 \sin\theta \, dr\, d\theta\, d\varphi$$

$$= 4\int_{\theta=0}^{\pi/2}\int_{\theta=0}^{\alpha}\left[\frac{r^3}{3}\sin\theta\right]_0^a d\theta\, d\varphi$$

$$= 4\int_{\theta=0}^{\pi/2}\int_{\theta=0}^{\alpha}\frac{a^3}{3}\sin\theta\, d\theta\, d\varphi$$

$$= 4\int_{\theta=0}^{\pi/2}\left[-\frac{a^3}{3}\cos\theta\right]_0^{\alpha} d\varphi$$

$$= 4\int_{\theta=0}^{\pi/2}\frac{a^3}{3}(1-\cos\alpha)d\varphi$$

$$= 4\frac{a^3}{3}(1-\cos\alpha).\frac{\pi}{2}$$

$$= \frac{2\pi a^3}{3}(1-\cos\alpha)$$

Example 1.4.7 Evaluate $\iint [xy(1-x-y)]^{1/2}\,dxdy$ taken over the triangle with sides $x = 0, y = 0, x + y = 1$.

Solution:

Substituting $x + y = u$, $y = uv$, we have $x = u(1-v)$, $y = uv$

$$\therefore J = \frac{\partial(x,y)}{\partial(u,v)} = \begin{vmatrix} \frac{\partial x}{\partial u} & \frac{\partial x}{\partial v} \\ \frac{\partial y}{\partial u} & \frac{\partial y}{\partial v} \end{vmatrix}$$

$$= \begin{vmatrix} 1-v & -u \\ v & u \end{vmatrix} = u$$

The triangle with sides $x = 0$, $y = 0$, $x + y = 1$ is transformed into a square with sides $u = 0, u = 1, v = 0, v = 1$.

$$\therefore \iint [xy(1-x-y)]^{1/2}\,dxdy = \int_0^1\int_0^1 [u(1-v)uv(1-u)]^{1/2}.u\,du\,dv$$

$$= \int_0^1\int_0^1 u(1-v)^{1/2} v^{1/2} (1-u)^{1/2}.u\,du\,dv$$

$$= \int_0^1 u^2(1-u)^{1/2}du \int_0^1 v^{1/2}(1-v)^{1/2}dv$$

$$= I_1.I_2 \qquad (1)$$

In I_1 put $u = \sin^2\theta$, $du = 2\sin\theta\cos\theta\, d\theta$

$$\therefore I_1 = \int_0^{\pi/2} \sin^4\theta \cos\theta . 2\sin\theta \cos\theta \, d\theta$$

$$= 2\int_0^{\pi/2} \sin^5\theta(1-\sin^2\theta)\, d\theta$$

$$= 2\int_0^{\pi/2} \sin^5\theta\, d\theta - 2\int_0^{\pi/2} \sin^7\theta\, d\theta$$

$$= 2.\frac{4}{5}.\frac{2}{3}.1 - 2.\frac{6}{7}.\frac{4}{5}.\frac{2}{3}.1$$

$$= \frac{16}{15}\left(1-\frac{6}{7}\right) = \frac{16}{105}$$

Similarly substituting $v = \sin^2\theta$ in I_2 we get,

$$I_2 = \int_0^{\pi/2} \sin\theta\cos\theta . 2\sin\theta\cos\theta\, d\theta$$

$$= 2\int_0^{\pi/2} \sin^2\theta(1-\sin^2\theta)\, d\theta$$

$$= 2\int_0^{\pi/2} \sin^2\theta\, d\theta - 2\int_0^{\pi/2} \sin^4\theta\, d\theta$$

$$= 2.\frac{1}{2}.\frac{\pi}{2} - 2.\frac{3}{4}.\frac{1}{2}.\frac{\pi}{2} = \frac{\pi}{8}$$

$\therefore$ From (1) we get, the given integral $= \dfrac{16}{105}\dfrac{\pi}{8} = \dfrac{2\pi}{105}$

Example 1.4.8 Evaluate $\iiint [xyz(1-x-y-z)]^{1/2} dx\, dy\, dz$ taken over the tetrahedron bounded by the planes $x = 0$, $y = 0$, $x + y + z = 1$.

Solution:

Substituting $x + y + z = u$, $y + z = uv$, $z = uvw$. Then $x = u(1 - v)$, $y = uv(1 - w)$, $z = uvw$.

$$J = \frac{\partial(x,y,z)}{\partial(u,v,w)} = \begin{vmatrix} 1-v & -u & 0 \\ v(1-w) & u(1-w) & -uv \\ vw & uw & uv \end{vmatrix} = u^2 v$$

The tetrahedron is transformed into a cube, whose sides are the planes u = 0, u = 1, v = 0, v = 1, w = 0 and w = 1.

$\therefore$ The given integral $= \int_0^1\int_0^1\int_0^1 [u^3v^2w(1-u)(1-v)(1-w)]^{1/2} u^2 v \, du\, dv\, dw$

$$= \int_0^1 u^{7/2}(1-u)^{1/2}\, du \int_0^1 v^2(1-v)^{1/2}\, dv \int_0^1 w^{1/2}(1-w)^{1/2}\, dw$$

$$= I_1 . I_2 . I_3$$

Substituting $u = \sin^2\theta$ in I_1, we get,

$$I_1 = \int_0^{\pi/2} \sin^7\theta \cos\theta . 2\sin\theta\cos\theta \, d\theta$$

$$= 2\int_0^{\pi/2} \sin^8\theta\,(1-\sin^2\theta)\, d\theta$$

$$= 2\left[\frac{7}{8}.\frac{5}{6}.\frac{3}{4}.\frac{1}{2}.\frac{\pi}{2} - \frac{9}{10}.\frac{7}{8}.\frac{5}{6}.\frac{3}{4}.\frac{1}{2}.\frac{\pi}{2}\right] = \frac{35\pi}{1280}$$

Similarly substituting $v = \sin^2\theta$ in I_2, we get,

$$I_2 = \int_0^{\pi/2} \sin^4\theta \cos\theta . 2\sin\theta\cos\theta \, d\theta$$

$$= 2\int_0^{\pi/2} \sin^5\theta \cos^2\theta \, d\theta = \frac{16}{105}$$

Similarly substituting $w = \sin^2\theta$ in I_3, we get,

$$I_3 = \int_0^{\pi/2} \sin\theta \cos\theta . 2\sin\theta\cos\theta \, d\theta$$

$$= 2\int_0^{\pi/2} \sin^2\theta \cos^2\theta \, d\theta = \frac{\pi}{8}$$

$\therefore$ The given integral $= \frac{35\pi}{1280}.\frac{16}{105}.\frac{\pi}{8}$

$$= \frac{\pi^2}{1920}$$

Example 1.4.9 Evaluate $\iiint xyz\left(\frac{x^2}{a^2}+\frac{y^2}{b^2}+\frac{z^2}{c^2}\right) dx\,dy\,dz$, over the positive octant of the ellipsoid $\frac{x^2}{a^2}+\frac{y^2}{b^2}+\frac{z^2}{c^2}=1$.

Solution:

First transform the given integral using the substitution

x = au, y = bv, z = cw. Then J = abc and dx dy dz = abc du dv dw and the ellipsoid is transformed into a sphere $u^2+v^2+w^2=1$. The integral becomes

$$I = \iiint au.bv.cw\,(u^2+v^2+w^2)abc\,du\,dv\,dw$$

$$= a^2b^2c^2 \iiint uvw\,(u^2+v^2+w^2)du\,dv\,dw$$

Now transform this integral to spherical polar coordinate by substituting $u = r\sin\theta\cos\varphi$, $v = r\sin\theta\sin\varphi$, $w = r\cos\theta$. Then $J = r^2\sin\theta$ and $du\,dv\,dw = r^2\sin\theta\,dr\,d\theta\,d\varphi$

The limits of integration are $0 \le r \le 1$, $0 \le \theta \le \pi/2$ and $0 \le \varphi \le \pi/2$

$$\therefore I = a^2b^2c^2 \int_0^1\int_0^{\pi/2}\int_0^{\pi/2} r^3\sin^2\theta.\cos\theta.\sin\varphi.\cos\varphi.r^2.r^2\sin\theta\,dr\,d\theta\,d\varphi$$

$$= a^2b^2c^2 \int_0^1 r^7dr \int_0^{\pi/2}\sin^3\theta.\cos\theta\,d\theta \int_0^{\pi/2}\sin\varphi.\cos\varphi\,d\varphi$$

$$= a^2b^2c^2\left[\frac{r^8}{8}\right]_0^1\left[\frac{\sin^4\theta}{4}\right]_0^{\pi/2}\left[\frac{\sin^2\varphi}{2}\right]_0^{\pi/2}$$

$$= a^2b^2c^2.\frac{1}{8}.\frac{1}{4}.\frac{1}{2} = \frac{a^2b^2c^2}{64}$$

Example 1.4.10 Find the volume of the solid surrounded by the surface

$$\left(\frac{x}{a}\right)^{2/3}+\left(\frac{y}{b}\right)^{2/3}+\left(\frac{z}{c}\right)^{2/3}=1.$$

Solution:

Change the variables x, y, z to u, v, w, where

$$\left(\frac{x}{a}\right)^{1/3}=u,\quad \left(\frac{y}{c}\right)^{1/3}=v,\quad \left(\frac{z}{c}\right)^{1/3}=w.$$

i.e., $x = au^3$, $y = bv^3$, $z = cw^3$. Then we get,

$$J = \frac{\partial(x,y,z)}{\partial(u,v,w)} = \begin{vmatrix} 3au^2 & 0 & 0 \\ 0 & 3bv^2 & 0 \\ 0 & 0 & 3cw^2 \end{vmatrix} = 27abcu^2v^2w^2$$

Hence the required volume $= \iiint dx\,dy\,dz = 27abc \iiint u^2v^2w^2\,du\,dv\,dw$
Integrated over the sphere $u^2 + v^2 + w^2 = 1$.
Now change the variables u, v, w to spherical polar coordinates r, θ, φ

$u = r\sin\theta\cos\varphi,\ v = r\sin\theta\sin\varphi,\ w = r\cos\theta$. Then we have, $J = \dfrac{\partial(u,v,w)}{\partial(r,\theta,\varphi)} = r^2\sin\theta$

Hence the required volume

$$= 8.27abc\int_0^1\int_0^{\pi/2}\int_0^{\pi/2} r^2\sin^2\theta\cos^2\varphi\, r^2\sin^2\theta\sin^2\varphi\, r^2\cos^2\theta\, r^2\sin\theta\, dr\, d\theta\, d\varphi$$

$$= 216abc\int_0^1 r^8 dr\int_0^{\pi/2}\sin^5\theta\cos^2\theta\, d\theta\int_0^{\pi/2}\sin^2\varphi\cos^2\varphi\, d\varphi$$

$$= 216abc.\, I_1.\, I_2.\, I_3$$

$$I_1 = \int_0^1 r^8 dr = \frac{1}{9}$$

$$I_2 = \int_0^{\pi/2}\sin^5\theta.\cos^2\theta\, d\theta = \int_0^{\pi/2}\sin^5\theta.(1-\sin^2\theta)\, d\theta$$

$$= \int_0^{\pi/2}(\sin^5\theta - \sin^7\theta)\, d\theta$$

$$= \frac{4}{5}.\frac{2}{3}.1 - \frac{6}{7}.\frac{4}{5}.\frac{2}{3}.1$$

$$= \frac{8}{105}$$

$$I_3 = \int_0^{\pi/2}\sin^2\varphi.\cos^2\varphi\, d\varphi$$

$$= \int_0^{\pi/2}(\sin^2\varphi - \sin^4\varphi)\, d\varphi$$

$$= \frac{1}{2}.\frac{\pi}{2} - \frac{3}{4}.\frac{1}{2}.\frac{\pi}{2}$$

$$= \frac{\pi}{16}$$

$\therefore$ Required volume $= 216abc.\dfrac{1}{9}\dfrac{8}{105}.\dfrac{\pi}{16}$

$$= \frac{4\pi abc}{35}$$

EXERCISE 1.4

PART-A

1. What is the Jocobian of the transformation $x = r\cos\theta$, $y = r\sin\theta$?
2. Transform the volume element dx dy dz to cylindrical coordinates.
3. Transform the volume element dx dy dz to spherical polar coordinates.
4. What is the Jacobian of the transformation from Cartesian coordinates to spherical polar coordinates?
5. What is the image of the sphere $x^2 + y^2 + z^2 = a^2$, when it is transformed into spherical polar coordinates.
6. State the transformation equations that transform the ellipsoid $\frac{x^2}{a^2} + \frac{y^2}{b^2} + \frac{z^2}{c^2} = 1$ into a sphere of unit radius.
7. State the triple integral in cylindrical coordinates for finding volume.
8. State the triple integral in spherical polar coordinates for finding volume.
9. Express the area element dx dy in terms of du dv when $x = au$, $y = bv$.
10. Express the area element dx dy in terms of du dv when $x = u\cos\alpha - v\sin\alpha$, $y = u\sin\alpha + v\cos\alpha$.

PART-B

Evaluate the following integrals by changing them into polar co-ordinates (Ex: 11 – 24)

11. $\int_0^{2a}\int_0^{\sqrt{2ax-x^2}} (x^2 + y^2)\, dx\, dy$

12. $\int_0^{\infty}\int_0^{\infty} e^{-(x^2+y^2)}\, dx\, dy$. Hence evaluate $\int_0^{\infty} e^{-x^2}\, dx$

13. $\int_0^{\infty}\int_0^{\infty} \frac{dx\, dy}{(x^2 + y^2 + a^2)^2}$

14. $\int_0^{2}\int_0^{\sqrt{2x-x^2}} \frac{x\, dx\, dy}{x^2 + y^2}$

15. $\int_0^{a}\int_{\sqrt{ax-x^2}}^{\sqrt{a^2-x^2}} \frac{dx\, dy}{\sqrt{a^2 - x^2 - y^2}}$

16. $\iint \frac{dx\, dy}{(1 + x^2 + y^2)^2}$, over one loop of the lemniscate $(x^2 + y^2)^2 = x^2 - y^2$.

17. $\iint_R \left(\frac{1 - x^2 - y^2}{1 + x^2 + y^2}\right)^{1/2} dx\, dy$, where R is the positive quadrant of the circle $x^2 + y^2 = 1$

18. $\int_0^a\int_0^a \frac{x^2\,dx\,dy}{(x^2+y^2)^{3/2}}$

19. $\int_0^1\int_0^{\sqrt{x-x^2}} \frac{4xy}{x^2+y^2}e^{-(x^2+y^2)}\,dy\,dx$

20. $\int_0^a\int_0^{\sqrt{a^2-x^2}} \sqrt{a^2-x^2-y^2}\,dy\,dx$

21. $\iint_R \sqrt{x^2+y^2}\,dx\,dy$, where R is the region bounded by $x^2+y^2=4$ and $x^2+y^2=9$.

22. $\int_0^a\int_0^{\sqrt{a^2-x^2}} (x^2+y^2)\,dy\,dx$

23. $\int_0^a\int_y^a \frac{x^2\,dy\,dx}{\sqrt{x^2+y^2}}$

24. $\int_0^{4a}\int_{y^2/4a}^{a} \frac{x^2-y^2}{x^2+y^2}\,dy\,dx$

Evaluate the following integrals by changing to spherical polar coordinates (Ex 25 – 34)

25. $\int_0^1\int_0^{\sqrt{1-x^2}}\int_{\sqrt{x^2+y^2}}^{1} \frac{1}{\sqrt{x^2+y^2+z^2}}\,dz\,dy\,dx$

26. $\int_0^1\int_0^{\sqrt{1-x^2}}\int_0^{\sqrt{1-x^2-y^2}} \frac{1}{\sqrt{1-x^2-y^2-z^2}}\,dx\,dy\,dz$

27. $\iiint z^2\,dx\,dy\,dz$, taken over the volume bounded by the surfaces $x^2+y^2=a^2$, $x^2+y^2=z$ and $z=0$.

28. $\iiint \frac{dx\,dy\,dz}{x^2+y^2+z^2}$, over the interior of the sphere $x^2+y^2+z^2=a^2$

29. $\iiint \frac{dx\,dy\,dz}{a^2+x^2+y^2+z^2}$, over the interior of the sphere $x^2+y^2+z^2=a^2$

30. $\iiint \frac{xyz}{\sqrt{x^2+y^2+z^2}}\,dx\,dy\,dz$, taken over the positive octant of the interior of the sphere $x^2+y^2+z^2=a^2$.

31. $\iiint (ax+by+cz)^2\,dx\,dy\,dz$, taken over the interior of the sphere $x^2+y^2+z^2=1$.

32. $\iiint\limits_R \sqrt{1-x^2-y^2-z^2}\,dx\,dy\,dz$, taken over the interior of the sphere $x^2+y^2+z^2=1$.

33. $\iiint\limits_R \left(\frac{1-x^2-y^2-z^2}{1+x^2+y^2+z^2}\right)dx\,dy\,dz$, where R is the positive octant of the interior of the sphere $x^2+y^2+z^2=1$.

Evaluate the following integrals by changing to cylindrical coordinates. (Ex: 34 -35)

34. $\iiint\limits_R z^2\,dx\,dy\,dz$, where R is the region common to the interior of the sphere $x^2+y^2+z^2=a^2$ and the cylinder $x^2+y^2=ax$.

35. $\iiint\limits_R \sqrt{x^2+y^2+z^2}\,dx\,dy\,dz$, where R is the region bounded by the plane $z=3$ and the cone $z=x^2+y^2$.

36. Find the volume of the portion of the sphere $x^2+y^2+z^2=a^2$ lying inside the cylinder $x^2+y^2=ax$ (use cylindrical coordinates).
37. Find the volume enclosed by the cylinder $x^2+y^2=2ax$ and $z^2=2ax$.
38. Find the volume of the cylinder $x^2+y^2-2ax=0$ intercepted between the paraboloid $x^2+y^2=2az$ and xy-plane.
39. Find the volume bounded by the cylinder $x^2+y^2=4$ and the hyperboloid $x^2+y^2+z^2=1$.
40. Evaluate $\iiint\limits_R x^2y\,dx\,dy\,dz$, where R is $x^2+y^2\le 1, 0\le z\le 1$.
41. Evaluate $\iiint\limits_R x^2y^2z\,dx\,dy\,dz$, where R is $x^2+y^2\le 1, 0\le z\le 1$.
42. Evaluate $\iint\limits_R (x-y)^2\cos^2(x+y)\,dx\,dy$, where R is the rhombus with successive vertices at $(\pi, 0)$, $(2\pi, \pi)$, $(\pi, 2\pi)$ and $(0, \pi)$ (substitute $u = y - x$, $v = y + x$).
43. Evaluate $\iint\limits_R (x+y)^2\,dx\,dy$, where R is the parallelogram in the xy-plane with vertices (1, 0), (3, 1), (2, 2), (0, 1) using the transformation $u = x + y$, $v = x - 2y$.
44. Evaluate $\iint\limits_R \sqrt{x+y}\,dx\,dy$, where R is the parallelogram bounded by the lines $x+y=0$, $x+y=1$, $2x-3y=0$, $2x-3y=4$.
45. Evaluate $\iint\limits_R \cos\left(\frac{x-y}{x+y}\right)dx\,dy$, where R is the region bounded by the lines $x+y=1$, $x=0$, $y=0$.(Put $x-y=u$, $x+y=v$)
46. Evaluate $\iint\limits_R (x-y)^4 e^{x+y}\,dx\,dy$, where R is the square with vertices (1, 0), (2, 1),

(1, 2) and (0, 1) using the transformation x + y = u, x-y = v.

47. Evaluate $\iint_R e^{\frac{x-y}{x+y}}\,dx\,dy$, where R is the triangle bounded by the x-axis, lines x = y and x + y = 1.

48. Use the substitution u = x + y, v = y to evaluate $\iint_R ye^{-(x+y)}\sin\left[\frac{\pi y^2}{(x+y)^2}\right]dx\,dy$, where R is the infinite positive quadrant.

49. Evaluate $\int_0^\infty\int_0^\infty e^{-(x+y)}\sin\left(\frac{\pi y}{x+y}\right)dx\,dy$, using the transformation u = x + y, v = y.

50. Evaluate $\iiint_R (x+y+z+1)^2\,dx\,dy\,dz$, where R is the tetrahedron formed by the coordinate planes and x + y + z = 1.

51. Evaluate $\iiint_R \sqrt{\frac{1-x-y-z}{xyz}}\,dx\,dy\,dz$, where R is the tetrahedron formed by the coordinate planes and x + y + z = 1.

52. Find the volume of the tetrahedron bounded by the coordinate planes and the plane $\frac{x}{a}+\frac{y}{b}+\frac{z}{c}=1$.

53. Find the volume of the ellipsoid $\frac{x^2}{a^2}+\frac{y^2}{b^2}+\frac{z^2}{c^2}=1$, using the transformation x = au, y = bv, z = cw.

54. Evaluate $\iiint_R xyzy\,dx\,dy\,dz$, where R is the positive octant of the ellipsoid $\frac{x^2}{a^2}+\frac{y^2}{b^2}+\frac{z^2}{c^2}=1$.

55. Evaluate $\iiint_R \left(1-\frac{x^2}{a^2}-\frac{y^2}{b^2}-\frac{z^2}{c^2}\right)^{1/2}dx\,dy\,dz$, where R is the interior of the ellipsoid $\frac{x^2}{a^2}+\frac{y^2}{b^2}+\frac{z^2}{c^2}=1$.

56. Evaluate the integral $\int_0^2\int_0^x \left[(x-y)^2+2(x+y)+1\right]^{-1/2}dx\,dy$, by using the transformation x = u(1 + v), y = v(1 + u).

57. Using the transformation $x=\frac{u}{1+v^2}$, $y=\frac{uv}{1+v^2}$, evaluate $\iint_R \frac{dx\,dy}{x+y}$ where R is region for which $x^2+y^2-x<0<y$.

58. Evaluate $\iiint_R xy^2z^3\,dx\,dy\,dz$, where R is the region bounded by x + y + z = 1 and

the coordinate planes.

59. Evaluate $\iiint_R \sqrt{xyz}\, dx\, dy\, dz$, where R is the region bounded by $x^3 + y^3 + z^3 = 8$ and the coordinate planes. (Use the transformation $u = (x/2)^3$, $v = (y/2)^3$, $w = (z/2)^3$.

ANSWERS

EXERCISE 1.1

(5) $\frac{a^3}{6}$ (6) $\int_0^1\int_0^{1-y} x\, dx\, dy$ (7) $\frac{ab}{3}(a^2+b^2)$

(8) $\frac{47}{30}$ (9) $\frac{67}{60}$ (10) $\frac{4\pi a^3}{3}$

(11) $\frac{32}{9}$ (12) $\frac{1}{3}$ (13) $\frac{3}{35}$

(14) $\frac{1}{2}$ (15) $\frac{1}{40}$ (16) $\frac{36}{5}$

(17) $\int_0^1\int_{x^2}^{2x} dy\, dx$ (18) $\frac{3}{2}$ (19) $\frac{1}{20}$

(20) 4 (21) 5 (22) $\frac{16\sqrt{2}}{3}$

(23) $\frac{8}{3}$ (24) $\frac{9}{2}$ (25) $\frac{9}{2}$

(26) $\frac{a^4}{8}$ (27) $\frac{44}{15}\sqrt{2}+\frac{5}{3}$ (28) πa^3

(29) $\frac{1286}{105}$ (30) $\frac{\pi}{24}$ (31) ½ (e-1)

(32) $\frac{9}{24}$ (33) 8 (34) $\frac{32}{3}$

(35) $\frac{8}{3}\pi + 4\sqrt{3} - 4$ (36) $8\left(\frac{2}{3}\pi + \sqrt{3}\right)$ (37) 6π

(38) $\frac{4a^2}{3}$ (39) $\frac{\pi a^2\sqrt{2}}{16}$ (40) $\frac{\pi a^4}{128}$

(41) $\frac{128}{3} + 8\pi$ (42) 3π (43) $\frac{1485}{16}$

(44) 8 (45) $\frac{1024}{3}$ (46) $\frac{14\sqrt{2}}{5}$

(47) π (48) $\dfrac{81\pi}{16}$ (49) $\dfrac{27\pi}{2}$

(50) $\dfrac{2a^3}{3}$ (51) $\dfrac{16a^3}{3}$ (52) 3π

(53) 4π.

EXERCISE 1.2

(1) $\int_3^4\int_1^2 f(x,y)\,dx\,dy$

(2) $\int_0^4\int_{x/4\sqrt{2}}^{\sqrt{x}} f(x,y)\,dx\,dy + \int_4^{8\sqrt{2}}\int_{x/4\sqrt{2}}^{2} f(x,y)\,dx\,dy$

(3) $\int_{-2}^{1}\int_{-y}^{\sqrt{2-y}} f(x,y)\,dx\,dy + \int_1^2\int_{-\sqrt{2-y}}^{\sqrt{2-y}} f(x,y)\,dx\,dy$

(4) $\int_{a^2/b}^{a}\int_{a^2/y}^{b} f(x,y)\,dx\,dy + \int_a^b\int_y^b f(x,y)\,dx\,dy$

(5) $\int_0^a\int_0^{\sqrt{a^2-y^2}} f(x,y)\,dy\,dx$

(6) $\int_0^1\int_{-\sqrt{1-y^2}}^{\sqrt{1-y^2}} f(x,y)\,dx\,dy$

(7) $\int_0^1\int_{x^{3/2}}^{x} dy\,dx$

(8) $\int_0^3\int_0^{x/2} e^{x^2}\,dy\,dx$

(9) $\int_0^1\int_{\sqrt{y}}^{1} xe^y\,dx\,dy$

(10) $\int_0^1\int_0^{\sqrt{y}} xy\,dy\,dx + \int_1^2\int_0^{2-y} xy\,dy\,dx$

(11) $\dfrac{\pi a}{4}$ (12) $\dfrac{31}{60}$ (13) $\text{Sinh}^{-1}\dfrac{b^2}{2}$

(14) $\dfrac{a^3}{6}$ (15) $\dfrac{3}{8}a^4$ (16) $\dfrac{\pi a}{2}$

(17) $\dfrac{511}{60}$ (18) $1-\dfrac{1}{\sqrt{2}}$ (19) $\dfrac{3}{8}$

(20) $\dfrac{16a^2}{3}$ (21) $\dfrac{1}{2}$

EXERCISE 1.3

(4) abc (5) 9 (6) $\dfrac{4}{35}$

(7) $\dfrac{13}{240}$ (8) $\dfrac{2}{3}$ (9) $\dfrac{1}{24}$

(10) $(2-\sqrt{2})\pi$ (11) 32 (12) 16

(13) ½ (14) $\frac{4}{3}a^b$ (15) $\frac{3\pi}{8}a^3$

(16) $\frac{a^6}{48}$ (17) $\frac{1}{2}abc(a+b+c)$ (18) 1

(19) $\frac{1}{4}$

(20) $\frac{a^6}{48}$ (21) $\frac{a^3b^2c^2}{2520}$ (22) $\frac{31}{60}$

(23) $\frac{16\pi}{3}$ (24) $\frac{\pi a^2h^2}{2}$ (25) $\frac{a^5}{20}$

(26) $\frac{a^8}{64}$ (27) πa^3 (28) $\frac{3\pi a^3}{2}$

(29) $\frac{\pi}{2}$ (30) $\frac{16\pi}{3}$ (31) $\frac{128a^3}{15}$

(32) $\frac{\pi a^4}{4}$

EXERCISE 1.4

(1) r (2) r dr dθ dz (3) $r^2\sin\theta\, dr\, d\theta\, d\varphi$

(4) $r^2\sin\theta$ (5) r = a (6) x = au, y = bv, z = cw

(7) $\iiint_R r\, dr\, d\theta\, dz$ (8) $\iiint_R r^2\sin\theta\, dr\, d\theta\, d\varphi$ (9) dx dy = ab dud v

(10) dx dy = du dv (11) $\frac{3}{4}\pi a^4$ (12) $\frac{\pi}{4}, \frac{\sqrt{\pi}}{4}$

(13) $\frac{\pi}{4}a^2$ (14) $\frac{\pi}{2}$ (15) $\frac{a^3}{8}(3\pi-4)$

(16) $\frac{1}{2}\left(1-\frac{\pi}{2}\right)$ (17) $\frac{\pi}{4}\left(\frac{\pi}{2}-1\right)$ (18) $\frac{a}{\sqrt{2}}$

(19) $\frac{1}{e}$ (20) $\frac{1}{6}\pi a^3$ (21) $\frac{38}{3}\pi$

(22) $\frac{\pi a^4}{8}$ (23) $\frac{a^3}{3}\log(1+\sqrt{2})$ (24) $\left(\frac{\pi}{2}-\frac{5}{3}\right)8a^2$

(25) $(\sqrt{2}-1)\frac{\pi}{4}$ (26) $\frac{\pi^2}{8}$ (27) $\frac{\pi a^8}{12}$

(28) $4\pi a$ (29) $\pi a(4-\pi)$ (30) $\frac{a^5}{40}$

(31) $\frac{4\pi}{15}(a^2+b^2+c^2)$ (32) $\frac{\pi^2}{4}$ (33) $\frac{\pi}{2}\left(\frac{5}{3}-\frac{\pi}{2}\right)$

(34) $\frac{2\pi a^5}{15}$ (35) $\frac{27\pi}{2}(2\sqrt{2}-1)$ (36) $\frac{2a^3}{3}(3\pi-4)$

(37) $\frac{128a^3}{15}$ (38) $3\pi a^3$ (39) $4\sqrt{3}\,\pi$

(40) 0 (41) $\frac{\pi}{48}$ (42) $\frac{\pi^4}{3}$ (43) 21

(44) $-\frac{8}{15}$ (45) $\frac{1}{2}\sin 1$ (46) $\frac{e^3-e}{5}$

(47) $\frac{e^2-1}{4e}$ (48) 2 (49) $\frac{2}{\pi}$

(50) $\frac{31}{60}$ (51) $\frac{\pi^2}{4}$ (52) $\frac{1}{6}abc$

(53) $\frac{4\pi}{3}abc$ (54) $\frac{a^2b^2c^2}{48}$ (55) $\frac{\pi^2 abc}{4}$

(56) $2\log 2 - \frac{1}{2}$ (57) $\frac{\pi}{4}$ (58) $\frac{12}{9!}$

(59) $\frac{64}{81}\sqrt{2}\,\pi$

CHAPTER 2

VECTOR CALCULUS

2.0 INTRODUCTION

In this chapter we shall study the differentiation and integration of vector functions and the applications of the theorems of Gauss, Stoke and Green in the evaluation of line, surface and volume integrals.

Definition 2.0.1 A **vector** is a mathematical quantity that possesses both a magnitude and a specific direction.

For example, quantities such as force, velocity and acceleration have a magnitude as well as a direction. Hence, they are vector quantities. Geometrically, a vector is a directed line segment, the length of the segment gives the magnitude and the direction of the segment gives the direction of the vector. If $\bar{a}$ (also denoted as **a**) is a vector quantity, its magnitude is denoted by $|\bar{a}|$ or a. If $\bar{a}$ is represented geometrically by the directed line segment $\overline{OP}$, then O is the **initial point** of the vector whose magnitude a = OP, denoted by $|\overline{OP}|$ and whose direction is from O to P. The line OP is called the **line of support** of the vector. A vector whose length is zero is called a **zero vector** or a **null vector** and is denoted by $\bar{0}$. A vector whose length is 1 is called a **unit vector** and is generally denoted by **e** or $\hat{n}$. The unit vectors along the coordinate axes OX, OY, OZ are denoted by **i, j, k** or $\mathbf{e_1}, \mathbf{e_2}, \mathbf{e_3}$ respectively.

Definition 2.0.2 If P(x, y, z) is a point in space and O is the origin, then the vector $\overline{OP}$ joining O to the point P is called the **position vector** of the point P and is denoted by $\underline{p}$.

Definition 2.0.3 If λ is a scalar and $\bar{a}$ is a nonzero vector then the **scalar multiplication** $\lambda\bar{a}$ is defined as a vector of length $|\lambda|.|\bar{a}|$, having the same direction as that of $\bar{a}$ if λ is positive, but the opposite direction if λ is negative. If $\lambda = 0$ or $\bar{a} = \bar{0}$, then $\lambda\bar{a}$ is the zero vector.

$(-\lambda)\bar{a}$ is denoted by $-\lambda\bar{a}$ and $(-1)\bar{a}$ by $-\bar{a}$.

Note:

Two non-zero vectors $\bar{a}$ and $\bar{b}$ are said to be collinear if they have the same or parallel lines of support. $\bar{a}$ and $\bar{b}$ are collinear if and only if $\bar{a} = \lambda\bar{b}$ for some scalar λ.

Definition 2.0.4 The **sum of two vectors** $\bar{a}$ and $\bar{b}$, written as $\bar{a}+\bar{b}$ is defined as a vector represented by the diagonal of a parallelogram, two of whose adjoining sides represent the vectors $\bar{a}$ and $\bar{b}$ (Fig. 2.1).

i.e., $\bar{a}+\bar{b} = \overline{OA}+\overline{OB} = \overline{OC}$

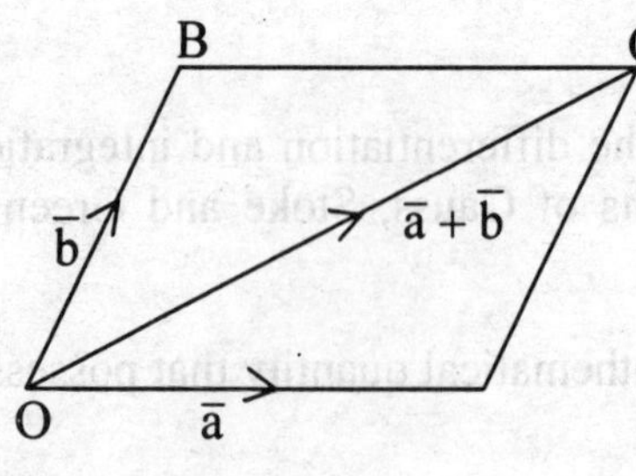

Fig. 2.1

Note:

(i) $\bar{a}+(-\bar{b})$ is denoted by $\bar{a}-\bar{b}$

(ii) $\bar{a}+\bar{b} = \bar{b}+\bar{a}$ (Commutative property)

(iii) $(\bar{a}+\bar{b})+\bar{c} = \bar{a}+(\bar{b}+\bar{c})$ (Associative property)

(iv) $\lambda(\bar{a}+\bar{b}) = \lambda\bar{a}+\lambda\bar{b}$ (Distributive property)

(v) $\bar{a}+\bar{0} = \bar{0}+\bar{a} = \bar{a}$ (Identity)

(vi) $|\bar{a}+\bar{b}| \le |\bar{a}|+|\bar{b}|$

(vii) $|\bar{a}|-|\bar{b}| \le |\bar{a}-\bar{b}|$

If P(x, y, z) is any point in space and O is the origin, then the position vector $\overline{OP}$ (also denoted as $\bar{r}$ or $\mathbf{r}$) of the point P can be expressed as $\bar{r} = x\mathbf{i}+y\mathbf{j}+z\mathbf{k}$ where **i, j, k** are the unit vectors along the coordinate axis OX, OY and OZ. Then $OP^2 = x^2+y^2+z^2$

$\therefore r = |\bar{r}| = |\overline{OP}| = \sqrt{x^2+y^2+z^2}$

If $\bar{a} = x_1\mathbf{i}+y_1\mathbf{j}+z_1\mathbf{k}$ and $\bar{b} = x_2\mathbf{i}+y_2\mathbf{j}+z_2\mathbf{k}$, then $\bar{b}-\bar{a} = (x_2-x_1)\mathbf{i}+(y_2-y_1)\mathbf{j}+(z_2-z_1)\mathbf{k}$.

Now, if A and B are any two points in space with position vectors $\bar{a}$ and $\bar{b}$ then

$\overline{AB} = \overline{OB}-\overline{OA} = \bar{b}-\bar{a} = (x_2-x_1)\mathbf{i}+(y_2-y_1)\mathbf{j}+(z_2-z_1)\mathbf{k}$

$\therefore |\overline{AB}| = AB = \sqrt{(x_2-x_1)^2+(y_2-y_1)^2+(z_2-z_1)^2}$

2.0.1 PRODUCT OF VECTORS

Definition 2.0.5 The **scalar product** $\bar{a}.\bar{b}$ of two vectors $\bar{a}$ and $\bar{b}$ is defined as the scalar $|\bar{a}||\bar{b}|\cos\theta$, where θ is the angle between the vectors.

i.e., $\bar{a}.\bar{b} = |\bar{a}||\bar{b}|\cos\theta = ab\cos\theta$

Note:

(i) $\bar{a}.\bar{b} = \bar{b}.\bar{a}$ (Commutative property).

(ii) $\bar{a}.\bar{a} = |\bar{a}||\bar{a}|\cos 0 = |\bar{a}|^2 = a^2$ ($\bar{a}.\bar{a}$ is written as $\bar{a}^2$)

(iii) $\mathbf{i}.\mathbf{i} = \mathbf{j}.\mathbf{j} = \mathbf{k}.\mathbf{k} = 1, \mathbf{i}.\mathbf{j} = \mathbf{j}.\mathbf{k} = \mathbf{k}.\mathbf{i} = 0$

(iv) $\bar{a}.(\bar{b}+\bar{c}) = \bar{a}.\bar{b} + \bar{a}.\bar{c}$ (Distributive property)

(v) If $\bar{a} = a_1\mathbf{i} + a_2\mathbf{j} + a_3\mathbf{k}$ and $\bar{b} = b_1\mathbf{i} + b_2\mathbf{j} + b_3\mathbf{k}$ then, $\bar{a}.\bar{b} = a_1b_1 + a_2b_2 + a_3b_3$

Definition 2.0.6 Let $\bar{a}$ and $\bar{b}$ be any two vectors and θ be the angle between them. Then the **vector product** $\bar{a}\times\bar{b}$ of $\bar{a}$ and $\bar{b}$ is defined as the vector whose magnitude is $|\bar{a}||\bar{b}|\sin\theta$ and whose direction is perpendicular to the plane of $\bar{a}$ and $\bar{b}$ such that $\bar{a}$, $\bar{b}$ and $\bar{a}\times\bar{b}$ form a right-handed system. (i.e., when we rotate a screw from $\bar{a}$ to $\bar{b}$, the top of the screw moves in the direction of $\bar{a}\times\bar{b}$)

i.e., $\bar{a}\times\bar{b} = |\bar{a}||\bar{b}|\sin\theta\,\hat{n}$ where $\hat{n}$ is the unit vector in the direction of $\bar{a}\times\bar{b}$.

Note:

(i) $\bar{a}\times\bar{b} = -\bar{b}\times\bar{a}$

(ii) $|\bar{a}\times\bar{b}| = |\bar{a}||\bar{b}|\sin\theta =$ Area of the parallelogram whose adjacent sides are $\bar{a}$ and $\bar{b}$.

(ii) $\mathbf{i}\times\mathbf{i} = \mathbf{j}\times\mathbf{j} = \mathbf{k}\times\mathbf{k} = 0, \mathbf{i}\times\mathbf{j} = \mathbf{k}, \mathbf{j}\times\mathbf{k} = \mathbf{i}$ and $\mathbf{k}\times\mathbf{i} = \mathbf{j}$

(iv) $\bar{a}\times(\bar{b}+\bar{c}) = \bar{a}\times\bar{b} + \bar{a}\times\bar{c}$ (Distributive property)

(v) If $\bar{a} = a_1\mathbf{i} + a_2\mathbf{j} + a_3\mathbf{k}$ and $\bar{b} = b_1\mathbf{i} + b_2\mathbf{j} + b_3\mathbf{k}$ then, $\bar{a}\times\bar{b} = \begin{vmatrix} \mathbf{i} & \mathbf{j} & \mathbf{k} \\ a_1 & a_2 & a_3 \\ b_1 & b_2 & b_3 \end{vmatrix}$

Definition 2.0.7 If $\bar{a}, \bar{b}, \bar{c}$ are any three given vectors, then their **scalar triple product** is $\bar{a}.(\bar{b}\times\bar{c})$ and is denoted by $[\bar{a}\,\bar{b}\,\bar{c}]$

Note:

(i) If $\bar{a} = a_1\mathbf{i} + a_2\mathbf{j} + a_3\mathbf{k}$, $\bar{b} = b_1\mathbf{i} + b_2\mathbf{j} + b_3\mathbf{k}$, $\bar{c} = c_1\mathbf{i} + c_2\mathbf{j} + c_3\mathbf{k}$, then,

$$[\bar{a}\,\bar{b}\,\bar{c}] = \begin{vmatrix} a_1 & a_2 & a_3 \\ b_1 & b_2 & b_3 \\ c_1 & c_2 & c_3 \end{vmatrix}$$

(ii) $[\bar{a}\,\bar{b}\,\bar{c}] = [\bar{b}\,\bar{c}\,\bar{a}] = [\bar{c}\,\bar{a}\,\bar{b}]$ = Volume of the parallelopiped formed by $\bar{a}, \bar{b}, \bar{c}$ as coterminous edges.

(iii) $\bar{a}.(\bar{b}\times\bar{c}) = (\bar{a}\times\bar{b}).\bar{c}$

(iv) $[\bar{a}\,\bar{b}\,\bar{c}] = -[\bar{a}\,\bar{c}\,\bar{b}]$

Definition 2.0.8 If $\bar{a}, \bar{b}, \bar{c}$ are any three given vectors, then their **vector triple** product is $(\bar{a}\times\bar{b})\times\bar{c}$ and is equal to $(\bar{a}.\bar{c})\bar{b} - (\bar{a}.\bar{b}).\bar{c}$

i.e., $\bar{a}\times(\bar{b}\times\bar{c}) = (\bar{a}.\bar{c})\bar{b} - (\bar{a}.\bar{b}).\bar{c}$

Note:

$$(\bar{a}\times\bar{b})\times\bar{c} = (\bar{a}.\bar{c})\bar{b} - (\bar{b}.\bar{c})\bar{a}$$

Definition 2.0.9 If $\bar{a}, \bar{b}, \bar{c}, \bar{d}$ are any four given vectors, then $(\bar{a}\times\bar{b}).(\bar{c}\times\bar{d})$ is called the **scalar product** and $(\bar{a}\times\bar{b})\times(\bar{c}\times\bar{d})$ is called the **vector product** of the four vectors

Note:

(i) $(\bar{a}\times\bar{b}).(\bar{c}\times\bar{d}) = \begin{vmatrix} \bar{a}.\bar{c} & \bar{a}.\bar{d} \\ \bar{b}.\bar{c} & \bar{b}.\bar{d} \end{vmatrix}$

(ii) $(\bar{a}\times\bar{b})\times(\bar{c}\times\bar{d}) = [\bar{a}\bar{b}\bar{d}]\bar{c} - [\bar{a}\bar{b}\bar{c}]\bar{d}$

$= [\bar{a}\bar{c}\bar{d}]\bar{b} - [\bar{b}\bar{c}\bar{d}]\bar{a}$

2.0.2 LIMIT, CONTINUITY AND DIFFERENTIABILITY OF VECTOR FUNCTIONS

The concept of function is fundamental in Mathematics. In this section we discuss vector-valued functions and introduce the concept of limit, continuity, and differentiability of such functions.

Definition 2.0.10 If to every point P(x, y, z) of a certain domain D in space, there corresponds a scalar (real number) f(x, y, z), then the correspondence f is called a **scalar point function** or simply a **scalar function** We say that a **scalar field** f is defined in D and write, f: D → ℝ.

Definition 2.0.11 If to every point P(x, y, z) of a certain domain D in space, there corresponds a vector $\bar{F}(x,y,z)$ in $\mathbb{R}^3$, then the correspondence $\bar{F}$ is called a **vector point function** or simply a **vector function**. We say that a **vector field** $\bar{F}$ is defined in D and we write $\bar{F}(x,y,z) = F_1(x, y, z)\mathbf{i} + F_2(x, y, z)\mathbf{j} + F_3(x, y, z)\mathbf{k}$

For example, the position vector function $\bar{F}(x,y,z) = x\mathbf{i} + y\mathbf{j} + z\mathbf{k}$., which associates with each point (x, y, z) of $\mathbb{R}^3$ a vector $x\mathbf{i} + y\mathbf{j} + z\mathbf{k}$ is a vector function.
Similarly $\bar{F}(x,y,z) = xy\mathbf{i} + yz\mathbf{j} + zx\mathbf{k}$ is a vector function.

Note:

(i) If the scalar and vector fields depend on time t also, then we denote them by $f(x, y, z, t)$ and $\overline{F}(x, y, z, t)$.

(ii) If $x = x(t)$, $y = y(t)$, $z = z(t)$, $a \le t \le b$ are the parametric equations of a curve in space, then the position vector $\overline{r}(t) = x(t)\mathbf{i} + y(t)\mathbf{j} + z(t)\mathbf{k}$, $a \le t \le b$ of any point on the curve defines a vector function.

(iii) A vector function $\overline{F}(t) = F_1(t)\mathbf{i} + F_2(t)\mathbf{j} + F_3(t)\mathbf{k}$ is said to be in the parametric form and $F_1(t)$, $F_2(t)$, $F_3(t)$ are called the **component functions**, which are scalar functions.

Definition 2.0.12 The vector function $\overline{F}(t)$ is said to have a **limit** $\overline{l}$ as $t \to a$ if $\overline{F}(t)$ is defined in some neighborhood of a, except possibly at $t = a$ and for every $\epsilon > 0$, there exists a $\delta > 0$ such that $|\overline{F}(t) - \overline{l}| < \epsilon$ when ever $0 < |t - a| < \delta$. We write $\lim\limits_{t \to a} \overline{F}(t) = \overline{l}$

Note:

If $\overline{F}(t) = F_1(t)\mathbf{i} + F_2(t)\mathbf{j} + F_3(t)\mathbf{k}$ and $\lim\limits_{t \to a} \overline{F}(t) = \overline{l} = l_1\mathbf{i} + l_2\mathbf{j} + l_3\mathbf{k}$, then the limit of the component functions exists and $\lim\limits_{t \to a} F_1(t) = l_1$, $\lim\limits_{t \to a} F_2(t) = l_2$ and $\lim\limits_{t \to a} F_3(t) = l_3$ and conversely.

Definition 2.0.13 A vector function $\overline{F}(t)$ is said to be **continuous at t = a** if $\overline{F}(t)$ is defined in some neighborhood of a and $\lim\limits_{t \to a} \overline{F}(t)$ exists and is equal to $\overline{F}(a)$.

Note:

$\overline{F}(t) = F_1(t)\mathbf{i} + F_2(t)\mathbf{j} + F_3(t)\mathbf{k}$ is continuous at $t = a$ if and only if the component functions $F_1(t)$, $F_2(t)$, and $F_3(t)$ are continuous at $t = a$.

Definition 2.0.14 A vector function $\overline{F}(t)$ is said to be **differentiable** at a point t, if $\lim\limits_{\Delta t \to 0} \dfrac{\overline{F}(t + \Delta t) - \overline{F}(t)}{\Delta t}$ exists. If the limit exists, then we write it as $\overline{F}'(t)$ or as $\dfrac{d\overline{F}}{dt}$.

Note:

(i) If $\overline{F}(t) = F_1(t)\mathbf{i} + F_2(t)\mathbf{j} + F_3(t)\mathbf{k}$ is differentiable at a point t, then the component functions are also differentiable at t and $\overline{F}'(t) = F_1'(t)\mathbf{i} + F_2'(t)\mathbf{j} + F_3'(t)\mathbf{k}$.

(ii) If $\overline{r}(t) = x(t)\mathbf{i} + y(t)\mathbf{j} + z(t)\mathbf{k}$ is the parametric equation of a curve C, then $\dfrac{d\overline{r}}{dt} = \overline{r}'(t) = x'(t)\mathbf{i} + y'(t)\mathbf{j} + z'(t)\mathbf{k}$. If $\dfrac{d\overline{r}}{dt} = \overline{r}'(t) \ne 0$, then $\overline{r}'(t)$ represents the tangent vector to the curve C at the point P whose position vector is $\overline{r}(t)$.

(iii) The unit vector in the direction of the tangent to the curve at the point P with position vector $\overline{r}(t)$ is $\dfrac{\overline{r}'(t)}{|\overline{r}'(t)|}$

(iv) If $\bar{r}(t)$ has continuous first derivative and $\bar{r}'(t) \neq 0$ for all $t \in (a, b)$, then $\bar{r}(t)$ defines a smooth function on (a, b) and the curve C traced by $\bar{r}(t)$ is called a **smooth curve**.

Example 2.0.1 Find the unit tangent vector at any point on the curve $x = t^2 + 2$, $y = 4t-5$, $z = 2t^2 - 6t$. Also determine the unit tangent vector at the point $t = 2$.

Solution:
The position vector of any point on the curve is $\bar{r}(t) = (t^2 + 2)\mathbf{i} + (4t - 5)\mathbf{j} + (2t^2 - 6t)\mathbf{k}$

$\therefore \bar{r}'(t) = 2t\mathbf{i} + 4\mathbf{j} + (4t-6)\mathbf{k}$

$$|\bar{r}'(t)| = \sqrt{(2t)^2 + 4^2 + (4t-6)^2}$$

$$= \sqrt{4t^2 + 16 + 16t^2 + 36 - 48t} = 2\sqrt{5t^2 - 12t + 13}$$

Hence the unit tangent vector at any point $= \dfrac{\bar{r}'(t)}{|\bar{r}'(t)|} = \dfrac{2t\mathbf{i} + 4\mathbf{j} + (4t-6)\mathbf{k}}{2\sqrt{5t^2 - 12t + 13}}$

$$= \frac{t\mathbf{i} + 2\mathbf{j} + (2t-3)\mathbf{k}}{\sqrt{5t^2 - 12t + 13}}$$

$\therefore$ The unit tangent vector at $t = 2$ is

$$= \frac{2\mathbf{i} + 2\mathbf{j} + \mathbf{k}}{\sqrt{5.4 - 12.2 + 13}}$$

$$= \frac{2\mathbf{i} + 2\mathbf{j} + \mathbf{k}}{3}$$

Example 2.0.2 A particle moves on the curve $x = 2t^2$, $y = t^2 - 4t$, $z = 3t - 5$, where t is the time. Find the components of velocity and acceleration at time $t = 1$, in the direction $i - 3j + 2k$.

Solution:
The parametric equation of the curve is $\bar{r}(t) = 2t^2\mathbf{i} + (t^2 - 4t)\mathbf{j} + (3t - 5)\mathbf{k}$ then r(t) is the position vector of the moving particle.

Velocity $= \bar{r}'(t) = 4t\mathbf{i} + (2t-4)\mathbf{j} + 3\mathbf{k}$

Velocity at $t = 1$ is $4\mathbf{i} - 2\mathbf{j} + 3\mathbf{k}$

The unit vector in the direction of $\mathbf{i} - 3\mathbf{j} + 2\mathbf{k} = \dfrac{\mathbf{i} - 3\mathbf{j} + 2\mathbf{k}}{\sqrt{14}}$

$\therefore$ Component of velocity in the direction $\mathbf{i} - 3\mathbf{j} + 2\mathbf{k}$ is $= (4\mathbf{i} - 2\mathbf{j} + 3\mathbf{k}) \cdot \dfrac{(\mathbf{i} - 3\mathbf{j} + 2\mathbf{k})}{\sqrt{14}}$

$$= \frac{4 + 6 + 6}{\sqrt{14}} = \frac{16}{\sqrt{14}}$$

Acceleration $\bar{r}''(t) = 4\mathbf{i} + 2\mathbf{j}$

$\therefore$ Acceleration at time $t = 1$ is $4\mathbf{i} + 2\mathbf{j}$

Component of acceleration in the direction of **i** - 3**j** + 2**k**

$$= (4\mathbf{i} + 2\mathbf{j}).\frac{(\mathbf{i} - 3\mathbf{j} + 2\mathbf{k})}{\sqrt{14}}$$

$$= \frac{4 - 6 + 0}{\sqrt{14}} = \frac{-2}{\sqrt{14}}$$

2.1 GRADIENT, DIVERGENCE AND CURL

The gradient operator when operated on a scalar field produces a vector field, whereas the divergence operator when operated on a vector field produces a scalar field. The curl of a vector field is again a vector field. In this section we define these three notions in terms of a vector differential operator called **del operator**, denoted by ∇. Further we discuss the application of these three notions.

Definition 2.1.1 The vector differential operator called **del operator** and denoted by ∇ is defined as $\nabla = \mathbf{i}\frac{\partial}{\partial x} + \mathbf{j}\frac{\partial}{\partial y} + \mathbf{k}\frac{\partial}{\partial z}$. In the case of two dimensions, $\nabla = \mathbf{i}\frac{\partial}{\partial x} + \mathbf{j}\frac{\partial}{\partial y}$

Definition 2.1.2 Let f(x, y, z) be a scalar field. The **gradient** of f(x, y z), denoted by **∇f** or **grad f** is defined as the vector

$$\nabla f = \mathbf{i}\frac{\partial f}{\partial x} + \mathbf{j}\frac{\partial f}{\partial y} + \mathbf{k}\frac{\partial f}{\partial z}$$

Theorem 2.1.1 Let f(x, y, z) and g(x, y, z) be two differentiable scalar functions and a and b be arbitrary constants. Then

(i) $\nabla f = 0$ if f is a constant function.

(ii) $\nabla(af + bg) = a\nabla f + b\nabla g$

(iii) $\nabla(fg) = f\,\nabla g + g\,\nabla f$

(iv) $\nabla\left(f/g\right) = \frac{g\,\nabla f - f\,\nabla g}{g^2}$, if $g \neq 0$.

Proof:

(i) If f(x, y, z) is a constant function, the partial derivatives w.r.t x, y and z are zero.

i.e., $\frac{\partial f}{\partial x} = \frac{\partial f}{\partial y} = \frac{\partial f}{\partial z} = 0$

$$\therefore \quad \nabla f = \mathbf{i}\frac{\partial f}{\partial x} + \mathbf{j}\frac{\partial f}{\partial y} + \mathbf{k}\frac{\partial f}{\partial z} = 0$$

(ii) $\nabla(af + bg) = \mathbf{i}\frac{\partial(af + bg)}{\partial x} + \mathbf{j}\frac{\partial(af + bg)}{\partial y} + \mathbf{k}\frac{\partial(af + bg)}{\partial z}$

$$= \mathbf{i}\left(a\frac{\partial f}{\partial x} + b\frac{\partial g}{\partial x}\right) + \mathbf{j}\left(a\frac{\partial f}{\partial y} + b\frac{\partial g}{\partial y}\right) + \mathbf{k}\left(a\frac{\partial f}{\partial z} + b\frac{\partial g}{\partial z}\right)$$

$$= a\left(\mathbf{i}\frac{\partial f}{\partial x}+\mathbf{j}\frac{\partial f}{\partial y}+\mathbf{k}\frac{\partial f}{\partial z}\right)+b\left(\mathbf{i}\frac{\partial g}{\partial x}+\mathbf{j}\frac{\partial g}{\partial y}+\mathbf{k}\frac{\partial g}{\partial z}\right)$$

$$= a\,\nabla f + b\,\nabla g.$$

(iii) $$\nabla(fg) = \mathbf{i}\frac{\partial}{\partial x}(fg)+\mathbf{j}\frac{\partial}{\partial y}(fg)+\mathbf{k}\frac{\partial}{\partial z}(fg)$$

$$= \mathbf{i}\left(f\frac{\partial g}{\partial x}+g\frac{\partial f}{\partial x}\right)+\mathbf{j}\left(f\frac{\partial g}{\partial y}+g\frac{\partial f}{\partial y}\right)+\mathbf{k}\left(f\frac{\partial g}{\partial z}+g\frac{\partial f}{\partial z}\right)$$

$$= f\left(\mathbf{i}\frac{\partial g}{\partial x}+\mathbf{j}\frac{\partial g}{\partial y}+\mathbf{k}\frac{\partial g}{\partial z}\right)+g\left(\mathbf{i}\frac{\partial f}{\partial x}+\mathbf{j}\frac{\partial f}{\partial y}+\mathbf{k}\frac{\partial f}{\partial z}\right)$$

$$= f\,\nabla g + g\,\nabla f$$

(iv) $$\nabla\left(f/g\right) = \mathbf{i}\frac{\partial}{\partial x}\left(f/g\right)+\mathbf{j}\frac{\partial}{\partial y}\left(f/g\right)+\mathbf{k}\frac{\partial}{\partial z}\left(f/g\right)$$

$$= \mathbf{i}\left(\frac{g\frac{\partial f}{\partial x}-f\frac{\partial g}{\partial x}}{g^2}\right)+\mathbf{j}\left(\frac{g\frac{\partial f}{\partial y}-f\frac{\partial g}{\partial y}}{g^2}\right)+\mathbf{k}\left(\frac{g\frac{\partial f}{\partial z}-f\frac{\partial g}{\partial z}}{g^2}\right)$$

$$= \frac{g\left(\mathbf{i}\frac{\partial f}{\partial x}+\mathbf{j}\frac{\partial f}{\partial y}+\mathbf{k}\frac{\partial f}{\partial z}\right)-f\left(\mathbf{i}\frac{\partial g}{\partial x}+\mathbf{j}\frac{\partial g}{\partial y}+\mathbf{k}\frac{\partial g}{\partial z}\right)}{g^2}$$

$$= \frac{g\,\nabla f - f\,\nabla g}{g^2}, \text{ if } g \neq 0$$

2.1.1 GEOMETRICAL MEANING OF THE GRADIENT

Let f(x, y, z) be a differentiable scalar field. Let f(x, y, z) = k be any **level surface (or equipotential surface)** and $P_0(x_0, y_0, z_0)$ be a point on it. A vector normal to the tangent plane to the surface at P_0 is called the **normal vector** to the surface at this point.

Let x = x(t), y = y(t), z = z(t) be the parametric equations of a smooth curve C on the surface f(x, y, z) = k, passing through a point P on the surface. Then the point P on C has position vector $\bar{r}(t) = x(t).\mathbf{i} + y(t).\mathbf{j} + z(t).\mathbf{k}$ and f(x(t), y(t), z(t)) = k.

$$\therefore \quad \frac{d}{dt}f(x(t), y(t), z(t)) = 0$$

i.e., $$\frac{\partial f}{\partial x}.\frac{dx}{dt}+\frac{\partial f}{\partial y}.\frac{dy}{dt}+\frac{\partial f}{\partial z}.\frac{dz}{dt} = 0 \qquad \text{(By chain rule)}$$

i.e., $$\left(\mathbf{i}\frac{\partial f}{\partial x}+\mathbf{j}\frac{\partial f}{\partial y}+\mathbf{k}\frac{\partial f}{\partial z}\right).\left(\mathbf{i}\frac{dx}{dt}+\mathbf{j}\frac{dy}{dt}+\mathbf{k}\frac{dz}{dt}\right) = 0$$

i.e., $$\nabla f\,.\,\bar{r}'(t) = 0 \qquad (1)$$

Let $\nabla f \neq 0$ at P and $\bar{r}'(t) \neq 0$.

$\bar{r}'(t)$ is a tangent vector to C at the point P and lies in the tangent plane to the surface at P. Hence (1) implies that ∇f is orthogonal to every tangent vector at P and then ∇f is the vector normal to the surface f(z, y ,z) = k at the point P(Fig. 2.2)

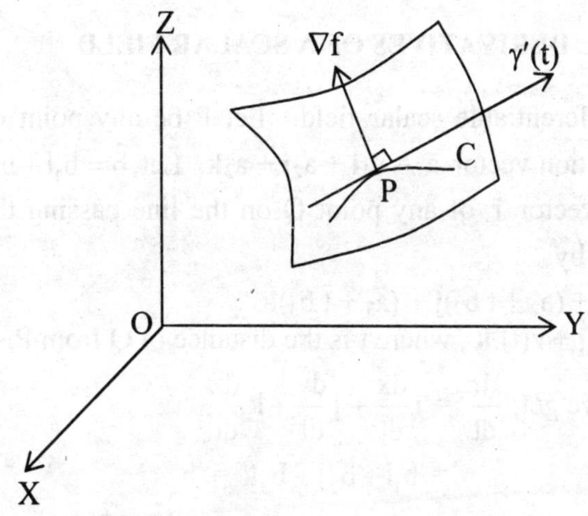

Fig. 2.2

Note:

(i) The unit normal vector to a surface f(x, y, z) = k is $\hat{n} = \dfrac{\nabla f}{|\nabla f|}$

(ii) Let $P_0(x_0, y_0, z_0)$ be any given point on the surface f(x, y, z) = k. Let (x, y, z) be any point on the tangent plane to the surface at P_0. Then the vector $(x - x_0)\mathbf{i} + (y - y_0)\mathbf{j} + (z - z_0)\mathbf{k}$ is a vector in the tangent plane and is then orthogonal to the normal vector ∇f at P_0. $\therefore$ $\nabla f . [(x - x_0)\mathbf{i} + (y - y_0)\mathbf{j} + (z - z_0)\mathbf{k}] = 0$

i.e., $$(x - x_0)\frac{\partial f}{\partial x} + (y - y_0)\frac{\partial f}{\partial x} + (z - z_0)\frac{\partial f}{\partial x} = 0$$

This is the equation of the tangent plane to the surface f(x, y, z) = k at the point $P_0(x_0, y_0, z_0)$ on it.

Definition 2.1.3 The operator $\nabla.\nabla$ or $\nabla^2 = \dfrac{\partial^2}{\partial x^2} + \dfrac{\partial^2}{\partial y^2} + \dfrac{\partial^2}{\partial x^2}$ is called the **Laplace operator.**

Note:

We have $\nabla^2 f = \nabla.\nabla f = \dfrac{\partial^2 f}{\partial x^2} + \dfrac{\partial^2 f}{\partial y^2} + \dfrac{\partial^2 f}{\partial z^2}$

i.e., $$\nabla^2 f = \frac{\partial^2 f}{\partial x^2} + \frac{\partial^2 f}{\partial y^2} + \frac{\partial^2 f}{\partial z^2}$$

Definition 2.1.4 If f(x, y, z) is any scalar function, then the equation $\nabla^2 f = 0$ is called the **Laplace equation** of f or the **Lapacian** of f.

Definition 2.1.5 A scalar function f(x,. y, z) is called a **harmonic function** if it satisfies the Laplace equation. i.e., $\nabla^2 f = 0$.

2.1.2 DIRECTIONAL DERIVATIVES OF A SCALAR FIELD

Let f(x, y, z) be a differentiable scalar field. Let P be any point on the level surface f(x, y, z) = k, with position vector $\bar{a} = a_1\mathbf{i} + a_2\mathbf{j} + a_3\mathbf{k}$. Let $\hat{b} = b_1\mathbf{i} + b_2\mathbf{j} + b_3\mathbf{k}$ be any unit vector. The position vector $\bar{r}$ of any point Q on the line passing through P and in the direction of $\hat{b}$ is given by

$$\bar{r} = a + t\hat{b} = (a_1 + t\,b_1)\mathbf{i} + (a_2 + t\,b_2)\mathbf{j} + (a_3 + t\,b_3)\mathbf{k}$$

$= x(t).\mathbf{i} + y(t)\mathbf{j} + z(t).\mathbf{k}$, where t is the distance of Q from P. (Fig. 2.3)

Differentiating w.r.t t we get, $\frac{d\bar{r}}{dt} = \mathbf{i}\frac{dx}{dt} + \mathbf{j}\frac{dy}{dt} + \mathbf{k}\frac{dz}{dt}$

$$= b_1\mathbf{i} + b_2\mathbf{j} + b_3\mathbf{k} = \hat{b}$$

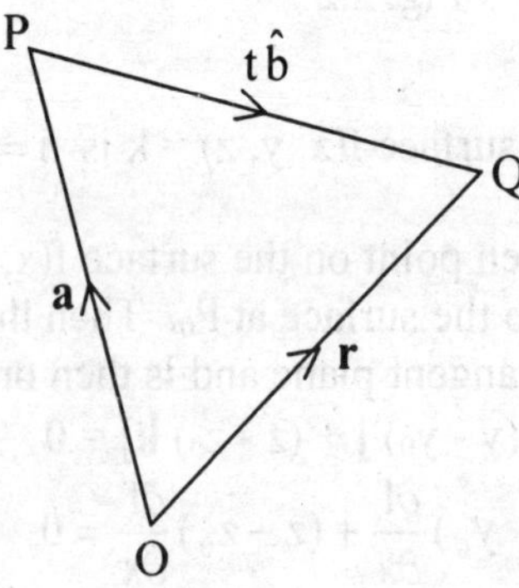

Fig. 2.3

$\frac{df}{dt} = \lim_{t\to 0} \frac{f(a_1 + tb_1, a_2 + tb_2, a_3 + tb_3) - f(a_1, a_2, a_3)}{t}$ is called the **directional derivative** of f at the point P in the direction of $\hat{b}$.

We have, $\frac{df}{dt} = \frac{\partial f}{\partial x}.\frac{dx}{dt} + \frac{\partial f}{\partial y}.\frac{dy}{dt} + \frac{\partial f}{\partial z}.\frac{dz}{dt}$

$$= \left(\mathbf{i}\frac{\partial f}{\partial x} + \mathbf{j}\frac{\partial f}{\partial y} + \mathbf{k}\frac{\partial f}{\partial z}\right).\left(\mathbf{i}\frac{dx}{dt} + \mathbf{j}\frac{dy}{dt} + \mathbf{k}\frac{dz}{dt}\right)$$

$$= \nabla f . \frac{d\bar{r}}{dt}$$

$$= \nabla f . \hat{b} \qquad (\text{Since } \frac{d\bar{r}}{dt} = \hat{b})$$

Note:

$\frac{df}{dt}$ is the rate of change of f w.r.t distance t in the direction of the unit vector $\hat{b}$.

Definition 2.1.6 The **directional derivative** of the scalar function f(x, y, z) in the direction of the unit vector $\hat{b}$ is $\nabla f . \hat{b}$ and is denoted by $D_b(f)$.

Note:

(i) The directional derivative of f(x, y, z) in the direction of the vector $\bar{b}$ is $D_b(f) = \nabla f . \frac{\bar{b}}{|\bar{b}|}$

(ii) $D_b(f) = \nabla f . \hat{b}$

$= |\nabla f| \, |\hat{b}| \cos\theta$, where θ is the angle between ∇f and $\hat{b}$.

$= |\nabla f| \cos\theta$

$|\nabla f|. \cos\theta$ is maximum when $\theta = 0$ and the **maximum value** is $|\nabla f|$. Thus the directional derivative is maximum when $\hat{b}$ has the direction of ∇f and this direction is the direction of the normal to the surface f(x, y, z) = k. The directional derivative is minimum when $\theta = -\pi$ and the **minimum value** is $-|\nabla f|$. Then $\hat{b}$ and ∇f have opposite directions.

(iii) $\frac{\partial f}{\partial x}, \frac{\partial f}{\partial y}, \frac{\partial f}{\partial z}$ are the directional derivatives of f in the directions of the coordinate axes.

Definition 2.1.7 A vector field $\bar{F}$ (x, y, z) is said to be **conservative** if it is the gradient of a scalar function f(x, y, z). i.e., $\bar{F} = \nabla f$. The function f(x, y, z) is called the **scalar potential** of $\bar{F}$ (x, y, z).

Example 2.1.1 Compute the gradient of the scalar function $f(x, y, z) = 3x^2y + 2y^3 - 5z$.

Solution:

$$\frac{\partial f}{\partial x} = 6xy, \quad \frac{\partial f}{\partial y} = 3x^2 + 6y^2, \quad \frac{\partial f}{\partial z} = -5$$

$$\therefore \; \nabla f = \mathbf{i}\frac{\partial f}{\partial x} + \mathbf{j}\frac{\partial f}{\partial y} + \mathbf{k}\frac{\partial f}{\partial z} = 6xy\,\mathbf{i} + (3x^2 + 6y^2)\,\mathbf{j} - 5\,\mathbf{k}.$$

Example 2.1.2 Compute $\nabla^2(\log r)$

Solution:

$$r = \sqrt{x^2 + y^2 + z^2}$$

$$\therefore\ f = \log r = \frac{1}{2}\log\left(x^2 + y^2 + z^2\right)$$

$$\frac{\partial f}{\partial x} = \frac{1}{2\left(x^2 + y^2 + z^2\right)}.2x = \frac{x}{\left(x^2 + y^2 + z^2\right)}$$

$$\frac{\partial^2 f}{\partial x^2} = \frac{x^2 + y^2 + z^2 - 2x^2}{\left(x^2 + y^2 + z^2\right)^2} = \frac{r^2 - 2x^2}{r^4} = \frac{1}{r^2} - \frac{2x^2}{r^4}$$

Similarly $\dfrac{\partial^2 f}{\partial y^2} = \dfrac{1}{r^2} - \dfrac{2y^2}{r^4}$

$$\frac{\partial^2 f}{\partial z^2} = \frac{1}{r^2} - \frac{2z^2}{r^4}$$

Hence $\nabla^2 f = \dfrac{\partial^2 f}{\partial x^2} + \dfrac{\partial^2 f}{\partial y^2} + \dfrac{\partial^2 f}{\partial z^2}$

$$= \frac{3}{r^2} - \frac{2}{r^4}\left(x^2 + y^2 + z^2\right) = \frac{3}{r^2} - \frac{2}{r^2} = \frac{1}{r^2}$$

Example 2.1.3 Find the normal vector and unit normal vector to the surface $x^2 + 2y^2 + z^2 = 4$ at the point (1, 1, 1).

Solution:

$$f(x, y, z) = x^2 + 2y^2 + z^2 - 4 = 0$$

$$\frac{\partial f}{\partial x} = 2x,\quad \frac{\partial f}{\partial y} = 4y,\quad \frac{\partial f}{\partial z} = 2z$$

$$\therefore\ \nabla f = \mathbf{i}\frac{\partial f}{\partial x} + \mathbf{j}\frac{\partial f}{\partial y} + \mathbf{k}\frac{\partial f}{\partial z} = 2x\,\mathbf{i} + 4y\,\mathbf{j} + 2z\,\mathbf{k}$$

At the point (1, 1, 1), $\nabla f = 2\mathbf{i} + 4\mathbf{j} + 2\mathbf{k}$.

$\therefore$ The normal vector to the surface at (1, 1, 1) is $2\mathbf{i} + 4\mathbf{j} + 2\mathbf{k}$

The unit normal vector is $\dfrac{2\mathbf{i} + 4\mathbf{j} + 2\mathbf{k}}{\sqrt{2^2 + 4^2 + 2^2}} = \dfrac{1}{\sqrt{6}}(\mathbf{i} + 2\mathbf{j} + \mathbf{k})$

Example 2.1.4 Find the directional derivative of $f(x, y, z) = \left(x^2 + y^2 + z^2\right)^{3/2}$ at (-1, 1, 2) in the direction $\mathbf{i} - 2\mathbf{j} + \mathbf{k}$.

Solution:

$$\frac{\partial f}{\partial x} = \frac{3}{2}\left(x^2 + y^2 + z^2\right)^{1/2}.2x = 3x\left(x^2 + y^2 + z^2\right)^{1/2}$$

$$\frac{\partial f}{\partial y} = 3y\left(x^2 + y^2 + z^2\right)^{1/2}$$

$$\frac{\partial f}{\partial z} = 3z\left(x^2 + y^2 + z^2\right)^{1/2}$$

At (-1, 1, 2), $\frac{\partial f}{\partial x} = (-3).6^{1/2} = -3\sqrt{6}$ $\quad \frac{\partial f}{\partial y} = 3\sqrt{6} \quad \frac{\partial f}{\partial z} = 6\sqrt{6}$

$\therefore \quad \nabla f = -3\sqrt{6}\,\mathbf{i} + 3\sqrt{6}\,\mathbf{j} + 6\sqrt{6}\,\mathbf{k}$

Unit vector in the direction of $\mathbf{i} - 2\mathbf{j} + \mathbf{k}$ is $\hat{b} = \frac{\mathbf{i} - 2\mathbf{j} + \mathbf{k}}{\sqrt{1^2 + (-2)^2 + 1^2}} = \frac{1}{\sqrt{6}}(\mathbf{i} - 2\mathbf{j} + \mathbf{k})$

$\therefore$ **The directional derivative at (-1, 1, 2) in the direction of $\mathbf{i} - 2\mathbf{j} + \mathbf{k}$**

$$= \nabla f . \hat{b}$$

$$= \left(-3\sqrt{6}\,\mathbf{i} + 3\sqrt{6}\,\mathbf{j} + 6\sqrt{6}\,\mathbf{k}\right)\frac{1}{\sqrt{6}}(\mathbf{i} - 2\mathbf{j} + \mathbf{k})$$

$$= \frac{1}{\sqrt{6}}\left(-3\sqrt{6} - 6\sqrt{6} + 6\sqrt{6}\right) = -3$$

Example 2.1.5 Find the angle between the surfaces $x^2 - y^2 - z^2 = 11$ and $xy + yz - zx - 18 = 0$ at the point (6, 4, 3).

Solution:

Let $f(x, y, z) = x^2 - y^2 - z^2 - 11 = 0$

$g(x, y, z) = xy + yz - zx - 18 = 0$

$$\nabla f = \mathbf{i}\frac{\partial f}{\partial x} + \mathbf{j}\frac{\partial f}{\partial y} + \mathbf{k}\frac{\partial f}{\partial z}$$

$$= \mathbf{i}\,(2x) + \mathbf{j}(-2y) + \mathbf{k}(-2z)$$

$$= 12\mathbf{i} - 8\mathbf{j} - 6\mathbf{k} \qquad \text{at } (6, 4, 3)$$

$$\nabla g = \mathbf{i}\frac{\partial g}{\partial x} + \mathbf{j}\frac{\partial g}{\partial y} + \mathbf{k}\frac{\partial g}{\partial z}$$

$$= \mathbf{i}\,(y - z) + \mathbf{j}(x + z) + \mathbf{k}(y - x)$$

$$= \mathbf{i} + 9\mathbf{j} - 2\mathbf{k} \qquad \text{at } (6, 4, 3)$$

The angle between the surfaces at (6, 4, 3) is equal to the angle between their normals at (6, 4, 3). i.e., the angle between the vectors ∇f and ∇g. If θ is the angle, then

$$\cos\theta = \frac{\nabla f . \nabla g}{|\nabla f| . |\nabla g|}$$

$$= \frac{(12\mathbf{i} - 8\mathbf{j} - 6\mathbf{k})(\mathbf{i} + 9\mathbf{j} - 2\mathbf{k})}{\sqrt{12^2 + (-8)^2 + (-6)^2}\sqrt{1^2 + 9^2 + (-2)^2}}$$

$$= \frac{12 - 72 + 12}{\sqrt{244}\sqrt{86}} = \frac{-48}{2\sqrt{61}\sqrt{86}} = \frac{-24}{\sqrt{61}\sqrt{86}}$$

$$\therefore\ \theta = \cos^{-1}\left(\frac{-24}{\sqrt{61}\sqrt{86}}\right)$$

Example 2.1.6 Find a vector that gives the direction of maximum rate of increase of $f(x, y, z) = x^2y^2z^2 + xz^2 + x^2y$ at (1, 2, -1). Find the maximum rate.

Solution:

The required vector is ∇f and the required maximum rate is $|\nabla f|$.

$$\nabla f = \mathbf{i}\frac{\partial f}{\partial x} + \mathbf{j}\frac{\partial f}{\partial y} + \mathbf{k}\frac{\partial f}{\partial z}$$

$$= \mathbf{i}(2y^2z^2.x + z^2 + 2xy) + \mathbf{j}(2x^2z^2y + x^2) + \mathbf{k}(2x^2y^2z + 2xz)$$

$$= 13\mathbf{i} + 5\mathbf{j} - 10\mathbf{k} \text{ at } (1, 2, -1)$$

$$|\nabla f| = \sqrt{13^2 + 5^2 + (-10)^2} = \sqrt{294}$$

Example 2.1.7 Show that the vector field defined by $\overline{F}(x, y, z) = e^{xyz}(yz\mathbf{i} + zx\mathbf{j} + xy\mathbf{k})$ is a conservative field.

Solution:

To show that there exists a scalar field f(x, y, z) such that $\nabla f = \overline{F}$.

i.e., $\mathbf{i}\frac{\partial f}{\partial x} + \mathbf{j}\frac{\partial f}{\partial y} + \mathbf{k}\frac{\partial f}{\partial z} = e^{xyz}(yz\mathbf{i} + zx\mathbf{j} + xy\mathbf{k})$

Comparing both sides we get,

$$\frac{\partial f}{\partial x} = e^{xyz}yz \qquad (1)$$

$$\frac{\partial f}{\partial y} = e^{xyz}zx \qquad (2)$$

$$\frac{\partial f}{\partial z} = e^{xyz}xy \qquad (3)$$

Integrating the equation (1), we get,

$f(x, y, z) = e^{xyz} + g(y, z)$

Then $\frac{\partial f}{\partial y} = e^{xyz}xz + \frac{\partial g}{\partial y}$

i.e., $e^{xyz}.zx = e^{xyz}zx + \frac{\partial g}{\partial y}$ (Using (2))

$\therefore\ \frac{\partial g}{\partial y} = 0$ i.e., $g = g(z)$

$\therefore\ f(x, y, z) = e^{xyz} + g(z)$

Again, $\frac{\partial f}{\partial z} = e^{xyz}xy + \frac{dg}{dz}$

i.e., $e^{xyz}.xy = e^{xyz}xy + \frac{dg}{dz}$ (Using (3))

i.e., $\frac{dg}{dz} = 0$ i.e, g = k, a constant

$\therefore$ $f(x, y, z) = e^{xyz} + k$

Thus there exists a scalar function f such that $\nabla f = \overline{F}$

$\therefore$ $\overline{F}$ is a conservative vector field.

Example 2.1.8 If $\nabla\varphi = (6r - 8r^2)\overline{r}$ and $\varphi(2) = 4$, find $\varphi(r)$.

Solution:

$$\nabla\varphi = \mathbf{i}\frac{\partial\varphi}{\partial x} + \mathbf{j}\frac{\partial\varphi}{\partial y} + \mathbf{k}\frac{\partial\varphi}{\partial z}$$

$$= \mathbf{i}\varphi'(r)\frac{\partial r}{\partial x} + \mathbf{j}\varphi'(r)\frac{\partial r}{\partial y} + \mathbf{k}\varphi'(r)\frac{\partial r}{\partial z}$$

$$= \varphi'(r)\left[\mathbf{i}\frac{\partial r}{\partial x} + \mathbf{j}\frac{\partial r}{\partial y} + \mathbf{k}\frac{\partial r}{\partial z}\right] \quad (1)$$

But $r^2 = x^2 + y^2 + z^2$

$\therefore$ $2r.\frac{\partial r}{\partial x} = 2x$ i.e., $\frac{\partial r}{\partial x} = \frac{x}{r}$

Similarly $\frac{\partial r}{\partial y} = \frac{y}{r}, \ \frac{\partial r}{\partial z} = \frac{z}{r}$

$\therefore$ (1) $\Rightarrow$ $\nabla\varphi = \varphi'(r)\left[\mathbf{i}\frac{\partial r}{\partial x} + \mathbf{j}\frac{\partial r}{\partial y} + \mathbf{k}\frac{\partial r}{\partial z}\right] = \varphi'(r).\frac{\overline{r}}{r}$

Given $\nabla\varphi = (6r - 8r^2)\overline{r} = (6r^2 - 8r^3)\frac{\overline{r}}{r}$

$\therefore$ $\varphi'(r) = 6r^2 - 8r^3$

Hence $\varphi(r) = 6.\frac{r^3}{3} - 8\frac{r^4}{4} + c = 2(r^3 - r^4) + c$

$\therefore$ $\varphi(2) = 2(2^3 - 2^4) + c$

$\therefore$ i.e., $4 = 2(8 - 16) + c$ i.e., c = 20

$\therefore$ $\varphi(r) = 2(r^3 - r^4 + 10)$

Example 2.1.9 Find the equation of tangent plane to the surface $x^2 - 4y^2 + 3z^2 + 4 = 0$ at (3, 2, 1).

Solution:

Let $f(x, y, z) = x^2 - 4y^2 + 3z^2 + 4$

The vector normal to the surface is ∇f.

$$\nabla f = \mathbf{i}\frac{\partial f}{\partial x} + \mathbf{j}\frac{\partial f}{\partial y} + \mathbf{k}\frac{\partial f}{\partial z} = \mathbf{i}\,2x + \mathbf{j}\,(-8y) + \mathbf{k}\,6z = 6\mathbf{i} - 16\mathbf{j} + 6\mathbf{k}, \text{ at } (3, 2, 1)$$

∴ Direction ratios of the normal to the tangent plane at (3, 2, 1) are [6, -16, 6].

∴ Equation of the tangent plane is $6x - 16y + 6z + c = 0$

As this plane passes through (3, 2, 1), $6.3 - 16.2 + 6.1 + c = 0 \Rightarrow c = 8$

∴ The equation of the tangent plane is

$$6x - 16y + 6z + 8 = 0$$

i.e., $3x - 8y + 3z + 4 = 0$

2.1.3 DIVERGENCE AND CURL OF A VECTOR FIELD

Definition 2.1.8 The **divergence** of a vector field $\overline{F}(x,y,z) = F_1(x,y,z)\mathbf{i} + F_2(x,y,z)\mathbf{j} + F_3(x,y,z)\mathbf{k}$ is defined as the scalar field $\frac{\partial F_1}{\partial x} + \frac{\partial F_2}{\partial y} + \frac{\partial F_3}{\partial z}$ and is denoted by **div** $\overline{F}$ or $\nabla.\overline{F}$

i.e., $\operatorname{div}\overline{F} = \nabla.\overline{F}$

$$= \left(\mathbf{i}\frac{\partial}{\partial x} + \mathbf{j}\frac{\partial}{\partial y} + \mathbf{k}\frac{\partial}{\partial z}\right).(F_1\mathbf{i} + F_2\mathbf{j} + F_3\mathbf{k})$$

$$= \frac{\partial F_1}{\partial x} + \frac{\partial F_2}{\partial y} + \frac{\partial F_3}{\partial z}$$

Also $\operatorname{div}\overline{F} = \left(\sum \mathbf{i}\frac{\partial}{\partial x}\right).\overline{F} = \sum \mathbf{i}.\frac{\partial \overline{F}}{\partial x}$

Definition 2.1.9 A vector field $\overline{F}(x,y,z)$ is said to be **solenoidal (or incompressible)** if $\operatorname{div}\overline{F} = 0$.

Definition 2.1.10 The **curl** of a vector field $\overline{F}(x,y,z) = F_1(x,y,z)\mathbf{i} + F_2(x,y,z)\mathbf{j} + F_3(x,y,z)\mathbf{k}$ is defined as the vector field $\left(\frac{\partial F_3}{\partial y} - \frac{\partial F_2}{\partial z}\right)\mathbf{i} + \left(\frac{\partial F_1}{\partial z} - \frac{\partial F_3}{\partial x}\right)\mathbf{j} + \left(\frac{\partial F_2}{\partial x} - \frac{\partial F_1}{\partial y}\right)\mathbf{k}$ and is denoted by **curl** $\overline{F}$ or $\nabla \times \overline{F}$.

i.e., $\operatorname{curl}\overline{F} = \nabla \times \overline{F} = \begin{vmatrix} \mathbf{i} & \mathbf{j} & \mathbf{k} \\ \frac{\partial}{\partial x} & \frac{\partial}{\partial y} & \frac{\partial}{\partial z} \\ F_1 & F_2 & F_3 \end{vmatrix} = \sum \left(\frac{\partial F_3}{\partial y} - \frac{\partial F_2}{\partial z}\right)\mathbf{i}$

Also $\operatorname{curl}\overline{F} = \left(\sum \mathbf{i}\frac{\partial}{\partial x}\right) \times \overline{F} = \sum \mathbf{i} \times \frac{\partial \overline{F}}{\partial x}$

Definition 2.1.11 A vector field $\overline{F}(x,y,z)$ is said to be **irrotational** if curl $\overline{F} = 0$. A vector field which is not irrotational is called a **vortex** field.

Theorem 2.1.2 If f is scalar function and $\overline{F}$ and $\overline{G}$ are vector functions, then

(i) $\nabla.(\overline{F} \pm \overline{G}) = \nabla.\overline{F} \pm \nabla.\overline{G}$

(ii) $\nabla.(f\overline{F}) = \nabla f.\overline{F} + f(\nabla.\overline{F})$

(iii) $\nabla.(\overline{F}\times\overline{G}) = (\nabla\times\overline{F}).\overline{G} - \overline{F}.(\nabla\times\overline{G})$

Proof:

(i) Let $\overline{F} = F_1\mathbf{i} + F_2\mathbf{j} + F_3\mathbf{k}$ and $\overline{G} = G_1\mathbf{i} + G_2\mathbf{j} + G_3\mathbf{k}$

Then $\nabla.(\overline{F} \pm \overline{G}) = \frac{\partial}{\partial x}(F_1 \pm G_1) + \frac{\partial}{\partial y}(F_2 \pm G_2) + \frac{\partial}{\partial z}(F_3 \pm G_3)$

$$= \frac{\partial F_1}{\partial x} \pm \frac{\partial G_1}{\partial x} + \frac{\partial F_2}{\partial y} \pm \frac{\partial G_2}{\partial y} + \frac{\partial F_3}{\partial z} \pm \frac{\partial G_3}{\partial z}$$

$$= \left(\frac{\partial F_1}{\partial x} + \frac{\partial F_2}{\partial y} + \frac{\partial F_3}{\partial z}\right) \pm \left(\frac{\partial G_1}{\partial x} + \frac{\partial G_2}{\partial y} + \frac{\partial G_3}{\partial z}\right)$$

$$= \nabla.\overline{F} \pm \nabla.\overline{G}$$

(i.e., div $(\overline{F} \pm \overline{G}) = \text{div}\,\overline{F} \pm \text{div}\,\overline{G}$)

(ii) $\nabla.(f\overline{F}) = \frac{\partial}{\partial x}(f\,F_1) + \frac{\partial}{\partial y}(f\,F_2) + \frac{\partial}{\partial z}(f\,F_3)$

$$= \left(\frac{\partial f}{\partial x}\right)F_1 + f\left(\frac{\partial F_1}{\partial x}\right) + \left(\frac{\partial f}{\partial y}\right)F_2 + f\left(\frac{\partial F_2}{\partial y}\right) + \left(\frac{\partial f}{\partial z}\right)F_3 + f\left(\frac{\partial F_3}{\partial z}\right)$$

$$= \left[\left(\frac{\partial f}{\partial x}\right)F_1 + + \left(\frac{\partial f}{\partial y}\right)F_2 + \left(\frac{\partial f}{\partial z}\right)F_3\right] + f\left(\frac{\partial F_1}{\partial x} + \frac{\partial F_2}{\partial y} + \frac{\partial F_3}{\partial z}\right)$$

$$= \left(\mathbf{i}\frac{\partial f}{\partial x} + \mathbf{j}\frac{\partial f}{\partial y} + \mathbf{k}\frac{\partial f}{\partial z}\right).(F_1\mathbf{i} + F_2 j + F_3 k) + f(\nabla.\overline{F})$$

$$= \nabla f.\overline{F} + f(\nabla.\overline{F})$$

(i.e., div $(f\overline{F}) = (\text{grad } f).\overline{F} + f\,(\text{div}\,\overline{F})$)

(iii) $\nabla.(\overline{F}\times\overline{G}) = \sum \mathbf{i}.\frac{\partial}{\partial x}(\overline{F}\times\overline{G})$

$$= \sum \mathbf{i}.\left[\frac{\partial \overline{F}}{\partial x}\times\overline{G} + \overline{F}\times\frac{\partial \overline{G}}{\partial x}\right]$$

$$= \sum \mathbf{i}.\left(\frac{\partial \overline{F}}{\partial x}\times\overline{G}\right) - \sum \mathbf{i}.\left(\overline{F}\times\frac{\partial \overline{G}}{\partial x}\right)$$

$$= \sum \left(\mathbf{i} \times \frac{\partial \overline{F}}{\partial x} \right) . \overline{G} - \sum \left(\mathbf{i} \times \frac{\partial \overline{G}}{\partial x} \right) . \overline{F}$$

$$= \left(\sum \mathbf{i} \times \frac{\partial \overline{F}}{\partial x} \right) . \overline{G} - \left(\sum \mathbf{i} \times \frac{\partial \overline{G}}{\partial x} \right) . \overline{F}$$

$$= (\nabla \times \overline{F}) . \overline{G} - (\nabla \times \overline{G}) . \overline{F}$$

$$= (\nabla \times \overline{F}) . \overline{G} - \overline{F} . (\nabla \times \overline{G})$$

(i.e., div $(\overline{F} \times \overline{G}) = (\text{curl } \overline{F}) . \overline{G} - \overline{F} . (\text{curl } \overline{G})$)

Theorem 2.1.3 If f is a scalar function, and $\overline{F}$ and $\overline{G}$ are vector functions then,

(i) $\nabla \times (\overline{F} \pm \overline{G}) = \nabla \times \overline{F} \pm \nabla \times \overline{G}$

(ii) $\nabla \times (f\overline{F}) = (\nabla f) \times \overline{F} + f(\nabla \times \overline{F})$

(iii) $\nabla \times (\overline{F} \times \overline{G}) = (\nabla . \overline{G})\overline{F} - (\nabla . \overline{F})\overline{G} + (\overline{G} . \nabla)\overline{F} - (\overline{F} . \nabla)\overline{G}$

Proof:

(i) $\nabla \times (\overline{F} \pm \overline{G}) = \sum \mathbf{i} \times \frac{\partial}{\partial x} (\overline{F} \pm \overline{G})$

$$= \sum \mathbf{i} \times \left(\frac{\partial \overline{F}}{\partial x} \pm \frac{\partial \overline{G}}{\partial x} \right)$$

$$= \sum \mathbf{i} \times \frac{\partial \overline{F}}{\partial x} \pm \sum \mathbf{i} \times \frac{\partial \overline{G}}{\partial x}$$

$$= \nabla \times \overline{F} \pm \nabla \times \overline{G}$$

(i.e., curl $(\overline{F} \pm \overline{G}) = \text{curl } \overline{F} \pm \text{curl } \overline{G}$)

(ii) $\nabla \times (f\overline{F}) = \sum \mathbf{i} \times \frac{\partial}{\partial x} (f\overline{F})$

$$= \sum \mathbf{i} \times \left(\frac{\partial f}{\partial x} \overline{F} + f \frac{\partial \overline{F}}{\partial x} \right)$$

$$= \sum \mathbf{i} \times \left(\frac{\partial f}{\partial x} \overline{F} \right) + \sum \mathbf{i} \times \left(f \frac{\partial \overline{F}}{\partial x} \right)$$

$$= \sum \mathbf{i} \frac{\partial f}{\partial x} \times \overline{F} + f \sum \mathbf{i} \times \frac{\partial \overline{F}}{\partial x}$$

$$= (\nabla f) \times \overline{F} + f(\nabla \times \overline{F})$$

(i.e., curl $(f\overline{F}) = (\text{grad } f) \times \overline{F} + f \, \text{curl } \overline{F}$)

(iii) $\nabla \times (\overline{F} \times \overline{G}) = \sum \mathbf{i} \times \frac{\partial}{\partial x} (\overline{F} \times \overline{G})$

$$= \sum \mathbf{i} \times \left(\frac{\partial \overline{F}}{\partial x} \times \overline{G} + \overline{F} \times \frac{\partial \overline{G}}{\partial x} \right)$$

$$= \sum \mathbf{i} \times \left(\frac{\partial \overline{F}}{\partial x} \times \overline{G} \right) + \sum \mathbf{i} \times \left(\overline{F} \times \frac{\partial \overline{G}}{\partial x} \right)$$

$$= \sum (\mathbf{i}.\overline{G}) \frac{\partial \overline{F}}{\partial x} - \sum \left(\mathbf{i}.\frac{\partial \overline{F}}{\partial x} \right) \overline{G} + \sum \left(\mathbf{i}.\frac{\partial \overline{G}}{\partial x} \right) \overline{F} - \sum (\mathbf{i}.\overline{F}) \frac{\partial \overline{G}}{\partial x}$$

$$= \sum \left(\overline{G}.\mathbf{i}\frac{\partial}{\partial x} \right) \overline{F} - \overline{G} \sum \left(\mathbf{i}.\frac{\partial \overline{F}}{\partial x} \right) + \overline{F} \sum \left(\mathbf{i}.\frac{\partial \overline{G}}{\partial x} \right) - \sum \left(\overline{F}.\mathbf{i}\frac{\partial}{\partial x} \right) \overline{G}$$

$$= \left(\overline{G}.\sum \mathbf{i}\frac{\partial}{\partial x} \right) \overline{F} - \overline{G} \sum \left(\mathbf{i}.\frac{\partial \overline{F}}{\partial x} \right) + \overline{F} \sum \left(\mathbf{i}.\frac{\partial \overline{G}}{\partial x} \right) - \left(\overline{F}.\sum \mathbf{i}\frac{\partial}{\partial x} \right) \overline{G}$$

$$= \left(\overline{G}.\sum \mathbf{i}\frac{\partial}{\partial x} \right) \overline{F} - \overline{G} \sum \left(\mathbf{i}.\frac{\partial \overline{F}}{\partial x} \right) + \overline{F} \sum \left(\mathbf{i}.\frac{\partial \overline{G}}{\partial x} \right) - \left(\overline{F}.\sum \mathbf{i}\frac{\partial}{\partial x} \right) \overline{G}$$

$$= (\overline{G}.\nabla)\overline{F} - \overline{G}(\nabla.\overline{F}) + \overline{F}(\nabla.\overline{G}) - (\overline{F}.\nabla)\overline{G}$$
$$= (\overline{G}.\nabla)\overline{F} - (\nabla.\overline{F})\overline{G} + (\nabla.\overline{G})\overline{F} - (\overline{F}.\nabla)\overline{G}$$
$$= (\nabla.\overline{G})\overline{F} - (\nabla.\overline{F})\overline{G} + (\overline{G}.\nabla)\overline{F} - (\overline{F}.\nabla)\overline{G}$$

(i.e., curl $(\overline{F} \times \overline{G}) = (\text{div}\,\overline{G})\,\overline{F} - (\text{div}\,\overline{F})\overline{G} + (\overline{G}.\nabla)\overline{F} - (\overline{F}.\nabla)\overline{G}$)

Theorem 2.1.4 If f is a scalar function and is a vector function then

(i) $\nabla.(\nabla f) = \nabla^2 f$.

(ii) $\nabla \times \nabla f = 0$

(iii) $\nabla.(\nabla \times \overline{F}) = 0$

(iv) $\nabla \times (\nabla \times \overline{F}) = \nabla(\nabla.\overline{F}) - \nabla^2 \overline{F}$

(v) $\nabla(\nabla.\overline{F}) = \nabla \times (\nabla \times \overline{F}) + \nabla^2 \overline{F}$

Proof:

(i) $\nabla.(\nabla f) = \left(\mathbf{i}\frac{\partial}{\partial x} + \mathbf{j}\frac{\partial}{\partial y} + \mathbf{k}\frac{\partial}{\partial z} \right) . \left(\mathbf{i}\frac{\partial f}{\partial x} + \mathbf{j}\frac{\partial f}{\partial y} + \mathbf{k}\frac{\partial f}{\partial z} \right)$

$$= \frac{\partial^2 f}{\partial x^2} + \frac{\partial^2 f}{\partial y^2} + \frac{\partial^2 f}{\partial z^2} = \nabla^2 f$$

(i.e., div (grad f) = ∇^2f)

(ii) $\nabla \times (\nabla f) = \sum \mathbf{i} \times \frac{\partial}{\partial x}(\nabla f)$

$$= \sum \mathbf{i} \times \frac{\partial}{\partial x} \left(\mathbf{i}\frac{\partial f}{\partial x} + \mathbf{j}\frac{\partial f}{\partial y} + \mathbf{k}\frac{\partial f}{\partial z} \right)$$

$$= \sum\left(\mathbf{k}\frac{\partial^2 f}{\partial x\partial y} - \mathbf{j}\frac{\partial^2 f}{\partial x\partial z}\right)$$

$$= \mathbf{k}\frac{\partial^2 f}{\partial x\partial y} - \mathbf{j}\frac{\partial^2 f}{\partial x\partial z} + \mathbf{i}\frac{\partial^2 f}{\partial y\partial z} - \mathbf{k}\frac{\partial^2 f}{\partial x\partial y} + \mathbf{j}\frac{\partial^2 f}{\partial z\partial x} - \mathbf{i}\frac{\partial^2 f}{\partial y\partial z}$$

$$= 0$$

(i.e., curl (grad f) = 0)

(iii) $\nabla.(\nabla\times\overline{F}) = \sum \mathbf{i}.\frac{\partial}{\partial x}(\nabla\times\overline{F})$

$$= \sum \mathbf{i}.\frac{\partial}{\partial x}\left(\sum \mathbf{i}\times\frac{\partial \overline{F}}{\partial x}\right)$$

$$= \sum \mathbf{i}.\frac{\partial}{\partial x}\left(\mathbf{i}\times\frac{\partial \overline{F}}{\partial x} + \mathbf{j}\times\frac{\partial \overline{F}}{\partial y} + \mathbf{k}\times\frac{\partial \overline{F}}{\partial z}\right)$$

$$= \sum \mathbf{i}.\left(\mathbf{i}\times\frac{\partial^2 \overline{F}}{\partial x^2} + \mathbf{j}\times\frac{\partial^2 \overline{F}}{\partial x\partial y} + \mathbf{k}\times\frac{\partial^2 \overline{F}}{\partial x\partial z}\right)$$

$$= \sum(\mathbf{i}\times\mathbf{i}).\frac{\partial^2 \overline{F}}{\partial x^2} + \sum(\mathbf{i}\times\mathbf{j}).\frac{\partial^2 \overline{F}}{\partial x\partial y} + \sum(\mathbf{i}\times\mathbf{k}).\frac{\partial^2 \overline{F}}{\partial x\partial z}$$

$$= \sum \mathbf{k}.\frac{\partial^2 \overline{F}}{\partial x\partial y} - \sum \mathbf{j}.\frac{\partial^2 f}{\partial x\partial z}$$

$$= \mathbf{k}.\frac{\partial^2 \overline{F}}{\partial x\partial y} - \mathbf{j}.\frac{\partial^2 \overline{F}}{\partial x\partial z} + \mathbf{i}.\frac{\partial^2 \overline{F}}{\partial y\partial z} - \mathbf{k}.\frac{\partial^2 \overline{F}}{\partial x\partial y} + \mathbf{j}.\frac{\partial^2 \overline{F}}{\partial z\partial x} - \mathbf{i}.\frac{\partial^2 \overline{F}}{\partial y\partial z}$$

$$= 0$$

(i.e., div (curl $\overline{F}$) = 0)

(iv) $\nabla\times(\nabla\times\overline{F}) = \begin{vmatrix} i & j & k \\ \frac{\partial}{\partial x} & \frac{\partial}{\partial y} & \frac{\partial}{\partial z} \\ \frac{\partial F_3}{\partial y} - \frac{\partial F_2}{\partial z} & \frac{\partial F_1}{\partial z} - \frac{\partial F_3}{\partial x} & \frac{\partial F_2}{\partial x} - \frac{\partial F_1}{\partial y} \end{vmatrix}$

$$= \mathbf{i}\left(\frac{\partial^2 F_2}{\partial x\partial y} - \frac{\partial^2 F_1}{\partial y^2} - \frac{\partial^2 F_1}{\partial z^2} + \frac{\partial^2 F_3}{\partial x\partial z}\right) + \mathbf{j}\left(\frac{\partial^2 F_3}{\partial y\partial z} - \frac{\partial^2 F_2}{\partial z^2} - \frac{\partial^2 F_2}{\partial x^2} + \frac{\partial^2 F_1}{\partial x\partial y}\right)$$

$$+ \mathbf{k}\left(\frac{\partial^2 F_1}{\partial x\partial z} - \frac{\partial^2 F_3}{\partial x^2} - \frac{\partial^2 F_3}{\partial y^2} + \frac{\partial^2 F_2}{\partial z\partial y}\right) \qquad (1)$$

$$\nabla(\nabla.\overline{F}) - \nabla^2\overline{F} = \left(\mathbf{i}\frac{\partial}{\partial x}+\mathbf{j}\frac{\partial}{\partial y}+\mathbf{k}\frac{\partial}{\partial z}\right).\left(\frac{\partial F_1}{\partial x}+\frac{\partial F_2}{\partial y}+\frac{\partial F_3}{\partial z}\right) - \nabla^2(F_1\mathbf{i}+F_2\mathbf{j}+F_3\mathbf{k})$$

$$= \mathbf{i}\left(\frac{\partial^2 F_1}{\partial x^2}+\frac{\partial^2 F_2}{\partial x\partial y}+\frac{\partial^2 F_3}{\partial x\partial z}\right)+\mathbf{j}\left(\frac{\partial^2 F_1}{\partial x\partial y}+\frac{\partial^2 F_2}{\partial y^2}+\frac{\partial^2 F_3}{\partial y\partial z}\right)+$$

$$\mathbf{k}\left(\frac{\partial^2 F_1}{\partial z\partial x}+\frac{\partial^2 F_2}{\partial z\partial y}+\frac{\partial^2 F_3}{\partial z^2}\right)-\mathbf{i}\left(\frac{\partial^2 F_1}{\partial x^2}+\frac{\partial^2 F_1}{\partial y^2}+\frac{\partial^2 F_1}{\partial z^2}\right)$$

$$-\mathbf{j}\left(\frac{\partial^2 F_2}{\partial x^2}+\frac{\partial^2 F_2}{\partial y^2}+\frac{\partial^2 F_2}{\partial z^2}\right)-\mathbf{k}\left(\frac{\partial^2 F_3}{\partial x^2}+\frac{\partial^2 F_3}{\partial y^2}+\frac{\partial^2 F_3}{\partial z^2}\right)$$

$$= \mathbf{i}\left(\frac{\partial^2 F_2}{\partial x\partial y}+\frac{\partial^2 F_3}{\partial x\partial z}-\frac{\partial^2 F_1}{\partial y^2}-\frac{\partial^2 F_1}{\partial z^2}\right)+\mathbf{j}\left(\frac{\partial^2 F_1}{\partial x\partial y}+\frac{\partial^2 F_3}{\partial y\partial z}-\frac{\partial^2 F_2}{\partial x^2}-\frac{\partial^2 F_2}{\partial z^2}\right)+$$

$$\mathbf{k}\left(\frac{\partial^2 F_1}{\partial z\partial x}+\frac{\partial^2 F_2}{\partial z\partial y}-\frac{\partial^2 F_3}{\partial x^2}-\frac{\partial^2 F_3}{\partial y^2}\right) \quad (2)$$

In (1) and (2) R.H.S are equal.

$\therefore \nabla\times(\nabla\times\overline{F}) = \nabla(\nabla.\overline{F}) - \nabla^2\overline{F}$

(i.e., curl (curl $\overline{F}$) = grad (div $\overline{F}$) - $\nabla^2\overline{F}$)

(v) From (iv) we have, $\nabla\times(\nabla\times\overline{F}) = \nabla(\nabla.\overline{F}) - \nabla^2\overline{F}$

$\therefore \nabla(\nabla.\overline{F}) = \nabla\times(\nabla\times\overline{F}) + \nabla^2\overline{F}$

Note:

(i) If the vector field $\overline{F}$ is **conservative**, then there exists a scalar function f such that $\overline{F}$ = grad f. Then curl($\overline{F}$) = curl(grad f) = 0. Therefore, if $\overline{F}$ is conservative then curl($\overline{F}$) = 0.

(ii) If a vector field $\overline{F}$ is irrotational, there exists a scalar field f such that $\overline{F} = \nabla f$. f is known as the **scalar potential** of $\overline{F}$.

Example 2.1.10 If $\overline{F} = xz^3\,\mathbf{i} - 2xyz\,\mathbf{j} + xz\,\mathbf{k}$, find div $\overline{F}$ and curl $\overline{F}$ at (1, 2, 3).

Solution:

$$\text{div}\,\overline{F} = \frac{\partial F_1}{\partial x}+\frac{\partial F_2}{\partial y}+\frac{\partial F_3}{\partial z}$$

$$= \frac{\partial}{\partial x}(xz^3)+\frac{\partial}{\partial y}(-2xyz)+\frac{\partial}{\partial z}(xz)$$

$$= z^3 - 2xz + x$$

div $\overline{F}$ at (1, 2, 3) = 3^3-2.1.3 + 1 = 22

$$\text{curl}\,\overline{F} = \begin{vmatrix} \mathbf{i} & \mathbf{j} & \mathbf{k} \\ \frac{\partial}{\partial x} & \frac{\partial}{\partial y} & \frac{\partial}{\partial z} \\ xz^3 & -2xyz & xz \end{vmatrix} = \mathbf{i}(0 + 2xy) - \mathbf{j}(z - 3xz^2) + \mathbf{k}(-2yz - 0)$$

$$= 2xy\,\mathbf{i} + (3xz^2 - z)\,\mathbf{j} - 2yz.\mathbf{k}$$

$\therefore$ curl $\overline{F}$ at (1, 2, 3) = 4**i** - 24**j** -12**k**

Example 2.1.11 Show that the vector field $\bar{a} = 5y^4z^3\,\mathbf{i} + 8xz^2\,\mathbf{j} - y^2x\,\mathbf{k}$ is solenoidal.

Solution:

$$\text{div a} = \frac{\partial}{\partial x}(5y^4z^3) + \frac{\partial}{\partial y}(8xz^2) + \frac{\partial}{\partial z}(-y^2x)$$

$$= 0 + 0 + 0 = 0$$

$\therefore$ a is solenoidal.

Example 2.1.12 Show that the vector field $\overline{A} = yz\,\mathbf{i} + zx\,\mathbf{j} + xy\,\mathbf{k}$ is irrotational.

Solution:

$$\text{curl}\,\overline{A} = \begin{vmatrix} \mathbf{i} & \mathbf{j} & \mathbf{k} \\ \frac{\partial}{\partial x} & \frac{\partial}{\partial y} & \frac{\partial}{\partial z} \\ yz & zx & xy \end{vmatrix} = \mathbf{i}(x - x) - \mathbf{j}(y - y) + \mathbf{k}(z - z) = \mathbf{i}(0) - \mathbf{j}(0) + \mathbf{k}(0) = 0$$

$\therefore \overline{A}$ is irrotational.

Example 2.1.13 If $\overline{F} = (3x^2y - z)\mathbf{i} + (xz^3 + y^4)\mathbf{j} - 2x^3z^2\mathbf{k}$, find grad div $\overline{F}$ at (2, -1, 0).

Solution:

$$\text{div}\,\overline{F} = \frac{\partial}{\partial x}(3x^2y - z) + \frac{\partial}{\partial y}(xz^3 + y^4) + \frac{\partial}{\partial z}(-2x^3z^2) = 6xy + 4y^3 - 4x^3z$$

$$\therefore \text{grad div}\,\overline{F} = \left(\mathbf{i}\frac{\partial}{\partial x} + \mathbf{j}\frac{\partial}{\partial y} + \mathbf{k}\frac{\partial}{\partial z}\right)(6xy + 4y^3 - 4x^3z)$$

$$= \mathbf{i}(6y - 12x^2z) + \mathbf{j}(6x + 12y^2) + \mathbf{k}(-4x^3)$$

At (2, -1, 0), grad div $\overline{F}$ = **i**(-6) + **j**(24) + **k**(-32) = -6**i** + 24**j** - 32**k**

Example 2.1.14 If the vector fields $\overline{F}$ and $\overline{G}$ are irrotational, show that $\overline{F} \times \overline{G}$ is solenoidal.

Solution:

Given $\overline{F}$ and $\overline{G}$ are irrotational

$\therefore$ curl $\overline{F} = 0$ and curl $\overline{G} = 0$.

Now div($\overline{F} \times \overline{G}$) = (curl $\overline{F}$).G - $\overline{F}$.(curl $\overline{G}$) = 0 . $\overline{G}$ - $\overline{F}$.0 = 0

$\therefore\ \overline{F}\times\overline{G}$ is solenoidal.

Example 2.1.15 If $\bar{a}$ is a constant vector and $\bar{r} = x\mathbf{i} + y\mathbf{j} + z\mathbf{k}$, prove that

(i) curl $(\bar{a}\times\bar{r}) = 2\bar{a}$

(ii) $\nabla.[(\bar{a}.\bar{r})r^n]\bar{r} = (n+4)(\bar{a}.\bar{r})r^n$

(iii) $\nabla\times[(\bar{a}\times\bar{r})r^n]\bar{r} = (\bar{a}\times\bar{r})r^n$ where $r = \sqrt{x^2+y^2+z^2}$

Solution:

(i) Let $\bar{a} = a_1\mathbf{i} + a_2\mathbf{j} + a_3\mathbf{k}$

$$\text{Then } \bar{a}\times\bar{r} = \begin{vmatrix} \mathbf{i} & \mathbf{j} & \mathbf{k} \\ a_1 & a_2 & a_3 \\ x & y & z \end{vmatrix} = \mathbf{i}(a_2z - a_3y) - \mathbf{j}(a_1z - a_3x) + \mathbf{k}(a_1y - a_2x)$$

$$\text{curl}(\bar{a}\times\bar{r}) = \begin{vmatrix} \mathbf{i} & \mathbf{j} & \mathbf{k} \\ \partial/\partial x & \partial/\partial y & \partial/\partial z \\ a_2z - a_3y & a_3x - a_1z & a_1y - a_2x \end{vmatrix} = 2(a_1\mathbf{i} + a_2\mathbf{j} + a_3\mathbf{k}) = 2\bar{a}$$

(ii) $\bar{a}.\bar{r} = a_1x + a_2y + a_3z$

$(\bar{a}.\bar{r})r^n = (a_1x + a_2y + a_3z)\,(x^2+y^2+z^2)^{n/2}$

$[(\bar{a}.\bar{r})r^n]\bar{r} = (a_1x + a_2y + a_3z)(x^2+y^2+z^2)^{n/2}\ x\mathbf{i} + y\mathbf{j} + z\mathbf{k} = F_1\mathbf{i} + F_2\mathbf{j} + F_3\mathbf{j}$ (say)

$$\text{Then } \nabla.[(\bar{a}.\bar{r})r^n]\bar{r} = \frac{\partial F_1}{\partial x} + \frac{\partial F_2}{\partial y} + \frac{\partial F_3}{\partial z} \tag{1}$$

$$\frac{\partial F_1}{\partial x} = \frac{\partial}{\partial x}\left[(a_1x + a_2y + a_3z)\,(x^2+y^2+z^2)^{n/2}x\right]$$

$$= (a_1x + a_2y + a_3z)x.\frac{n}{2}(x^2+y^2+z^2)^{n/2-1}.2x + (2a_1x + a_2y + a_3z)\,(x^2+y^2+z^2)^{n/2}$$

$$= (\bar{a}.\bar{r})n\,x^2 r^{n-2} + (a_1x + \bar{a}.\bar{r})r^n$$

$$\text{Similarly, } \frac{\partial F_2}{\partial y} = (\bar{a}.\bar{r})n\,y^2 r^{n-2} + (a_2y + \bar{a}.\bar{r})r^n$$

$$\frac{\partial F_3}{\partial z} = (\bar{a}.\bar{r})n\,z^2 r^{n-2} + (a_3z + \bar{a}.\bar{r})r^n$$

$\therefore$ From (1) $\nabla.[(\bar{a}.\bar{r})r^n]\bar{r} = (\bar{a}.\bar{r})n\,r^2 r^{n-2} + (\bar{a}.\bar{r} + 3\bar{a}.\bar{r})r^n = (n+4)(\bar{a}.\bar{r})r^n$

(iii) We have $[(\bar{a}.\bar{r})r^n]\bar{r} = F_1\mathbf{i} + F_2\mathbf{j} + F_3\mathbf{j}$

$$\nabla\times[(\bar{a}\times\bar{r})\,r^n]\bar{r} = \begin{vmatrix} \mathbf{i} & \mathbf{j} & \mathbf{k} \\ \frac{\partial}{\partial x} & \frac{\partial}{\partial y} & \frac{\partial}{\partial z} \\ F_1 & F_2 & F_3 \end{vmatrix}$$

coefficient of $\mathbf{i} = \frac{\partial F_3}{\partial y} - \frac{\partial F_2}{\partial z}$

$$= \frac{\partial}{\partial y}\left[(a_1x + a_2y + a_3z)\, z\, (x^2 + y^2 + z^2)^{n/2}\right]$$

$$- \frac{\partial}{\partial z}\left[(a_1x + a_2y + a_3z)\, y\, (x^2 + y^2 + z^2)^{n/2}\right]$$

$$= a_2 z (x^2 + y^2 + z^2)^{n/2} + (a_1x + a_2y + a_3z)\, z.n\, (x^2 + y^2 + z^2)^{n/2 - 1}.y$$

$$- a_3 y (x^2 + y^2 + z^2)^{n/2} - (a_1x + a_2y + a_3z)\, y.n\, (x^2 + y^2 + z^2)^{n/2 - 1}.z$$

$$= (a_2z - a_3y)\, r^n$$

Similarly coefficient of $\mathbf{j} = (a_3x - a_1z)\, r^n$

Similarly coefficient of $\mathbf{k} = (a_1y - a_2x)\, r^n$

$$\therefore \nabla\times[(\bar{a}\times\bar{r})r^n]\bar{r} = [(a_2z - a_3y)\,\mathbf{i} + (a_3x - a_1z)\,\mathbf{j} + (a_1y - a_2x)\,\mathbf{k}]r^n$$

$$= (\bar{a}\times\bar{r})r^n$$

Example 2.1.16 Let f(x, y, z) be a solution of the Laplace equation $\nabla^2 f = 0$. Then show that ∇f is both irroational and solenoidal.

Solution:

Given $\nabla^2 f = 0$(i.e., f is harmonic)

i.e., we have $\nabla.\nabla f = 0$

i.e., div $\nabla f = 0$ $\quad\therefore$ ∇f is solenoidal.

For any scalar function f we have

curl (grad f) = 0

i.e., $\nabla\times(\nabla f) = 0$ $\quad\therefore \nabla f$ is irrotational

Example 2.1.17 If f and g are differentiable scalar fields, prove that $\nabla f\times\nabla g$ and $f\,\nabla g \times g\,\nabla f$ are solenoidal.

Solution:

We have $\nabla.(\bar{F}\times\bar{G}) = (\nabla\times\bar{F}).\bar{G} - \bar{F}.(\nabla\times\bar{G})$

$\therefore\ \nabla.(\nabla f\times\nabla g) = (\nabla\times\nabla f).\nabla g - \nabla f.(\nabla\times\nabla g)$

$= 0.\nabla g - \nabla f.0 = 0$

$\therefore\ \nabla f\times\nabla g$ is solenodal.

$\nabla.(f\nabla g\times g\nabla f) = [\nabla\times(f\nabla g)].g\nabla f - f\nabla g.[\nabla\times(g\nabla f)]$

$= [\nabla f\times\nabla g + f(\nabla\times\nabla g)].g\nabla f - f\nabla g.[\nabla g\times\nabla f + g(\nabla\times\nabla f)]$

$= (\nabla f\times\nabla g).g\nabla f - f\nabla g.(\nabla g\times\nabla f)$ $\quad$ (Since $\nabla\times\nabla f = \nabla\times\nabla g = 0$)

$= (g\nabla f\times\nabla f)\times\nabla g - (f\nabla g\times\nabla g).\nabla f$

$= 0$ $\quad$ (Using $\bar{a}\times\bar{a} = 0$)

$\therefore\ f\,\nabla g\times g\,\nabla f$ is solenoidal.

Example 2.1.18 Prove that the vector $\bar{a} = r^n \bar{r}$ is irrotational. Find n when $\bar{a}$ is solenoidal.

Solution:

$$\nabla \times \bar{a} = \nabla \times (r^n . \bar{r})$$
$$= \nabla (r^n) \times \bar{r} + r^n (\nabla \times \bar{r}) \qquad (1)$$

We have
$$\nabla(r^n) = \mathbf{i}\frac{\partial}{\partial x}(r^n) + \mathbf{j}\frac{\partial}{\partial y}(r^n) + \mathbf{k}\frac{\partial}{\partial z}(r^n)$$
$$= \mathbf{i}\, nr^{n-1}.\frac{\partial r}{\partial x} + \mathbf{j}\, nr^{n-1}.\frac{\partial r}{\partial y} + \mathbf{k}\, nr^{n-1}.\frac{\partial r}{\partial z} \quad (\text{Since } r^2 = x^2 + y^2 + z^2)$$
$$= n.r^{n-1}\left(\mathbf{i}.\frac{x}{r} + \mathbf{j}\frac{y}{r} + \mathbf{k}\frac{z}{r}\right)$$
$$= n.r^{n-2}.\bar{r} \qquad (2)$$

Also
$$\nabla \times \bar{r} \begin{vmatrix} \mathbf{i} & \mathbf{j} & \mathbf{k} \\ \frac{\partial}{\partial x} & \frac{\partial}{\partial y} & \frac{\partial}{\partial z} \\ x & y & z \end{vmatrix} = 0 \qquad (3)$$

Using (2) and (3) in (1), we get,

$$\nabla \times \bar{a} = n.r^{n-2}\ \bar{r} \times \bar{r} + r^n (0) = 0 \qquad (\text{Since } \bar{r} \times \bar{r} = 0)$$

$\therefore$ $\bar{a}$ irrotational

When $\bar{a}$ is solenoiial, $\nabla . \bar{a} = 0$

Now, $\nabla . \bar{a} = \nabla (r^n \bar{r})$

$$= \nabla r^n . \bar{r} + r^n (\nabla . \bar{r}) = nr^{n-2} \bar{r} . \bar{r} + r^n (3) = n.r^{n-2} r^2 + 3r^n = (n+3) r^n$$

$\therefore$ $\bar{a}$ is solenoidal implies $(n + 3)r^n = 0$ i.e., $n = -3$

Example 2.1.19 Show that the vector field $\bar{F} = (2xye^z)\mathbf{i} + (x^2 e^z)\mathbf{j} + (x^2ye^z)\mathbf{k}$ is irroational and find its scalar potential.

Solution:

$$\nabla \times \bar{F} = \begin{vmatrix} \mathbf{i} & \mathbf{j} & \mathbf{k} \\ \frac{\partial}{\partial x} & \frac{\partial}{\partial y} & \frac{\partial}{\partial z} \\ 2xye^z & x^2 e^z & x^2 ye^z \end{vmatrix} = \mathbf{i}(x^2e^z - x^2e^z) - \mathbf{j}(2xye^z - 2xye^z) + \mathbf{k}(2xe^z - 2xe^z) = 0$$

$\therefore$ $\bar{F}$ is irrotational

Let
$$\bar{F} = \nabla f = \mathbf{i}\frac{\partial f}{\partial x} + \mathbf{j}\frac{\partial f}{\partial y} + \mathbf{k}\frac{\partial f}{\partial z}$$

$$\frac{\partial f}{\partial x} = 2xye^z, \frac{\partial f}{\partial y} = x^2 e^z \text{ and } \frac{\partial f}{\partial z} = x^2 ye^z$$

Integrating, $\frac{\partial f}{\partial x} = 2xye^z$, we get,

$$f = x^2ye^z + g(y, z) \qquad (1)$$

Differentiating (1) w.r.t y, we get,

$$\frac{\partial f}{\partial y} = x^2e^z + \frac{\partial g(y,z)}{\partial y}$$

i.e., $x^2e^z = x^2e^z + \frac{\partial g(y,z)}{\partial y} \qquad \therefore \frac{\partial g(y,z)}{\partial y} = 0$

Hence $g = g(z)$ $\quad\therefore$(1) becomes,

$$f = x^2ye^z + g(z) \qquad (2)$$

Differentiating (2) w.r.t z, we get,

$$\frac{\partial f}{\partial z} = x^2ye^z + g'(z)$$

i.e., $x^2ye^z = x^2ye^z + g'(z)$

Hence $g'(z) = 0$ $\qquad$ i.e., $g(z) = c$, a constant

$\therefore$ Equation (2) gives, $f(x, y, z) = x^2ye^z + c$

Example 2.1.20 If $\overline{F}$ is a solenoidal field, show that curl curl curl curl $\overline{F} = \nabla^4\overline{F}$

Solution:

We have $\nabla \times (\nabla \times \overline{F}) = \nabla(\nabla.\overline{F}) - \nabla^2\overline{F}$ (1)

Since $\overline{F}$ is solenidal, $\nabla.\overline{F} = 0$

$\therefore$(1) becomes, $\qquad \nabla \times(\nabla\times \overline{F}) = -\nabla^2\overline{F}$

i.e., curl curl $\overline{F} = -\nabla^2\overline{F}$

$\therefore$ curl curl curl curl $\overline{F} = \nabla \times (\nabla\times(-\nabla^2\overline{F}))$

$= \nabla(\nabla.(-\nabla^2\overline{F})) - \nabla^2(-\nabla^2\overline{F})$

$= -\nabla(\nabla^2(\nabla.\overline{F})) + \nabla^2\nabla^2\overline{F} = \nabla^4\overline{F}$ $\quad$ (Since $\nabla.\overline{F} = 0$)

EXERCISE 2.1

PART –A

1. Define gradient of a scalar function. Find grad f when $f(x, y, z) = 3x^2y + 2y^3-5z$.
2. What is the geometrical meaning of ∇f?
3. What is the directional derivative of f in the direction of the vector $\overline{a}$?
4. What are the maximum and minimum values of the directional derivatives of f?
5. When is a vector function said to be conservative?
6. Define divergence of a vector field. Given $\overline{r} = x\mathbf{i} + y\mathbf{j} + z\mathbf{k}$, find div $\overline{r}$?
7. Define curl of a vector field. Find curl $\overline{F}$ when $\overline{F}(x, y, z) = xe^{-y}\mathbf{i} + 2ze^{-y}\mathbf{j} + xy^2\mathbf{k}$.
8. Prove $\nabla \times (\nabla f) = 0$ for any scalar field f.
9. Prove $\nabla.(\nabla\times \overline{F}) = 0$ for any vector field $\overline{F}$.

10. Define the terms solenoidal and irrotational.
11. If $\overline{F}$ is a conservative vector field, prove that curl($\overline{F}$) = 0.
12. If $\overline{F}$ and $\overline{G}$ are irrotational vector fields, show that $\overline{F} \times \overline{G}$ is a solenoidal vector field.
13. If f(x, y, z) is harmonic, prove ∇f is solenoidal.
14. Prove that ∇f is irrotational.
15. If $f(x, y, z) = x^2y - 2y^2z^3$, find ∇f at the point (1, -1, 2)
16. Find grad f when $f = 2xz^4 - x^2y$ at (2, -2, -1).
17. Compute the gradient of $(x^2 + y^2 + z^2)^{1/2}$ at (1, 1, 1).
18. Find the unit vector normal to the surface $z^2 = x^2 - y^2$ at $(2, 1, \sqrt{3})$.
19. If $\overline{r} = x\mathbf{i} + y\mathbf{j} + z\mathbf{k}$, show that $(\overline{a}.\nabla)\overline{r} = \overline{a}$.
20. Find the unit normal vector to the surface $xy^2 + 2yz = 8$ at the point (3, -2, 1).
21. Find the divergence of $x^2\mathbf{i} + y^2\mathbf{j} + z^2\mathbf{k}$.
22. If $\overline{F} = (x^2 + yz)\mathbf{i} + (y^2 + zx)\mathbf{j} + (z^2 + xy)\mathbf{k}$. find div $\overline{F}$.
23. If $\overline{V} = xy\mathbf{i} + yz\mathbf{j} + zx\mathbf{k}$, find curl $\overline{V}$.
24. If $f(x, y, z) = 16xy^3z^2$, prove that curl(grad f) = 0.
25. If $\overline{a} = x^2\mathbf{i} + y^2\mathbf{j} + z^2\mathbf{k}$, prove that div(curl $\overline{a}$) = 0.
26. If $\overline{a}$ is a constant vector and $\overline{r} = x\mathbf{i} + y\mathbf{j} + z\mathbf{k}$, prove that grad($\overline{a}.\overline{r}$) = $\overline{a}$.
27. If $\overline{a}$ is a constant vector and $\overline{r} = x\mathbf{i} + y\mathbf{j} + z\mathbf{k}$, prove that div($\overline{a} \times \overline{r}$) = 0 and $\overline{a} \times (\text{curl}\,\overline{r}) = 0$.
28. Show that $(x + 3y)\mathbf{i} + (y-3z)\mathbf{j} + (x-2z)\mathbf{k}$ is solenoidal.
29. If $\overline{F} = \text{grad}(x^3 + y^3 + z^3 - 3xyz)$, find div $\overline{F}$.
30. Show that the vector function $\overline{V} = (6xy + z^3)\mathbf{i} + (3x^2-z)\mathbf{j} + (3xz^2-y)\mathbf{k}$ is irrotational.

PART-B

31. Prove that (i) $\nabla r = \frac{\overline{r}}{r}$ (ii) $\nabla f(r) = f'(r)\,\nabla r = f'(r)\hat{r}$ (iii) $\nabla(\log r) = \frac{\overline{r}}{r^2}$ where $\overline{r} = x\mathbf{i} + y\mathbf{j} + z\mathbf{k}$ and $r = |\overline{r}|$.
32. If $\overline{r} = x\mathbf{i} + y\mathbf{j} + z\mathbf{k}$ and $r = |\overline{r}|$, prove that (i) $\nabla\left(\frac{1}{r}\right) = \frac{-\overline{r}}{r^3}$ (ii) $\nabla(r^n) = n\,r^{n-2}\,\overline{r}$ (iii) $\nabla f(r) \times \overline{r} = 0$ (iv) $\nabla e^{r^2} = 2e^{r^2}\,\overline{r}$
33. If $\overline{a}$ and $\overline{b}$ are constant vectors and $\overline{r} = x\mathbf{i} + y\mathbf{j} + z\mathbf{k}$, prove that
    (i) $\nabla(\overline{r}.\overline{a}) = \overline{a}$ (ii) $\nabla(\overline{r}.(\overline{a} \times \overline{b})) = \overline{a} \times \overline{b}$.
34. Find the scalar function φ(r), given that ∇φ is (i) $5r^3\overline{r}$ (ii) $(6r - 8r^2)\overline{r}$ and φ(2) = 4.
35. Find the scalar function φ(x, y, z), given that ∇φ is
    (i) $2xyz\,\mathbf{i} + (x^2z + 1)\mathbf{j} + x^2y\mathbf{k}$
    (ii) $(y + y^2 + z^2)\,\mathbf{i} + (x + z + 2xy)\mathbf{j} + (y + 2zx)\mathbf{k}$ and φ(1, 1, 1) = 0
    (iii) $(6xy + z^3)\,\mathbf{i} + (3x^2 - z)\mathbf{j} + (3xz^2 - y)\mathbf{k}$
    (iv) $(y + \sin z)\,\mathbf{i} + x\mathbf{j} + x\cos z\,\mathbf{k}$
36. Find the unit vector normal to the following surfaces.

(i) $xy^3z^2 = 4$ at (-1, -1, 2)
(ii) $x^3 - xyz + z^3 = 1$ at (1, 1, 1)
(iii) $(x-1)^2 + y^2 + (z+2)^2 = 9$ at (3, 1, -4)
(iv) $xy + yz + zx = 7$ at (1, 1, 3)

37. Prove that the directional derivative of $f = x^3y^2z$ at (1, 2, 3) is maximum along the direction $9\mathbf{i} + 3\mathbf{j} + \mathbf{k}$. Also find the maximum directional derivative.
38. Find the directional derivative of $f = x^2yz+4xz^2$ at (1,-2,-1) in the direction $2\mathbf{i} - \mathbf{j} - 2\mathbf{k}$.
39. Find the maximum directional derivative of the following functions
(i) $f(x, y, z) = x^2 + y^2 + z^2$ at (1, 1, 1)
(ii) $f(x, y, z) = 2xz^4 - x^2y$ at (2, -2, -1)
40. Find the direction and magnitude of the greatest directional derivative of $f(x, y, z) = 3x^2y - y^3z^3$ at (1, -2, -1)
41. Find the angle between the normals to the surface $z^2 - xy = 0$ at points (1, 1, 1) and (4, 1, 2)
42. Find the equation of the tangent plane to the surface $x^2-4y^2 + 3z^2 + 4 = 0$ at (3, 2, 1)
43. Find the values of a, b and c such that the maximum value of the directional derivative of $f(x, y, z) = axy^2 + byz + cx^2z^2$ at (1, -1, 1) is in the direction pararlled to the axis of y and has magnitude 6.
44. Find the angle between the surfaces $x^2 + y^2 + z^2 = 9$ and $x^2 + y^2 = z + 3$ at (-2, 1, 2).
45. Find the equation of the normal to the surface $x^2 + 2y^2 + 4z^2 = 10$ at (2, 1, -1).
46. Determine the values of a and b so that the surface $5x^2-2yz-9x = 0$ is orthogonal to the surface $ax^2y + bz^3 = 4$ at (1, -1, 2).
47. Show that (i) $\nabla^2(r^n) = n(n+1)r^{n-2}$ (ii) $\nabla^2(e^r) = e^r + \frac{2}{r}e^r$.
48. Show that $\nabla^2(f(r)) = f''(r) + \frac{2}{r}f'(r)$.
49. If $f = x^2yz$ and $g = xy - 3z^2$, find (i) $\nabla(\nabla f.\nabla g)$ (ii) $\nabla.(\nabla f\times\nabla g)$
50. Determine f(r) so that the vector $f(r)\bar{r}$ is solenoidal
51. Prove that $f(r)\bar{r}$ is irrotational
52. Show that $r^n\bar{r}$ is irrotational for any value of n, but is solenoidal only when n = -3.
53. If $\bar{a} = \bar{w}\times\bar{r}$ and $\bar{w}$ is a constant vector, prove that $\bar{w} = \frac{1}{2}\text{curl}\,\bar{a}$.
54. Determine the constant a so that the vector $(z + 3y)\mathbf{i} + (x - 2z)\mathbf{j} + (x + az)\mathbf{k}$ is solenoidal.
55. Find the constant a, b and c so that the vector $(x + 2y + az)\mathbf{i} + (bx - 3y - z)\mathbf{j} + (4x + cy + 2z)\mathbf{k}$ is irrotational.
56. Show that the vector field $\bar{F} = yz^2\mathbf{i} + (xz^2 - 1)\mathbf{j} + 2(xyz - 1)\mathbf{k}$ is irrotational and find its scalar potential.
57. Show that the $\bar{F} = (y^2\cos x + z^3)\mathbf{i} + (2y\sin x - 4)\mathbf{j} + 3xz^2\,\mathbf{k}$ is irrotational and find its scalar potential.
58. Show that the $\bar{F} = (6xy + z^3)\mathbf{i} + (3x^2 - z)\mathbf{j} + (3xz^2 - y)\mathbf{k}$ is irrotational. Find f such that $\bar{F} = \nabla f$.

59. Prove that $\overline{F} = (2x + yz)\mathbf{i} + (4y + zx)\mathbf{j} - (6z - xy)\mathbf{k}$ is solenoidal as well as irrotational. Also find the scalar potential of $\overline{F}$.
60. If $\overline{F} = 2yz\mathbf{i} - x^2y\mathbf{j} + xz\mathbf{k}$ and $\varphi = 2xyz$, verify that
    (i) $\text{curl}(\varphi\nabla\varphi) = 0$ (ii) $(\overline{F}.\nabla)\,\varphi = \overline{F}.\nabla\varphi$
61. If $\bar{a}$ is a constant vector, show that
    (i) $\nabla.(\bar{a}\times\bar{r}) = 0$
    (ii) $\nabla\times(\bar{a}\times\bar{r}) = 2\bar{a}$
    (iii) $\nabla.\left(\dfrac{\bar{a}\times\bar{r}}{r}\right) = 0$
    (iv) $\text{grad}\left(\dfrac{\bar{a}.\bar{r}}{r^3}\right) = \text{curl}\left(\dfrac{\bar{r}\times\bar{a}}{r^3}\right)$
    (v) $\text{curl}(\bar{a}\times\bar{r})r^n = (n+2)r^n\bar{a} - n\,r^{n-2}(\bar{a}.\bar{r})\bar{r}$
62. If $\bar{a}$ and $\bar{b}$ are constants vectors and $\overline{F} = \bar{r} - \bar{a}, \overline{G} = \bar{r} - \bar{b}$, show that
    (i) $\nabla.(\overline{F}\times\overline{G}) = 0$
    (ii) $\nabla(\overline{F}.\overline{G}) = \overline{F} + \overline{G}$
    (iii) $\nabla\times(\overline{F}\times\overline{G}) = 2(\bar{b} - \bar{a})$
63. If φ is any scalar function and $\bar{a}$ is any vector function show that
    (i) $\text{curl}\,(\varphi\,\text{grad}\,\varphi) = 0$
    (ii) $\text{div}\,(\varphi\,\text{curl}\,\bar{a}) = \text{grad}\,\varphi.\text{curl}\,\bar{a}$

2.2 VECTOR INTEGRATION

In this section we extend the concept of integration to vector valued functions. We define line, surface and volume integrals of vector functions and consider the evaluation of these integrals. We also discuss some simple applications of line, surface and volume integrals. Let $\overline{F}(t) = F_1(t)\mathbf{i} + F_2(t)\mathbf{j} + F_3(t)\mathbf{k}$. be a vector function of a scalar variable t. Then we define $\int\overline{F}(t)dt = \mathbf{i}\int F_1(t)dt + \mathbf{j}\int F_2(t)dt + \mathbf{k}\int F_3(t)dt$.

If the vector function $\overline{F}(t)$ is defined in the interval $a \le t \le b$, then the definite integral

$$\int_a^b \overline{F}(t)dt = \mathbf{i}\int_a^b F_1(t)dt + \mathbf{j}\int_a^b F_2(t)dt + \mathbf{k}\int_a^b F_3(t)dt$$

2.2.1 LINE INTEGRAL OF VECTOR FIELDS

Let $\overline{F}(x, y, z) = F_1(x, y, z)\mathbf{i} + F_2(x, y, z)\mathbf{j} + F_3(x, y, z)\mathbf{k}$ be a vector function over a region in space. Let C be any simple smooth curve in the above region and the parametric equation of C be

$x = x(t), y = y(t), z = z(t), a \le t \le b$ (1)

Then the position vector of any point P on C is $\bar{r} = \bar{r}(t) = x(t)\mathbf{i} + y(t)\mathbf{j} + z(t)\mathbf{k}$.

Let Q be a neighboring point on the curve C with position vector $\bar{r} + d\bar{r}$, so that $\overline{PQ} = d\bar{r}$ (Fig. 2.4). If $\bar{F}$ (x, y, z) at P has a direction that makes an angel θ with $\overline{PQ}$, then

$\bar{F}.d\bar{r} = F\cos\theta\, dr$ (2)

Also $d\bar{r} = dx\mathbf{i} + dy\mathbf{j} + dz\mathbf{k}$

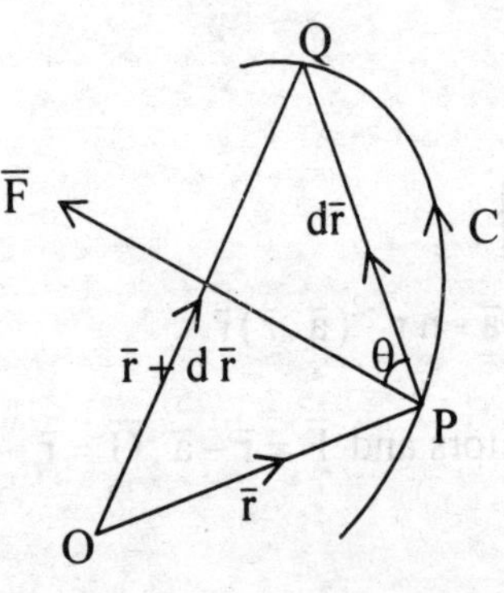

Fig. 2.4

Definition 2.2.1 Let C be a simple smooth curve given by $\bar{r} = \bar{r}(t)$, $a \le t \le b$ and $\bar{F}$ be a vector field, continuous on C. Then the **line integral** of $\bar{F}$ over C is defined as $\int_C \bar{F}.d\bar{r}$

Note:

(i) $\int_C \bar{F}.d\bar{r}$ is also equal to $\int_C F\cos\theta\, dr$ (Using (2))

(ii) If $\bar{F}(x, y, z) = F_1(x, y, z)\mathbf{i} + F_2(x, y, z)\mathbf{j} + F_3(x, y, z)\mathbf{k}$, then

$$\int_C \bar{F}.d\bar{r} = \int_C (F_1 dx + F_2 dy + F_3 dz)$$

(iii) If the curve C has the parametric equation given by (1), then

$$\int_C \bar{F}.d\bar{r} = \int_a^a \left(\bar{F}(x(t), y(t), z(t)).\frac{d\bar{r}}{dt} \right) dt$$

(iv) If A and B are the initial and terminal points of a curve C and $\int_C \bar{F}.d\bar{r}$ is independent of the path joining A and B, then we write $\int_C \bar{F}.d\bar{r} = \int_A^B \bar{F}.d\bar{r}$

Theorem 2.2.1 The line integral $\int_C \bar{F}.d\bar{r}$ is independent of the path joining the initial and terminal points of C if and only if $\bar{F} = \nabla f$ for a scalar function f (i.e., $\bar{F}$ is conservative).

Proof:

Let $\overline{F} = F_1\mathbf{i} + F_2\mathbf{j} + F_3\mathbf{k}$

Assume that the line integral $\int_C \overline{F}.d\overline{r}$ is independent of path. We have to prove that there exists a scalar function f such that $\overline{F} = \nabla f$. Let $A(x_0, y_0, z_0)$ be any point in the region R where $\overline{F}$ is defined. Then for any arbitrary point P with position vector $\overline{r} = x\mathbf{i} + y\mathbf{j} + z\mathbf{k}$

define $f(x, y, z) = \int_A^P \overline{F}.d\overline{r}$

$$= \int_{(x_0,y_0,z_0)}^{(x,y,z)} F_1 dx + F_2 dy + F_3 dz$$

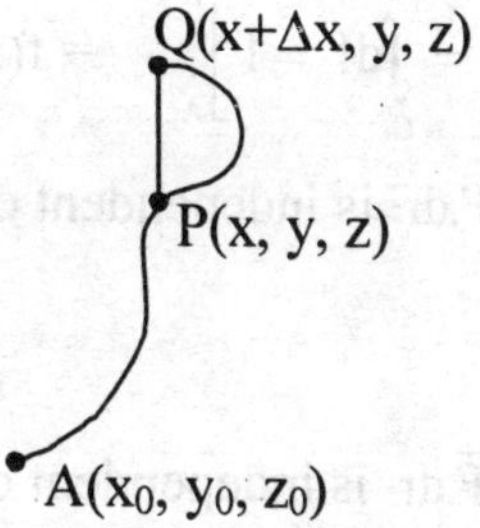

This scalar function f is well-defined in the region R, since the integral in (1) is independent of the path joining A and P. Let $Q(x + \Delta x, y, z)$ be a point in the region R lying in the neighborhood of $P(x, y, z)$ (Fig. 2.5).

Q(x+Δx, y, z)
P(x, y, z)
A(x₀, y₀, z₀)

Fig. 2.5

Then $f(x + \Delta x, y, z) - f(x, y, z) = \int_{(x_0,y_0,z_0)}^{(x+\Delta x,y,z)} \overline{F}.d\overline{r} - \int_{(x_0,y_0,z_0)}^{(x,y,z)} \overline{F}.d\overline{r}$

$$= \int_{(x,y,z)}^{(x+\Delta x,y,z)} \overline{F}.d\overline{r}$$

$$= \int_{(x,y,z)}^{(x+\Delta x,y,z)} (F_1 dx + F_2 dy + F_3 dz)$$

Choosing the path of integration from P to Q as the straight line parallel to x-axis, we get $dy = dz = 0$.

$f(x + \Delta x, y, z) - f(x, y, z) = \int_{(x,y,z)}^{(x+\Delta x,y,z)} F_1 dx$

$= \Delta x.F_1(x + \theta\,\Delta x, y, z), \quad 0 < \theta < 1$ (By Mean-Value theorem)

$$\therefore \quad \lim_{\Delta x \to 0} \frac{f(x + \Delta x, y, z) - f(x, y, z)}{\Delta x} = F_1(x, y, z)$$

i.e., $\frac{\partial f}{\partial x} = F_1$

In a similar manner, we can prove that $\frac{\partial f}{\partial y} = F_2$ and $\frac{\partial f}{\partial z} = F_3$

$$\therefore \quad \nabla f = \mathbf{i}\frac{\partial f}{\partial x} + \mathbf{j}\frac{\partial f}{\partial y} + \mathbf{k}\frac{\partial f}{\partial z}$$

$$= F_1\mathbf{i} + F_2\mathbf{j} + F_3\mathbf{k} = \overline{F}$$

Conversely, assume that there exists a scalar function f such that $\overline{F} = \nabla f$.
Let $A(x_0, y_0, z_0)$ be any point in the region R and P(x, y, z), an arbitrary point in R. Let C be any simple curve joining A and P. Then

$$\int_C \overline{F}.d\overline{r} = \int_C \nabla f.d\overline{r}$$

$$= \int_C \left(\frac{\partial f}{\partial x} dx + \frac{\partial f}{\partial y} dy + \frac{\partial f}{\partial z} dz \right)$$

$$= \int_C df = f \Big]_A^P = f(x, y, z) - f(x_0, y_0, z_0)$$

$\therefore \int_C \overline{F}.d\overline{r}$ is independent of path.

Note:

(i) $\int_C \overline{F}.d\overline{r}$ is independent of path if and only if curl $\overline{F} = 0$

(ii) If $\int_C \overline{F}.d\overline{r}$ is independent of path, then $\int \overline{F}.d\overline{r}$ over any simple closed path is zero.

(iii) A vector field $\overline{F}$ is conservative if and only if $\int_C \overline{F}.d\overline{r}$ is independent of path.

(iv) If a force $\overline{F}$ acts on a particle and moves it from the position A to the position B along the curve C, then $\int_C \overline{F}.d\overline{r}$ represents the **work done** by the force.

(v) If the vector field $\overline{F}$ represents the velocity of air or fluid, then $\int_C \overline{F}.d\overline{r}$ is called the **circulation** of $\overline{F}$ about C.

(vi) $\overline{F}.d\overline{r} = F\cos\theta\, dr$ is the tangential component of $\overline{F}$ along the curve C. Hence the line integral $\int_C \overline{F}.d\overline{r}$ is referred to as the **tangential line integral** of $\overline{F}$ along C.

Definition 2.2.2 Let $\overline{F} = F_1\mathbf{i} + F_2\mathbf{j} + F_3\mathbf{k}$. Then line integrals $\int_C \overline{F} \times d\overline{r}$ and $\int_C f\, dr$ are defined as follows.

$$\int_C \overline{F} \times d\overline{r} = \mathbf{i} \int_C (F_2 dz - F_3 dy) + \mathbf{j} \int_C (F_3 dx - F_1 dz) + \mathbf{k} \int_C (F_1 dy - F_2 dx)$$

$$\int_C f\, d\overline{r} = \mathbf{i} \int_C f\, dx + \mathbf{j} \int_C f\, dy + \mathbf{k} \int_C f\, dz$$

Definition 2.2.3 Let f(x,y,z) be a scalar function defined and continuous on a simple smooth curve C given by x = x(t), y = y(t), z = z(t), $a \le t \le b$. Let s be the arc length measured along C from a fixed point on C. Then the **line integral with respect to arc length s** is defined by $\int_C f(x,y,z)\,ds = \int_a^b f(x(t),y(t),z(t)).\left(\frac{ds}{dt}\right)dt$

where $\frac{ds}{dt} = \sqrt{\left(\frac{dx}{dt}\right)^2 + \left(\frac{dy}{dt}\right)^2 + \left(\frac{dz}{dt}\right)^2}$

If $\bar{r}(t) = x(t)\mathbf{i} + y(t)\mathbf{j} + z(t)\mathbf{k}$, then $\frac{ds}{dt} = \left|\frac{d\bar{r}}{dt}\right|$

Example 2.2.1 If $\bar{F} = x\mathbf{i} + (\sin y)\mathbf{j} + \mathbf{k}$ and C is given by $x = t^2, y = t, z = 2t, 0 \le t \le 1$, evaluate the line integral $\int_C \bar{F}.d\bar{r}$.

Solution:

We have $F_1 = x$, $F_2 = \sin y$, $F_3 = 1$.

$$\therefore\ \bar{F}.d\bar{r} = F_1dx + F_2dy + F_3dz$$
$$= x.2tdt + \sin y.dt + 1.2dt$$
$$= t^2.2t\,dt + \sin t\,dt + 2dt = (2t^3 + \sin t + 2)dt$$

$$\therefore\ \int_C \bar{F}.d\bar{r} = \int_0^1 (2t^3 + \sin t + 2)dt$$

$$= \left.2\frac{t^4}{4} - \cos t + 2t\right]_0^1 = \left(\frac{1}{2} - \cos 1 + 2\right) - (0 - 1 + 0) = \frac{7}{2} - \cos 1$$

Example 2.2.2 Evaluate $\int_C \bar{F}.d\bar{r}$ when $\bar{F} = (y^2+z^2)\mathbf{i} + (z^2+x^2)\mathbf{j} + (x^2+y^2)\mathbf{k}$ and C is $\bar{r}(t) = t\mathbf{i} + t^2\mathbf{j} + t^3\mathbf{k},\ 0 \le t \le 1$.

Solution:

The parametric equation of C is $x = t, y = t^2, z = t^3$.

$\therefore\ dx = dt$, $dy = 2tdt$ and $dz = 3t^2dt$.

Now $\int_C \bar{F}.d\bar{r} = \int_C (y^2+z^2)dx + (z^2+x^2)dy + (x^2+y^2)dz$

$$= \int_0^1 (t^4+t^6)dt + (t^6+t^2)2tdt + (t^2+t^4)3t^2dt$$

$$= \int_0^1 (2t^3 + 4t^4 + 4t^6 + 2t^7)dt = \left[2\frac{t^4}{4} + 4\frac{t^5}{5} + 4\frac{t^7}{7} + 2\frac{t^8}{8}\right]_0^1$$

$$= 2\frac{1}{4} + 4\frac{1}{5} + 4\frac{1}{7} + 2\frac{1}{8} = \frac{297}{140}$$

Example 2.2.3 Let F_1, F_2, F_3 be continuous functions having continuous first order partial derivatives. Then prove that the line integral $\int_C F_1dx + F_2dy + F_3dz$ is independent of path iff $\frac{\partial F_1}{\partial y} = \frac{\partial F_2}{\partial x}$, $\frac{\partial F_1}{\partial z} = \frac{\partial F_3}{\partial x}$ and $\frac{\partial F_2}{\partial z} = \frac{\partial F_3}{\partial y}$.

Solution:

Let $\overline{F} = F_1\mathbf{i} + F_2\mathbf{j} + F_3\mathbf{k}$

Then $\int_C \overline{F}.d\overline{r} = \int_C F_1dx + F_2dy + F_3dz$ is independent of path iff curl $\overline{F} = 0$.

i.e., $\begin{vmatrix} \mathbf{i} & \mathbf{j} & \mathbf{k} \\ \frac{\partial}{\partial x} & \frac{\partial}{\partial y} & \frac{\partial}{\partial z} \\ F_1 & F_2 & F_3 \end{vmatrix} = 0$

i.e., $\mathbf{i}\left(\frac{\partial F_3}{\partial y} - \frac{\partial F_2}{\partial z}\right) + \mathbf{j}\left(\frac{\partial F_1}{\partial z} - \frac{\partial F_3}{\partial x}\right) + \mathbf{k}\left(\frac{\partial F_2}{\partial x} - \frac{\partial F_1}{\partial y}\right) = 0$

i.e., $\frac{\partial F_3}{\partial y} = \frac{\partial F_2}{\partial z}, \frac{\partial F_1}{\partial z} = \frac{\partial F_3}{\partial x}, \frac{\partial F_2}{\partial x} = \frac{\partial F_1}{\partial y}$

i.e., $\frac{\partial F_1}{\partial y} = \frac{\partial F_2}{\partial x}, \frac{\partial F_1}{\partial z} = \frac{\partial F_3}{\partial x}, \frac{\partial F_2}{\partial z} = \frac{\partial F_3}{\partial y}$

Note:

The line integral $\int_C F_1\,dx + F_2\,dy$ is independent of path iff $\frac{\partial F_1}{\partial y} = \frac{\partial F_2}{\partial x}$

(Here F_1 and F_2 are functions of x and y only)

Example 2.2.4 Prove that $\int_A^B 2xy^2dx + (2x^2y + 1)dy$ is independent of path and hence evaluate the integral, given A(-1, 2), b(2, 3)

Solution:

$F_1(x, y) = 2xy^2$, $F_2(x, y) = 2x^2y + 1$

$\therefore \quad \frac{\partial F_1}{\partial y} = 4xy$ and $\frac{\partial F_2}{\partial x} = 4xy \qquad$ i.e., $\frac{\partial F_1}{\partial y} = \frac{\partial F_2}{\partial x}$

$\therefore$ The given integral is independent of path.

Choosing the path of integration as the line joining A(-1, 2) and B(2, 3), the equation of the line is $\quad \frac{y-2}{x+1} = \frac{3-2}{2+1}$

i.e., $y - 2 = \frac{1}{3}(x + y)$

i.e., $\frac{y-2}{1} = \frac{x+1}{3} = t$ (say)

$x = 3t - 1,\ y = t + 2$

$\therefore dx = 3\,dt, \quad dy = dt$

$$\therefore \int_A^B 2xy^2 dx + (2x^2y + 1)dy = \int_{t=0}^{1} 2(3t-1)(t+1)^2\, 3\,dt + (2(3t-1)^2(t+2) + 1)dt$$

$$= \int_0^1 6(3t^3 + 11t^2 + 8t - 4)\,dt + (18t^3 + 24t^2 - 22t + 5)\,dt$$

$$= \int_0^1 (36t^3 + 90t^2 + 26t - 19)\,dt$$

$$= \left[36.\frac{t^4}{4} + 90.\frac{t^3}{3} + 26.\frac{t^2}{2} - 19t\right]_0^1 = 33$$

Example 2.2.5 If $\overline{F} = (e^x z - 2xy)\mathbf{i} + (1 - x^2)\mathbf{j} + (e^x + z)\mathbf{k}$, show that $\int_C \overline{F}.d\overline{r}$ is independent of path. Hence evaluate the integral along a curve from (0, 1, -1) to (2, 3, 0).

Solution:

$$\nabla \times \overline{F} = \begin{vmatrix} \mathbf{i} & \mathbf{j} & \mathbf{k} \\ \frac{\partial}{\partial x} & \frac{\partial}{\partial y} & \frac{\partial}{\partial z} \\ e^x z - 2xy & 1 - x^2 & e^x + z \end{vmatrix}$$

$$= \mathbf{i}(0 - 0) + \mathbf{j}(e^x - e^x) + \mathbf{k}(-2x + 2x) = 0$$

$\int_C \overline{F}.d\overline{r}$ is independent of path.

Then there exists a scalar function f such that $\overline{F} = \nabla f = \mathbf{i}\frac{\partial f}{\partial x} + \mathbf{j}\frac{\partial f}{\partial y} + \mathbf{k}\frac{\partial f}{\partial z}$

$$\therefore \frac{\partial f}{\partial x} = e^x z - 2xy \quad (1)$$

$$\frac{\partial f}{\partial y} = 1 - x^2 \quad (2)$$

$$\frac{\partial f}{\partial z} = e^x + z \quad (3)$$

Integrating (1) w.r.t x, we get, $f = e^x z - x^2 y + g(y, z)$

Differentiating w.r.t y,

$$\frac{\partial f}{\partial y} = -x^2 + \frac{\partial g}{\partial y} \quad \text{i.e., } 1 - x^2 = -x^2 + \frac{\partial g}{\partial y} \qquad \text{(Using (2))}$$

$$\therefore \frac{\partial g}{\partial y} = 1$$

$g = y + h(z)$

Hence $f = e^x z - x^2 y + y + h(z)$

Differentiating w.r.t z, $\frac{\partial f}{\partial z} = e^x + h'(z)$

i.e., $e^x + z = e^x + h'(z)$ (Using (3))

$\therefore h'(z) = z$

$h(z) = \frac{z^2}{2} + c$ where c is a constant

Hence $f = e^x z - x^2 y + y + \frac{z^2}{2} + c$

$$\text{Now } \int_C \bar{F}.d\bar{r} = \int \nabla f.d\bar{r} = \int df = \left[f \right]_{(0,1,-1)}^{(2,3,0)}$$

$$= \left[e^x z - x^2 y + y + \frac{z^2}{2} + c \right]_{(0,1,-1)}^{(2,3,0)}$$

$$= (-12 + 3 + c) - (-1 + 1 + \frac{1}{2} + c) = -\frac{19}{2}.$$

Example 2.2.6 A force $\bar{F} = xy\mathbf{i} + (y - z)\mathbf{j} + 2x\mathbf{k}$ acts on a particle and moves it along the curve $x = t, y = t^2, z = t^3$ from $t = 1$ to $t = 2$. Find the work done by the force.

Solution:

The work done by the force is given by $\int_C \bar{F}.d\bar{r} = \int_C xy\,dx + (y - z)\,dy + 2x\,dz$

where $x = t$, $dy = 2t\,dt$, $dz = 3t^2\,dt$

$$\therefore \text{Work done} = \int_{t=1}^{2} tt^2 dt + (t^2 - t^3)\, 2t\,dt + 2t.3t^2 dt$$

$$= \int_{t=1}^{2} (t^3 + 2t^3 - 2t^4 + 6t^3) dt$$

$$= \int_{t=1}^{2} (9t^3 - 2t^4) dt$$

$$= \int_{t=1}^{2} \left[\frac{9t^4}{4} - \frac{2t^5}{5} \right]_1^2 = \frac{427}{20}$$

Example 2.2.7 Find the work done by the force $\bar{F} = (2x - y - z)\mathbf{i} + (x + y - z^2)\mathbf{j} + (3x - 2y + 4z)\mathbf{k}$ when it moves a particle once round the circle $x^2 + y^2 = 16$ in the x - y plane in anti-clockwise direction.

Solution:

$$\text{Work done} = \int_C \bar{F}.d\bar{r}$$

$$= \int_C (2x - y - z)\,dx + (x + y - z^2)\,dy + (3x - 2y + 4z)\,dz$$

Along the circle C, $x = 4\cos\theta$, $y = 4\sin\theta$, $z = 0$

$\therefore dx = -4\sin\theta\,d\theta$, $dy = 4\cos\theta\,d\theta$, $dz = 0 \quad 0 \le \theta \le 2\pi$

$$\therefore \text{Work done} = \int_{\theta=0}^{2\pi} (8\cos\theta - 4\sin\theta)(-4\sin\theta)\,d\theta + (4\cos\theta + 4\sin\theta)4\cos\theta\,d\theta$$

$$= \int_0^{2\pi} (-16\sin\theta\cos\theta + 16)\,d\theta$$

$$= \left[-16\frac{\sin^2\theta}{2} + 16\theta\right]_0^{2\pi} = 32\pi$$

Example 2.2.8 Evaluate $\int_C f(x,y,z)\,ds$, when $f(x, y, z) = x + y^2 + yz$, C is the curve $y = 2x$, $z = 2$ from (1, 2, 2) to (3, 6, 2)

Solution:

$$\int_C f(x,y,z)\,ds = \int_C (x + y^2 + yz).\frac{ds}{dt}.dt$$

The parametric equation of the curve C is $x = t$, $y = 2t$, $z = 2$

$$\therefore \frac{dx}{dt} = 1, \quad \frac{dy}{dt} = 2, \quad \frac{dz}{dt} = 0$$

$$\frac{ds}{dt} = \sqrt{\left(\frac{dx}{dt}\right)^2 + \left(\frac{dy}{dt}\right)^2 + \left(\frac{dz}{dt}\right)^2} = \sqrt{1^2 + 2^2} = \sqrt{5}$$

$$\int_C f(x,y,z)\,ds = \int_{t=1}^{3} (t + (2t)^2 + 2t.2)\sqrt{5}.dt$$

$$= \int_1^3 (4t^2 + 5t)\,dt$$

$$= \sqrt{5}\left[4\frac{t^3}{3} + 5\frac{t^2}{2}\right]_1^3 = \sqrt{5}\left(36 + \frac{45}{2} - \frac{4}{3} - \frac{5}{2}\right) = \frac{164\sqrt{5}}{3}$$

2.2.2 SURFACE INTEGRAL

Let $\overline{F}(x, y, z)$ be a vector function defined over a region R in space and S be a surface whose equation is f(x, y, z) = 0. Let P(x, y, z) be any arbitrary point on S. Let $\hat{n}$ be the unit normal vector drawn outwards to S at P. Then $\hat{n} = \frac{\nabla f}{|\nabla f|}$ and $\overline{F}.\hat{n}$ represents the component of $\overline{F}$ along the normal to S at P. The surface integral of $\overline{F}$ over the surface S is defined as follows.

Definition 2.2.4 Divide the surface S into a number of elementary areas ΔS_i $(1 \le i \le m)$. Let $P_i(x_i, y_i, z_i)$ be an arbitrary point in ΔS_i. Let $\hat{n}_i$ be the unit normal vector at P_i to ΔS_i. Then the limit of the sum $\sum_{i=1}^{m} \overline{F}(x_i, y_i, z_i).\hat{n}_i \, \Delta S_i$ as $\text{Max}|\Delta S_i| \to 0$ is called the surface integral of $\overline{F}$ over S and is written as $\iint_S \overline{F}.d\overline{S}$ or $\iint_S \overline{F}.\hat{n}\, dS$.

$$\text{i.e.,} \quad \iint_S \overline{F}.d\overline{S} = \iint_S \overline{F}.\hat{n}\, dS = \lim_{\substack{m\to\infty \\ \text{Max}|\Delta S_i|\to 0}} \sum_{i=1}^{m} \overline{F}(x_i, y_i, z_i).\hat{n}_i \, \Delta S_i$$

Let $\overline{F} = F_1\mathbf{i} + F_2\mathbf{j} + F_3\mathbf{k}$ and $\hat{n} = \cos\alpha\mathbf{i} + \cos\beta\mathbf{j} + \cos\gamma\mathbf{k}$, where α, β, γ are the angles made by the unit normal vector $\hat{n}$ with the x, y and z-axes. Then we have $\cos\alpha = \hat{n}.\mathbf{i}$, $\cos\beta = \hat{n}.\mathbf{j}$ and $\cos\gamma = \hat{n}.\mathbf{k}$

$$\iint_S \overline{F}.\hat{n}\, dS = \iint_S (F_1\cos\alpha + F_2\cos\beta + F_3\cos\gamma)dS$$

$$= \iint_S \left(F_1(\cos\alpha)dS + F_2(\cos\beta)dS + F_3(\cos\gamma)dS\right)$$

$$= \iint_S \left(F_1 dydz + F_2 dzdx + F_3 dxdy\right), \text{ where}$$

$$(\cos\alpha)dS = dydz \text{ or } dS = \frac{dydz}{\hat{n}.\mathbf{i}} \qquad (1)$$

$$(\cos\beta)dS = dzdx \text{ or } dS = \frac{dzdx}{\hat{n}.\mathbf{j}} \qquad (2)$$

$$(\cos\gamma)dS = dxdy \text{ or } dS = \frac{dxdy}{\hat{n}.\mathbf{k}} \qquad (3)$$

Note:

(i) If the vector field $\overline{F}$ represents the velocity field of a moving fluid, then the surface integral $\iint_S \overline{F}.\hat{n}\, d\overline{S}$ gives the total volume of fluid flowing through S per unit times and is called the **flux of $\overline{F}$ through S.**

(ii) $\iint_S dS$ gives the area of the surface S.

(iii) Let R_1, R_2, R_3 be the projections of S in the xy, yz and zx planes respectively. Then by using the equations (1), (2) and (3), we can write,

$$\iint_S \overline{F}.\hat{n}\, dS = \iint_{R_1} \overline{F}.\hat{n}\frac{dxdy}{(\hat{n}.\mathbf{k})} \quad (4)$$

$$= \iint_{R_2} \overline{F}.\hat{n}\frac{dydz}{(\hat{n}.\mathbf{i})} \quad (5)$$

$$= \iint_{R_3} \overline{F}.\hat{n}\frac{dzdx}{(\hat{n}.\mathbf{j})} \quad (6)$$

Equation (4), (5) or (6) can be used for the evaluation of the surface integral $\iint_S \overline{F}.\hat{n}\, d\overline{S}$.

(iv) The other forms of surface integral are $\iint_S \overline{F} dS$, $\iint_S \overline{F}\times\hat{n}\, dS$ and $\iint_S f.\hat{n}\, dS$, where $\overline{F}$ is a vector field and f is a scalar field and $\hat{n}$ is the unit normal vector to the surface.

Example 2.2.9 Evaluate $\iint_S \overline{F}.d\overline{S}$ where $\overline{F} = xy\mathbf{i} - x^2\mathbf{j} + (x+z)\mathbf{k}$ and S is the region of the plane $2x + 2y + z = 6$ bounded in the first octant.

Solution:

$$\iint_S \overline{F}.d\overline{S} = \iint_S \overline{F}.\hat{n}\, dS = \iint_{R_1} \overline{F}.\hat{n}\frac{dxdy}{(\hat{n}.\mathbf{k})} \quad \text{(By projecting S on the xy-plane)} \quad (1)$$

Let $f(x, y, z) = 2x + 2y + z - 6 = 0$.

Then $\nabla f = \nabla(2x + 2y + z - 6) = 2\mathbf{i} + 2\mathbf{j} + \mathbf{k}$

$$\therefore\ \hat{n} = \frac{\nabla f}{|\nabla f|} = \frac{1}{3}(2\mathbf{i} + 2\mathbf{j} + \mathbf{k})$$

Hence $\overline{F}.\hat{n} = \left[xy\mathbf{i} - x^2\mathbf{j} + (x+z)\mathbf{k}\right].\frac{1}{3}(2\mathbf{i} + 2\mathbf{j} + \mathbf{k}) = \frac{2}{3}xy - \frac{2}{3}x^2 + \frac{1}{3}(x+z)$

i.e., $\overline{F}.\hat{n} = \frac{2}{3}xy - \frac{2}{3}x^2 + \frac{1}{3}(6 - x - 2y)$ (Since $z = 6 - 2x - 2y$)

Also $\hat{n}.\mathbf{k} = \frac{1}{3}$. Hence equation (1) becomes,

$$\iint_S \overline{F}\, d\overline{S} = \int_{x=0}^{3}\int_{y=0}^{3-x}\left[\frac{2}{3}xy - \frac{2}{3}x^2 + \frac{1}{3}(6 - x - 2y)\right].\frac{dxdy}{(1/3)} \quad \text{(Refer Fig. 2.6)}$$

$$= \int_0^3\int_0^{3-x}\left[2xy - 2x^2 + 6 - x - 2y\right]dydx$$

$$= \int_0^3 \left[xy^2 - 2x^2y + (6-x)y - y^2 \right]_0^{3-x} dx$$

$$= \int_0^3 3\left(x^3 - 4x^2 + 2x + 3\right)dx = 3\left(\frac{x^4}{4} - 4\frac{x^3}{3} + x^2 + 3x\right)_0^3 = \frac{27}{4}$$

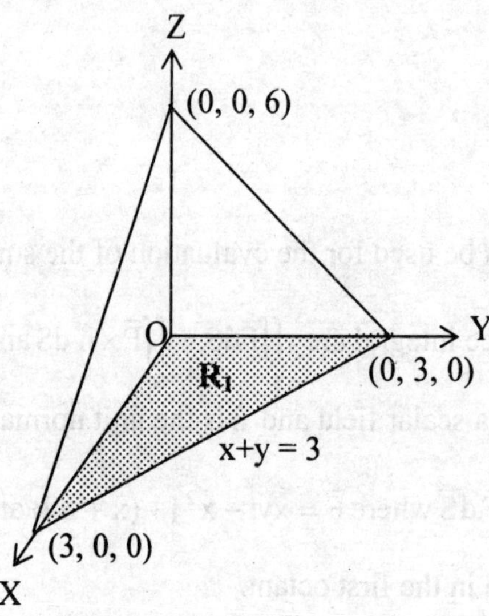

Fig. 2.6

Example 2.2.10 Evaluate the surface integral $\iint_S \overline{F}.\hat{n}\, dS$ where $\overline{F} = z^2\mathbf{i} + xy\mathbf{j} - y^2\mathbf{k}$ and S is the surface of the cylinder $x^2 + y^2 = 25$, $0 \le z \le 3$ included in the first octant.

Solution:

Let $f(x, y, z) = x^2 + y^2 - 25 = 0$

Then $\nabla f = 2x\mathbf{i} + 2y\mathbf{j}$ and $\hat{n} = \frac{\nabla f}{|\nabla f|} = \frac{2(x\mathbf{i} + y\mathbf{j})}{\sqrt{4(x^2 + y^2)}} = \frac{1}{5}(x\mathbf{i} + y\mathbf{j})$

$$\overline{F}.\hat{n} = (z^2\mathbf{i} + xy\mathbf{j} - y^2\mathbf{k}).\frac{1}{5}(x\mathbf{i} + y\mathbf{j})$$

$$= \frac{1}{5}(xz^2 + xy^2)$$

$$\therefore \iint_S \overline{F}.\hat{n}\, dS = \iint_{R_2} \overline{F}.\hat{n}\frac{dydz}{(\hat{n}.\mathbf{i})} \qquad \text{(By projecting S on the yz-plane)}$$

$$= \iint_{R_2} \frac{1}{5}\left(xz^2 + xy^2\right)\frac{dydz}{(x/5)}$$

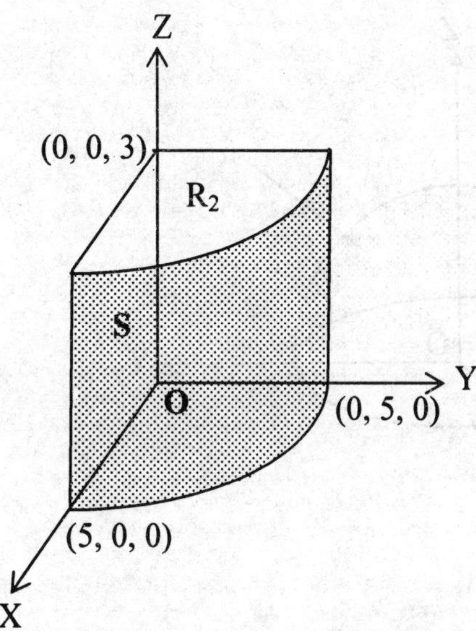

Fig. 2.7

$$= \int_{z=0}^{3}\int_{y=0}^{5}(y^2 + z^2)dydz \quad \text{(As } R_2 \text{ is a rectangle)}$$

$$= \int_0^3 \left(\frac{y^3}{3} + z^2 y\right)_0^5 dz$$

$$= \int_0^3 \left(\frac{125}{3} + 5z^2\right) dz$$

$$= \left(\frac{125}{3} z + 5\frac{z^3}{3}\right)_0^3 = 170$$

Example 2.2.11 Evaluate $\iint_S \overline{F}.\hat{n}\, dS$, given $\overline{F} = x\mathbf{i} + y\mathbf{j} - 2z\mathbf{k}$ and S is the surface of the sphere $x^2 + y^2 + z^2 = a^2$ above the xy-plane.

Solution:

Let $f(x, y, z) = x^2 + y^2 + z^2 - a^2 = 0$

Then $\nabla f = 2x\mathbf{i} + 2y\mathbf{j} + 2z\mathbf{k}$ and

$$\hat{n} = \frac{\nabla f}{|\nabla f|} = \frac{2(x\mathbf{i} + y\mathbf{j} + z\mathbf{k})}{2\sqrt{(x^2 + y^2 + z^2)}} = \frac{1}{a}(x\mathbf{i} + y\mathbf{j} + z\mathbf{k})$$

$$\overline{F}.\hat{n} = (xi + yj - 2zk).\frac{1}{a}(xi + yj + zk)$$

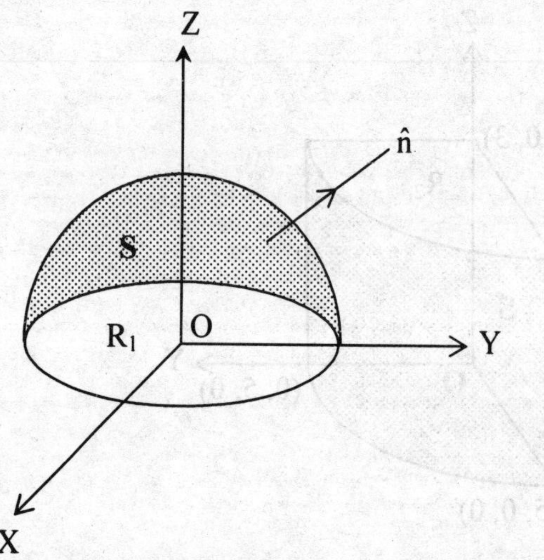

Fig. 2.8

$$= \frac{1}{a}(x^2 + y^2 - 2z^2) = \frac{1}{a}\left[x^2 + y^2 - 2(a^2 - x^2 - y^2)\right] = \frac{1}{a}\left[3x^2 + 3y^2 - 2a^2\right]$$

$$\therefore \iint_S \bar{F}.\hat{n}\, dS = \iint_{R_1} \bar{F}.\hat{n} \frac{dxdy}{(\hat{n}.\mathbf{k})} \qquad \text{(By projecting S on the xy-plane)}$$

$$= \iint_{R_1} \frac{1}{a}\left[3x^2 + 3y^2 - 2a^2\right] \frac{dxdy}{z/a}$$

$$= \iint_{R_1} \frac{3x^2 + 3y^2 - 2a^2}{\sqrt{a^2 - x^2 - y^2}}\, dx\, dy \qquad (1)$$

R_1 is the circular area bounded by $x^2 + y^2 = a^2$.

$\therefore$ Transforming the integral (1) to polar coordinates, $x = r\cos\theta$, $y = r\sin\theta$, we get,

$$\iint_S \bar{F}.\hat{n}\, dS = \int_{r=0}^{a} \int_{\theta=0}^{2\pi} \frac{3r^2 - 2a^2}{\sqrt{a^2 - r^2}}\, rdr\, d\theta$$

$$= \int_0^a \frac{3r^2 - 2a^2}{\sqrt{a^2 - r^2}}\, r.dr \int_0^{2\pi} d\theta$$

$$= 2\pi \int_a^0 \frac{3(a^2 - t^2) - 2a^2}{t}(-t)dt \qquad \text{(Substituting } a^2 - r^2 = t^2;\ -2rdr = 2tdt)$$

$$= 2\pi \int_0^a (a^2 - 3t^2)dt = 2\pi(a^2 t - t^3)_0^a = 2\pi(a^3 - a^3) = 0$$

Example 2.2.12 Evaluate $\iint_S \overline{F}.\hat{n}\, dS$ where $\overline{F} = 4x\mathbf{i} - 2y^2\mathbf{j} + z^2x^2\mathbf{k}$ and S is the outer surface of the cylinder $x^2 + y^2 = 9$, $0 \le z \le 4$.

Solution:

Let $f(x, y, z) = x^2 + y^2 - 9$. Then $\nabla f = 2x\mathbf{i} + 2y\mathbf{j}$

$$\hat{n} = \frac{\nabla f}{|\nabla f|} = \frac{2(x\mathbf{i} + y\mathbf{j})}{2\sqrt{x^2 + y^2}} = \frac{1}{3}(x\mathbf{i} + y\mathbf{j})$$

$$\overline{F}.\hat{n} = \frac{1}{3}(4x^2 - 2y^3) = \frac{2}{3}(2x^2 - y^3)$$

$$\iint_S \overline{F}.\hat{n}\, dS = \iint_S \frac{2}{3}(2x^2 - y^3)dS$$

Using cylindrical coordinates $x = 3\cos\theta$, $y = 3\sin\theta$, $z = z$, we have $dS = 3d\theta\, dz$

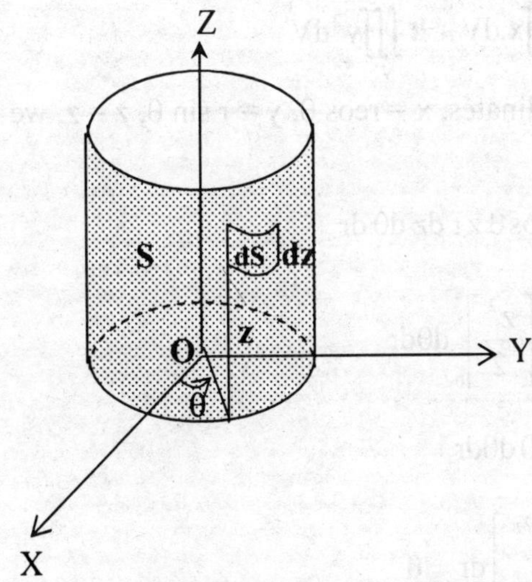

Fig. 2.9

$$\iint_S \overline{F}.\hat{n}\, dS = \int_{\theta=0}^{2\pi}\int_{z=0}^{4} \frac{2}{3}(18\cos^2\theta - 27\sin^3\theta)3dzd\theta$$

$$= 2\int_0^{2\pi}(18\cos^2\theta - 27\sin^3\theta)d\theta\int_0^4 dz$$

$$= 72\int_0^{2\pi}(2\cos^2\theta - 3\sin^3\theta)d\theta$$

$$= 72(2\pi)$$

$$= 144\pi.$$

2.2.3 VOLUME INTEGRAL

Definition 2.2.5 Consider a closed surface in space that encloses a volume V. Let $\overline{F}(x,y,z)$ be a vector function defined over V and $\overline{F} = F_1\mathbf{i} + F_2\mathbf{j} + F_3\mathbf{k}$. Then volume integral $\iiint_V \overline{F}dV$ is defined as $\iiint_V \overline{F}dV = \mathbf{i}\iiint_V F_1 dV + \mathbf{j}\iiint_V F_2 dV + \mathbf{k}\iiint_V F_3 dV$

Note:

The other forms of volume integral are $\iiint_V \nabla.\overline{F}dV$, $\iiint_V \nabla\times\overline{F}dV$ and $\iiint_V f.\hat{n}dV$

Example 2.2.13 Evaluate $\iiint_V \overline{F}dV$ where $\overline{F} = 2xz\mathbf{i} - x\mathbf{j} + y^2\mathbf{k}$ and V is the volume of the region enclosed by the cylinder $x^2 + y^2 = a^2$ between $z = 0$ and $z = 4$.

Solution:

$$\iiint_V \overline{F}dV = 2\mathbf{i}\iiint_V xz dV - \mathbf{j}\iiint_V x.dV + \mathbf{k}\iiint_V y^2 dV \qquad (1)$$

Using the cyclindrical coordinates, $x = r\cos\theta$, $y = r\sin\theta$, $z = z$, we have,
$dV = dx\,dy\,dz = r dr\,d\theta\,dz$

$$\therefore \quad \iiint_V xz dV = \int_{r=0}^{a}\int_{\theta=0}^{2\pi}\int_{z=0}^{4} r\cos\theta\, z\, r\, dz\, d\theta\, dr$$

$$= \int_0^a\int_0^{2\pi} r^2\cos\theta\left[\frac{z^2}{2}\right]_0^4 d\theta dr$$

$$= 8\int_0^a\int_0^{2\pi} r^2\cos\theta\, d\theta dr$$

$$= 8\int_0^a r^2\left[\sin\theta\right]_0^{2\pi} dr = 0 \qquad (2)$$

$$\iiint_V x dV = \int_{r=0}^{a}\int_{\theta=0}^{2\pi}\int_{z=0}^{4} r\cos\theta\, r\, dz\, d\theta\, dr$$

$$= \int_0^a\int_0^{2\pi} r^2\cos\theta\left[z\right]_0^4 d\theta\, dr$$

$$= 4\int_0^a r^2\left[\sin\theta\right]_0^{2\pi} dr = 0 \qquad (3)$$

$$\iiint_V y^2 dV = \int_{r=0}^{a}\int_{\theta=0}^{2\pi}\int_{z=0}^{4} r^2\sin^2\theta\, r\, dz\, d\theta\, dr$$

$$= \int_0^a \int_0^{2\pi} r^3 \sin^2\theta \Big[z\Big]_0^4 \, d\theta\, dr$$

$$= 4\int_0^a \int_0^{2\pi} r^3 \sin^2\theta \; d\theta\, dr$$

$$= 4\int_0^a r^3 \left[\frac{\theta}{2} - \frac{\sin 2\theta}{4}\right]_0^{2\pi} dr$$

$$= 4\int_0^a r^3 \,[\pi]\, dr = 4\pi\left[\frac{r^4}{4}\right]_0^a = 4\pi.\frac{a^4}{4} = \pi a^4 \qquad (4)$$

$$\therefore \quad \iiint_V \overline{F}\, dV = \pi a^4 \mathbf{k}. \qquad \text{(Using (2), (3) and (4) in (1))}$$

Example 2.2.14 If $\overline{F} = y\mathbf{i} + 2x\mathbf{j} - z\mathbf{k}$ and V be the closed region bounded by x = 0, x = 2, y = 0, y = 6, z = x^2, z = 4, show that $\iiint_V \nabla \times \overline{F} \; dV = 32\,\mathbf{k}$

Solution:

$$\nabla \times \overline{F} = \begin{vmatrix} \mathbf{i} & \mathbf{j} & \mathbf{k} \\ \frac{\partial}{\partial x} & \frac{\partial}{\partial y} & \frac{\partial}{\partial z} \\ y & 2x & -z \end{vmatrix} = \mathbf{k}$$

$$\therefore \quad \iiint_V \nabla \times \overline{F} \; dV = \iiint_V \mathbf{k}\, dV$$

$$= \mathbf{k} \int_{x=0}^{2} \int_{y=0}^{6} \int_{z=x^2}^{4} dz\, dy\, dx$$

$$= \mathbf{k} \int_0^2 \int_0^6 \Big[z\Big]_{x^2}^4 \, dy\, dx$$

$$= \mathbf{k} \int_0^2 \int_0^6 (4 - x^2)\, dy\, dx$$

$$= \mathbf{k} \int_0^2 \Big[(4 - x^2)\, y\Big]_0^6 dx$$

$$= 6\mathbf{k} \int_0^2 (4 - x^2) dx$$

$$= 6\mathbf{k}\left[4x - \frac{x^3}{3}\right]_0^2 = 6\mathbf{k}.(8 - 8/3)$$

$$= 32\,\mathbf{k}$$

Example 2.2.15 If $\overline{F} = (2x^2 - z)\mathbf{i} - 2xy\mathbf{j} - 4x\mathbf{k}$ and V is the region bounded by the coordinate planes and the plane 2x + 2y + z = 4, show that $\iiint_V \nabla.\overline{F}\ dV = \frac{8}{3}$.

Solution:

$$\nabla.\overline{F} = \left(\mathbf{i}\frac{\partial}{\partial x} + \mathbf{j}\frac{\partial}{\partial y} + \mathbf{k}\frac{\partial}{\partial k}\right).\left((2x^2 - z)\mathbf{i} - 2xy\mathbf{j} - 4x\mathbf{k}\right) = 4x - 2x - 0 = 2x.$$

$$\therefore \iiint_V \nabla.\overline{F}\ dV = 2\int_0^2 \int_{y=0}^{2-x} \int_{z=0}^{4-2x-2y} x\ dz\ dy\ dx$$

$$= 2\int_0^2 \int_0^{2-x} x\Big[z\Big]_0^{4-2x-2y} dy\ dx$$

$$= 2\int_0^2 \int_0^{2-x} x\Big[4 - 2x - 2y\Big] dy\ dx$$

$$= 2\int_0^2 \int_0^{2-x} \left(4x - 2x^2 - 2xy\right) dy\ dx$$

$$= 2\int_0^2 \Big[4xy - 2x^2y - xy^2\Big]_0^{2-x} dx$$

$$= 2\int_0^2 \Big[4x(2-x) - 2x^2(2-x) - x(2-x)^2\Big]_0^{2-x} dx$$

$$= 2\int_0^2 \Big[4x - 4x^2 + x^3\Big]dx = 2\left(2x^2 - \frac{4}{3}x^3 + \frac{x^4}{4}\right)_0^2 = 2\left(\frac{4}{3}\right) = \frac{8}{3}$$

EXERCISE 2.2

PART-A

1. Define line integral of a vector function.
2. What is the necessary and sufficient condition for the line integral $\int_C \overline{F}.d\overline{r}$ to be independent of the path of integration?
3. What is the physical meaning of $\int_C \overline{F}.d\overline{r}$?
4. Why $\int_C \overline{F}.d\overline{r}$ is called tangential line integral of $\overline{F}$ along C?
5. Define circulation of $\overline{F}$ about the curve C.

6. What is the condition for $\int_C F_1 dx + F_2 dy$ to be independent of path?
7. Evaluate $\int_C \bar{F}.d\bar{r}$, where $\bar{F} = xy^2\mathbf{i} + y^3\mathbf{j}$ and C is $y^2 = x$ from (0, 0) to (1, 1).
8. If $\bar{F} = zx\mathbf{i} + zy\mathbf{j} + z^2\mathbf{k}$, evaluate $\int_C \bar{F}.d\bar{r}$ along the curve $\bar{r} = t\mathbf{i} + t^2\mathbf{j} + t^3\mathbf{k}$ from (0, 0, 0) to (1, 1, 1).
9. If $f(x, y, z) = 2x^2y$, then evaluate $\int_C f.d\bar{r}$ where C is the portion of the curve $x = t$, $y = 2t^2$, $z = t^3$ from $t = 0$ to $t = 1$.
10. Show that $\int_C \bar{r}.d\bar{r}$ is independent of path.
11. Find the work done by $\bar{F} = x^2\mathbf{i} + yz\mathbf{j} + z\mathbf{k}$ in moving a particle along the line segment from (1, 2, 2) to (3, 4, 2).
12. Define the surface integral $\iint_S \bar{F}.d\bar{S}$.
13. Define flux of the vector field $\bar{F}$ across the surface S.
14. Define volume integral.
15. Evaluate $\iiint_V \bar{F} dV$ where $\bar{F} = xy\mathbf{i} - y\mathbf{j} + 2z\mathbf{k}$ and V is the region bounded by the surfaces $x = 0$, $x = 1$, $y = 0$, $y = 2$, $z = x^2$, $z = 1$.
16. Find the work done in moving a particle in a force field given by $\bar{F} = 3xy\mathbf{i} - 5z\mathbf{j} + 10x\mathbf{k}$ along the curve $x = t^2 + 1$, $y = 2t^2$, $z = t^3$ from $t = 1$ to $t = 2$.
17. Evaluate $\int_C \bar{F}.d\bar{r}$, given $\bar{F}$ is irrotational and C is a closed curve in space.
18. If $\bar{r}(t) = 5t^2\mathbf{i} + t\mathbf{j} - t^3\mathbf{k}$, show that $\int_2^3 \left(\bar{r}.\frac{d\bar{r}}{dt}\right)dt = \frac{2295}{2}$.
19. If $f = 15xy$ and V denotes the closed region bounded by the planes $x + 4y + z = 2$, $x = 0$, $y = 0$, $z = 0$, show that $\iiint_V f dV = \frac{1}{4}$.
20. Evaluate $\iiint_V \nabla.\bar{F} dV$ if $\bar{F} = x^2\mathbf{i} + y^2\mathbf{j} + z^2\mathbf{k}$ and if V is the volume of the region enclosed by the cube $0 \leq x, y, z \leq 1$.

PART-B

21. If $\bar{F} = y\mathbf{i} - z\mathbf{j} + x\mathbf{k}$ and C is the curve $\bar{r} = \cos\theta\,\mathbf{i} + \sin\theta\,\mathbf{j} + \theta\,\mathbf{k}$, $0 \leq \theta \leq 2\pi$ then find $\int_C \bar{F} dS$, $\int_C \bar{F}.d\bar{r}$ and $\int_C \bar{F} \times d\bar{r}$.

22. If f = 3y - z, $\overline{F} = 2xy\mathbf{i} + y\mathbf{j} + xz\mathbf{k}$ and C is the arc of the parabola $y^2 = 2x$, z = 0 from (0, 0, 0) to $\left(\frac{1}{2}, 1, 0\right)$, then evaluate $\int_C f dS$, $\int_C \overline{F} dS$, $\int_C \overline{F}.d\overline{r}$ and $\int_C \overline{F} \times d\overline{r}$.

23. Show that if C is the circle x = 3cos t, y = 3 sin t, z = 0, then
$$\int_C \left[(2x - y + z)\mathbf{i} + (x + y - z^2)\mathbf{j} + (3x - 2y + 4z)\mathbf{k}\right].d\overline{r} = 18\pi.$$

24. Show that $\int \overline{F} \times d\overline{r} = -\left(2 + \frac{\pi}{4}\right)\mathbf{i} + \left(\frac{\pi}{2} - \frac{1}{2}\right)\mathbf{j} + \frac{3}{2}\mathbf{k}$ where $\overline{F} = y\mathbf{i} + z\mathbf{j} + x\mathbf{k}$ and the integral is taken along the curve x = cos t, y = sin t, z = 2 cos t from t = 0 to $t = \frac{\pi}{2}$.

25. If $\overline{F} = xy\mathbf{i} - z\mathbf{j} + x^2\mathbf{k}$ and C is the curve $x = t^2$, y = 2t, $z = t^3$ from t = 0 to t = 1, show that $\int_C \overline{F} \times d\overline{r} = -\frac{9}{10}\mathbf{i} - \frac{2}{3}\mathbf{j} + \frac{7}{5}\mathbf{k}$.

26. If $\overline{F} = (y^2 + z^2)\mathbf{i} + (z^2 + x^2)\mathbf{j} + (x^2 + y^2)\mathbf{k}$ evaluate $\int_C \overline{F}.d\overline{r}$ where C is rectilinear path from (0, 0, 0) to (1, 1, 1) through the points (1, 0, 0) and (1, 1, 0).

27. Prove that the integral $\int_P^Q (2xz + y)dx + (x + z)dy + (x^2 + y)dz$, where P = (-1, 2, 3) and Q = (2, 2, 4) is independent of path of integration and hence evaluate the integral.

28. IF $\overline{F} = (2xy^2 + yz)\mathbf{i} + (2x^2y + xz + 2yz^2)\mathbf{j} + (2y^2z + xy)\mathbf{k}$, prove that $\int_P^Q \overline{F}.d\overline{r}$ is independent of the path jointing P and Q.

29. The force $\overline{F} = xy\mathbf{i} + (y - z)\mathbf{j} + 2x\mathbf{k}$ acting on a particle moves it along the curve $x = t, y = t^2, z = t^3$ from t = 1 to t = 2. Find the work done by $\overline{F}$.

30. Find the work done by a force $\overline{F} = 3xy\mathbf{i} + (x + y)\mathbf{j} - z\mathbf{k}$ while moving a particle along the curve x = t + 1, y = t-1, $z = t^2$ from the point (2, 0, -1) to (4, 2, 9).

31. Find the work done by the force $\overline{F} = (2x - y + 2z)\mathbf{i} + (x + y - z^2)\mathbf{j} + (3x - 2y - 5z)\mathbf{k}$ in moving a particle once around a circle in the xy-plane, which has its centre at the origin and has radius 2.

32. Show that $\overline{F} = (2xy + z^3)\mathbf{i} + x^2\mathbf{j} + 3xz^2\mathbf{k}$ is conservative and evaluate $\int \overline{F}.d\overline{r}$ along any curve joining (1, -2, 1) and (3, 1, 4).

33. Evaluate $\iint_S \overline{F}.\hat{n}\, dS$ if $F = (x + y^2)\mathbf{i} - 2x\mathbf{j} + 2yz\mathbf{k}$ and S is the surface of the plane 2x + y + 2z = 6 in the first octant.

34. Evaluate $\iint_S \overline{F}.\hat{n}\, dS$ if $F = 4y\mathbf{i} + 18z\mathbf{j} - x\mathbf{k}$ and S is the surface of the portion of the plane 3x + 2y + 6z = 6 contained in the first octant.

35. Evaluate $\iint_S \bar{F}.\hat{n}\, dS$ where $\bar{F} = z\mathbf{i} + x\mathbf{j} - y^2 z\mathbf{k}$ and S is the surface of the cylinder $x^2 + y^2 = 1$ included in the first octant between the planes z = 0 and z = 2.
36. Evaluate $\iint_S \bar{F}.\hat{n}\, dS$ if $\bar{F} = yz\mathbf{i} + 2y^2\mathbf{j} + xz^2\mathbf{k}$ and S is the surface of the cylinder $x^2 + y^2 = 9$ contained in the first octant between the planes z = 0 and z = 2.
37. Find the values of the following integrals, given $\bar{F} = 2x\mathbf{j} - xz\mathbf{k}$ and S is the surface of the portion of the plane 2x + 2y + z = 4 in the first octant.
(i) $\iint_S \bar{F}.\hat{n}\, dS$ (ii) $\iint_S \bar{r}.\hat{n}\, dS$ (iii) $\iint_S \bar{F} dS$ (iv) $\iint_S \bar{F} \times \hat{n}\, dS$.
38. If $\bar{F} = 2zx\mathbf{i} + y^2\mathbf{j} + yz\mathbf{k}$ and S is the cube bounded by x = 0, x = 1, y = 0, y = 1, z = 0, z = 1, show that $\iint_S \bar{F}.\hat{n}\, dS = \frac{5}{2}$.
39. Evaluate $\iint_S \bar{F}.dS$ where $\bar{F} = x\mathbf{i} - y\mathbf{j} + z\mathbf{k}$ and S is the surface of the cylinder $x^2 + y^2 = a^2$ bounded by the planes z = 0 and z = b.
40. Evaluate $\iiint_V 45x^2 y dV$ where V is the region bounded by the planes x = 0, y = 0, z = 0, 4x + 2y + z = 8.
41. Evaluate $\iiint_V \bar{F} dV$ where V is the region enclosed by the cylinder $x^2 + y^2 = a^2$ between z = 0 and z = c and $\bar{F} = 2xz\mathbf{i} - x\mathbf{j} + y^2\mathbf{k}$.
42. Evaluate $\iiint_V \nabla.\bar{F}\, dV$ if $\bar{F} = 2x^2y\mathbf{i} - y^2\mathbf{j} + 4xz^2\mathbf{k}$ and V is the region in the first octant bounded by the cylinder $y^2 + z^2 = 9$ and the plane x = 2.
43. If V is the region common to the cylinders $x^2 + y^2 = a^2$ and $x^2 + z^2 = a^2$ contained in the first octant and $\bar{F} = xy\mathbf{i} - 3y^2 z\mathbf{j}$, evaluate $\iiint_V \nabla \times \bar{F}\, dV$.

2.3 INTEGRAL THEOREMS OF GREEN, GAUSS AND STOKE

The theory of vector fields and vector integration provide three important theorems – Green's theorem (by George Green: 1793-1841), Divergence theorem of Gauss (by C.F. Gauss: 1777-1855) and Stoke's theorem (by G.G Stokes: 1819-1903). The Green's theorem provides a relationship between a double integral over a region R and the line integral over the closed curve C, the boundary of R. Green's theorem is also known as the first fundamental theorem of vector calculus. The Divergence theorem of Gauss provides a relationship between the volume integral over a closed and bounded region V in the three dimensional space and the double integral over the surface S which is the boundary of the region V. The Stoke's theorem relates the surface integral over a surface S bounded by a simple closed curve C with the line integral over C.

2.3.1 GREEN'S THEOREM

Theorem 2.3.1 (Green's Theorem) Let C be a piecewise smooth simple closed curve, which is the boundary of a region R in the plane. Let $F_1(x, y)$, $F_2(x, y)$, $\frac{\partial F_1}{\partial y}$ and $\frac{\partial F_2}{\partial x}$ be continuous on R. Then $\int_C F_1(x,y)dx + F_2(x,y)dy = \iint_R \left(\frac{\partial F_2}{\partial x} - \frac{\partial F_1}{\partial y}\right) dxdy$, when integrated in the positive direction (anti-clockwise direction) of C

Proof:
Assume that the region R in the xy-plane can be expressed simultaneously in the following forms

$a \le x \le b,\ f_1(x) \le y \le f_2(x)$ (1)

$c \le y \le d,\ g_1(y) \le x \le g_2(y)$ (2)

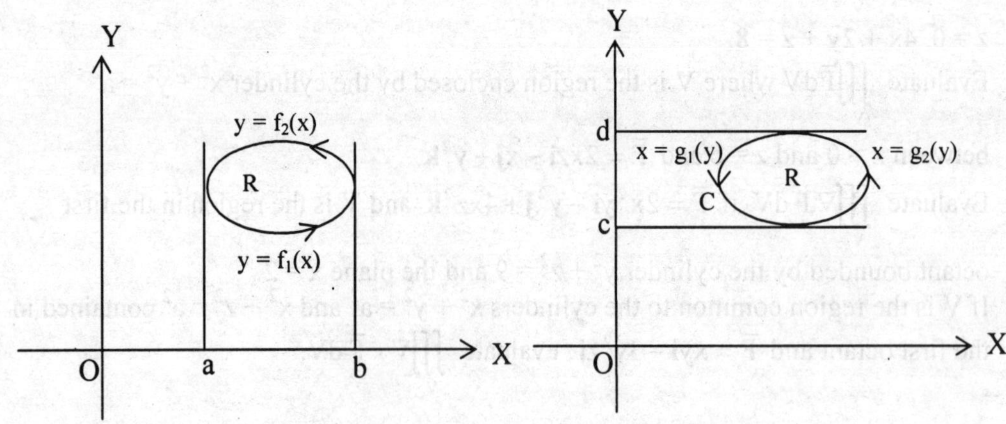

Fig. 2.10 Fig. 2.11

Using (1) (Fig. 2.10) we get, $\iint_R \frac{\partial F_1}{\partial y} dx\, dy = \int_a^b \left[\int_{f_1(x)}^{f_2(x)} \frac{\partial F_1}{\partial y} .dy\right] dx$

$$= \int_a^b [F_1(x, f_2(x)) - F_1(x, f_1(x))] dx$$

$$= \int_a^b F_1(x, f_2(x))dx + \int_b^a F_1(x, f_1(x))dx$$

$$= -\int_C F_1(x,y)dx \tag{3}$$

Using (2) (Fig. 2.11) we get,

$$\iint_R \frac{\partial F_2}{\partial x} dx\, dy = \int_c^d \left[\int_{g_1(y)}^{g_2(y)} \frac{\partial F_2}{\partial x} .dx \right] dy$$

$$= \int_c^d [F_2(g_2(y), y) - F_2(g_1(y), y)] dy$$

$$= \int_c^d F_2(g_2(y), y)dy + \int_d^c F_2(g_1(y), y)dy$$

$$= \int_C F_2(x,y)dy \tag{4}$$

From (3) and (4), we get

$$\iint_R \left(\frac{\partial F_2}{\partial x} - \frac{\partial F_1}{\partial y} \right) dxdy = \int_C F_1(x,y)dx + F_2(x,y)dy$$

The proof can be extended to other regions R by decomposing R into a finite number of subregions R_1, R_2, …, R_n such that each of these subregions can be expressed in both the forms (1) and (2). Hence the theorem.

Note:

Green's theorem can be expressed in the vector form as

$$\int_C \overline{F}.d\overline{r} = \iint_R (\nabla \times \overline{F}).k\, dx\, dy, \text{ where } \overline{F} = F_1\mathbf{i} + F_2\mathbf{j} \tag{1}$$

$$\int_C \overline{F}.d\overline{r} = \int_C (F_1\mathbf{i} + F_2\mathbf{j}).(dx\mathbf{i} + dy\mathbf{j})$$

$$= \int_C F_1 dx + F_2 dy \tag{2}$$

$$\iint_R (\nabla \times \overline{F}).\mathbf{k}\, dx\, dy = \iint_R (\nabla \times (F_1\mathbf{i} + F_2\mathbf{j})).\mathbf{k}\, dx\, dy$$

$$= \iint_R \left(\frac{\partial F_2}{\partial x} - \frac{\partial F_1}{\partial y} \right) \mathbf{k}.\mathbf{k}\, dx\, dy$$

$$= \iint_R \left(\frac{\partial F_2}{\partial x} - \frac{\partial F_1}{\partial y} \right) dx\, dy \tag{3}$$

From (2) and (3), we get the result (1).

Example 2.3.1 Verify Green's theorem for $\int_C (x-2y)dx + xdy$ where C is the circle $x^2 + y^2 = 1$.

Solution:
Let $F_1 = x - 2y$ and $F_2 = x$.
Along the circle, $x = \cos\theta$, $y = \sin\theta$ and $0 \le \theta \le 2\pi$.

$$\therefore \int_C (x-2y)dx + xdy = \int_0^{2\pi} (\cos\theta - 2\sin\theta)(-\sin\theta)d\theta + \cos\theta.\cos\theta d\theta$$

$$= \int_0^{2\pi} (-\cos\theta\sin\theta + 2\sin^2\theta + \cos^2\theta)d\theta$$

$$= \int_0^{2\pi} (-\cos\theta\sin\theta + \sin^2\theta + 1)d\theta$$

$$= \int_0^{2\pi} \left(-\frac{1}{2}\sin 2\theta + \frac{1-\cos 2\theta}{2} + 1\right)d\theta$$

$$= \left(\frac{1}{4}\cos 2\theta + \frac{3}{2}\theta - \frac{1}{4}\sin 2\theta\right)_0^{2\pi}$$

$$= 3\pi \qquad (1)$$

$$\iint_R \left(\frac{\partial F_2}{\partial x} - \frac{\partial F_1}{\partial y}\right) dx\,dy = \iint_R (1 - (-2))dx\,dy$$

$$= 3\iint_R dx\,dy$$

$$= 3 \text{ (area of the circle)}$$

$$= 3\pi \qquad (2)$$

In (1) and (2) R.H.S are equal.
Hence Green's theorem is verified.

Example 2.3.2 Verify Green Theorem for $\int_C (x-y)dx + 3xydy$ where C is the boundary of the region enclosing $x^2 = 4y$ and $y^2 = 4x$.

Solution:
The two curves $y^2 = 4x$ and $x^2 = 4y$ intersect at A(4, 4).
Along $x^2 = 4y$, $x = 2t$ and $y = t^2$.

$$\therefore \int_{OBA} (x-y)dx + 3xydy = \int_{t=0}^{2} (2t - t^2)2.dt + 3.2t.t^2.2t.dt$$

$$= \int_0^2 (4t - 2t^2 + 12t^4)dt$$

$$= \left[4\frac{t^2}{2} - 2\frac{t^3}{3} + 12.\frac{t^5}{5}\right]_0^2 = \frac{1192}{15}$$

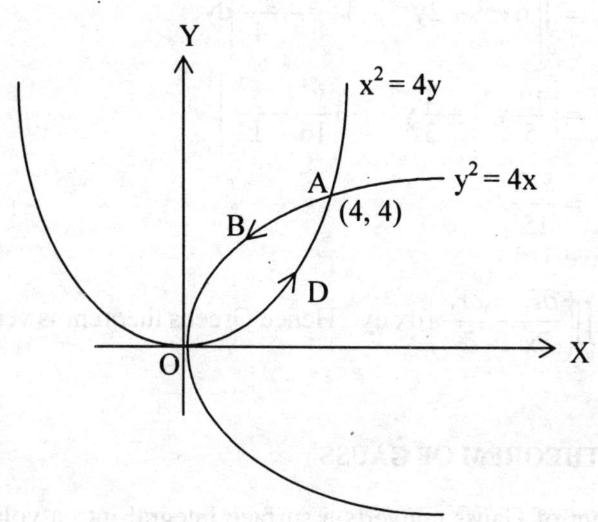

Fig. 2.12

Along $y^2 = 4x$, $x = t^2$ and $y = 2t$

$$\therefore \int_{ADO} (x-y)dx + 3xydy = \int_{t=2}^{0} (t^2 - 2t)2t.dt + 3.t^2.2t.2dt$$

$$= \int_{2}^{0} (14t^3 - 4t^2)dt = \left[14\frac{t^4}{4} - 4\frac{t^3}{3}\right]_2^0 = -\frac{136}{3}$$

$$\therefore \int_C (x-y)dx + 3xydy = \frac{1192}{15} - \frac{136}{15} = \frac{512}{15} \qquad (1)$$

Let $F_1 = x - y$ and $F_2 = 3xy$ then $\frac{\partial F_1}{\partial y} = -1$ and $\frac{\partial F_2}{\partial x} = 3y$

$$\therefore \iint_R \left(\frac{\partial F_2}{\partial x} - \frac{\partial F_1}{\partial y}\right) dx\,dy = \iint_R (3y+1)dx\,dy$$

$$= \int_{y=0}^{4} \int_{x=\frac{y^2}{4}}^{2\sqrt{y}} (3y+1)dx\,dy$$

$$= \int_0^4 \left[(3y+1)x\right]_{\frac{y^2}{4}}^{2\sqrt{y}} dy$$

$$= \int_0^4 \left[(3y+1)2\sqrt{y} - (3y+1)\frac{y^2}{4}\right] dy$$

$$= \int_0^4 \left[6y^{3/2} + 2y^{1/2} - 3\frac{y^3}{4} - \frac{y^2}{4} \right] dy$$

$$= \left[\frac{12}{5} y^{5/2} + \frac{4}{3} y^{3/2} - 3\frac{y^4}{16} - \frac{y^3}{12} \right]_0^4$$

$$= \frac{512}{15} \tag{2}$$

From (1) and (2) we have

$$\int_C (x-y)dx + 3xydy = \iint_R \left(\frac{\partial F_2}{\partial x} - \frac{\partial F_1}{\partial y} \right) dx\, dy$$ Hence Greens theorem is verified.

2.3.2 DIVERGENCE THEOREM OF GAUSS

The Divergence theorem of Gauss converts a surface integral into a volume integral and consequently this conversion simplifies the evaluation of several integrals.

Theorem 2.3.2 (Divergence Theorem of Gauss)

Let D be a closed and bounded region in the three dimensional space whose boundary is a piecewise smooth surface S. Let $\overline{F}(x, y, z) = F_1(x, y, z)\mathbf{i} + F_2(x, y, z)\mathbf{j} + F_3(x, y, z)\mathbf{k}$ be a vector field in D where F_1, F_2, F_3 have continuous first order partial derivatives. Then $\iint_S \overline{F}.\hat{n}\, dS = \iiint_D \text{div}\overline{F}\, dV$ where $\hat{n}$ is the outer unit normal vector to S.

Proof:

We prove the theorem for a special case of the region D whose bounding surface S is such that any line parallel to any coordinate axis (say z-axis) cuts S in at most two points. Let the equations of the lower and upper portions, S_1 and S_2 be $z = f_1(x, y)$ and $z = f_2(x, y)$ respectively (Fig. 2.13). Let the projection of S on the xy-plane be R. $\hat{n}_1$ and $\hat{n}_2$ be the outward unit normal to the surfaces S_1 and S_2 respectively.

$$\iiint_D \text{div}\overline{F}\, dV = \iiint_D \frac{\partial F_1}{\partial x} dV + \iiint_D \frac{\partial F_2}{\partial y} dV + \iiint_D \frac{\partial F_3}{\partial z} dV \tag{1}$$

$$\iiint_D \frac{\partial F_3}{\partial z} dV = \iiint_D \frac{\partial F_3}{\partial z} dx\, dy\, dz$$

$$= \iint_R \left[\int_{z=f_1(x,y)}^{f_2(x,y)} \frac{\partial F_3}{\partial z} dz \right] dy\, dx$$

$$= \iint_R [F_3(x, y, f_2) - F_3(x, y, f_1)] dy\, dx \tag{2}$$

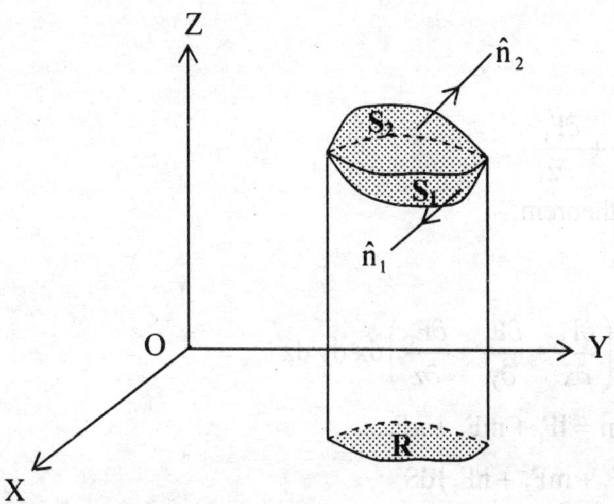

Fig. 2.13

On the surface S_1, the normal $\hat{n}_1$ makes an obtuse angle γ_1 with the z-axis and on the surface S_2, the normal $\hat{n}_2$ makes an acute angle γ_2 with the z-axis

On S_1, $dy\,dx = -dS_1 \cos\gamma_1 = -k.\hat{n}_1\,dS_1$

On S_2, $dy\,dx = dS_2 \cos\gamma_2 = k.\hat{n}_2\,dS_2$ Hence (2) becomes,

$$\iiint_D \frac{\partial F_3}{\partial z}\,dV = \iint_{S_2} F_3\,k.\hat{n}_2 dS_2 + \iint_{S_1} F_3\,k.\hat{n}_1 dS_1 = \iint_S F_3\,k.\hat{n}dS \tag{3}$$

Similarly projecting S on the other coordinate planes, we get,

$$\iiint_D \frac{\partial F_1}{\partial x}\,dV = \iint_S F_1\,i.\hat{n}dS \tag{4}$$

and $$\iiint_D \frac{\partial F_2}{\partial y}\,dV = \iint_S F_2\,j.\hat{n}dS \tag{5}$$

Now adding (3), (4) and (5) and using (1) we get,

$$\iiint_D \text{div}\bar{F}\,dV = \iint_S (F_1\mathbf{i} + F_2\mathbf{j} + F_3\mathbf{k}).\hat{n}\,dS = \iint_S \bar{F}.\hat{n}\,dS$$ Hence the theorem.

Theorem 2.3.3 (Green's theorem in space)

If F_1, F_2, F_3 are any three functions defined in a region containing a volume V, bounded by a surface S and have continuous first order partial derivatives then,

$$\iiint_V \left(\frac{\partial F_1}{\partial x} + \frac{\partial F_2}{\partial y} + \frac{\partial F_3}{\partial z}\right) dx\,dy\,dz = \iint_S (F_1 dydz + F_2 dzdx + F_3 dxdy) \tag{1}$$

(This is only a special form of divergence theorem of Gauss)

Proof:

Let $\overline{F} = F_1\mathbf{i} + F_2\mathbf{j} + F_3\mathbf{k}$

Then $\nabla.\overline{F} = \dfrac{\partial F_1}{\partial x} + \dfrac{\partial F_2}{\partial y} + \dfrac{\partial F_3}{\partial z}$

By Gauss divergence theorem,

$$\iiint_V \nabla.\overline{F}\, dV = \iint_S \overline{F}.\hat{n}\, dS \qquad (2)$$

$$\text{Now } \iiint_V \nabla.\overline{F}\, dV = \iiint_V \left(\frac{\partial F_1}{\partial x} + \frac{\partial F_2}{\partial y} + \frac{\partial F_3}{\partial z}\right) dx\, dy\, dz \qquad (3)$$

If $\hat{n} = l\mathbf{i} + m\mathbf{j} + n\mathbf{k}$, $\overline{F}.\hat{n} = lF_1 + mF_2 + nF_3$

$$\therefore \quad \iint_S \overline{F}.\hat{n}\, dS = \iint_S (lF_1 + mF_2 + nF_3)\, dS$$

$$= \iint_S F_1 l dS + F_2 m dS + F_3 n dS$$

$$= \iint_S F_1 dydz + F_2 dzdx + F_3 dxdy \qquad (4)$$

$$\left(l\, dS = (\hat{n}.i)\frac{dy\, dz}{(\hat{n}.i)} = dy\, dz, \quad m\, dS = (\hat{n}.j)\frac{dz\, dx}{(\hat{n}.j)} = dz\, dx, \quad n\, dS = (\hat{n}.k)\frac{dx\, dy}{(\hat{n}.k)} = dx\, dy \right)$$

From (2), (3) and (4) we get,

$$\iiint_V \left(\frac{\partial F_1}{\partial x} + \frac{\partial F_2}{\partial y} + \frac{\partial F_3}{\partial z}\right) dx\, dy\, dz = \iint_S (F_1 dydz + F_2 dzdx + F_3 dxdy)$$

Theorem 2.3.4 (Green's Identities)

If f and g are any two scalar functions possessing continuous partial derivatives of first order on a region D bounded by a closed surface S, then

(i) $\iiint_D (f\nabla^2 g + \nabla f.\nabla g)\, dV = \iint_S f(\nabla g.\hat{n})\, dS$

(ii) $\iiint_D (f\nabla^2 g - g\nabla^2 f)\, dV = \iint_S (f\nabla g - g\nabla f).\hat{n}\, dS$

((i) and (ii) are called Green's first and second identities respectively).

Proof:

By Gauss divergence theorem, we have

$$\iiint_V \nabla.\overline{F}\, dV = \iint_S \overline{F}.\hat{n}\, dS \qquad (1)$$

Let $\overline{F} = f\nabla g$

Then $\nabla.(f\nabla g) = f(\nabla.\nabla g) + \nabla f.\nabla g = f\nabla^2 g + \nabla f.\nabla g$

$\therefore$ (1) becomes,

$$\iiint_D \left(f\nabla^2 g + \nabla f.\nabla g\right)dV = \iint_S f\nabla g\,.\,\hat{n}\,dS = \iint_S f\left(\nabla g\,.\,\hat{n}\right)dS \qquad (2)$$

Interchanging f and g, we get,

$$\iiint_D \left(g\nabla^2 f + \nabla g.\nabla f\right)dV = \iint_S g\left(\nabla f\,.\,\hat{n}\right)dS \qquad (3)$$

Subtracting (3) from (2), we get,

$$\iiint_D \left(f\nabla^2 g - g\nabla^2 f\right)dV = \iint_S \left[f\left(\nabla g\,.\,\hat{n}\right) - g\left(\nabla f\,.\,\hat{n}\right)\right]dS$$

$$= \iint_S \left(f\nabla g - g\nabla f\right).\hat{n}\,dS$$

Example 2.3.3 Verify the divergence theorem for $\overline{F} = 4xz\mathbf{i} - y^2\mathbf{j} + yz\mathbf{k}$ over the cube bounded by x = 0, x = 1, y = 0, y = 1, z = 0, z = 1.

Solution:

Let D be the region bounded by the cube.

$\overline{F} = 4xz\mathbf{i} - y^2\mathbf{j} + yz\mathbf{k}$

$$\therefore\quad \nabla.\overline{F} = \frac{\partial}{\partial x}(4xz) + \frac{\partial}{\partial y}(-y^2) + \frac{\partial}{\partial z}(yz)$$

$$= 4z - 2y + y = 4z - y$$

$$\iiint_D \nabla.\overline{F}\,dV = \int_0^1\int_0^1\int_0^1 (4z - y)dx\,dy\,dz$$

$$= \int_0^1\int_0^1 \Big[(4z - y)x\Big]_0^1 dy\,dz$$

$$= \int_0^1\int_0^1 (4z - y)\,dy\,dz$$

$$= \int_0^1 \left[4yz - \frac{y^2}{2}\right]_0^1 dz$$

$$= \int_0^1 \left[4z - \tfrac{1}{2}\right]dz$$

$$= \left[4\frac{z^2}{2} - \frac{1}{2}z\right]_0^1 = \frac{3}{2} \qquad (1)$$

$\iint_S \overline{F}.\hat{n}\,dS = \iint_{S_1} + \iint_{S_2} + \iint_{S_3} + \iint_{S_4} + \iint_{S_5} + \iint_{S_6}$, where S_1, S_2, S_3, S_4, S_5, S_6 are the six outer faces of the cube whose outward unit normals are **i, -i, j, -j, k** and **–k** respectively.

$$\iint_{S_1} \overline{F}.\hat{n}\, dS = \iint_{S_1} \left(4xz\mathbf{i} - y^2\mathbf{j} + yz\mathbf{k}\right).\mathbf{i}\, dy\, dz$$

$$= \int_0^1\int_0^1 4z\, dy\, dz \qquad \text{Since } x = 1 \text{ on } S_1$$

$$= \int_0^1 4z dz = \left[4.\frac{z^2}{2}\right]_0^1 = 2$$

$$\iint_{S_2} \overline{F}.\hat{n}\, dS = \iint_{S_2} \left(4xz\mathbf{i} - y^2\mathbf{j} + yz\mathbf{k}\right).(-\mathbf{i})\, dy\, dz$$

$$= \iint_{S_2} (-4xz).\, dy\, dz = 0 \qquad \text{Since } x = 0 \text{ on } S_2.$$

$$\iint_{S_3} \overline{F}.\hat{n}\, dS = \iint_{S_3} \left(-y^2\right) dz\, dx = \int_0^1\int_0^1 (-1)\, dz\, dx = -1 \qquad \text{Since } y = 1 \text{ on } S_3$$

$$\iint_{S_4} \overline{F}.\hat{n}\, dS = \iint_{S_4} \left(y^2\right) dz\, dx = 0 \qquad \text{Since } y = 0 \text{ on } S_4$$

$$\iint_{S_5} \overline{F}.\hat{n}\, dS = \iint_{S_5} yz.\, dx\, dy$$

$$= \iint_{S_5} y.\, dx\, dy \qquad \text{Since } z = 1 \text{ on } S_5$$

$$= \int_0^1\int_0^1 y\, dx\, dy$$

$$= \int_0^1 \left[\frac{y^2}{2}\right]_0^1 dx$$

$$= \int_0^1 \frac{1}{2} dx = \frac{1}{2}$$

$$\iint_{S_6} \overline{F}.\hat{n}\, dS = \iint_{S_6} (-yz).\, dx\, dy = 0 \qquad \text{Since } z = 0 \text{ on } S_6.$$

Hence adding all the integrals, we get,

$$\iint_S \overline{F}.\hat{n}\, dS = 2 + 0 + (-1) + 0 + \left(\tfrac{1}{2}\right) + 0 = \frac{3}{2} \qquad (2)$$

From (1) and (2), we have, $\iiint_D \nabla.\overline{F}\, dV = \iint_S \overline{F}.\hat{n}\, dS$

Thus Divergence theorem is verified.

Example 2.3.4 If D is the region bounded by the closed cylinder $x^2 + y^2 = 16$, $z = 0$ and $z = 4$ and $\overline{F} = 3x^2\mathbf{i} + 6y^2\mathbf{j} + z\mathbf{k}$, verify the divergence theorem

Solution:

$$\nabla.\bar{F} = \frac{\partial}{\partial x}(3x^2) + \frac{\partial}{\partial y}(6y^2) + \frac{\partial}{\partial z}(z)$$

$$= 6x + 12y + 1$$

$$\therefore \iiint_D \nabla.\bar{F}\, dV = \int_{z=0}^{4}\int_{-4}^{4}\int_{y=-\sqrt{16-x^2}}^{\sqrt{16-x^2}} (6x + 12y + 1)dy\, dx\, dz$$

$$= \int_{z=0}^{4} dz. \int_{-4}^{4}\left(12x\sqrt{16-x^2} + 2\sqrt{16-x^2}\right)dx$$

$$= 4.\int_{-4}^{4} 2\sqrt{16-x^2}dx$$

$$= 4.2.2\int_{0}^{4}\sqrt{16-x^2}dx$$

$$= 16\left[\frac{x}{2}\sqrt{16-x^2} + \frac{16}{2}\sin^{-1}(x/4)\right]_0^4 = 16.8.\frac{\pi}{2} = 64\pi$$

The surface S of the cylinder consists of S_1 (top), S_2 (bottom) and S_3 (side).

On S_1, $z = 4$ and $\hat{n} = k$

$$\therefore \iint_{S_1} F.\hat{n}\, dS = \iint_{S_1} F.\hat{k}\, dS$$

$$= \iint_{S_1} z\, dS$$

$$= \iint_{S_1} 4\, dS$$

$$= 4(\text{area of } S_1) = 4.\pi.4^2 = 64\pi \qquad (1)$$

On S_2, $z = 0$ and $\hat{n} = -k$

$$\iint_{S_2} F.\hat{n}\, dS = \iint_{S_2} - zdS = 0 \qquad (2)$$

On S_3, $x^2 + y^2 = 16$ and $\hat{n} = \dfrac{2x\mathbf{i} + 2y\mathbf{j}}{\sqrt{4x^2 + 4y^2}} = \dfrac{1}{4}(x\mathbf{i} + y\mathbf{j})$

$$\therefore \iint_{S_3} F.\hat{n}\, dS = \iint_{S_3} \left(3x^2\mathbf{i} + 6y^2\mathbf{j} + z\mathbf{k}\right).\frac{1}{4}(x\mathbf{i} + y\mathbf{j})\, dS$$

$$= \frac{1}{4}\iint_{S_3}\left(3x^3 + 6y^3\right)dS$$

Using cylindrical coordinates we have,

$x = 4\cos\theta$, $y = 4\sin\theta$, $z = z$ and $dS = 4\, d\theta\, dz$.

$$\therefore \iint_{S_3} \overline{F}.\hat{n}\, dS = \frac{3}{4}\int_{z=0}^{4}\int_{\theta=0}^{2\pi}(64\cos^3\theta + 128\sin^3\theta)4d\theta dz$$

$$= 192.\ 4\int_0^{2\pi}(\cos^3\theta + 2\sin^3\theta)d\theta$$

$$= 768\int_0^{2\pi}\left[\frac{1}{4}(\cos 3\theta + 3\cos\theta) + \frac{1}{2}(3\sin\theta - \sin 3\theta)\right]d\theta$$

$$= 768.\frac{1}{2}\left(-3\cos\theta + \frac{1}{3}\cos 3\theta\right)_0^{2\pi} = 0 \qquad (3)$$

From (1), (2) and (3), we get

$$\iint_S \overline{F}.\hat{n}\, dS = 64\pi.$$

Hence $\iint_S \overline{F}.\hat{n}\, dS = \iiint_D \nabla.\overline{F}\, dV$

2.3.3 STOKE'S THEOREM

Stoke's theorem converts an integral over a surface into a line integral over a closed curve, which is the boundary of the surface.

Theorem 2.3.5 (Stoke's Theorem)

Let S be a piecewise smooth surface bounded by a piecewise smooth simple closed curve C. Let $\overline{F}(x, y, z) = F_1(x, y, z)\mathbf{i} + F_2(x, y, z)\mathbf{j} + F_3(x, y, z)\mathbf{k}$ be a vector function having continuous first order partial derivatives in a domain containing S. Then $\int_C \overline{F}.d\overline{r} = \iint_S \text{curl}\,\overline{F}.\hat{n}\, dS$, where $\hat{n}$ is the unit outward normal vector to S and C is the oriented in the anti-clockwise direction.

Proof:

$\overline{F} = F_1\mathbf{i} + F_2\mathbf{j} + F_3\mathbf{k}$ and $\overline{r} = x\mathbf{i} + y\mathbf{j} + z\mathbf{k}$

$$\therefore \text{curl}\,\overline{F} = \nabla \times \overline{F} = \begin{vmatrix} \mathbf{i} & \mathbf{j} & \mathbf{k} \\ \frac{\partial}{\partial x} & \frac{\partial}{\partial y} & \frac{\partial}{\partial z} \\ F_1 & F_2 & F_3 \end{vmatrix}$$

$$= \mathbf{i}\left(\frac{\partial F_3}{\partial y} - \frac{\partial F_2}{\partial z}\right) + \mathbf{j}\left(\frac{\partial F_1}{\partial z} - \frac{\partial F_3}{\partial x}\right) + \mathbf{k}\left(\frac{\partial F_2}{\partial x} - \frac{\partial F_1}{\partial y}\right)$$

$$= \mathbf{j}\frac{\partial F_1}{\partial z} - \mathbf{k}\frac{\partial F_1}{\partial y} + \text{other terms}$$

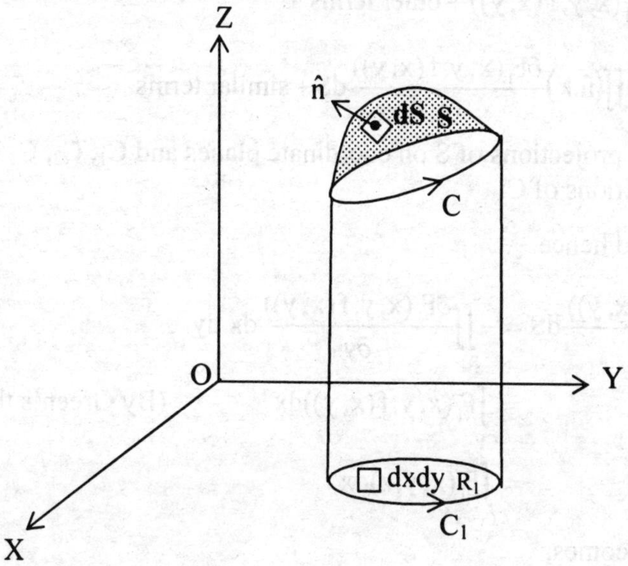

Fig. 2.14

$$\text{curl}\,\bar{F}.\hat{n} = \hat{n}.\mathbf{j}\frac{\partial F_1}{\partial z} - \hat{n}.\mathbf{k}\frac{\partial F_1}{\partial y} + \text{other terms} \qquad (1)$$

Let the equation of the surface S be given by $z = f(x, y)$

Then the position vector of any point on S is given by $\bar{r} = x\mathbf{i} + y\mathbf{j} + z\mathbf{k}$ where $z = f(x, y)$

$$\therefore \quad \frac{\partial \bar{r}}{\partial y} = \mathbf{j} + \mathbf{k}\frac{\partial z}{\partial y}$$

$$\hat{n}.\frac{\partial \bar{r}}{\partial y} = \hat{n}.\mathbf{j} + \hat{n}.\mathbf{k}\frac{\partial z}{\partial y} \qquad (2)$$

$\frac{\partial \bar{r}}{\partial y}$ is a vector tangent to the surface S and thus is perpendicular to $\hat{n}$

$$\therefore \quad \hat{n}.\frac{\partial \bar{r}}{\partial y} = 0$$

Hence $\hat{n}.\mathbf{j} = -\hat{n}.\mathbf{k}\frac{\partial z}{\partial y}$ (3)

Using (3) in (1), we get

$$\text{curl}\,\bar{F}.\hat{n} = -\hat{n}.\mathbf{k}\left(\frac{\partial F_1}{\partial y} + \frac{\partial F_1}{\partial z}.\frac{\partial z}{\partial y}\right) + \text{other terms}$$

$$= -\hat{n}.k \frac{\partial}{\partial y} F_1(x, y, f(x, y)) + \text{other terms}$$

$$\therefore \iint_S \text{curl}\, \overline{F}.\hat{n}\, dS = -\iint_S (\hat{n}.k) \frac{\partial F_1(x, y, f(x, y))}{\partial y} dS + \text{similar terms} \qquad (4)$$

Let R_1, R_2, R_3 be the projections of S on coordinate planes and C_1, C_2, C_3 be the corresponding projections of C.

Then $dS = \dfrac{dx\, dy}{\hat{n}.k}$ and hence

$$-\iint_S (\hat{n}.k) \frac{\partial F_1(x, y, f(x, y))}{\partial y} dS = -\iint_{R_1} \frac{\partial F_1(x, y, f(x, y))}{\partial y} dx\, dy$$

$$= \int_{C_1} F_1(x, y, f(x, y)) dx \qquad \text{(By Green's theorem)}$$

$$= \int_C F_1(x, y, z) dx$$

Thus equation (4) becomes,

$$\iint_S \text{curl}\, \overline{F}.\hat{n}\, dS = \int_C F_1 dx + \int_C F_2 dy + \int_C F_3 dz = \int_C \overline{F}.d\overline{r}$$ Hence the theorem.

Note:

Green's theorem in plane can be deduced from Stoke's theorem as follows.

We have $\displaystyle\int_C \overline{F}.d\overline{r} = \iint_S (\nabla \times \overline{F}).\hat{n}\, dS \qquad (1)$

Let $\overline{F} = F_1\mathbf{i} + F_2\mathbf{j}$ and S be a region R in the xy-plane

Then $\overline{F}.d\overline{r} = F_1 dx + F_2 dy$ and $\hat{n} = k$

$$\nabla \times \overline{F} = \begin{vmatrix} \mathbf{i} & \mathbf{j} & \mathbf{k} \\ \frac{\partial}{\partial x} & \frac{\partial}{\partial y} & \frac{\partial}{\partial z} \\ F_1 & F_2 & 0 \end{vmatrix} = \left(\frac{\partial F_2}{\partial x} - \frac{\partial F_1}{\partial y} \right) \mathbf{k}$$

$$(\nabla \times \overline{F}).\hat{n} = \left(\frac{\partial F_2}{\partial x} - \frac{\partial F_1}{\partial y} \right) \mathbf{k}.\mathbf{k} = \frac{\partial F_2}{\partial x} - \frac{\partial F_1}{\partial y}$$

$\therefore$ (1) becomes,

$$\int_C F_1 dx + F_2 dy = \iint_R \left(\frac{\partial F_2}{\partial x} - \frac{\partial F_1}{\partial y} \right) dx\, dy$$, which is Green's theorem.

Note:

$\overline{F} = F_1\mathbf{i} + F_2\mathbf{j} + F_3\mathbf{k}$ and $d\overline{r} = \mathbf{i}dx + \mathbf{j}dy + \mathbf{k}dz$

$\therefore \overline{F}.d\overline{r} = F_1 dx + F_2 dy + F_3 dz$

$$\nabla \times \overline{F} = \mathbf{i}\left(\frac{\partial F_3}{\partial y} - \frac{\partial F_2}{\partial z}\right) + \mathbf{j}\left(\frac{\partial F_1}{\partial z} - \frac{\partial F_3}{\partial x}\right) + \mathbf{k}\left(\frac{\partial F_2}{\partial x} - \frac{\partial F_1}{\partial y}\right)$$

$$(\nabla \times \overline{F}).\hat{n}\, dS = \left(\frac{\partial F_3}{\partial y} - \frac{\partial F_2}{\partial z}\right) dy\, dz + \left(\frac{\partial F_1}{\partial z} - \frac{\partial F_3}{\partial x}\right) dz\, dx + \left(\frac{\partial F_2}{\partial x} - \frac{\partial F_1}{\partial y}\right) dx\, dy$$

Thus the Stoke's theorem takes the form

$$\int_C F_1 dx + F_2 dy + F_3 dz = \iint_S \left[\left(\frac{\partial F_3}{\partial y} - \frac{\partial F_2}{\partial z}\right) dy\, dz + \left(\frac{\partial F_1}{\partial z} - \frac{\partial F_3}{\partial x}\right) dz\, dx + \left(\frac{\partial F_2}{\partial x} - \frac{\partial F_1}{\partial y}\right) dx\, dy\right]$$

Example 2.3.5 Verify Stokes theorem for $\overline{F} = xy\mathbf{i} + yz\mathbf{j} + zx\mathbf{k}$, taken over the triangular surface S in the plane x + y + z = 1 bounded by the planes x = 0, y = 0 and z = 0 and over its boundary.

Solution:

The vertices of the triangular surface are L(1, 0, 0), M(0, 1, 0), N(0, 0, 1).

Along LM, $\overline{r} = (1-t)\mathbf{i} + t\mathbf{j}$, $0 \le t \le 1$.

$\therefore d\overline{r} = -\mathbf{i}dt + \mathbf{j}dt$

$\overline{F}.d\overline{r} = (xy\mathbf{i} + yz\mathbf{j} + zx\mathbf{k}).(-\mathbf{i} + \mathbf{j})dt$

$= (-xy + yz)dt$

$= -(1-t)t\, dt$ Since x = 1-t, y = t, z = 0.

$$\therefore \int_L^M \overline{F}.d\overline{r} = \int_{t=0}^{1} -(1-t)t\,dt$$

$$= \left[\frac{-t^2}{2} + \frac{t^3}{3}\right]_0^1 = -\frac{1}{2} + \frac{1}{3} = -\frac{1}{6}$$

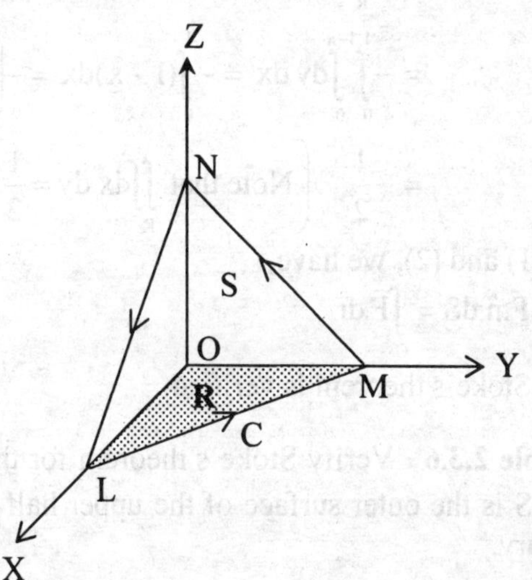

Fig. 2.15

Similarly, by symmetry,

$$\int_M^N \overline{F}.d\overline{r} = \int_N^L \overline{F}.d\overline{r} = -\frac{1}{6}$$

$$\therefore \int_C \overline{F}.dr = -\frac{1}{6} - \frac{1}{6} - \frac{1}{6} = -\frac{1}{2} \qquad (1)$$

$$\text{curl}\, \overline{F} = \begin{vmatrix} \mathbf{i} & \mathbf{j} & \mathbf{k} \\ \frac{\partial}{\partial x} & \frac{\partial}{\partial y} & \frac{\partial}{\partial z} \\ xy & yz & zx \end{vmatrix} = \mathbf{i}(0 - y) - \mathbf{j}(z - 0) + \mathbf{k}(0 - x) = -(y\mathbf{i} + z\mathbf{j} + x\mathbf{k})$$

$$\hat{n} = \frac{\nabla f}{|\nabla f|}, \text{ where } f(x, y, z) = x + y + z - 1$$

$$= \frac{(\mathbf{i}+\mathbf{j}+\mathbf{k})}{\sqrt{3}}$$

Then $\text{curl}\,\overline{F}.\hat{n} = \dfrac{-(x+y+z)}{\sqrt{3}}$ and $\hat{n}.\mathbf{k} = \dfrac{1}{\sqrt{3}}$

$$\iint_S \text{curl}\,\overline{F}.\hat{n}\,dS = \iint_S \frac{-(x+y+z)}{\sqrt{3}}\,dS$$

$$= \iint_R \frac{-(x+y+z)}{\sqrt{3}}\frac{dx\,dy}{\hat{n}.\mathbf{k}} \qquad \text{(Where R is the projection of S on xy-plane)}$$

$$= -\iint_R (x+y+z)dx\,dy$$

$$= -\iint_R dx\,dy$$

$$= -\int_0^1\int_0^{1-x} dy\,dx = -\int_0^1 (1-x)dx = -\left(x-\frac{x^2}{2}\right)_0^1$$

$$= -\frac{1}{2} \quad \left(\text{Note that } \iint_R dx\,dy = \frac{1}{2}, \text{ area of the region R}\right) \qquad (2)$$

From (1) and (2), we have,

$$\iint_S \text{curl}\,\overline{F}.\hat{n}\,dS = \int_C \overline{F}.d\bar{r}$$

Hence Stoke's theorem is verified.

Example 2.3.6 Verify Stoke's theorem for the vector field $\overline{F} = (2x-y)\mathbf{i} - y^2z\mathbf{k} - y^2z\mathbf{k}$ where S is the outer surface of the upper half of the sphere $x^2 + y^2 + z^2 = 1$ and C is its boundary.

Proof:

Along the boundary C, $x = \cos t$, $y = \sin t$ and $z = 0$ where $0 \le t \le 2\pi$.

On C, $\overline{F}.d\bar{r} = (2x-y)dx - y^2z\,dy - y^2z\,dz$

$= (2\cos t - \sin t)(-\sin t)\,dt$

$$\therefore \int_C \overline{F}.d\bar{r} = \int_0^{2\pi} (2\cos t - \sin t)(-\sin t)dt$$

$$= -2\int_0^{2\pi} \cos t \sin t\,dt + \int_0^{2\pi} \sin^2 t\,dt = 0 + \pi = \pi$$

$$\text{Now curl}\,\overline{F} = \begin{vmatrix} \mathbf{i} & \mathbf{j} & \mathbf{k} \\ \frac{\partial}{\partial x} & \frac{\partial}{\partial y} & \frac{\partial}{\partial z} \\ 2x-y & -yz^2 & -y^2z \end{vmatrix} = \mathbf{i}(-2yz+2yz) + \mathbf{j}(0-0) + \mathbf{k}(0+1) = \mathbf{k}$$

$$\iint_S \text{curl}\,\overline{F}.\hat{n}\,dS = \iint_S \mathbf{k}.\hat{n}\,dS$$

$$= \iint_S \mathbf{k}.\hat{n}\frac{dx\,dy}{\mathbf{k}.\hat{n}}$$, where R is the projection of S on xy-plane.

$$= \iint_R dx\,dy = \text{Area of the circular region R of radius } 1 = \pi.$$

Hence $\iint_S \text{curl}\,\overline{F}.\hat{n}\,dS = \int_C \overline{F}.d\overline{r}$ Thus Stoke's theorem is verified.

EXERCISE 2.3

PART-A

1. State Green's theorem in plane and give its vector form.
2. State Green's theorem in space.
3. State Green's first identity.
4. State Green's second identity.
5. State the Divergence theorem of Gauss.
6. State Stoke's theorem
7. Prove that Green's theorem is a special case of Stoke's theorem.

PART-B

Verify Green's theorem in plane for the following integrals (Ex: 8-15)

8. $\int_C (x-2y)dx + ydy$ taken around the unit circle $x^2 + y^2 = 1$.
9. $\int_C (x^2 - y^2)dx + 2xydy$, where C is the closed boundary of the region enclosed by $y = x^2$ and $y^2 = x$.
10. $\int_C (x^3 - x^2y)dx + xy^2dy$, where C is the boundary of the region enclosed by the circles $x^2 + y^2 = 4$ and $x^2 + y^2 = 16$.
11. $\int_C (x^2 + y)dx - xy^2dy$ taken around the boundary of the square whose vertices are (0, 0), (1, 0), (1, 1) and (0, 1).
12. $\int_C (x^2 - xy^2)dx + (y^2 - 2xy)dy$, where C is the square in the xy-plane given by $x = 0, x = 3, y = 0, y = 3$ described in anticlockwise direction.
13. $\int_C (x^2 - 2xy)dx + (x^2y + 3)dy$ along the boundary of the region defined by $y^2 = 8x$ and $x = 2$.

14. $\int_C (xy + y^2)dx + x^2 dy$, where C is the curve enclosing the region R bounded by $y = x^2$ and $y = x$.

15. $\int_C (3x^2 - 8y^2)dx + (4y - 6xy)dy$, where C is the boundary of the region R enclosed by $x = 0, y = 0$ and $x + y = 1$.

Verify Gauss' divergence theorem for the following functions and regions (or surfaces) (Ex: 16-23)

16. $\overline{F} = 3x^2\mathbf{i} + 6y^2\mathbf{j} + z\mathbf{k}$ over the region bounded by the closed cylinder $x^2 + y^2 = 16$, $z = 0$ and $z = 4$.
17. $\overline{F} = x\mathbf{i} + y\mathbf{j} + z\mathbf{k}$ over the region bounded by the sphere $x^2 + y^2 + z^2 = 16$.
18. $\overline{F} = 2xy\mathbf{i} + 6yz\mathbf{j} + 3zx\mathbf{k}$ over the region bounded by the coordinate planes and the plane $x + y + z = 2$.
19. $\overline{F} = (x^2 - 2yz)\mathbf{i} + (y^2 - 3xz)\mathbf{j} + (z^2 - xy)\mathbf{k}$ over the region bounded by the planes $x = 0, x = a, y = 0, y = b, z = 0, z = c$.
20. $\overline{F} = 2xz\mathbf{i} + yz\mathbf{j} + z^2\mathbf{k}$ over the upper half of the sphere $x^2 + y^2 + z^2 = a^2$
21. $\overline{F} = a(x + y)\mathbf{i} + a(y - x)\mathbf{j} + z^2\mathbf{k}$ over the hemisphere bounded by the xy-plane and the upper half of the sphere $x^2 + y^2 + z^2 = a^2$.
22. $\overline{F} = y\mathbf{i} + x\mathbf{j} + z^2\mathbf{k}$ over the cylindrical region bounded by $x^2 + y^2 = a^2$, $z = 0$, $z = h$.
23. $\overline{F} = 2x^2y\mathbf{i} - y^2\mathbf{j} + 4xz^2\mathbf{k}$ over the region in the first quadrant bounded by the cylinder $y^2 + z^2 = 9$ and the planes $x = 0$ and $x = 2$.
24. Verify the divergence theorem when $\overline{F} = \alpha x\mathbf{i} + \beta y\mathbf{j} + \gamma z\mathbf{j}$ and the volume enclosed is the interior of the cylinder $x^2 + y^2 = a^2$, bounded by the planes $z = \pm h$.

Verify Stoke's Theorem for the following functions and surfaces. (Ex: 25-33)

25. $\overline{F} = (2x - y)\mathbf{i} - yz^2\mathbf{j} - y^2z\mathbf{k}$, where S is the upper hemisphere of the unit sphere $x^2 + y^2 + z^2 = 1$ and C is its boundary.
26. $\overline{F} = (x + y)\mathbf{i} + (2x - z)\mathbf{j} + (y + z)\mathbf{k}$ and S is the surface of the triangle formed by the plane $3x + 2y + z = 6$ and the coordinate planes.
27. $\overline{F} = x^2\mathbf{i} + xy\mathbf{j}$ taken over the square surface S in the xy-plane whose vertices are O(0, 0, 0), A(a, 0, 0), B(a, a, 0), C(0, a, 0) and over its boundary.
28. $\overline{F} = (y - z + 2)\mathbf{i} + (yz + 4)\mathbf{j} - xz\mathbf{k}$ where S is the surface of the cube $x = 0, x = 2, y = 0, y = 2, z = 0, z = 2$ above the xy-plane. (i.e., excluding the face on the xy-plane).
29. $\overline{F} = (x^2 - y^2)\mathbf{i} + 2x\mathbf{j}$ around the rectangle bounded by the straight lines $x = 0, x = a, y = 0, y = b$.
30. $\overline{F} = y^2\mathbf{i} + xy\mathbf{j} - xz\mathbf{k}$ over the upper half of the sphere $x^2 + y^2 + z^2 = a^2$.

31. $\bar{F} = (x^2 + y^2 - 4)\mathbf{i} + 3xy\mathbf{j} + (2xz + z^2)\mathbf{k}$, where S is the surface of the hemisphere $x^2 + y^2 + z^2 = 16$ above the xy-plane and C is its boundary.
32. $\bar{F} = xz\mathbf{i} - y\mathbf{j} + x^2y\mathbf{k}$ and S is the region bounded by the planes x = 0, y = 0, z = 0, 2x + y + 2z = 8 not included in the xy-plane.
33. $\bar{F} = z\mathbf{i} + (2x + z)\mathbf{j} + x\mathbf{k}$ and C is the boundary of the triangle with vertices at (1, 0, 0), (0, 2, 0) and (0, 0, 3).

2.4 EVALUATION OF INTEGRALS USING INTEGRAL THEOREMS

In this section, we will discuss methods of evaluating certain types of line, surface and volume integrals using the integral theorems of Green, Gauss and Stoke.

2.4.1 Area using Green's Theorem

Let R be a region bounded by a simple closed curve C. By Green's theorem in plane, we have, $\int_C F_1 dx + F_2 dy = \iint_R \left(\frac{\partial F_2}{\partial x} - \frac{\partial F_1}{\partial y} \right) dx\, dy$

Put F_1 = -y and F_2 = x

Then $\int_C x dy - y dx = \iint_R 2 dx\, dy$

$= 2.$ Area of R

$\therefore$ Area of R $= \frac{1}{2} \int_C (x dy - y dx)$

Note:

In polar coordinates x = r cos θ, y = r sin θ,

Area of R $= \frac{1}{2} \int_C (r^2 \cos^2\theta + r^2 \sin^2\theta) d\theta \; = \frac{1}{2} \int_C r^2 d\theta$

Example 2.4.1 Find the area between the curve $y^2 = 9x$ and the straight line y = 3x.

Solution:

Area between the curve and line $= \frac{1}{2} \int_C x dy - y dx$

$$= \frac{1}{2} \int_{y=3x} (x dy - y dx) - \frac{1}{2} \int_{y^2=9x} (x dy - y dx)$$

When y = 3x, dy = 3dx

When $y^2 = 9x,\ y = 3\sqrt{x}$ and $dy = \frac{3}{2\sqrt{x}} dx$

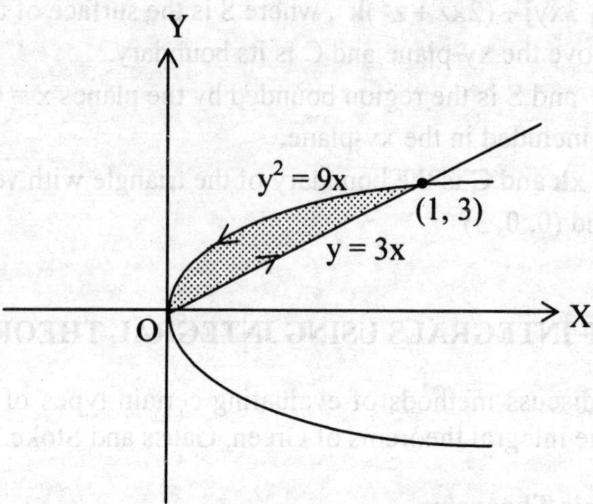

Fig. 2.16

$$\therefore \text{ Area } = \frac{1}{2}\int_{x=0}^{1}(x.3dx - 3x.dx) - \frac{1}{2}\int_{x=0}^{1}(x.\frac{3}{2\sqrt{x}}dx - 3\sqrt{x}.dx)$$

$$= 0 + \frac{3}{4}\int_{0}^{1}\sqrt{x}dx = \frac{3}{4}.\left[\frac{x^{3/2}}{3/2}\right]_0^1 = \frac{1}{2}$$

Example 2.4.2 Find the area bounded by the four cusped hypocycloid $x = a\cos^3\theta$, $y = a\sin^3\theta$.

Solution:

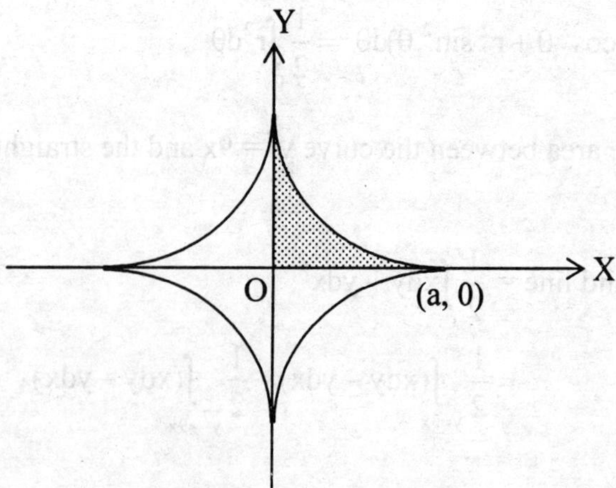

Fig. 2.17

Area $= 4.\frac{1}{2}\int_C (x\,dy - y\,dx)$ where C is the curve in the first quadrant where $0 \le \theta \le \pi/2$

$dx = -3a\cos^2\theta \sin\theta\, d\theta$
$dy = 3a\sin^2\theta \cos\theta\, d\theta$

$$\therefore \text{Area} = 2.\int_0^{\pi/2}\left(3a^2\cos^4\theta\sin^2\theta + 3a^2\sin^4\theta\cos^2\theta\right)d\theta$$

$$= 6a^2\int_0^{\pi/2}\cos^2\theta\sin^2\theta\, d\theta$$

$$= 6a^2\int_0^{\pi/2}\cos^2\theta(1-\cos^2\theta)\,d\theta$$

$$= 6a^2\int_0^{\pi/2}(\cos^2\theta - \cos^4\theta)\,d\theta = 6a^2\left(\frac{1}{2}.\frac{\pi}{2} - \frac{3}{4}.\frac{1}{2}.\frac{\pi}{2}\right) = \frac{3}{8}\pi a^2$$

Example 2.4.3 Evaluate $\int_C\left[(3x^2 - 8y^2)dx + (4y - 6xy)dy\right]$ where C is the boundary of the rectangular area enclosed by the lines x = 0, x = 1, y = 0, y = 2.

Solution:

Let $F_1 = 3x^2 - 8y^2$, $F_2 = 4y - 6xy$

Then $\frac{\partial F_1}{\partial y} = -16y$ and $\frac{\partial F_2}{\partial x} = -6y$

By Green's theorem,

$$\int_C F_1dx + F_2dy = \iint_R\left(\frac{\partial F_2}{\partial x} - \frac{\partial F_1}{\partial y}\right)dx\,dy$$

$$= \iint_R(-6y + 16y)\,dx\,dy$$

$$= \int_{x=0}^{1}\int_{y=0}^{2}10y\,dy\,dx$$

$$= \int_0^1\left[5y^2\right]_0^2 dx = 20\int_0^1 dx = 20$$

Example 2.4.4 Show that the volume V of a region bounded by surface S is

$$V = \iint_S x\,dy\,dz = \frac{1}{3}\iint_S(x\,dydz + y\,dzdx + x\,dydx)$$

Solution:

By Green's theorem in space, we have,

$$\iiint_V \left(\frac{\partial F_1}{\partial x}+\frac{\partial F_2}{\partial y}+\frac{\partial F_3}{\partial z}\right)dx\,dy\,dz = \iint_S \left(F_1 dydz + F_2 dzdx + F_3 dxdy\right) \qquad (1)$$

Put $F_1 = x$, $F_2 = 0$, $F_3 = 0$ in (1).

Then $\iint_S x\,dy\,dz = \iiint_V dx\,dy\,dz = V$

Again put $F_1 = x$, $F_2 = y$, $F_3 = z$ in (1)

Then $\iint_S (x.dydz + y.dzdx + z.dxdy) = \iiint_V (1+1+1)dx\,dy\,dz = 3V$

$$\therefore\ V = \frac{1}{3}\iint_S (x.dydz + y.dzdx + z.dxdy)$$

Example 2.4.5 Evaluate $\iint_S (2xy^2.dydz + y^3.dzdx - y^2z.dxdy)$ where S is the closed surface formed from the circular cylinder $x^2 + y^2 = a^2$, $z = 0$, $z = a$.

Solution:

Let $F_1 = 2xy^2$, $F_2 = y^3$, $F_3 = -y^2z$.

Then $\frac{\partial F_1}{\partial x} = 2y^2$, $\frac{\partial F_2}{\partial y} = 3y^2$, $\frac{\partial F_3}{\partial z} = -y^2$

By Green's theorem in space,

$$\iint_S (2xy^2.dydz + y^3.dzdx - y^2z.dxdy) = \iiint_V \left(\frac{\partial F_1}{\partial x}+\frac{\partial F_2}{\partial y}+\frac{\partial F_3}{\partial z}\right)dx\,dy\,dz$$

$$= \iiint_V 4y^2 dx\,dy\,dz$$

$$= \int_{z=0}^{a}\int_{\theta=0}^{2\pi}\int_{r=0}^{a} 4.r^2\sin^2\theta\, r\,dr\,d\theta\,dz$$

(Using the cylindrical coordinates $x = r\cos\theta$, $y = r\sin\theta$, $z = z$)

$$= \int_0^a 4dz \int_0^{2\pi}\sin^2\theta\,d\theta \int_0^a r^3 dr$$

$$= 4a.\pi.\frac{a^4}{4}$$

$$= \pi a^5.$$

Example 2.4.6 Evaluate using Green's theorem in space,

$\iint_S (x^3.dydz + y^3.dzdx + z^3.dxdy)$, where S is the spherical surface $x^2 + y^2 + z^2 = 1$.

Solution:

Let $F_1 = x^3$, $F_2 = y^3$, $F_3 = z^3$.

Then $\frac{\partial F_1}{\partial x} = 3x^2$, $\frac{\partial F_2}{\partial y} = 3y^2$, $\frac{\partial F_3}{\partial z} = 3z^2$

$$\therefore \iint_S \left(x^3.dydz + y^3.dzdx + z^3.dxdy\right) = \iiint_V \left(\frac{\partial F_1}{\partial x} + \frac{\partial F_2}{\partial y} + \frac{\partial F_3}{\partial z}\right) dx\, dy\, dz$$

$$= \iiint_V 3\left(x^2 + y^2 + z^2\right) dx\, dy\, dz \quad (1)$$

Transforming to spherical polar coordinates $x = r \sin\theta \cos\varphi$, $y = r \sin\theta \sin\varphi$, $z = r\cos\theta$ $0 \le r \le 1$, $0 \le \theta \le \pi$, $0 \le \varphi \le 2\pi$, we have $x^2 + y^2 + z^2 = r^2$ and $dx\, dy\, dz = r^2 \sin\theta\, dr\, d\theta\, d\varphi$

$$\therefore \iiint_V 3\left(x^2 + y^2 + z^2\right) dx\, dy\, dz = 3 \int_{\varphi=0}^{2\pi} \int_{\theta=0}^{\pi} \int_{r=0}^{1} r^2.r^2 \sin\theta\, dr\, d\theta\, d\varphi$$

$$= 3\int_0^{2\pi} d\varphi \int_0^{\pi} \sin\theta\, d\theta \int_0^1 r^4 dr$$

$$= 3.2\pi.2.\frac{1}{5} = \frac{12\pi}{5}.$$

2.4.2 Integration using Gauss' divergence theorem

The following results are proved using divergence theorem.

1. $\iint_S f.d\bar{S} = \iiint_V \nabla f\, dV$ (1)

Proof:

By divergence theorem $\iint_S \bar{F}.\hat{n}dS = \iiint_V \nabla.\bar{F}\, dV$ (2)

Put $\bar{F} = f\,\bar{a}$ where $\bar{a}$ is an arbitrary constant vector.

Then $\nabla.(f.\bar{a}) = \nabla f.\bar{a} + f(\nabla.\bar{a})$

$= \nabla f.\bar{a}$ (Since $\nabla.\bar{a} = 0$)

$$\therefore \iint_S (f.\bar{a}).\hat{n}dS = \iiint_V \nabla.(f.\bar{a})\, dV$$

i.e., $\iint_S f.\bar{a}\, dS = \iiint_V \nabla f.\bar{a}\, dV$

i.e., $\left(\iint_S f\, dS\right).\bar{a} = \left(\iiint_V \nabla f\, dV\right).\bar{a}$

Since $\bar{a}$ is arbitrary, we get, $\iint_S f.d\bar{S} = \iiint_V \nabla f\, dV$

2. $\iint_S \bar{F} \times d\bar{S} = -\iiint_V (\nabla \times \bar{F})\, dV$ (3)

Proof:

Let $\bar{a}$ be an arbitrary constant vector

Replacing $\bar{F}$ by $\bar{a} \times \bar{F}$ in (2), we get, $\iint_S (\bar{a} \times \bar{F}).d\bar{S} = \iiint_V \nabla.(\bar{a} \times \bar{F})\, dV$

i.e., $\iint_S \bar{a}.(\bar{F} \times d\bar{S}) = \iiint_V [\bar{F}.(\nabla \times \bar{a}) - \bar{a}.(\nabla \times \bar{F})]\, dV$

Now $\nabla \times \bar{a} = 0$ (Since $\bar{a}$ is constant)

$\therefore \iint_S \bar{a}.(\bar{F} \times d\bar{S}) = -\iiint_V \bar{a}.(\nabla \times \bar{F})\, dV$

i.e., $\bar{a}.\left(\iint_S \bar{F} \times d\bar{S}\right) = -\bar{a}.\left(\iiint_V (\nabla \times \bar{F})\, dV\right)$

Since $\bar{a}$ is arbitrary, we get, $\iint_S \bar{F} \times d\bar{S} = -\iiint_V (\nabla \times \bar{F})\, dV$

3. If S is a closed surface, then

(i) Area of S $= \iiint_V (\nabla.\hat{n})\, dV$ (4)

(ii) $\iint_S d\bar{S} = 0$ (5)

(iii) $\iint_S \bar{r} \times d\bar{S} = 0$ (6)

(iv) $\iiint_V (\nabla \times \hat{n})\, dV = 0$ (7)

Proof:

(i) We have $\iint_S \bar{F}.\hat{n}\, dS = \iiint_V \nabla.\bar{F}\, dV$

Put $\bar{F} = \hat{n}$

Then $\iint_S \hat{n}.\hat{n}\, dS = \iiint_V (\nabla.\hat{n})\, dV$

i.e., $\iint_S dS = \iiint_V (\nabla.\hat{n})\, dV$

$\iint_S dS$ is the area of surface S.

(ii) By (1) we have $\iint_S f\, d\bar{S} = \iiint_V \nabla f\, dV$

Put f = 1, so that $\nabla f = 0$.

Then $\iint_S d\overline{S} = \iiint_V 0.dV = 0$

(iii) By (3) we have, $\iint_S \overline{F} \times d\overline{S} = -\iiint_V (\nabla \times \overline{F}) dV$

Put $\overline{F} = \overline{r}$ so that $\nabla \times \overline{r} = \begin{vmatrix} \mathbf{i} & \mathbf{j} & \mathbf{k} \\ \frac{\partial}{\partial x} & \frac{\partial}{\partial y} & \frac{\partial}{\partial z} \\ x & y & z \end{vmatrix} = 0$

$\therefore \iint_S \overline{r} \times d\overline{S} = -\iiint_V (\nabla \times \overline{r}) dV = 0$

(iv) Put $\overline{F} = \hat{n}$ in $\iint_S \overline{F} \times d\overline{S} = -\iiint_V (\nabla \times \overline{F}) dV$

Then $\iint_S \overline{F} \times d\overline{S} = \iint_S \hat{n} \times \hat{n} dS = 0$ (Since $\hat{n} \times \hat{n} = 0$)

i.e., $-\iiint_V (\nabla \times \hat{n}) dV = 0$

i.e., $\iiint_V (\nabla \times \hat{n}) dV = 0$

4. If S is a closed surface then,

(i) $\iint_S (\nabla r^2) . d\overline{S} = 6V$ (8)

(ii) $\iint_S (\nabla \times \overline{u}) . d\overline{S} = 0$ (9)

(iii) $\iint_S (\overline{r} \times \overline{a}) \times d\overline{S} = 2\overline{a}V$ where $\overline{a}$ is a constant. (10)

Proof:

(i) $\iint_S \overline{F} \times d\overline{S} = \iiint_V (\nabla \times \overline{F}) dV$

Put $\overline{F} = \nabla r^2$ where $\overline{r} = x\mathbf{i} + y\mathbf{j} + z\mathbf{k}$ Then $r^2 = \overline{r}.\overline{r} = x^2 + y^2 + z^2$

$\nabla r^2 = 2x\mathbf{i} + 2y\mathbf{j} + 2z\mathbf{k}$

$$\nabla.(\nabla r^2) = \frac{\partial}{\partial x}(2x) + \frac{\partial}{\partial y}(2y) + \frac{\partial}{\partial z}(2z) = 2 + 2 + 2 = 6$$

$$\iint_S .(\nabla r^2) . d\overline{S} = \iiint_V \nabla.(\nabla r^2) dV$$

$$= \iiint_V 6\, dV = 6V$$

(ii) Again substituting $\overline{F} = \nabla \times \overline{u}$, we have, $\nabla.\overline{F} = \nabla(\nabla \times \overline{u}) = 0$

$$\therefore \iint_S (\nabla\times\bar{u}).d\bar{S} = \iiint_V \nabla.(\nabla\times\bar{u})\,dV$$

$$= \iiint_V 0\,dV = 0$$

(iii) Put $\bar{F} = \bar{r}\times\bar{a}$ in $\iint_S \bar{F}\times d\bar{S} = -\iiint_V (\nabla\times\bar{F})dV$

Then $\iint_S (\bar{r}\times\bar{a})\times d\bar{S} = -\iiint_V \nabla\times(\bar{r}\times\bar{a})dV$

$$\nabla\times(\bar{r}\times\bar{a}) = \begin{vmatrix} \mathbf{i} & \mathbf{j} & \mathbf{k} \\ \frac{\partial}{\partial x} & \frac{\partial}{\partial y} & \frac{\partial}{\partial z} \\ a_3y - a_2z & a_1z - a_3x & a_2x - a_1y \end{vmatrix}$$

Since $\bar{r} = x\mathbf{i} + y\mathbf{j} + a\mathbf{k}$ and $\bar{a} = a_1\mathbf{i} + a_2\mathbf{j} + a_3\mathbf{k}$

$$= \mathbf{i}(-2a_1) + \mathbf{j}(-2a_2) + \mathbf{k}(-2a_3)$$

$$= -2\bar{a}$$

$$\iint_S (\bar{r}\times\bar{a})\times d\bar{S} = -\iiint_V (-2\bar{a})dV = 2\bar{a}\,V$$

Example 2.4.7 If a vector $\bar{F}$ is always normal to a given surface S, show that $\iiint_V \text{curl}\,\bar{F}\,dV = 0$.

Solution:

We have, by equation (3), $\iint_S \bar{F}\times d\bar{S} = -\iiint_V (\nabla\times\bar{F})dV$

i.e., $\iiint_V \text{curl}\,\bar{F}\,dV = -\iint_S \bar{F}\times\hat{n}\,dS$

Since $\bar{F}$ is always normal to S, $\bar{F}$ and $\hat{n}$ have the same direction. $\therefore \bar{F}\times\hat{n} = 0$

Hence $\iiint_V \text{curl}\,\bar{F}\,dV = 0$.

Example 2.4.8 If S is a closed surface enclosing a volume and $\hat{n}$ is the unit normal to S, prove $\iint_S \left(\frac{\hat{n}.\bar{r}}{r^3}\right)dS = 0$

Solution:

By Gauss theorem, we have,

$$\iint_S \left(\frac{\hat{n}.\bar{r}}{r^3}\right)dS = \iint_S \left(\frac{.\bar{r}}{r^3}\right).\hat{n}\,dS = \iiint_V \nabla.\left(\frac{\bar{r}}{r^3}\right)dV$$

Where $\nabla.\left(\frac{\bar{r}}{r^3}\right) = \frac{1}{r^3}\nabla.\bar{r} + \bar{r}.\nabla\left(\frac{1}{r^3}\right)$

Now $\nabla.\bar{r} = \nabla.(x\mathbf{i} + y\mathbf{j} + z\mathbf{k}) = 3$

$$\nabla.\left(\frac{1}{r^3}\right) = \mathbf{i}\frac{\partial}{\partial x}\left(\frac{1}{r^3}\right) + \mathbf{j}\frac{\partial}{\partial y}\left(\frac{1}{r^3}\right) + \mathbf{k}\frac{\partial}{\partial z}\left(\frac{1}{r^3}\right)$$

$$\frac{\partial}{\partial x}\left(\frac{1}{r^3}\right) = \frac{\partial}{\partial x}\left(\frac{1}{(x^2+y^2+z^2)^{3/2}}\right) = \frac{-3x}{r^5}$$

Similarly $\frac{\partial}{\partial y}\left(\frac{1}{r^3}\right) = \frac{-3y}{r^5}$ and $\frac{\partial}{\partial y}\left(\frac{1}{r^3}\right) = \frac{-3z}{r^5}$

$$\therefore\ \nabla.\left(\frac{1}{r^3}\right) = \frac{-3(x\mathbf{i} + y\mathbf{j} + z\mathbf{k})}{r^5} = \frac{-3\bar{r}}{r^5}$$

$$\nabla.\left(\frac{\bar{r}}{r^3}\right) = \frac{3}{r^3} - \frac{3\,\bar{r}.\bar{r}}{r^5} = \frac{3}{r^3} - \frac{3}{r^3} = 0$$

$$\iint_S \left(\frac{\hat{n}.\bar{r}}{r^3}\right) dS = \iiint_V 0\, dV = 0$$

Example 2.4.9 Evaluate $\iint_S \bar{F}.d\bar{S}$ where $\bar{F} = 2xy\mathbf{i} + yz^2\mathbf{j} + xz\mathbf{k}$ and S is the surface of the rectangular parallelopiped bounded by x = 0, y = 0, z = 0, x = 2, y = 1 and z = 3

Solution:

By divergence theorem,

$$\iint_S \bar{F}.d\bar{S} = \iiint_D \nabla.\bar{F}\, dV$$, where D is the closed region enclosed by the parallelopiped

$$= \iiint_D \left[\frac{\partial}{\partial x}(2xy) + \frac{\partial}{\partial y}(yz^2) + \frac{\partial}{\partial z}(xz)\right] dV$$

$$= \iiint_D (2y + z^2 + x)\, dV$$

$$= \int_{x=0}^{2}\int_{y=0}^{1}\int_{z=0}^{3} (2y + z^2 + x)\, dz\, dy\, dx$$

$$= \int_{x=0}^{2}\int_{y=0}^{1} \left[2yz + \frac{z^3}{3} + xz\right]_0^3 dy\, dx$$

$$= \int_{x=0}^{2}\int_{y=0}^{1} (6y + 9 + 3x)\, dy\, dx$$

$$= \int_{x=0}^{2} \left[6\frac{y^2}{2} + 9y + 3xy\right]_0^1 dx = \int_{x=0}^{2} (3 + 9 + 3x) dx = 6 + 18 + 6 = 30$$

Example 2.4.10 Given $\overline{F} = \mathbf{i}x - \mathbf{j}y + \mathbf{k}(z^2 - 1)$, find the value of $\iint_S \overline{F}.\hat{n}\ dS$ over the closed surface bounded by the planes z = 0, z = 1, and the cylinder $x^2 + y^2 = a^2$.

Solution:

$$\nabla.\overline{F} = \frac{\partial}{\partial x}(x) + \frac{\partial}{\partial y}(-y) + \frac{\partial}{\partial z}(z^2 - 1) = 1 - 1 + 2z = 2z$$

$$\therefore \quad \iint_S \overline{F}.\hat{n}\ dS = \iiint_D \nabla.\overline{F}\ dV \qquad \text{(By divergence theorem)}$$

$$= \iiint_D 2z\ dx\ dy\ dz$$

$$= \int_{r=0}^{a} \int_{\theta=0}^{2\pi} \int_{z=0}^{1} 2z\ r\ dr\ d\theta\ dz \qquad \text{(By using cylindrical coordinates } x = r\cos\theta,\ y = r\sin\theta,\ z = z.)$$

$$= \int_0^a \int_0^{2\pi} \left[2r.\frac{z^2}{2} \right]_0^1 d\theta\ dr$$

$$= \int_0^a \int_0^{2\pi} r\ d\theta\ dr = \int_0^a r.2\pi\ dr = \left[2\pi.\frac{r^2}{2} \right]_0^a = \pi a^2.$$

Example 2.4.11 Evaluate $\iint_S (\nabla \times \overline{F}).\hat{n}\ ds$ where $\overline{F} = (x + 2y)\mathbf{i} - 3z\mathbf{j} + x\mathbf{k}$ where S is the surface bounded by x = 0, y = 0, z = 0, x + y + z = 1.

Solution:

$$\overline{G} = \nabla \times \overline{F} = \begin{vmatrix} \mathbf{i} & \mathbf{j} & \mathbf{k} \\ \frac{\partial}{\partial x} & \frac{\partial}{\partial y} & \frac{\partial}{\partial z} \\ x + 2y & -3z & x \end{vmatrix} = \mathbf{i}(3) + \mathbf{j}(-1) + \mathbf{k}(-2) = 3\mathbf{i} - \mathbf{j} - 2\mathbf{k}$$

By divergence theorem, we have,

$$\iint_S (\nabla \times \overline{F}).\hat{n}\ dS = \iiint_S \overline{G}.\hat{n}\ dS$$

$$= \iiint_D \nabla.\overline{G}\ dV \quad \text{, where D is the volume enclosed by S}$$

$$= \iiint_D \nabla.(3\mathbf{i} - \mathbf{j} - 2\mathbf{k})\ dV = \iiint_D 0.\,dV = 0$$

Example 2.4.12 Evaluate $\iint_S \overline{F}.d\overline{S}$ where $\overline{F} = xz\mathbf{i} - yz\mathbf{j} + 2z^2\mathbf{k}$ and S is the surface of the region bounded by x = 0, y = 0, z = 0, $x^2 + y^2 = a^2$, $x^2 + z^2 = a^2$

Solution:

$$\nabla.\overline{F} = \frac{\partial}{\partial x}(xz) + \frac{\partial}{\partial y}(-yz) + \frac{\partial}{\partial z}(2z^2) = z - z + 4z = 4z$$

By divergence theorem,

$$\iint_S \overline{F}.d\overline{S} = \iiint_D \nabla.\overline{F}\, dV \quad \text{, where D is the volume enclosed by S}$$

$$= \int_{x=0}^{a} \int_{y=0}^{\sqrt{a^2-x^2}} \int_{z=0}^{\sqrt{a^2-x^2}} 4z\, dx\, dy\, dz$$

$$= \int_{x=0}^{a} \int_{y=0}^{\sqrt{a^2-x^2}} \left[4\frac{z^2}{2}\right]_0^{\sqrt{a^2-x^2}} dx\, dy$$

$$= \int_{x=0}^{a} \int_{y=0}^{\sqrt{a^2-x^2}} 2(a^2 - x^2)\, dy\, dx$$

$$= \int_{x=0}^{a} \left[2(a^2 - x^2)y\right]_0^{\sqrt{a^2-x^2}} dx$$

$$= \int_{x=0}^{a} 2(a^2 - x^2)^{3/2} dx$$

$$= \int_{\theta=0}^{\pi/2} 2a^3 \cos^3\theta . a\cos\theta\, d\theta \qquad \text{(By substituting } x = a\sin\theta)$$

$$= 2a^4 \int_0^{\pi/2} \cos^4\theta\, d\theta = 2a^4 . \frac{3}{4}.\frac{1}{2}.\frac{\pi}{2} = \frac{3\pi a^4}{8}$$

2.4.3 Integration using Stoke's theorem

The following results are proved using Stoke's theorem.

1. $\int_C f\, d\overline{r} = -\iint_S (\nabla f) \times d\overline{S} = \iint_S \hat{n} \times \nabla f\, dS$

Proof:

By Stoke's theorem $\int_C \overline{F}.d\overline{r} = -\iint_S (\nabla \times \overline{F}).d\overline{S}$

Replace $\overline{F}$ by $f\overline{a}$, where $\overline{a}$ is an arbitrary constant vector.

$$\int_C (f\overline{a}).d\overline{r} = -\iint_S [\nabla \times (f\overline{a})].d\overline{S}$$

$$= \iint_S [\nabla f \times \overline{a} + f(\nabla \times \overline{a})].d\overline{S}$$

$$= \iint_S (\nabla f \times \bar{a}).d\bar{S} + \iint_S f(\nabla \times \bar{a}).d\bar{S}$$

$$= -\iint_S (\nabla f \times d\bar{S}).\bar{a} + 0 \qquad \text{(Since } \nabla \times \bar{a} = 0\text{)}$$

i.e., $\int_C f\, d\bar{r}.\bar{a} = -\iint_S (\nabla f \times d\bar{S}).\bar{a}$

$$\left(\int_C f\, d\bar{r} + \iint_S \nabla f \times d\bar{S}\right).\bar{a} = 0$$

i.e., $\int_C f\, d\bar{r} + \iint_S \nabla f \times d\bar{S} = 0$ (Since $\bar{a}$ is arbitrary)

i.e., $\int_C f\, d\bar{r} = -\iint_S (\nabla f) \times d\bar{S} = -\iint_S \nabla f \times \hat{n}\; dS = \iint_S \hat{n} \times \nabla f\; dS$

2. (i) $\nabla \times (\nabla f) = 0$ and (ii) $\nabla.(\nabla \times \bar{F}) = 0$

Proof:

(i) Let S be any surface bounded by a simple closed curve C.

$\iint_S (\nabla \times \bar{F}).d\bar{S} = \int_C \bar{F}.d\bar{r}$ (By Stoke's theorem)

Put $\bar{F} = \nabla f$

Then $\iint_S (\nabla \times \nabla f).d\bar{S} = \int_C \nabla f.d\bar{r}$

$$= \int_C \frac{\partial}{\partial x}dx + \frac{\partial}{\partial y}dy + \frac{\partial}{\partial z}dz$$

$$= \int_C df = 0 \qquad \text{(Since C is closed)}$$

i..e, $\iint_S (\nabla \times \nabla f).d\bar{S} = 0$ for every surface S with a simple closed curve as boundary.

$\therefore\ \nabla \times \nabla f = 0.$

(ii) If S is any closed surface, divide S into two portions S_1 and S_2 by a closed curve C. As C is common boundary to S_1 and S_2, we have,

$$\iint_S (\nabla \times \bar{F}).d\bar{S} = \iint_{S_1} (\nabla \times \bar{F}).d\bar{S} + \iint_{S_2} (\nabla \times \bar{F}).d\bar{S}$$

$$= \int_C \bar{F}.d\bar{r} - \int_C \bar{F}.d\bar{r} = 0 \qquad \text{(By Stoke's theorem)}$$

$\iiint_V \nabla.(\nabla \times \bar{F})dV = \iint_S (\nabla \times \bar{F}).d\bar{S} = 0$ (By divergence theorem)

Since S is arbitrary the volume V enclosed by S is also arbitrary.

$\therefore\ \nabla.(\nabla \times \bar{F}) = 0$

3. $\iint_C (\bar{a}\times\bar{r}).d\bar{r} = 2\iint_S \bar{a}.\hat{n}\, d\bar{S}$, where $\bar{a}$ is a constant vector.

Proof:

By Stoke's theorem, $\int_C \bar{F}.d\bar{r} = \iint_S (\nabla\times\bar{F}).\hat{n}\, dS$

Put $\bar{F} = \bar{a}\times\bar{r}$

Then $\int_C (\bar{a}\times\bar{r}).d\bar{r} = \iint_S (\nabla\times(\bar{a}\times\bar{r})).\hat{n}\, dS$ (3)

But $\nabla\times(\bar{a}\times\bar{r}) = 2\bar{a}$ (Prove this!)

$\therefore$ (3) becomes, $\int_C (\bar{a}\times\bar{r}).d\bar{r} = -\iint_S 2\bar{a}.\hat{n}\, dS = 2\iint_S \bar{a}.\hat{n}\, dS$

4. $\int_C \bar{F}\times d\bar{r} = -\iint_S (\hat{n}\times\nabla)\times\bar{F}\, dS$

Proof:

By Stoke's theorem, $\int_C \bar{F}.d\bar{r} = \iint_S (\nabla\times\bar{F}).d\bar{S}$

Replace $\bar{F}$ by $\bar{a}\times\bar{F}$, where $\bar{a}$ is an arbitrary constant vector.

$$\text{Then } \int_C (\bar{a}\times\bar{F}).d\bar{r} = \iint_S (\nabla\times(\bar{a}\times\bar{F})).d\bar{S}$$

$$= \iint_S [(\bar{F}.\nabla)\bar{a} - (\bar{a}.\nabla)\bar{F} + (\nabla.\bar{F})\bar{a} - (\nabla.\bar{a})\bar{F}]d\bar{S}$$

$$= \iint_S [(\nabla.\bar{F})\bar{a} - (\bar{a}.\nabla)\bar{F}].d\bar{S} \quad (\text{Since } (\bar{F}.\nabla)\bar{a} = 0 \text{ and } \nabla.\bar{a} = 0)$$

$$= \iint_S [(\nabla.\bar{F})\bar{a} - (\bar{a}.\nabla)\bar{F}]\hat{n}\, dS$$

$$= \iint_S [(\nabla.\bar{F})(\bar{a}.\hat{n}) - (\bar{a}.\nabla)(\bar{F}.\hat{n})]dS$$

i.e., $\int_C \bar{a}.(\bar{F}\times d\bar{r}) = \iint_S [(\bar{a}.\hat{n})(\nabla.\bar{F}) - (\bar{a}.\nabla)(\bar{F}.\hat{n})]dS$

i.e., $\bar{a}.\left(\int_C (\bar{F}\times d\bar{r})\right) = \bar{a}.\left(\iint_S [\hat{n}(\nabla.\bar{F}) - \nabla(\bar{F}.\hat{n})]dS\right)$

Since $\bar{a}$ is arbitrary constant,

$$\int_C (\bar{F}\times d\bar{r}) = -\iint_S [\nabla(\bar{F}.\hat{n}) - \hat{n}(\nabla.\bar{F})]dS$$

$$= -\iint_S (\hat{n}\times\nabla)\times\bar{F}\, dS$$

Example 2.4.13 Evaluate $\int_C f_1(x)dx + f_2(y)dy + f_3(z)dz$ where C is any closed curve.

Solution:
Let S be any surface having C as its boundary.
Let $\overline{F}(x,y,z) = f_1(x)\mathbf{i} + f_2(y)\mathbf{j} + f_3(z)\mathbf{k}$

$$\nabla \times \overline{F} = \begin{vmatrix} \mathbf{i} & \mathbf{j} & \mathbf{k} \\ \frac{\partial}{\partial x} & \frac{\partial}{\partial y} & \frac{\partial}{\partial z} \\ f_1(x) & f_2(y) & f_3(z) \end{vmatrix} = \mathbf{i}(0) + \mathbf{j}(0) + \mathbf{k}(0) = 0$$

By Stoke's theorem, $\int_C \overline{F}.d\overline{r} = \iint_S (\nabla \times \overline{F}).\hat{n}\ dS = 0$

i.e., $\int_C f_1(x)dx + f_2(y)dy + f_3(z)dz = 0$

Example 2.4.14 Evaluate $\int_C yz\,dx + zx\,dy + xy\,dz$ where C is the circle $x^2 + y^2 + z^2 = a^2$, $z = 0$.

Solution:
Let S be any surface with C as boundary and $\overline{F} = yz\mathbf{i} + zx\mathbf{j} + xy\mathbf{k}$

Then $$\nabla \times \overline{F} = \begin{vmatrix} \mathbf{i} & \mathbf{j} & \mathbf{k} \\ \frac{\partial}{\partial x} & \frac{\partial}{\partial y} & \frac{\partial}{\partial z} \\ yz & zx & xy \end{vmatrix} = \mathbf{i}(x\text{-}x) + \mathbf{j}(y\text{-}y) + \mathbf{k}(z\text{-}z) = 0$$

By Stoke's theorem,
$\int_C \overline{F}.d\overline{r} = \iint_S (\nabla \times \overline{F}).\hat{n}\ dS = 0$

i.e., $\int_C yz\,dx + zx\,dy + xy\,dz = 0$

Example 2.4.15 Evaluate $\int_C \overline{F}.d\overline{r}$ where $\overline{F} = (3x + 2z)\mathbf{i} + (x + 3y)\mathbf{j} + (2y - 3z)\mathbf{k}$ and C is the curve of intersection of the plane 6x + 3y + 4z = 12 with the coordinate planes.

Solution: Let S be the surface of cone cutoff by the plane 6x + 3y + 4z = 12 in the first octant.

By Stoke's theorem, $\int_C \overline{F}.d\overline{r} = \iint_S (\nabla \times \overline{F}).\hat{n}\ dS$

$$\nabla \times \overline{F} = \begin{vmatrix} \mathbf{i} & \mathbf{j} & \mathbf{k} \\ \frac{\partial}{\partial x} & \frac{\partial}{\partial y} & \frac{\partial}{\partial z} \\ 3x + 2z & x + 3y & 2y - 3z \end{vmatrix} = \mathbf{i}(2 - 0) + \mathbf{j}(2 - 0) + \mathbf{k}(1 - 0) = 2\mathbf{i} + 2\mathbf{j} + \mathbf{k}$$

$\therefore \int_C \overline{F}.d\overline{r} = \iint_S (2\mathbf{i} + 2\mathbf{j} + \mathbf{k}).\hat{n}\ dS$

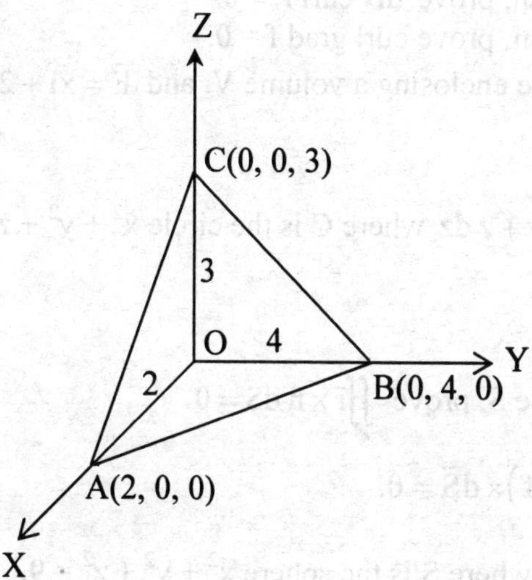

Fig. 2.18

$$= \iint_S (2dy\,dz + 2\,dz\,dx + dx\,dy).\hat{n}\ dS$$

$$= 2(\Delta OBC) + 2(\Delta OAC) + \Delta OAB = 2(6) + 2(3) + 4 = 22$$

EXERCISE 2.4

PART-A

1. Prove that $\iint_S \bar{r}.\hat{n}\ dS = 3V$ where V is the volume bounded by S.
2. Show that $\iint_S \bar{a}.\hat{n}\ dS = 0$ where $\bar{a}$ is a constant vector.
3. Show that $\iint_S (\nabla \times \bar{F}).\hat{n}\ dS = 0$.
4. Prove that $\iint_S (\nabla r^2).\hat{n}\ dS = 6V$ where V is the volume bounded by S.
5. If $\bar{G} = \text{curl}\,\bar{F}$, prove $\iint_S \bar{G}.\hat{n}\ dS = 0$ for any closed surface S.
6. If S is any closed surface enclosing a volume V and if $\bar{F} = ax\mathbf{i} + by\mathbf{j} + cz\mathbf{k}$, prove $\iint_S \bar{F}.\hat{n}\ dS = (a+b+c)V$

7. Using Stoke's theorem, prove div curl $\overline{F} = 0$.
8. Using Stoke's theorem, prove curl grad f = 0.
9. If S is a closed surface enclosing a volume V, and $\overline{F} = x\mathbf{i} + 2y\mathbf{j} + 3z\mathbf{k}$, show that $\iint_S \overline{F}.\hat{n}\, dS = 6V$.
10. Evaluate $\int_C x\, dx + y dy + z\, dz$ where C is the circle $x^2 + y^2 + z^2 = a^2, z = 0$.
11. Prove $\iint_C \bar{r}.d\bar{r} = 0$.
12. For any closed surface S, prove $\iint_S \bar{r} \times \hat{n}\, dS = 0$
13. Prove $\int_C f\, d\bar{r} = -\iint_S (\nabla f) \times d\overline{S} = 0$.
14. Evaluate $\iint_S \bar{r}.d\overline{S} = 0$ where S is the sphere $x^2 + y^2 + z^2 = 9$.
15. For a closed surface S, show that $\iint_S \frac{\bar{r}}{r^3}.d\overline{S} = 0$

PART-B

Use Green's theorem to solve the following problems (Ex: 16-25)

16. Find the area between the parabolas $y^2 = 4x$ and $x^2 = 4y$.
17. Find the area of the ellipse $x = a \cos \theta$, $y = b \sin \theta$, $0 \le \theta \le 2\pi$.
18. Evaluate $\int_C (3x^2 - 8y^2)dx + (4y - 6xy)dy$ where C is the boundary of the region bounded by $y = x^2$ and $x = y^2$.
19. Evaluate $\int_C (x^2 + xy)dx + (x^2 + y^2)dy$ where C is the square formed by the lines $y = \pm 1$ and $x = \pm 1$.
20. Evaluate $\int_C (\sin x - y)dx - \cos x\, dy$ where C is the triangle whose vertices are (0, 0), $(\pi/2, 0)$ and $(\pi/2, 1)$.
21. Evaluate $\int_C (3x^2 + 2y)dx + (x + 3\cos y)dy$ where C is the perimeter of the parallelogram whose vertices are (0, 0), (2, 0), (3, 1), (1, 1).
22. Evaluate $\int_C (xy + y^2)dx + x^2 dy$ where C is the closed curve of the region bounded by $y = x$ and $y = x^2$.
23. Evaluate $\int_C xy^2 dy - x^2 y\, dx$, where C is the cardioid $r = a(1 + \cos \theta)$.

24. Evaluate $\int_C (3x^2 - 8y^2)dx + (4y - 6xy)dy$ where C is the boundary of the triangular area enclosed by the lines $y = 0, x + y = 1, x = 0$.
25. Evaluate $\int_C \left[x^{-1}e^y dx + \left(e^y \log x + 2x\right)dy\right]$ where C is the boundary of the region bounded by $y = x^4 + 1$ and $y = 2$.

Use Gauss's divergence theorem to solve the following problems (Ex : 26-30)

26. Evaluate $\iint_S \overline{F}.\hat{n}\, dS$ if $\overline{F} = yz\mathbf{i} + 2y^2\mathbf{j} + xz^2\mathbf{k}$ and S is the surface of the cylinder $x^2 + y^2 = 9$ contained in the first octant between the planes $z = 0$ and $z = 2$.
27. Evaluate $\iint_S \overline{F}.\hat{n}\, dS$ where $\overline{F} = \alpha x\mathbf{i} + \beta y\mathbf{j} + \gamma z\mathbf{k}$ and S is the surface of the cylinder $x^2 + y^2 = a^2, z = \pm h$.
28. Evaluate $\iint_S \overline{r}.d\overline{S}$ where S is the cube bounded by $x = \pm 1, y = \pm 1, z = \pm 1$.
29. If $\overline{F} = x^3\mathbf{i} + y^3\mathbf{j} + z^3\mathbf{k}$ and S is the surface of the sphere $x^2 + y^2 + z^2 = a^2$, show that $\iint_S \overline{F}.d\overline{S} = \frac{12}{5}\pi a^5$.
30. Evaluate $\iint_S \overline{F}.d\overline{S}$ if $\overline{F} = 2x^2y\mathbf{i} - y^2\mathbf{j} + 4xz^2\mathbf{k}$ and S is the closed surface bounded by the region in the first quadrant bounded by the cylinder $y^2 + z^2 = 9, x = 0, x = 2$.

Use Stoke's theorem to solve the following problems (Ex: 31-37).

31. Evaluate $\int (\cos x\, dx + 2y^2 dy + z\, dz)$ where C is the curve $x^2 + y^2 = 1, z = 1$.
32. Evaluate $\iint_C \overline{F}.d\overline{r}$ where $\overline{F} = (2x - y)\mathbf{i} - yz^2\mathbf{j} - y^2z\mathbf{k}$ and C is the boundary of the surface S of the upper half of the sphere $x^2 + y^2 + z^2 = 1$.
33. Evaluate $\iint_S (\nabla \times F).d\overline{S}$ if $\overline{F} = y\mathbf{i} + z\mathbf{j} + x\mathbf{k}$ and S is the paraboloidal surface $x^2 + y^2 = 1 - z, z \geq 0$.
34. Evaluate $\int_C [(1+y)z\mathbf{i} + (1+z)x\mathbf{j} + (1+x)y\mathbf{k}].d\overline{r}$ where C is the circle $x^2 + y^2 = 1$, $z = 1$.
35. Evaluate $\iint_S \text{curl}\, \overline{F}.\hat{n}\, dS$ where S is the open hemispherical surface $x^2 + y^2 + z^2 = a^2$, $z \geq 0$ and $\overline{F} = (1 - ay)\mathbf{i} + 2y^2\mathbf{j} + (x^2 + 1)\mathbf{k}$

36. Evaluate $\iint_S \text{curl}\,\overline{F}.\hat{n}\,dS$ if $\overline{F} = (x^2 - y^2)\mathbf{i} + 2xy\mathbf{j}$ and S is the rectangular parallelopiped bounded by the planes x = 0, x = a, y = 0, y = b, z = 0, z = c and the face z = 0 missing.
37. Evaluate $\iint_S \text{curl}\,\overline{F}.\hat{n}\,dS$ where $\overline{F} = x^2\mathbf{i} - xy\mathbf{j}$ and S is the surface of the square in the xy-plane bounded by the lines x = 0, x = a, y = 0 and y = a.

ANSWERS

EXERCISE 2.1

PART-A

(15) $-2\mathbf{i} + 33\mathbf{j} - 24\mathbf{k}$ (16) $10\mathbf{i} - 4\mathbf{j} - 16\mathbf{k}$ (17) $(\mathbf{i}+\mathbf{j}+\mathbf{k})/\sqrt{3}$

(18) $(2\mathbf{i} - \mathbf{j} - \sqrt{3}\mathbf{k})/\sqrt{8}$ (20) $(2\mathbf{i}-5\mathbf{j}-2\mathbf{k})/\sqrt{33}$ (21) $2(x+y+z)$

(22) $2(x+y+z)$ (23) $-(\mathbf{i}y + \mathbf{j}z + \mathbf{k}x)$ (29) $6(x+y+z)$

PART-B

(34) (i) $\varphi = r^5 + c$ (ii) $\varphi = 2(r^3 - r^4 + 10)$

(35) (i) $\varphi = x^2yz + y + c$ (ii) $\varphi = xy + xy^2 + xz^2 + yz - 1$

(iii) $3x^2y + xz^3 - yz + c$ (iv) $xy + x\sin z + c$

(36) (i) $(-\mathbf{i}-3\mathbf{j}+\mathbf{k})/\sqrt{11}$ (ii) $(2\mathbf{i}-\mathbf{j}+2\mathbf{k})/3$

(iii) $(2\mathbf{i}+\mathbf{j}-2\mathbf{k})/3$ (iv) $(2\mathbf{i}+2\mathbf{j}+\mathbf{k})/3$

(37) $4\sqrt{91}$ (38) $\frac{37}{3}$

(39) (i) $2\sqrt{3}$ (ii) $2\sqrt{93}$ (40) $-12\mathbf{i} + 15\mathbf{j} + 24\mathbf{k}$, $3\sqrt{105}$

(41) $\cos^{-1}\left(\dfrac{13}{\sqrt{6}\sqrt{33}}\right)$ (42) $3x - 8y + 3z + 4 = 0$

(43) a = -2, b = 2, c = 1 (44) $\cos^{-1}\left(\dfrac{8}{3\sqrt{21}}\right)$

(45) $\dfrac{x-2}{4} = \dfrac{y-1}{4} = \dfrac{z+1}{-8}$ (46) a = -12, b = -1

(49) (i) $2(y^3 + 3x^2y - 6xy^2)z\mathbf{i} + 2(3xy^2 + x^3 - 6x^2y)z\mathbf{j} + 2(xy^2 + x^3 - 3x^2y)y\mathbf{k}$ (ii) 0

(50) $f(r) = \dfrac{c}{r^3}$, where c is a constant.

(54) $a = 0$ (55) $a = 4, b = 2, c = -1$

(56) $f(x, y, z) = xyz^2 - y - 2z + c$ (57) $y^2 \sin x + z^3 x - 4y + c$

(58) $f = 3x^2y + xz^3 - yz + c$ (59) $x^2 + 2y^2 - 3z^2 + xyz$

EXERCISE 2.2

PART-A

(7) $\frac{7}{12}$ (8) $\frac{86}{105}$ (9) $\frac{4}{5}\mathbf{i} + \frac{8}{3}\mathbf{j} + \frac{12}{7}\mathbf{k}$

(11) $\frac{62}{3}$ (15) $-\frac{4}{3}\mathbf{j} + \frac{12}{5}\mathbf{k}$ (16) 303

(17) zero (20) 3

PART-B

(21) $-2\sqrt{2}\pi^2\mathbf{j},\ -\pi,\ -(2\pi^2 + \pi)\mathbf{i} + 2\pi\mathbf{k}$

(22) $2\sqrt{2} - 1,\ \frac{2\sqrt{2}+2}{15}\mathbf{i} + \frac{2\sqrt{2}-1}{3}\mathbf{j},\ \frac{7}{10}$ and $\frac{-k}{12}$.

(26) 3 (27) 21 (29) $21\frac{7}{20}$

(30) -12 (31) 8π (32) 202

(33) 81 (34) 10 (35) 3

(36) 81

(37) (i) 4 (ii) 8 (iii) $8\mathbf{j} - 4\mathbf{k}$ (iv) $\frac{16}{3}\mathbf{i} - \frac{8}{3}\mathbf{j} - \frac{16}{3}\mathbf{k}$

(39) $\pi a^2 b$ (40) 128 (41) $\frac{a^4 c\pi}{4}$

(42) 180 (43) $\frac{8}{15}a^5\mathbf{i} - \frac{a^4}{4}\mathbf{k}$

EXERCISE 2.3

PART-B

(8) 3π (9) $\frac{3}{5}$ (10) 120π

(11) $-\frac{4}{3}$ (14) $-\frac{1}{20}$ (15) $\frac{5}{3}$

(16) 64π (17) 256π (18) $\frac{22}{3}$

(19) $abc(a + b + c)$ (20) $\frac{5\pi a^4}{4}$ (21) $\left(\frac{11}{6}\right)\pi a^4$

(22) $h^2a^2\pi$ (23) 180 (24) $2\pi ha^2(\alpha + \beta + \gamma)$

(25) π (27) $\dfrac{a^3}{2}$ (28) -4

(29) $2ab^2$ (30) zero (31) $\dfrac{256}{3}$

(32) zero (33) -1

EXERCISE 2.4

PART-A

(10) Zero (14) 108π

PART-B

(16) $\dfrac{16}{3}$ (17) πab (18) $\dfrac{3}{2}$

(19) zero (20) $\dfrac{2}{\pi} + \dfrac{\pi}{4}$ (21) -6

(22) $-\dfrac{1}{20}$ (23) $\dfrac{35}{16}\pi a^4$ (24) $\dfrac{5}{3}$

(25) $\dfrac{16}{5}$ (26) 108 (27) $2\pi ha^2(\alpha + \beta + \gamma)$

(28) 24 (30) 18 (31) zero

(32) π (33) $-\pi$ (34) π

(35) πa^2 (36) $2ab^2$ (37) $\dfrac{a^3}{2}$

CHAPTER 3

ANALYTIC FUNCTIONS

3.0 INTRODUCTION

Functions of a complex variable and analytic functions are widely used while solving certain types of differential equations and integrals in Physics, Fluid Dynamics, Space Mechanics and many other areas in Applied Mathematics. The theory of analytic functions provides a very efficient tool for the evaluation of definite integrals, which can not be evaluated by usual methods. In this Chapter we study the applications of Cauchy-Riemann equations, properties of analytic functions, conformal mappings and bilinear transformations.

3.0.1 COMPLEX NUMBERS

The equation $x^2 +1 = 0$ has no solution in the set $\mathbb{R}$ of real numbers, for $a^2 +1$ is always positive for any real number a. Suppose that we can find a larger set in which the equation $x^2 +1 = 0$ can be solved. We denote such a solution of the quadratic equation $x^2 +1 = 0$ by i. i.e., $i = \sqrt{-1}$ and $x^2 +1 = (x + i)(x - i)$. i is called the **imaginary unit**. The equation $x^2 +1 = 0$ has roots i and –i.

Definition 3.0.1 Let x and y be real numbers. Then the symbol x + iy is called a **complex number** and is denoted by z. x is called the real part of z, denoted by Re(z) or R(z) and y is called the imaginary part of z, denoted by Im(z) or I (z).

Note:

(i) Let $C = \{ x + iy \,/\, x, y \in \mathbb{R} \}$. C is called the **set of complex numbers**.

(ii) If x = 0, the complex number z is called **purely imaginary**. If y = 0, then z is **real**.

(iii) When two complex numbers are equal, then their real and imaginary parts are equal.

(iv) If $z_1 = x_1 + iy_1$, $z_2 = x_2 + iy_2$, then we define $z_1 + z_2 = (x_1 + x_2) + i(y_1 + y_2)$ and $z_1.z_2 = (x_1x_2 - y_1y_2) + i(x_1y_2 + x_2y_1)$

Definition 3.0.2 Let z = x + iy be any complex number. The complex number $\bar{z} = x - iy$ is called the **conjugate** of z.

Note:

(i) $\overline{\overline{z}} = z$

(ii) $z + \overline{z} = 2\text{Re}(z)$ and $z - \overline{z} = 2i\,\text{Im}(z)$

(iii) $\overline{z_1 + z_2} = \overline{z}_1 + \overline{z}_2$, $\overline{z_1 - z_2} = \overline{z}_1 - \overline{z}_2$ and $\overline{z_1 z_2} = \overline{z}_1 . \overline{z}_2$

(iv) $\overline{\left(\dfrac{z_1}{z_2}\right)} = \dfrac{\overline{z}_1}{\overline{z}_2}$ if $z_2 \neq 0$

Definition 3.0.3 Let $z = x + iy$ be any complex number. Then $|z| = +\sqrt{x^2 + y^2}$ is called the **modulus** or the **magnitude** (or **absolute value**) of z.

Note:

(i) $|z| \geq 0$ and $|z| = 0$ iff $z = 0$.

(ii) $z\overline{z} = |z|^2 = x^2 + y^2$

(iii) $|z_1 z_2| = |z_1|.|z_2|$

(iv) $|z_1 + z_2| \leq |z_1| + |z_2|$ (Triangle inequality).

(v) $|z_1 - z_2| \geq \big|\, |z_1| - |z_2| \,\big|$.

(vi) $\dfrac{1}{z} = \dfrac{\overline{z}}{z\overline{z}} = \dfrac{\overline{z}}{|z|^2} = \dfrac{x - iy}{x^2 + y^2} = \dfrac{x}{x^2 + y^2} - i\dfrac{y}{x^2 + y^2}$

(vii) $\left|\dfrac{z_1}{z_2}\right| = \dfrac{|z_1|}{|z_2|}$ $(z_2 \neq 0)$

3.0.2 THE ARGAND'S PLANE

The complex number $z = x + iy$ can be geometrically represented by a point (x, y) in the Cartesian plane $\mathbb{R} \times \mathbb{R}$. Conversely every point (x, y) in $\mathbb{R} \times \mathbb{R}$ represents a complex number $z = x + iy$. Hence this plane $\mathbb{R} \times \mathbb{R}$ is called the **complex plane or Argand's plane**. The x-axis is called the **real axis** and y-axis is called the **imaginary axis**.

If P(x, y) is the point in this plane corresponding to $z = x + iy$ and O is the origin, then $OP = \sqrt{x^2 + y^2} = |z|$.

If (r, θ) are the polar coordinates of the point P(x, y), then $x = r\cos\theta$ and $y = r\sin\theta$.

$\therefore z = x + iy$

$\quad = r(\cos\theta + i\sin\theta) = r\,e^{i\theta}$.

$\therefore r^2 = x^2 + y^2$ and $\tan\theta = y/x$

$r = |z| = \sqrt{x^2 + y^2}$ is the **magnitude** of z. θ is called the **amplitude or argument** of z and is denoted by arg(z) or amp(z), $0 \leq \theta < 2\pi$ The value of θ lying between $-\pi$ and π is called the **principal value** of arg(z).

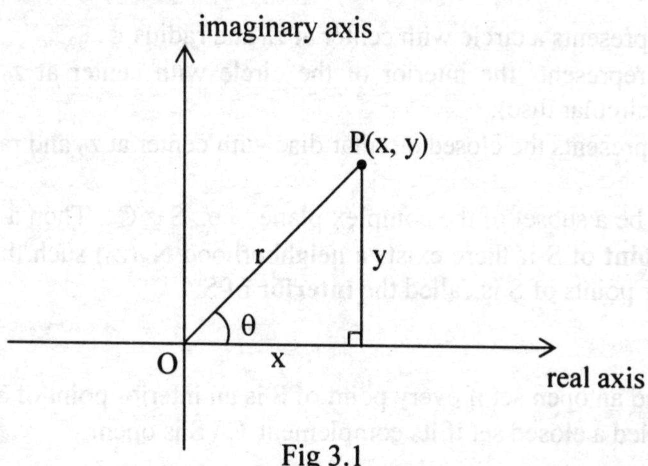

Fig 3.1

Note:

(i) $|z_1 - z_2|$ is the distance between z_1 and z_2 in the Argand's plane.

(ii) $\arg(\bar{z}_1) = -\arg(z_1)$

(iii) $\arg(z_1 z_2) = \arg(z_1) + \arg(z_2)$

(iv) $\arg\left(\frac{z_1}{z_2}\right) = \arg(z_1) - \arg(z_2)$

(v) If $z = r(\cos\theta + i\sin\theta)$ and n is any integer, $z^n = r^n(\cos n\theta + i\sin n\theta)$

(vi) (De-Movire's Theorem) If n is any integer or fraction, then the value or one of the values of $(\cos\theta + i\sin\theta)^n$ is $\cos n\theta + i\sin n\theta$

(vii) If $z = r(\cos\theta + i\sin\theta)$ and n is a positive integer, then the n values of $z^{1/n}$ are $r^{1/n}\left[\cos\left(\frac{\theta + 2k\pi}{n}\right) + i\sin\left(\frac{\theta + 2k\pi}{n}\right)\right]$ where k = 0, 1, 2, …, n-1.

(viii) The nth roots of unity are the n values of $1^{1/n}$. i.e., $(\cos 0 + i\sin 0)^{1/n}$.

They are $\cos\left(\frac{2k\pi}{n}\right) + i\sin\left(\frac{2k\pi}{n}\right)$ where k = 0, 1, 2, …, n-1.

If $\omega = \cos\left(\frac{2\pi}{n}\right) + i\sin\left(\frac{2\pi}{n}\right)$, then the nth roots of unity are given by $1, \omega, \omega^2, \ldots, \omega^{n-1}$. Also $\omega^n = 1$ and $1 + \omega + \omega^2 + \ldots + \omega^{n-1} = 0$.

3.0.3 REGIONS IN THE COMPLEX PLANE

Definition 3.0.4 Let z_0 be any complex number and $\in$ be a positive real number. Then the set $\{z \,/\, |z - z_0| < \in\}$ is called a **neighborhood** or **$\in$-neighborhood** of z_0 and is denoted by $N_\in(z_0)$ or $S(z_0, \in)$.

Note:

(i) $|z - z_0| = \in$ represents a circle with center at z_0 and radius $\in$.

(ii) $|z - z_0| < \in$ represents the interior of the circle with center at z_0 and radius $\in$ (called open circular disc).

(iii) $|z - z_0| \le \in$ represents the closed circular disc with center at z_0 and radius $\in$.

Definition 3.0.5 Let S be a subset of the complex plane. i.e., $S \subseteq \mathbb{C}$. Then a point $z_0 \in S$ is called an **interior point** of S if there exists a neighborhood $N_\in(z_0)$ such that $N_\in(z_0) \subseteq S$. The set of all interior points of S is called the **interior** of S

Note:

(i) $S \subseteq \mathbb{C}$ is called an open set if every point of S is an interior point of S.

(ii) A set S is called a closed set if its complement $\mathbb{C} \setminus S$ is open.

Definition 3.0.6 Let $S \subseteq \mathbb{C}$. S is called a **connected set** if every pair of points in S can be joined by a polygon, which lies in S.

Definition 3.0.7 A nonempty open connected subset of $\mathbb{C}$ is called a **region** or a **domain** in $\mathbb{C}$.

Definition 3.0.8 A set $S \subseteq \mathbb{C}$ is called a **bounded set** if there exists a real number K such that $|z| \le K$ for all $z \in S$.

Definition 3.0.9 Let x(t) and y(t) be real valued continuous functions of a real variable t, defined in the range $\alpha \le t \le \beta$. Then $\{z(t) = x(t) + iy(t) \,/\, \alpha \le t \le \beta\}$ is called a **continuous arc** or a **curve** in the Argand's plane. This is also referred to as a **path** in the complex plane. $z(\alpha)$ is called the **initial point** and $z(\beta)$ is called the **terminal point** of the curve. If $z(\alpha) = z(\beta)$, the curve is said to be a **closed curve.**

A point z in the complex plane is called a **multiple point** of the curve if $z = x(t) + iy(t)$ for more than one value of t in the given range $\alpha \le t \le \beta$.

Definition 3.0.10 A curve $z(t) = x(t) + iy(t)$, $\alpha \le t \le \beta$ is called a **Jordan arc** or a **simple curve** if $t_1 \ne t_2$ implies $z(t_1) \ne z(t_2)$. In other words, a Jordan curve is one without multiple points.

Definition 3.0.11 A simple curve $z(t) = x(t) + iy(t)$, $\alpha \le t \le \beta$ in which $z(\alpha) = z(\beta)$ is called a **simple closed Jordan curve** (or a simple closed curve).

The circle $z(t) = \cos t + i \sin t$; $0 \le t \le 2\pi$ is a simple closed Jordan curve since the values of z(t) coincide only at the end points $t = 0$ and $t = 2\pi$.

Theorem 3.0.1 (The Jordan Curve Theorem) A Simple closed Jordan curve divides the Argand's Plane into two regions, which have the curve as common boundary.

Note:

(i) Of the two regions determined by a simple closed Jordan curve, one is bounded and is called the **interior domain** and the other is unbounded, called the **exterior domain**.

Definition 3.0.12 A region S in the complex plane is said to be **simply connected** if every simple closed curve lying in S encloses only points of S. A region, which is not simply connected, is called a **multiply connected** region.

Note:

(i) In a simple connected region S, any simple closed curve, which lies in S, can be continuously shrunk to a point without leaving the region.

(ii) The complement $\mathbb{C} \setminus S$ of simple connected region S is connected where as the complement of a multiply connected region is not connected.

The annulus $S = \{ z / r_1 < |z - a| < r_2 \}$ is a multiply connected (Fig 3.2)

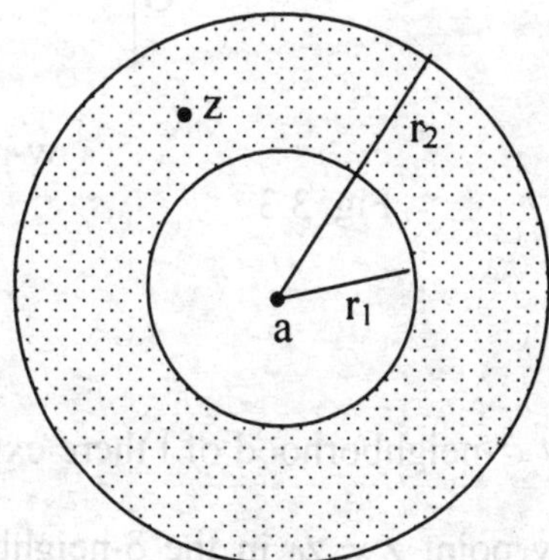

Fig. 3.2

3.0.4 FUNCTIONS OF A COMPLEX VARIABLE

Let $z = x + iy$ and $w = u + iv$ be complex variables. A complex valued function of a complex variable z, denoted by $w = f(z)$ associates with each value of z in a given region R, a unique value of w.
For example, $w = z^2$ is a function of the complex variable z.
Then $u + iv = (x + iy)^2 = x^2 - y^2 + i\,2xy$.
Thus $u(x, y) = x^2 - y^2$ and $v(x, y) = 2xy$. The functions u and v are real-valued functions of the real variables x and y.

If w = f(z) associates with each value of z, more than one value of w, then the relation w = f(z) is called a **multiple-valued** function.
For example, $w = z^2$ is a function. $w = z^{1/2}$ is a multiple-valued function.

Note:

(i) If w = f(z) = u(x, y) + i v(x,y), then u(x, y) and v(x, y) are called the real and imaginary part of the function f(z).

Definition 3.0.13 A function w = f(z) defined in the region R is said to have a limit l as z tends to $z_0 \in R$ if given $\in > 0$ there exists a $\delta > 0$ such that $0 < |z - z_0| < \delta \Rightarrow |f(z) - l| < \in$.

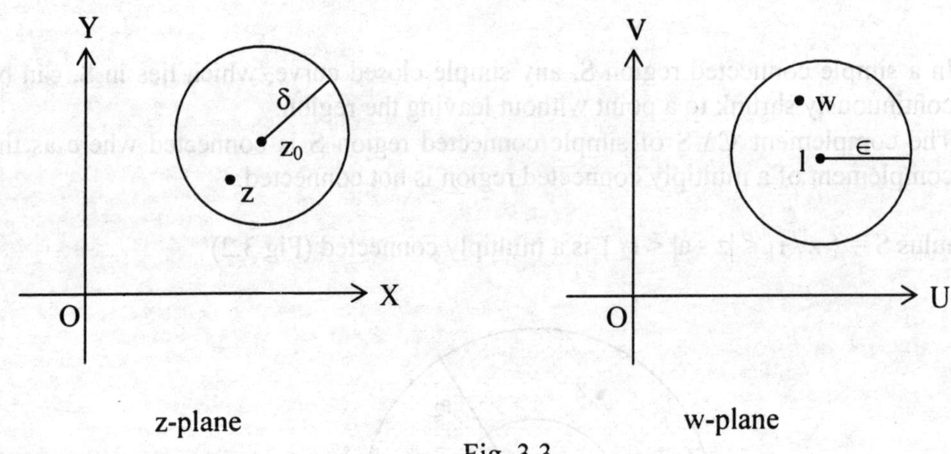

Fig. 3.3

We write $\lim_{z \to z_0} f(z) = l$

Note:

(i) $\lim_{z \to z_0} f(z) = l$ if given any $\in$-neighborhood of l there exists a δ-neighborhood of z_0 such that image of every point $z \neq z_0$ in the δ-neighborhood of z_0 is in the $\in$-neighborhood of l.

(ii) The condition $0 < |z - z_0| < \delta$ excludes the point $z = z_0$ and hence $\{z / 0 < |z - z_0| < \delta\}$ is called the **deleted neighborhood of z_0.**

(iii) To find $\lim_{z \to z_0} f(z)$, the function w = f(z) need not be defined at z_0.

For example, consider the function $f(z) = \dfrac{z^2 - 4}{z - 2} \quad z \neq 2$

$$\lim_{z \to 2} f(z) = \lim_{z \to 2} \frac{z^2 - 4}{z - 2} = \lim_{z \to 2} \frac{(z+2)(z-2)}{z-2}$$
$$= \lim_{z \to 2} (z+2) \text{ when } z \neq 2 = 4$$

Note that the function f(z) is defined for all z in the complex plane except at z = 2. But the limit as z tends to 2 exists.

Definition 3.0.14 Let w = f(z) be a function defined in a region R in the complex plane and $z_0 \in R$. Then f is said to be **continuous** at z_0 if given $\epsilon > 0$, there exists a $\delta > 0$ such that $|z - z_0| < \delta \Rightarrow |f(z) - f(z_0)| < \epsilon$.

Thus f is said to be continuous at the point z_0 if $\lim\limits_{z \to z_0} f(z) = f(z_0)$.

Also f is said to be continuous in the region R if it is continuous at each point of R.

Note:

(i) If f and g are continuous at z_0, then f + g, fg, |f| and $\bar{f}$ are continuous at z_0 and f/g is continuous at z_0 if $g(z_0) \neq 0$.

(ii) f is continuous at z_0 iff Re(f) and Im(f) are continuous at z_0.

(iii) Any polynomial $P(z) = a_0 + a_1 z + a_2 z^2 + \ldots + a_n z^n$ is continuous in the entire complex plane.

Definition 3.0.15 Let w = f(z) be defined in a region R and $z_0 \in R$. Then f is said to be **differentiable at z_0** if $\lim\limits_{z \to z_0} \dfrac{f(z) - f(z_0)}{z - z_0}$ exists.

This limit is called the **derivative** of f(z) at z_0 and is denoted by $f'(z_0)$ or $\dfrac{df}{dz}$ at z_0.

Also putting $z - z_0 = \Delta z$, we get, $f'(z_0) = \lim\limits_{\Delta z \to 0} \dfrac{f(z_0 + \Delta z) - f(z_0)}{\Delta z}$

f is said to be **differentiable in R** if it is differentiable at all points of R.

Note:

(i) If f(z) and g(z) are differentiable at z_0, then $(f + g)'(z_0) = f'(z_0) + g'(z_0)$

$(fg)'(z_0) = f'(z_0).g(z_0) + f(z_0).g'(z_0)$

$$\left(f/g\right)'(z_0) = \frac{f'(z_0).g(z_0) - f(z_0).g'(z_0)}{[g(z_0)]^2} \quad \text{if } g(z_0) \neq 0$$

(ii) Every differentiable function is continuous. The converse is not true.

Let f(z) be differentiable at z_0.

$$\text{Then } \lim_{z \to z_0} [f(z) - f(z_0)] = \lim_{z \to z_0} \frac{f(z) - f(z_0)}{(z - z_0)}.(z - z_0)$$

$$= \lim_{z \to z_0} \frac{f(z) - f(z_0)}{z - z_0}. \lim_{z \to z_0} (z - z_0) = f'(z_0).0 = 0.$$

$\lim_{\Delta z \to 0} f(z) = f(z_0)$

$\therefore$ f(z) is continuous at z_0.

Also note that $f(z) = \bar{z}$ is continuous at all points in the complex plane, but it is nowhere differentiable (prove this).

3.1 CAUCHY-RIEMANN EQUATIONS

3.1.1 ANALYTIC FUNCTIONS

Definition 3.1.1 Let the function w = f(z) be defined in a region R of the complex plane. The function f is said to be **analytic at a point** $z_0 \in R$ if its is differentiable at every point of some neighborhood of z_0.

If f is analytic at z_0 there exists $\in > 0$ such that f is differentiable at every point of the disc $s(z_0, \in) = \{ z / |z-z_0| < \in \}$. If f is analytic at every point of R, then f is said to be analytic in R.

Definition 3.1.2 A function w = f(z) that is analytic at every point of the complex plane is called an **entire function** or **integral function**. For example, any polynomial is an entire function.

Note:

(i) If f(z) is analytic at $a \in R$, there exists an $\in > 0$ such that f(z) is differentiable at each point of the $\in$-neighborhood $s(a, \in)$. If $z \in S(a, \in)$, we can find a $\delta > 0$ such that $S(z, \delta) \subseteq S(a, \in)$. Since f(z) is differentiable at each point of $S(z, \delta)$, it is analytic at z. Thus f(z) is analytic at every point of $S(a, \in)$. $\therefore$ f(z) is analytic at a iff it is analytic at each point of some neighborhood of a.

(ii) If f(z) is analytic at a, then it is differentiable at a. But the converse is not true. For example, $f(z) = |z|^2$ is differentiable at z = 0, but not analytic at z = 0 (Prove this!)

Theorem 3.1.1 (Necessary conditions for f(z) to be analytic)

If a function f(z) = u(x, y) + i v(x, y) is differentiable (or analytic) at any point z = x + iy, then the first order partial derivatives u_x; u_y, v_x, v_y exist and satisfy the **Cauchy-Riemann** partial differential equations $\mathbf{u_x = v_y}$ and $\mathbf{u_y = -v_x}$ at the point z.

Proof:

$$w = f(z) = u(x, y) + i\, v(x, y) \quad (1)$$

$$z = x + iy \text{ and } \Delta z = \Delta x + i\Delta y \quad (2)$$

since f(z) is differentiable at z, $f'(z) = \lim_{\Delta z \to 0} \dfrac{f(z + \Delta z) - f(z)}{\Delta z}$ (3)

must exist uniquely as $\Delta z \to 0$ along any path.

Using (1) and (2) in (3) we get

$$f'(z) = \lim_{\Delta z \to 0}\left[\frac{u(x+\Delta x, y+\Delta y) - u(x,y)}{\Delta x + i\,\Delta y} + i\frac{v(x+\Delta x, y+\Delta y) - v(x,y)}{\Delta x + i\,\Delta y}\right] \quad (4)$$

Let $\Delta z \to 0$ along the real axis so that $\Delta y = 0$Then (4) becomes,

$$f'(z) = \lim_{\Delta x \to 0}\left[\frac{u(x+\Delta x, y) - u(x,y)}{\Delta x} + i\frac{v(x+\Delta x, y) - v(x,y)}{\Delta x}\right]$$

i.e., $f'(z) = u_x + iv_x$ at z (5)

Since f′(z) exists, u_x and v_x also exist.

Again taking $\Delta z \to 0$ along the imaginary axis, we have $\Delta x = 0$. Then (4) becomes,

$$f'(z) = \lim_{\Delta y \to 0}\left[\frac{u(x, y+\Delta y) - u(x,y)}{i\Delta y} + i\frac{v(x, y+\Delta y) - v(x,y)}{i\Delta y}\right]$$

$$= \frac{1}{i}(u_y + iv_y) \text{ at z.}$$

i.e., $f'(z) = -i\,u_y + v_y$

$= v_y - i\,u_y$. (6)

Again f′(z) exists implies u_y and v_y also exist.

Equating the real and imaginary parts of (5) and (6) we get $u_x = v_y$ and $u_y = -v_x$.

Theorem 3.1.2 (Sufficient conditions for f(z) to be analytic)

The single valued continuous function f(z) is analytic in a region R if the four partial derivatives u_x, u_y, v_x, v_y exist, are continuous and satisfy the Cauchy-Riemann equations at each point of R.

Note:

(i). When f(z) is analytic at z, it is differentiable at z and hence the real and imaginary parts of f(z) = u(x, y) + i v(x, y) satisfy the Cauchy-Riemann equations, $u_x = v_y$ and $u_y = -v_x$.

(ii). If f(z) = u + iv is analytic, $f'(z) = u_x + iv_x$.

Hence $|f'(z)| = u_x^2 + v_x^2$

$= u_xv_y - v_xu_y$ (Since $u_x = v_y$, $u_y = -v_x$)

$$= \frac{\partial(u, v)}{\partial(x, y)}$$

(iii) If w = f(z) = u+iv is analytic, then $\frac{dw}{dz} = f'(z) = u_x + iv_x$

$$= \frac{\partial}{\partial x}(u + iv)$$

$$\text{i.e., } \frac{dw}{dz} = \frac{\partial w}{\partial x}$$

Example 3.1.1 Prove that $f(z) = \bar{z}$ is not differentiable at any point in the complex plane.

Solution:

Method 1.

$f(z) = \bar{z} = x - iy$

$\therefore u(x, y) = x$ and $v(x, y) = -y$

$u_x = 1, u_y = 0$ and $v_x = 0, v_y = -1$

$\therefore u_x \neq v_y$.

i.e., Cauchy-Riemann equations are not satisfied at any point.

$\therefore$ f(z) is not analytic and hence not differentiable.

Method 2.

$$f'(z) = \lim_{\Delta z \to 0} \frac{f(z+\Delta z) - f(z)}{\Delta z}$$

$$= \lim_{\Delta z \to 0} \frac{\overline{z+\Delta z} - \bar{z}}{\Delta z}$$

$$= \lim_{\Delta z \to 0} \frac{\bar{z}+\overline{\Delta z} - \bar{z}}{\Delta z}$$

$$= \lim_{\Delta z \to 0} \frac{\overline{\Delta z}}{\Delta z}$$

$$= \lim_{r \to 0} \frac{r(\cos\theta - i\sin\theta)}{r(\cos + i\sin\theta)} \qquad (\text{As } \Delta z \to 0 \text{ iff } r \to 0)$$

$$= \lim_{r \to 0}(\cos 2\theta - i\sin 2\theta) = \cos 2\theta - i\sin 2\theta,$$

which does not tend to unique limit as it dependents on θ.

Example 3.1.2 Prove that $f(z) = |z|^2$ is differentiable but not analytic at the origin.

Solution:

$$f'(z) = \lim_{\Delta z \to 0} \frac{f(z+\Delta z) - f(z)}{\Delta z}$$

$$= \lim_{\Delta z \to 0} \frac{|z+\Delta z|^2 - |z|^2}{\Delta z}$$

$$= \lim_{\Delta z \to 0} \frac{(z+\Delta z)(\bar{z}+\Delta\bar{z}) - z\bar{z}}{\Delta z}$$

$$= \lim_{\Delta z \to 0} \frac{z\Delta\bar{z} + \bar{z}.\Delta z - \Delta z.\Delta\bar{z}}{\Delta z}$$

$$= \lim_{\Delta z \to 0}\left(z.\frac{\Delta\bar{z}}{\Delta z} + \bar{z} + \Delta\bar{z}\right)$$

$$= \lim_{\Delta z \to 0}\left(z.\frac{\Delta \bar{z}}{\Delta z} + \bar{z} \right) \quad \text{Since } \Delta\bar{z} \to 0 \text{ as } \Delta z \to 0$$

$\therefore$ $f'(0) = 0$ (Since $z = 0$ implies $\bar{z} = 0$)

When $z \neq 0$, let $\Delta z = r(\cos\theta + i\sin\theta)$

Then $\Delta\bar{z} = r(\cos\theta - i\sin\theta)$

$$\frac{\Delta\bar{z}}{\Delta z} = \frac{\cos\theta - i\sin\theta}{\cos\theta + i\sin\theta} = \cos 2\theta - i\sin 2\theta$$

$\therefore$ $\frac{\Delta\bar{z}}{\Delta z}$ does not tend to a unique limit as $\Delta z \to 0$, since this limit depends on θ, $\arg(\Delta z)$.

$\therefore$ For $z \neq 0$ $f'(z)$ does not exist.

i.e., f(z) is differentiable only at the origin. We can not find a neighborhood of origin where f(z) is differentiable. Hence f(z) is not analytic at the origin.

Example 3.1.3 Show that the function $f(z) = \sqrt{|xy|}$ is not analytic at the origin although Cauchy-Riemann equation are satisfied at the point.

Solution:

Let $f(z) = u+iv = \sqrt{|xy|}$

Then $u(x, y) = \sqrt{|xy|}$ and $v(x, y) = 0$

At the origin, we have

$$u_x = \lim_{x \to 0} \frac{u(x,0) - u(0,0)}{x} = \lim_{x \to 0} \frac{0-0}{x} = 0$$

$$u_y = \lim_{y \to 0} \frac{u(0,y) - u(0,0)}{y} = \lim_{y \to 0} \frac{0-0}{y} = 0$$

$$v_x = \lim_{x \to 0} \frac{v(x,0) - v(0,0)}{x} = \lim_{x \to 0} \frac{0-0}{x} = 0$$

$$v_y = \lim_{y \to 0} \frac{v(0,y) - v(0,0)}{y} = \lim_{y \to 0} \frac{0-0}{y} = 0$$

$u_x = v_y$ and $u_y = -v_x$

Hence Cauchy-Riemann equations are satisfied at the origin.

$$f'(z) = \lim_{\Delta z \to 0} \frac{f(z+\Delta z) - f(z)}{\Delta z} = \lim_{z \to 0} \frac{f(z) - f(0)}{z}$$

$$= \lim_{z \to 0} \frac{\sqrt{|xy|}}{x+iy}$$

Let $z \to 0$ along the line $y = mx$, then we have

$$f'(0) = \lim_{z\to 0}\frac{\sqrt{|mx^2|}}{x+i\,mx} == \lim_{z\to 0}\frac{\sqrt{|m|}}{1+i\,m} = \frac{\sqrt{|m|}}{1+i\,m}$$

This limit is not unique as it depends on m.

$\therefore$ f'(0) does not exist.

Example 3.1.4 Show that the function f(z) = u + iv where $f(z) = \dfrac{x^3(1+i) - y^3(1-i)}{x^2+y^2}$, $z \neq 0$ and f(0) = 0 is continuous and Cauchy-Riemann equations are satisfied at the origin, yet f'(0) does not exist.

Solution:

$$f(z) = \frac{x^3 - y^3}{x^2+y^2} + i\frac{x^3+y^3}{x^2+y^2},\quad z \neq 0$$

$$\therefore u(x, y) = \frac{x^3 - y^3}{x^2+y^2},\quad v(x,y) = \frac{x^3+y^3}{x^2+y^2},\quad x, y \neq 0$$

Both u and v are rational and finite for $z \neq 0$. $\therefore$ u and v are continuous functions for $z \neq 0$.

$f(0) = 0 \Rightarrow u(0, 0) = 0, v(0, 0) = 0$

$$\therefore u(x, y) = \frac{x^3 - y^3}{x^2+y^2} = \frac{r^3\cos^3\theta - r^3\sin^3\theta}{r^2\cos^2\theta + r^2\sin^2\theta},\text{ substituting } x = r\cos\theta,\ y = r\sin\theta$$

$$= r(\cos^3\theta - \sin^3\theta) \to 0 \text{ as } r \to 0 \text{ irrespective of the value of } \theta.$$

i.e., $\lim_{\substack{x\to 0\\ y\to 0}} u(x, y) = 0 = u(0,0)$

$\therefore$ u(x, y) is continuous at the origin.

Similarly $v(x, y) = r(\cos^3\theta + \sin^3\theta) \to 0$ as $r \to 0$ irrespective of the value of θ.

$\therefore$ v(x, y) is continuous at the origin.

Hence u and v are continuous at all points, which implies f(z) is continuous.

At the origin,

$$u_x = \lim_{x\to 0}\frac{u(x,0) - u(0,0)}{x} = \lim_{x\to 0}\frac{x-0}{x} = 1$$

$$u_y = \lim_{y\to 0}\frac{u(0,y) - u(0,0)}{y} = \lim_{y\to 0}\frac{-y-0}{y} = -1$$

$$v_x = \lim_{x\to 0}\frac{v(x,0) - v(0,0)}{x} = \lim_{x\to 0}\frac{x-0}{x} = 1$$

$$v_y = \lim_{y\to 0}\frac{v(0,y) - v(0,0)}{y} = \lim_{y\to 0}\frac{y-0}{y} = 1$$

Since $u_x = v_y$ and $u_y = -v_x$, u and v satisfy the Cauchy-Riemann equations at the origin.

$$\text{Now } f'(0) = \lim_{z\to 0} \frac{f(z) - f(0)}{z}$$

$$= \lim_{z\to 0} \frac{(x^3 - y^3) + i(x^3 + y^3)}{(x^2 + y^2)(x + iy)}$$

Let $z \to 0$ along $y = x$, we get

$$f'(0) = \lim_{x\to 0} \frac{i2x^3}{2x^2(x + i\,x)} = \frac{i}{1+i}$$

Again taking $z \to 0$ along x –axis (i.e., y = 0) we get,

$$f'(0) = \lim_{x\to 0} \frac{x^3 + ix^3}{x^2 . x} = 1 + i$$

i.e., f'(0) has different values along different curves. ∴ f'(0) is not unique and hence f'(0) does not exist.

Example 3.1.5 Examine the nature of the function $f(z) = \dfrac{x^2y^5(x+iy)}{x^4 + y^{10}}$, $z \neq 0$, $f(0) = 0$ in a region including the origin.

Solution:

$$f(z) = u + iv = \frac{x^2y^5(x+iy)}{x^4 + y^{10}}$$

$$\therefore\ u(x, y) = \frac{x^3y^5}{x^4 + y^{10}}, \quad v(x, y) = \frac{x^2y^6}{x^4 + y^{10}}$$

At the origin,

$$u_x = \lim_{x\to 0} \frac{u(x,0) - u(0,0)}{x} = \lim_{x\to 0} \frac{0-0}{x} = 0$$

$$u_y = \lim_{y\to 0} \frac{u(0,y) - u(0,0)}{y} = \lim_{y\to 0} \frac{0-0}{y} = 0$$

Similarly $v_x = 0$, $v_y = 0$

$\therefore\ u_x = v_y$, $u_y = -v_x$

Hence Cauchy-Riemann equations are satisfied at the origin.

$$f'(0) = \lim_{z\to 0} \frac{f(z) - f(0)}{z}$$

$$= \lim_{z\to 0} \frac{x^2y^5(x+iy)}{(x^4 + y^{10})(x+iy)} \quad \text{since } f(0) = 0$$

$$= \lim_{z\to 0} \frac{x^2y^5}{x^4 + y^{10}}$$

Taking $z \to 0$ along the radius vector $y = mx$,

$$f'(0) = \lim_{x\to 0} \frac{x^2 m^5 x^5}{x^4 + m^{10} x^{10}}$$

$$= \lim_{x\to 0} \frac{m^5 x^3}{1 + m^{10} x^6} = 0$$

Taking $z \to 0$ along the curve $y^5 = x^2$,

$$f'(0) = \lim_{x\to 0} \frac{x^2 . x^2}{x^4 + x^4} = \frac{1}{2}$$

$\therefore f'(0)$ is not unique and hence it does not exist.

i.e., f(z) is not analytic at the origin, although Cauchy-Riemann equations are satisfied.

Example 3.1.6 Show that the function $f(z) = e^{-z^{-4}}$, $z \neq 0$ and $f(0) = 0$ is not analytic at $z = 0$ although the Cauchy-Riemann equations are satisfied at that point.

Solution:

Let $f(z) = u(x, y) + i\, v(x, y)$

$$\text{Now } f(z) = e^{-z^{-4}} = e^{-\frac{1}{(x+iy)^4}} = e^{-\frac{(x-iy)^4}{(x^2+y^2)^4}}$$

$$= e^{-(x^4+y^4-6x^2y^2-4ix^3y+4ixy^3)/r^8} \qquad (\text{Since } r^2 = x^2+y^2)$$

$$= e^{-(x^4+y^4-6x^2y^2)/r^8}\, e^{i.4xy(x^2-y^2)/r^8}$$

$$\therefore \text{ For } z \neq 0 \quad u(x, y) = e^{-(x^4+y^4-6x^2y^2)/r^8} .\cos\left(\frac{4xy(x^2 - y^2)}{r^8}\right)$$

$$\text{and } v(x, y) = e^{-(x^4+y^4-6x^2y^2)/r^8} .\sin\left(\frac{4xy(x^2 - y^2)}{r^8}\right)$$

Since $f(0) = 0$, $u(0, 0) = 0$ and $v(0, 0) = 0$.

At the origin, we have

$$u_x = \lim_{x\to 0} \frac{u(x,0) - u(0,0)}{x} = \lim_{x\to 0} \frac{e^{-x^{-4}} - 0}{x}$$

$$= \lim_{x\to 0} \frac{1}{x} . \frac{1}{e^{1/x^4}}$$

$$= \lim_{x\to 0} \frac{1}{x\left[1 + \frac{1}{x^4} + \frac{1}{2x^8} + ...\right]}$$

$$= \lim_{x\to 0} \frac{1}{x + \frac{1}{x^3} + \frac{1}{2x^7} + ...}$$

$$= \frac{1}{\infty} = 0$$

$$u_y = \lim_{y \to 0} \frac{u(0,y) - u(0,0)}{y} = \lim_{y \to 0} \frac{e^{-y^{-4}} - 0}{y} = 0$$

$$v_x = \lim_{x \to 0} \frac{v(x,0) - v(0,0)}{x} = \lim_{x \to 0} \frac{0-0}{x} = 0$$

$$v_y = \lim_{y \to 0} \frac{v(0,y) - v(0,0)}{y} = \lim_{y \to 0} \frac{0-0}{y} = 0$$

$u_x = v_y$ and $u_y = -v_x$

$\therefore$ Cauchy-Riemann equations are satisfied at the origin.

$$f'(0) = \lim_{z \to 0} \frac{f(z) - f(0)}{z} = \lim_{z \to 0} \frac{e^{-z^{-4}}}{z}$$

$$= \lim_{r \to 0} \frac{e^{-r^{-4}} . e^{-i\pi}}{re^{i\pi/4}} \quad (\text{taking } z \to 0 \text{ along } z = re^{i\pi/4})$$

$$= \lim_{r \to 0} \frac{e^{r^{-4}}}{re^{i\pi/4}} \quad \text{since } e^{-i\pi} = -1$$

$$= \lim_{r \to 0} \frac{1}{e^{i\pi/4}} . \frac{1}{r}\left[1 + \frac{1}{r^4} + \frac{1}{2r^8} + \dots\right]$$

$$= \lim_{r \to 0} \frac{1}{e^{i\pi/4}} . \left[\frac{1}{r} + \frac{1}{r^5} + \frac{1}{2r^9} + \dots\right]$$

$$= \infty$$

$\therefore$ f'(0) does not exist.

i.e., f(z) is not analytic at z = 0.

Example 3.1.7 Show that the function $e^x(\cos y + i \sin y)$ is analytic and find its derivative.

Solution:

Let $f(x) = u + iv = e^x(\cos y + i \sin y)$

Then $u = e^x \cos y$ and $v = e^x \sin y$.

$u_x = e^x \cos y$ and $u_y = -e^x \sin y$.

$v_x = e^x \sin y$ and $v_y = e^x \cos y$.

$u_x = v_y$ and $u_y = -v_x$.

$\therefore$ Cauchy-Riemann equations are satisfied by f(z).

Further u_x, u_y, v_x, v_y are continuous everywhere

$\therefore f(z) = e^x (\cos y + i \sin y)$ is analytic in the entire complex plane.

$f'(z) = u_x + i\, v_x$

$\quad = e^x \cos y + i\, e^x \sin y$

$= e^x (\cos y + i \sin y)$
$= e^x . e^{iy}$
$= e^{x+iy}$
$= e^z.$

Example 3.1.8 Find the constants a and b so that the function $f(z) = a(x^2 - y^2) + i\, bxy + c$ is analytic at all points.

Solution:
Let $f(z) = u + iv$
Then $u = a(x^2-y^2) + c$, $v = bxy$
$u_x = 2ax$, $u_y = -2ay$
$v_x = by$, $v_y = bx$
For f(z) to be analytic, $u_x = v_y$ and $u_y = - v_x$
i.e., $2ax = bx$ and $-2ay = -by$ for all x and y.
i.e., $2a = b$.
$\therefore$ f(z) is analytic for all values of a and b with $2a = b$.

3.1.2 ALTERNATE FORMS OF CAUCHY-RIEMANN EQUATIONS

The Cauchy-Riemann equations can be expressed in complex and polar forms.

Theorem 3.1.3 (Complex form of C-R equations)
Let $f(z) = u(x, y) + i\, v(x, y)$ be analytic. Then f(z) satisfies the equation $f_x = -i\, f_y$.

Proof:
$f(z) = u(x, y) + i\, v(x, y)$
Since f(z) is analytic, u and v satisfy the C-R equations.
$u_x = v_y,\ u_y = -v_x$ (1)
Now $f_x = u_x + i\, v_x$
and $f_y = u_y + i\, v_y$
$\therefore f_x = u_x + i\, v_x$
$= v_y + i\,(-u_y)$ (Using (1))
$= -i\,[u_y + i\, v_y]$
$= -i\, f_y$

Theorem 3.1.4 (C-R equations in polar coordinates)
Let $f(z) = u(r, \theta) + i\, v(r, \theta)$ be analytic at $z = re^{i\theta} \neq 0$. Then $\frac{\partial u}{\partial r} = \frac{1}{r}.\frac{\partial v}{\partial \theta}$ and $\frac{\partial v}{\partial r} = -\frac{1}{r}.\frac{\partial u}{\partial \theta}$

Also $f'(z) = \frac{r}{z}\left(\frac{\partial u}{\partial r} + i\frac{\partial v}{\partial r}\right)$

Proof:
We have $u_x = v_y$, $u_y = -v_x$ and $f'(z) = u_x + i\, v_x$ (1)

Put x = r cos θ and y = r sin θ

Then $\frac{\partial u}{\partial r} = \frac{\partial u}{\partial x}.\frac{\partial x}{\partial r} + \frac{\partial u}{\partial y}.\frac{\partial y}{\partial r} = \frac{\partial u}{\partial x}.\cos\theta + \frac{\partial u}{\partial y}.\sin\theta$ (2)

Also $\frac{\partial v}{\partial \theta} = \frac{\partial v}{\partial x}.\frac{\partial x}{\partial \theta} + \frac{\partial v}{\partial y}.\frac{\partial y}{\partial \theta} = \frac{\partial v}{\partial x}.(-r\sin\theta) + \frac{\partial v}{\partial y}.(r\cos\theta)$

$$\therefore \frac{1}{r}.\frac{\partial v}{\partial \theta} = -\frac{\partial v}{\partial x}.\sin\theta + \frac{\partial v}{\partial y}.\cos\theta$$

$$= \frac{\partial u}{\partial y}.\sin\theta + \frac{\partial u}{\partial x}.\cos\theta \quad \text{(Using (1))}$$

$$= \frac{\partial u}{\partial r} \quad \text{(Using (2))}$$

$$\frac{\partial u}{\partial r} = \frac{1}{r}.\frac{\partial v}{\partial \theta}$$

Now, $\frac{\partial v}{\partial r} = \frac{\partial v}{\partial x}.\frac{\partial x}{\partial r} + \frac{\partial v}{\partial y}.\frac{\partial y}{\partial r}$

$$= \frac{\partial v}{\partial x}.\cos\theta + \frac{\partial v}{\partial y}.\sin\theta \quad (3)$$

$$\frac{\partial u}{\partial \theta} = \frac{\partial u}{\partial x}.\frac{\partial x}{\partial \theta} + \frac{\partial u}{\partial y}.\frac{\partial y}{\partial \theta}$$

$$= \frac{\partial u}{\partial x}.(-r\sin\theta) + \frac{\partial u}{\partial y}.(r\cos\theta)$$

$$\therefore \frac{1}{r}.\frac{\partial u}{\partial \theta} = -\frac{\partial u}{\partial x}.\sin\theta + \frac{\partial u}{\partial y}.\cos\theta$$

$$= -\frac{\partial v}{\partial y}.\sin\theta - \frac{\partial v}{\partial x}.\cos\theta \quad \text{(Using (1))}$$

$$= -\frac{\partial v}{\partial r} \quad \text{(Using (3))}$$

$$\frac{\partial v}{\partial r} = -\frac{1}{r}.\frac{\partial u}{\partial \theta}$$

$$r\left(\frac{\partial u}{\partial r} + i\frac{\partial v}{\partial r}\right) = r\left[\left(\frac{\partial u}{\partial x}.\cos\theta + \frac{\partial u}{\partial y}.\sin\theta\right) + i\left(\frac{\partial v}{\partial x}.\cos\theta + \frac{\partial v}{\partial y}.\sin\theta\right)\right], \text{ (Using (2) and (3))}$$

$$= r\cos\theta\left(\frac{\partial u}{\partial x}. + i\frac{\partial v}{\partial x}\right) + r\sin\theta\left(\frac{\partial u}{\partial y} + i\frac{\partial v}{\partial y}\right)$$

$= x(u_x + iv_x) + i\,y(v_y - i\,u_y)$

$= x(u_x + iv_x) + i\,y(u_x + iv_x)$ (Using (1))

$= (x + iy)\,(u_x + iv_x)$

$$= z.f'(z) \qquad \text{(Using (1))}$$

$$\therefore f'(z) = \frac{r}{z}\left(\frac{\partial u}{\partial r} + i\frac{\partial v}{\partial r}\right)$$

Note:

$$\text{If } w = f(z),\ f'(z) = \frac{r}{z}\left(\frac{\partial u}{\partial r} + i\frac{\partial v}{\partial r}\right) = \frac{r}{re^{i\theta}}.\frac{\partial w}{\partial r} = e^{-i\theta}.\frac{\partial w}{\partial r}$$

Example 3.1.9 If f(z) is analytic, then $\frac{\partial f}{\partial \bar{z}} = 0$.

Solution:

$$f(z) = u(x, y) + i\,v(x, y) \qquad (1)$$

Now $x = \frac{z+\bar{z}}{2}$, $y = \frac{z-\bar{z}}{2i}$

$$\therefore f(z) = u\left(\frac{z+\bar{z}}{2}, \frac{z-\bar{z}}{2i}\right) + iv\left(\frac{z+\bar{z}}{2}, \frac{z-\bar{z}}{2i}\right)$$

$$\therefore \frac{\partial f}{\partial \bar{z}} = \frac{\partial f}{\partial x}.\frac{\partial x}{\partial \bar{z}} + \frac{\partial f}{\partial y}.\frac{\partial y}{\partial \bar{z}}$$

$$= \frac{\partial f}{\partial x}.\left(\frac{1}{2}\right) + \frac{\partial f}{\partial y}.\left(-\frac{1}{2i}\right)$$

$$\text{i.e., } \frac{\partial f}{\partial \bar{z}} = \frac{1}{2}\left(\frac{\partial f}{\partial x} + i\frac{\partial f}{\partial y}\right) \qquad (2)$$

Since f(z) is analytic, $f_x = -i\,f_y$.

$$\therefore \frac{\partial f}{\partial x} + i\frac{\partial f}{\partial y} = 0$$

Hence (2) becomes, $\frac{\partial f}{\partial \bar{z}} = 0$

Note:

$\frac{\partial f}{\partial \bar{z}} = 0$ is only another form of Cauchy-Riemann equations (in complex form).

Example 3.1.10 Show that $f(z) = \sqrt{r}\left(\cos \theta/2 + i \sin \theta/2\right)$ where $r > 0$ and $0 < \theta < 2\pi$ is analytic and find f'(z).

Solution:

Let f(z) = u(r, θ) + i v(r, θ).

Then $u(r, \theta) = \sqrt{r}.\cos \theta/2$ and $v(r, \theta) = \sqrt{r}.\sin \theta/2$

$$\therefore \frac{\partial u}{\partial r} = \frac{1}{2\sqrt{r}}.\cos\theta/2 \text{ and } \frac{\partial v}{\partial r} = \frac{1}{2\sqrt{r}}.\sin\theta/2$$

$$\frac{\partial u}{\partial \theta} = -\frac{\sqrt{r}}{2}.\sin \theta/2 \quad \text{and} \quad \frac{\partial v}{\partial \theta} = \frac{\sqrt{r}}{2}.\cos \theta/2$$

$$\frac{1}{r}.\frac{\partial v}{\partial \theta} = \frac{1}{r}.\frac{\sqrt{r}}{2}\cos \theta/2 = \frac{1}{2\sqrt{r}}\cos \theta/2 = \frac{\partial u}{\partial r}$$

i.e., $\dfrac{\partial u}{\partial r} = \dfrac{1}{r}.\dfrac{\partial v}{\partial \theta}$

$$-\frac{1}{r}.\frac{\partial u}{\partial \theta} = -\frac{1}{r}.\left(\frac{-\sqrt{r}}{2}\right)\sin \theta/2 = \frac{1}{2\sqrt{r}}\sin \theta/2 = \frac{\partial v}{\partial r}$$

i.e., $\dfrac{\partial v}{\partial r} = -\dfrac{1}{r}.\dfrac{\partial u}{\partial \theta}$

Hence Cauchy-Riemann equations in polar form are satisfied. Also all the first order partial derivatives are continuous for $r > 0$ and $0 < \theta < 2\pi$

$\therefore f(z)$ is analytic.

$$f'(z) = \frac{r}{z}\left(\frac{\partial u}{\partial r} + i\frac{\partial v}{\partial r}\right)$$

$$f'(z) = \frac{r}{z}\left(\frac{1}{2\sqrt{r}}\cos \theta/2 + i\frac{1}{2\sqrt{r}}\sin \theta/2\right)$$

$$f'(z) = \frac{1}{2z}.\sqrt{r}\left(\cos \theta/2 + i\sin \theta/2\right) = \frac{1}{2z}\sqrt{z} = \frac{1}{2\sqrt{z}}$$

Example 3.1.11 For what values of z, the function w defined by $z = \log \rho + i\,\phi$ where $w = \rho(\cos \phi + i \sin \phi)$ ceases to be analytic?

Solution:

$w = \rho\, e^{i\varphi} \Rightarrow \log w = \log \rho + i\,\phi = z$

$\therefore w = e^z$

$\therefore \dfrac{dw}{dz} = e^z$

Now, w is analytic at all points where $\dfrac{dw}{dz}$ is finite. i.e., at all points where z is finite

Hence, in any finite region in the complex plane, w is an analytic function.

Example 3.1.12 For what values of z the function w defined by $z = e^{-v}(\cos u + i \sin u)$ where $w = u + i\,v$ ceases to be analytic?

Solution:

We have $\dfrac{dw}{dz} = u_x + i\,v_x$ (1)

Method 1.

Now $z = x + iy = e^{-v}(\cos u + i \sin u)$

$\therefore\ x = e^{-v}\cos u$ and $y = e^{-v} \sin u$

Hence $u = \tan^{-1}(y/x)$ and $x^2 + y^2 = e^{-2v}$ or $v = -\frac{1}{2}\log(x^2 + y^2)$

i.e., $\therefore\ u_x = \dfrac{1}{1+(y/x)^2}\left(-\,y/x^2\right) = \dfrac{-y}{x^2 + y^2}$

$$v_x = -\frac{1}{2}.\frac{1}{x^2 + y^2}.2x = \frac{-x}{x^2 + y^2}$$

From (1) we get,

$$\frac{dw}{dz} = u_x + i\,v_x$$

$$= \frac{-y}{x^2 + y^2} - i\frac{x}{x^2 + y^2}$$

$$= \frac{-i(x - iy)}{x^2 + y^2}$$

$$= \frac{-i\bar{z}}{z\bar{z}} = \frac{1}{iz}$$

When $z = 0$, $\dfrac{dw}{dz}$ is not finite. Hence w ceases to be analytic at $z = 0$.

Method 2.

$z = e^{-v} e^{iu} = e^{i(u + iv)} = e^{iw}$

$\therefore\ w = \dfrac{1}{i}\log z$ i.e., $\dfrac{dw}{dz} = \dfrac{1}{iz}$

$\therefore$ w ceases to be analytic at $z = 0$.

Example 3.1.13 If n is real, show that $r^n(\cos n\theta + i \sin n\theta)$ is analytic except possibly when $r = 0$, and that its derivative is $n\, r^{n-1}[\cos (n-1)\theta + i \sin (n-1)\theta]$

Solution:

Let $w = f(z) = u + iv = r^n(\cos n\theta + i \sin n\theta)$

Then $u = r^n\cos n\theta$ and $v = r^n \sin n\theta$

$$\therefore\ \frac{\partial u}{\partial r} = n.r^{n-1}\cos n\theta \text{ and } \frac{\partial v}{\partial r} = n.r^{n-1}\sin n\theta$$

$$\frac{\partial u}{\partial \theta} = -\, n.r^{n}\sin n\theta \text{ and } \frac{\partial v}{\partial \theta} = n.r^{n}\cos n\theta$$

$$\therefore\ \frac{\partial u}{\partial r} = n.r^{n-1}\cos n\theta = \frac{1}{r}.n.r^n\cos n\theta = \frac{1}{r}.\frac{\partial v}{\partial \theta}$$

$$\frac{\partial v}{\partial r} = n.r^{n-1}\sin n\theta = \frac{1}{r}.n.r^n \sin n\theta = -\frac{1}{r}.\frac{\partial u}{\partial \theta}$$

i.e., Cauchy-Riemann equations are satisfied.

Further the four partial derivatives $\frac{\partial u}{\partial r}, \frac{\partial u}{\partial \theta}, \frac{\partial v}{\partial r}, \frac{\partial v}{\partial \theta}$ are continuous

$\therefore$ f(z) is analytic.

Now $\frac{dw}{dz} = e^{-i\theta}\frac{\partial w}{\partial r}$

$$= (\cos\theta - i\sin\theta.)\,n.r^{n-1}(\cos n\theta + i\sin n\theta)$$

$$= n.r^{n-1}[\cos(n-1)\theta + i\sin(n-1)\theta]$$

$\frac{dw}{dz}$ exists for all finite values of r except when r = 0

Example 3.1.14 If f(z) and $\overline{f(z)}$ are analytic in a region R, show that f(z) is constant in that region.

Solution:

Let f(z) = u(x, y) + i v(x, y).

Then $\overline{f(z)}$ = u(x, y) - i v(x, y)

Since f(z) is analytic in R, u and v satisfy the Cauchy-Riemann equations.

$\therefore u_x = v_y$ and $u_y = -v_x$ (1)

Similarly, since $\overline{f(z)}$ is analytic in R, u and -v satisfy the Cauchy-Riemann equations.

$\therefore u_x = -v_y$ and $u_y = v_x$ (2)

Adding (1) and (2) we get,

$u_x = 0$ and $v_x = 0$.

Now $f'(z) = u_x + i\,v_x = 0$ in R

$\therefore$ f(z) is analytic in R.

EXERCISE 3.1

PART – A

1. Define 'analytic function'
2. Define 'entire function'
3. Give an example of a function that is continuous, but not differentiable.
4. Give an example of a function that is differentiable at a point, but not analytic at that point.
5. Give an example of a function that is not differentiable at any point in the complex plane.
6. Give an example of a function that is differentiable at all points in the complex plane.

7. Give an example of an entire function.
8. State the Cauchy-Riemann equations.
9. State the necessary conditions for f(z) = u + iv to be analytic.
10. State the sufficient conditions for f(z) = u + iv to be analytic.
11. What is the complex form of Cauchy-Riemann equations?
12. State Cauchy-Riemann equations in polar form.
13. If f(z) is analytic prove $\frac{\partial f}{\partial \bar{z}} = 0$.
14. If w = f(z) is analytic, prove $\frac{dw}{dz} = \frac{\partial w}{\partial x} = -i\frac{\partial w}{\partial y}$
15. If f(z) = u + iv is analytic, what is f'(z) ?
16. Show that f(z) = xy + iy is everywhere continuous but not analytic.
17. Show that $w = |z|^2$ is continuous everywhere but differentiable nowhere, except at the origin.
18. If w = u(r, θ) + i v(r, θ) is an analytic function of z, prove that
$$e^{i\theta}\frac{dw}{dz} = \frac{\partial w}{\partial r} = \frac{-i}{r}\frac{\partial w}{\partial \theta}$$
19. Prove that $f(z) = \bar{z}$ is not analytic.
20. Prove that $f(z) = e^x(\cos y - i \sin y)$ is not analytic.
21. If u + iv is analytic, show that v + iu is not analytic.
22. If u + iv is analytic, show that v - iu is also analytic.
23. Find the points where C-R equations are satisfied by $f(z) = x^2 + iy^2$.
24. Find the points where C-R equations are satisfied by $f(z) = \frac{z-1}{z+1}$
25. Find constants a, b and c so that f(z) = x + ay – i (bx +cy) is analytic at all points.

PART - B

26. If $f(z) = \frac{x^3 y(y - ix)}{x^6 + y^2}$, $z \neq 0$ and f(0) = 0, show that $\frac{f(z) - f(0)}{z} \to 0$ as $z \to 0$ along any radius vector but not as $z \to 0$ in any manner.
27. If $f(z) = \frac{xy^2(x + iy)}{x^2 + y^4}$, $z \neq 0$ and f(0) = 0, show that $\frac{f(z) - f(0)}{z} \to 0$ as $z \to 0$ along any radius vector but not as $z \to 0$ in any manner.
28. Show that the function f(z) defined by $f(z) = \frac{xy(y - ix)}{x^2 + y^2}$, $z \neq 0$ and f(0) = 0 is not analytic at the origin, though it satisfies Cauchy-Riemann equations there.
29. Show that the function f(z) defined by $f(z) = \frac{x^2 y^3 (x - iy)}{x^6 + y^{10}}$, $z \neq 0$ and f(0) = 0 is not analytic at the origin, though it satisfies Cauchy-Riemann equations there.

30. Find the values of a, b, c, d so that the function
$f(z) = (x^2 + axy + by^2) + i\,(cx^2 + dxy + y^2)$ may be analytic.

31. Find the values of p for which the function $f(z) = \frac{1}{2}\log(x^2 + y^2) + i\tan^{-1}\left(\frac{px}{y}\right)$ is analytic.

32. Find the values of a and b for which
$f(z) = \cos x(\cos hy + a \sin hy) + i \sin x\,(\cos hy + b \sin hy)$ is analytic.

33. Prove that the functions f(z) and $\overline{f(\overline{z})}$ are simultaneously analytic.

34. Determine the constants a and b so that the function
$f(z) = (x^2 + ay^2 - 2xy) + i(bx^2 - y^2 + 2xy)$ is analytic. Find $f'(z)$.

35. Test whether $f(z) = e^{-x}(\cos y - i \sin y)$ is analytic.

36. For what values of z do the function w = u + iv defined by
$z = \sin u \cos hv + i \cos u \sin hv$ ceases to be analytic?

3.2 PROPERTIES OF ANALYTIC FUNCTIONS

3.2.1 HARMONIC FUNCTIONS

Definition 3.2.1 A real function $\phi(x, y)$ of two real variables x and y is said to satisfy the **Laplace equation** if $\frac{\partial^2\phi}{\partial x^2} + \frac{\partial^2\phi}{\partial y^2} = 0$

Note:

(i) The operator $\frac{\partial^2}{\partial x^2} + \frac{\partial^2}{\partial y^2}$ is called the **Laplace operator** and is denoted by ∇^2.
Then $\phi(x, y)$ is said to satisfy the Laplace equation if $\nabla^2\phi = 0$.

(ii) The Laplace equation in polar coordinates is

$$\frac{\partial^2\phi}{\partial r^2} + \frac{1}{r}\frac{\partial\phi}{\partial r} + \frac{1}{r^2}\frac{\partial^2\phi}{\partial\theta^2} = 0.$$

Definition 3.2.2 A real function of two variables x and y possessing continuous second order partial derivatives and satisfying the Laplace equation is called a **harmonic function.**

Definition 3.2.3 If u and v are harmonic functions such that f(z) = u + iv is analytic, then v is called the **conjugate harmonic function** of u or conjugate of u. Also u and v are called conjugate harmonic functions.

Theorem 3.2.1 The real and imaginary parts of an analytic function f(z) = u + iv are harmonic functions.

Proof:
Let f(z) = u + iv be analytic in some region in the complex plane. Then u and v satisfy the Cauchy-Riemann equations.

i.e., $\frac{\partial u}{\partial x} = \frac{\partial v}{\partial y}$ (1)

and $\frac{\partial u}{\partial y} = -\frac{\partial v}{\partial x}$ (2)

Differentiating (1) partially w.r.t x, we get

$$\frac{\partial^2 u}{\partial x^2} = \frac{\partial^2 v}{\partial x \partial y} \quad (3)$$

Differentiating (2) partially w.r.t y, we get

$$\frac{\partial^2 u}{\partial y^2} = -\frac{\partial^2 v}{\partial y \partial x} \quad (4)$$

Since u and v are the real and imaginary parts of an analytic function, partial derivatives of u and v of all orders exist and are continuous functions of x and y.

$$\therefore \frac{\partial^2 v}{\partial x \partial y} = \frac{\partial^2 v}{\partial y \partial x}$$

Hence from (3) and (4) we get,

$$\frac{\partial^2 u}{\partial x^2} + \frac{\partial^2 u}{\partial y^2} = 0$$

i.e., u satisfies the Laplace equation.
Further u possesses continuous second order partial derivatives.
∴ u is a harmonic function.
Similarly we can prove that v satisfies the Laplace equation and that v is a harmonic function.

Note:
(i) When f(z) = u + iv is analytic, u and v are harmonic function. Hence the real and imaginary parts of an analytic function are conjugate harmonic function.
(ii) When u and v are any two harmonic functions f(z) = u+iv need not be an analytic function.

Theorem 3.2.2 Every analytic function w = u(x, y) + i v(x, y) can be expressed as a function of z alone. i.e., an analytic function is independent of $\bar{z}$.

Proof:
w = u(x, y) + iv(x, y) (1)

$z = x + iy$ and $\bar{z} = x - iy$

$\therefore x = \frac{z+\bar{z}}{2}$ and $y = \frac{z-\bar{z}}{2i}$

Hence u, v and w can be expressed as functions of z and $\bar{z}$.

$$\text{Now } \frac{\partial w}{\partial \bar{z}} = \frac{\partial u}{\partial \bar{z}} + i\frac{\partial v}{\partial \bar{z}}$$

$$= \left(\frac{\partial u}{\partial x}.\frac{\partial x}{\partial \bar{z}} + \frac{\partial u}{\partial y}.\frac{\partial y}{\partial \bar{z}}\right) + i\left(\frac{\partial v}{\partial x}.\frac{\partial x}{\partial \bar{z}} + \frac{\partial v}{\partial y}.\frac{\partial y}{\partial \bar{z}}\right)$$

$$= \left[u_x.\frac{1}{2} + u_y.\left(-\frac{1}{2i}\right)\right] + i\left[v_x.\frac{1}{2} + v_y.\left(-\frac{1}{2i}\right)\right]$$

$$= \frac{1}{2}\left(u_x + iu_y\right) + i\frac{1}{2}\left(v_x + iv_y\right)$$

$$= \frac{1}{2}\left(u_x - v_y\right) + \frac{i}{2}\left(u_y + v_x\right)$$

$= 0$ (Since $u_x = v_y$ and $u_y = -v_x$)

$\therefore$ w is independent of $\bar{z}$.

i.e., w is a function of z alone.

3.2.2 ORTHOGONAL SYSTEMS

The equation $u(x, y) = c$ where c is an arbitrary constant (parameter) represent a family of curves

Definition 3.2.4 Two families of curves are said to form an orthogonal system if they intersect at right angles at each of their points of intersection. Two such families of curves are said to be **orthogonal trajectories** of each other.

Theorem 3.2.3 If $f(z) = u(x, y) + iv(x, y)$ is an analytic function, then the families of curves $u(x, y) = c_1$ and $v(x, y) = c_2$ form an **orthogonal system**, where c_1 and c_2 are parameters.

Proof:

Since $f(z) = u + iv$ is analytic, u and v satisfy the Cauchy-Riemann equations.

i.e., $\frac{\partial u}{\partial x} = \frac{\partial v}{\partial y}$ and $\frac{\partial u}{\partial y} = -\frac{\partial v}{\partial x}$ (1)

Now $u(x, y) = c_1 \Rightarrow$ $du = \frac{\partial u}{\partial x}dx + \frac{\partial u}{\partial y}dy = 0$

i.e., $\frac{dy}{dx} = -\frac{\partial u/\partial x}{\partial u/\partial y} = m_1$ (say)

m_1 is the slope of the curve $u(x, y) = c_1$ at (x, y). (2)

Similarly $v(x, y) = c_2 \Rightarrow dv = \frac{\partial v}{\partial x}dx + \frac{\partial v}{\partial y}dy = 0$

$$\text{i.e., } \frac{dy}{dx} = -\frac{\partial v/\partial x}{\partial v/\partial y} = m_2 \text{ (say)} \quad (3)$$

m_2 is the slope of the curve $v(x, y) = c_2$ at (x, y). From (2) and (3) we get,

$$m_1.m_2 = -\frac{\partial u/\partial x}{\partial u/\partial y}.\frac{\partial v/\partial x}{\partial v/\partial y} = -1 \quad \text{(Using equation (1))}$$

Hence, at the point of intersection (x, y) of the two curves, $u(x, y) = c_1$ and $v(x, y) = c_2$, $m_1.m_2 = -1$.

$\therefore$ The two families of curves (treating c_1 and c_2 as parameters) form an orthogonal system.

Note:

(i) In polar coordinates, the condition for two curves to be orthogonal is

$$\left(r.\frac{d\theta}{dr}\right)_1.\left(r.\frac{d\theta}{dr}\right)_2 = -1 \quad \text{(i.e., product of the slopes of the two curves is -1)}$$

Example 3.2.1 Prove that if $u = x^2 - y^2$, $v = \frac{-y}{x^2+y^2}$, then both u and v are satisfying the Laplace equation, but u + iv is not an analytic function.

Solution:

$u = x^2 - y^2$

$\therefore u_x = 2x, u_y = -2y$

$u_{xx} = 2, u_{yy} = -2.$

$\therefore u_{xx} + u_{yy} = 0$

i.e., u satisfies the Laplace equation.

$$v = \frac{-y}{x^2+y^2}$$

$$\therefore v_x = \frac{2xy}{(x^2+y^2)^2}, \quad v_y = \frac{(x^2+y^2)(-1)+y.2y}{(x^2+y^2)^2} = \frac{y^2-x^2}{(x^2+y^2)^2}$$

$$\therefore v_{xx} = \frac{(x^2+y^2)^2.2y - 2xy.2(x^2+y^2).2x}{(x^2+y^2)^4} = \frac{2y^3-6x^2y}{(x^2+y^2)^3}$$

$$\therefore v_{yy} = \frac{(x^2+y^2)^2.2y - (y^2-x^2).2(x^2+y^2).2y}{(x^2+y^2)^4} = \frac{6x^2y-2y^3}{(x^2+y^2)^3}$$

$\therefore v_{xx} + v_{yy} = 0$

i.e., v satisfies the Laplace equation.

Since $u_x \neq v_y$ and $u_y \neq -v_x$, u and v do not satisfy the Cauchy-Riemann equations.

$\therefore$ u + iv is not an analytic function.

Example 3.2.2 Show that a harmonic function u satisfies the differential equation $\dfrac{\partial^2 z}{\partial z \partial \bar{z}} = 0$.

Solution:

We have $z = x + iy$ and $\bar{z} = x - iy$

Then $x = \dfrac{1}{2}(z + \bar{z})$ and $y = \dfrac{1}{2i}(z - \bar{z})$

Since u is harmonic, u satisfies the Laplace equation.

$$\therefore \frac{\partial^2 u}{\partial x^2} + \frac{\partial^2 u}{\partial y^2} = 0 \qquad (1)$$

$$\text{Now } \frac{\partial u}{\partial \bar{z}} = \frac{\partial u}{\partial x} \cdot \frac{\partial x}{\partial \bar{z}} + \frac{\partial u}{\partial y} \cdot \frac{\partial y}{\partial \bar{z}}$$

$$= \frac{1}{2}\frac{\partial u}{\partial x} - \frac{1}{2i}\frac{\partial u}{\partial y} = \frac{1}{2}\frac{\partial u}{\partial x} + \frac{i}{2}\frac{\partial u}{\partial y}$$

$$\therefore \frac{\partial^2 u}{\partial z \partial \bar{z}} = \frac{\partial}{\partial z}\left(\frac{\partial u}{\partial \bar{z}}\right)$$

$$= \frac{\partial}{\partial z}\left(\frac{1}{2}\frac{\partial u}{\partial x} + \frac{i}{2}\frac{\partial u}{\partial y}\right)$$

$$= \frac{\partial}{\partial x}\left(\frac{1}{2}\frac{\partial u}{\partial x} + \frac{i}{2}\frac{\partial u}{\partial y}\right)\frac{\partial x}{\partial z} + \frac{\partial}{\partial y}\left(\frac{1}{2}\frac{\partial u}{\partial x} + \frac{i}{2}\frac{\partial u}{\partial y}\right)\frac{\partial y}{\partial z}$$

$$= \left(\frac{1}{2}\frac{\partial^2 u}{\partial x^2} + \frac{i}{2}\frac{\partial^2 u}{\partial x \partial y}\right)\frac{1}{2} + \left(\frac{1}{2}\frac{\partial^2 u}{\partial y \partial x} + \frac{i}{2}\frac{\partial^2 u}{\partial y^2}\right)\frac{1}{2i}$$

$$= \frac{1}{4}\left[\frac{\partial^2 u}{\partial x^2} + \frac{\partial^2 u}{\partial y^2} + \frac{\partial^2 u}{\partial x \partial y}\left(i + \frac{1}{i}\right)\right]$$

$$= \frac{1}{4}\left(\frac{\partial^2 u}{\partial x^2} + \frac{\partial^2 u}{\partial y^2}\right) + \frac{\partial^2 u}{\partial x \partial y}(i - i) = 0 \qquad \textbf{(Using (1))}$$

Note:

$$\frac{\partial^2}{\partial x^2} + \frac{\partial^2}{\partial y^2} = 4\frac{\partial^2 z}{\partial z \partial \bar{z}}$$

Example 3.2.3 If u and v are harmonic functions of x and y, show that s + it is analytic, where $s = \frac{\partial u}{\partial y} - \frac{\partial v}{\partial x}$ and $t = \frac{\partial u}{\partial x} + \frac{\partial v}{\partial y}$

Solution:

To prove that s + it is analytic, we have to prove s and t satisfy the Cauchy-Riemann equations.

$$\frac{\partial s}{\partial x} = \frac{\partial t}{\partial y} \text{ and } \frac{\partial s}{\partial y} = -\frac{\partial t}{\partial x} \quad (1)$$

Since u and v are harmonic they have continuous second order partial derivatives and satisfy the Laplace equation.

$$\frac{\partial^2 u}{\partial x^2} + \frac{\partial^2 u}{\partial y^2} = 0 \text{ and } \frac{\partial^2 v}{\partial x^2} + \frac{\partial^2 v}{\partial y^2} = 0 \quad (2)$$

Now $s = \frac{\partial u}{\partial y} - \frac{\partial v}{\partial x}$ and $t = \frac{\partial u}{\partial x} + \frac{\partial v}{\partial y}$

$$\therefore \frac{\partial s}{\partial x} = \frac{\partial^2 u}{\partial x \partial y} - \frac{\partial^2 v}{\partial x^2} \text{ and } \frac{\partial t}{\partial y} = \frac{\partial^2 u}{\partial y \partial x} + \frac{\partial^2 v}{\partial y^2}$$

$$\frac{\partial s}{\partial x} - \frac{\partial t}{\partial y} = \frac{\partial^2 u}{\partial x \partial y} - \frac{\partial^2 v}{\partial x^2} - \frac{\partial^2 u}{\partial y \partial x} - \frac{\partial^2 v}{\partial y^2}$$

$$= -\left(\frac{\partial^2 v}{\partial x^2} + \frac{\partial^2 v}{\partial y^2}\right) \quad \left(\text{Since } \frac{\partial^2 u}{\partial x \partial y} = \frac{\partial^2 u}{\partial y \partial x}\right)$$

$$= 0 \quad \text{(Using (2))}$$

$$\therefore \frac{\partial s}{\partial x} = \frac{\partial t}{\partial y} \quad (3)$$

Similarly $\frac{\partial s}{\partial y} = \frac{\partial^2 u}{\partial y^2} - \frac{\partial^2 v}{\partial y \partial x}$ and $\frac{\partial t}{\partial x} = \frac{\partial^2 u}{\partial x^2} + \frac{\partial^2 v}{\partial x \partial y}$

$$\frac{\partial s}{\partial y} + \frac{\partial t}{\partial x} = \frac{\partial^2 u}{\partial x^2} + \frac{\partial^2 u}{\partial y^2} = 0 \quad \text{(Using (2))}$$

$$\therefore \frac{\partial s}{\partial y} = -\frac{\partial t}{\partial x} \quad (4)$$

Equations (3) and (4) imply that s and t satisfy the Cauchy-Riemann equations.

∴ s + it is analytic.

Example 3.2.4 Prove that the real (and imaginary) parts of an analytic function w = u(r, θ) + i v(r, θ) satisfy the Laplace equation in polar coordinates.

i.e., $\frac{\partial^2 u}{\partial r^2} + \frac{1}{r}\frac{\partial u}{\partial r} + \frac{1}{r^2}\frac{\partial^2 u}{\partial \theta^2} = 0$

Solution:

Since w = u(r, θ) + i v(r, θ) is analytic, u and v satisfy the Cauchy-Riemann equations in polar form

i.e., $\frac{\partial u}{\partial r} = \frac{1}{r}.\frac{\partial v}{\partial \theta}$ (1)

And $\frac{\partial v}{\partial r} = -\frac{1}{r}.\frac{\partial u}{\partial \theta}$ (2)

Differentiating (1) partially w.r.t r, we get

$$\frac{\partial^2 u}{\partial r^2} = \frac{1}{r}\frac{\partial^2 v}{\partial r \partial \theta} - \frac{1}{r^2}\frac{\partial v}{\partial \theta} \quad (3)$$

Differentiating (2) partially w.r.t θ, we get

$$\frac{\partial^2 v}{\partial \theta \partial r} = -\frac{1}{r}\frac{\partial^2 u}{\partial \theta^2} \quad (4)$$

$$\therefore \frac{1}{r}.\frac{\partial^2 v}{\partial \theta \partial r} = -\frac{1}{r^2}\frac{\partial^2 u}{\partial \theta^2} \quad (5)$$

Since v is the imaginary part of an analytic function, it has continuous second order partial derivatives and hence $\frac{\partial^2 v}{\partial \theta \partial r} = \frac{\partial^2 v}{\partial r \partial \theta}$.

∴ Using (5) in (3), we get,

$$\frac{\partial^2 u}{\partial r^2} = \frac{1}{r^2}\frac{\partial^2 u}{\partial \theta^2} - \frac{1}{r^2}\frac{\partial v}{\partial \theta}$$

$$= \frac{1}{r^2}\frac{\partial^2 u}{\partial \theta^2} - \frac{1}{r}\frac{\partial u}{\partial r} \quad \text{(Using (1))}$$

i.e., $\frac{\partial^2 u}{\partial r^2} + \frac{1}{r}\frac{\partial u}{\partial r} + \frac{1}{r^2}\frac{\partial^2 u}{\partial \theta^2} = 0$

∴ u satisfies the Laplace equation in polar coordinates. Similarly we can prove that v alsio satisfies the Laplace equation in polar coordinates.

Example 3.2.5 Prove that an analytic function in a region with constant modulus is constant.

Solution:

Let f(z) = u + iv be analytic in a region R. Then u and v satisfy the Cauchy-Riemann equations. $u_x = v_y$, $u_y = -v_x$ (1)

Given $|f(z)| = \sqrt{u^2 + v^2}$ = constant.

Let $u^2 + v^2 = c$.

Differentiating partially w.r.t x, we get $2u.u_x + 2v.v_x = 0$

i.e., $u.u_x + v.v_x = 0$ (2)

Similarly, differentiating partially w.r.t y, we get,

$2u.u_y+2v.v_y = 0$

i.e., $u.u_y + v.v_y = 0$ (3)

using (1) in (2) and (3), we get,

$u.u_x - v.v_y = 0$ and $u.u_y + v.v_x = 0$

i.e., $u.u_x - v.u_y = 0$ and $v.u_x + u.u_y = 0.$

These equations in u_x and u_y have only a trivial solution if $\begin{vmatrix} u & -v \\ v & u \end{vmatrix} = u^2 + v^2 = c \neq 0.$

$\therefore$ When $c \neq 0$, $u_x = 0$, $u_y = 0$.

Then from (1), $v_x = 0$, $v_y = 0$.

$\therefore$ $f'(z) = u_x + i\, v_x = 0$.

Hence f(z) is a constant in the region R.

When c = 0, $u^2 + v^2 = 0$.

i.e., u = 0 and v = 0 at all points in R.

$\therefore$ f(z) = 0, a constant.

Example 3.2.6 If f(z) is analytic, prove that $\left(\frac{\partial^2}{\partial^2 x} + \frac{\partial^2}{\partial^2 y}\right)|f(z)|^2 = 4|f'(z)|^2$

Solution:

Let f(z) = u + iv be analytic.

Then $u_x = v_y$ and $u_y = -v_x$ (1)

Also $u_{xx} + u_{yy} = 0$ and $v_{xx} + v_{yy} = 0$ (2)

Now $|f(z)|^2 = u^2 + v^2$ and $f'(z) = u_x + iv_x$

$$\therefore \frac{\partial}{\partial x}|f(z)|^2 = 2u.u_x + 2v.v_x$$

and $$\frac{\partial^2}{\partial x^2}|f(z)|^2 = 2\left[u_x^2 + u.u_{xx} + v_x^2 + v.v_{xx}\right] \quad (3)$$

Similarly $$\frac{\partial^2}{\partial y^2}|f(z)|^2 = 2\left[u_y^2 + u.u_{yy} + v_y^2 + v.v_{yy}\right] \quad (4)$$

Adding (3) and (4)

$$\left(\frac{\partial^2}{\partial x^2} + \frac{\partial^2}{\partial y^2}\right)|f(z)|^2 = 2\left[u_x^2 + u_y^2 + u(u_{xx} + u_{yy}) + v_x^2 + v_y^2 + v(v_{xx} + v_{yy})\right]$$

$$= 2\left[u_x^2 + v_x^2 + u(0) + v_x^2 + u_x^2 + v(0)\right]$$

$$= 4\left[u_x^2 + v_x^2\right]$$

$$= 4|f'(z)|^2$$

Example 3.2.7 If f(z) = u + iv is analytic and f(z) ≠ 0, prove that

(i) $\nabla^2 (\log |f(z)|) = 0$

(ii) $\nabla^2(\text{amp } f(z)) = 0.$

Solution:

$\log f(z) = \log |f(z)| + i \text{ amp } f(z)$

Since $f(z) \neq 0$ and $f(z)$ is analytic, $\log f(z)$ exists and is analytic.

$\therefore$ $\log |f(z)|$ and amp $f(z)$ are real and imaginary parts of the analytic function $\log f(z)$.

$\therefore$ $\log |f(z)|$ and amp $f(z)$ are harmonic functions and satisfy the Laplace equation.

$\therefore$ (i) $\nabla^2 (\log |f(z)|) = 0$ and (ii) $\nabla^2(\text{amp } f(z)) = 0$.

Example 3.2.8 If $f(z) = u + iv$ is an analytic function in any region, prove that

(i) $\left(\dfrac{\partial^2}{\partial x^2} + \dfrac{\partial^2}{\partial y^2}\right) | f(z) |^p = p^2 \, | f(z) |^{p-2} . | f'(z) |^2$

(ii) $\left(\dfrac{\partial^2}{\partial x^2} + \dfrac{\partial^2}{\partial y^2}\right) | u |^p = p(p-1) \, | u |^{p-2} . | f'(z) |^2$

Solution:

We have $\dfrac{\partial^2}{\partial x^2} + \dfrac{\partial^2}{\partial y^2} = 4\dfrac{\partial^2 z}{\partial z \partial \bar{z}}$ (1)

(i) $\left(\dfrac{\partial^2}{\partial x^2} + \dfrac{\partial^2}{\partial y^2}\right) | f(z) |^p = 4.\dfrac{\partial^2}{\partial z \partial \bar{z}} | f(z) |^p$ (Using (1))

$$= 4.\frac{\partial^2}{\partial z \partial \bar{z}} \left[f(z).\overline{f(z)} \right]^{p/2} \quad \text{(Since } z\bar{z} = |z|^2 \text{)}$$

$$= 4.\frac{\partial^2}{\partial z \partial \bar{z}} \left(f(z) \right)^{p/2} \left(\overline{f(z)} \right)^{p/2}$$

$$= 4.\frac{\partial}{\partial z} \left[\left(f(z) \right)^{p/2} . \frac{p}{2} \left(\overline{f(z)} \right)^{p/2 - 1} \overline{f'(z)} \right]$$

$$\left(\text{Since } f(z) \text{ is analytic} \left(\overline{f(z)} \right)' = \overline{f'(z)} \right)$$

$$= 4.\frac{p}{2} \left(f(z) \right)^{p/2 - 1} f'(z).\frac{p}{2} \left(\overline{f(z)} \right)^{p/2 - 1} \overline{f'(z)}$$

(Since $\overline{f(z)}$ amd $\overline{f'(z)}$ are functions of $\bar{z}$)

$$= p^2\left(f(z).\overline{f(z)}\right)^{p/2-1} f'(z).\overline{f'(z)}$$

$$= p^2\left(|f(z)|^2\right)^{p/2-1} |f'(z)|^2 \quad = p^2|f(z)|^{p-2}|f'(z)|^2$$

(ii) $f(z) = u + iv$

$$\therefore \overline{f(z)} = u - iv \text{ and } u = \frac{1}{2}\left[f(z)+\overline{f(z)}\right]$$

$$\therefore |u|^p = \left|\frac{1}{2}\left[f(z)+\overline{f(z)}\right]\right|^p$$

$$= \frac{1}{2^p}\left[\left|f(z)+\overline{f(z)}\right|^2\right]^{p/2}$$

$$= \frac{1}{2^p}\left[\left(f(z)+\overline{f(z)}\right)\left(\overline{f(z)+\overline{f(z)}}\right)\right]^{p/2}$$

$$= \frac{1}{2^p}\left[\left(f(z)+\overline{f(z)}\right)\left(\overline{f(z)}+f(z)\right)\right]^{p/2} = \frac{1}{2^p}\left[f(z)+\overline{f(z)}\right]^p$$

$$\therefore \left(\frac{\partial^2}{\partial x^2}+\frac{\partial^2}{\partial y^2}\right)|u|^p = 4.\frac{\partial^2}{\partial z\partial\bar{z}}\left[\frac{1}{2^p}\left(f(z)+\overline{f(z)}\right)^p\right]$$

$$= \frac{4}{2^p}.\frac{\partial}{\partial z}\left[p.\left(f(z)+\overline{f(z)}\right)^{p-1}.\overline{f'(z)}\right]$$

$$= \frac{4}{2^p}.p.(p-1).\left(f(z)+\overline{f(z)}\right)^{p-2} f'(z).\overline{f'(z)}$$

$$= p.(p-1).\left[\frac{1}{2}\left(f(z)+\overline{f(z)}\right)\right]^{p-2} |f'(z)|^2 \quad (\text{Since } \overline{f'(z)} \text{ is a function of } \bar{z} \text{ only})$$

$$= p.(p-1).u^{p-2}|f'(z)|^2$$

$$= p.(p-1).\left(u^2\right)^{\frac{p-2}{2}} |f'(z)|^2$$

$$= p.(p-1).\left(|u|^2\right)^{\frac{p-2}{2}} |f'(z)|^2 = p.(p-1).|u|^{p-2}|f'(z)|^2$$

3.2.3 CONSTRUCTION OF ANALYTIC FUNCTIONS FROM THEIR REAL OR IMAGINARY PARTS

Method 1.

Let u(x, y) be the real part of an unknown analytic function. If the imaginary part v(x, y) is found, then f(z) = u(x, y) + i v(x, y) is the analytic function. Since u(x, y) is known, $\frac{\partial u}{\partial x}$ and $\frac{\partial u}{\partial y}$ are known.

$$\text{Now } dv = \frac{\partial v}{\partial x}.dx + \frac{\partial v}{\partial y}.dy$$

$$= \left(-\frac{\partial u}{\partial y}\right).dx + \left(\frac{\partial u}{\partial x}\right).dy \qquad \text{(Using Cauchy-Riemann equations)}$$

$$= Mdx + Ndy$$

$$\frac{\partial M}{\partial y} - \frac{\partial N}{\partial x} = \frac{\partial}{\partial y}\left(-\frac{\partial u}{\partial y}\right) - \frac{\partial}{\partial x}\left(\frac{\partial u}{\partial x}\right)$$

$$= -\left(\frac{\partial^2 u}{\partial x^2} + \frac{\partial^2 u}{\partial y^2}\right)$$

$$= 0 \qquad \text{(Since u satisfies the Laplace equation)}$$

$$\therefore \frac{\partial M}{\partial y} = \frac{\partial N}{\partial x}$$

Hence dv = Mdx + Ndy is an exact differential equation. Integrating, we get

$$v = \int (Mdx + Ndy) + c$$

$$\text{i.e., } v(x, y) = \int\left[\left(-\frac{\partial u}{\partial y}\right)dx + \left(\frac{\partial u}{\partial x}\right)dy\right] + c \qquad \text{where c is an arbitrary constant}$$

This gives f(z) = u(x, y) + i v(x, y).

Note:

If v(x, y) is known, then f(z) = u(x, y) + i v(x, y) can be found after computing

$$u(x, y) = \int\left[\frac{\partial v}{\partial y}dx - \frac{\partial v}{\partial x}dy\right] + c$$

Method 2. (Milne-Thomson's Method)

Let the real part of the analytic function f(z) = u(x, y) + i v(x, y) is known. In this method, we first find f′(z) as a function of z and then find f(z) by integration.

Since u(x, y) is known, $\frac{\partial u}{\partial x}$ and $\frac{\partial u}{\partial y}$ are known.

Now $f'(z) = \frac{\partial u}{\partial x} + i\frac{\partial v}{\partial x}$

i.e., $f'(z) = \frac{\partial u}{\partial x} - i\frac{\partial u}{\partial y}$ (using C-R equation $u_y = -v_x$)

i.e., $f'(z) = u_x(x, y) - iu_y(x, y)$ (1)

Now substituting $x = \frac{z+\bar{z}}{2}$, $y = \frac{z-\bar{z}}{2i}$ in (1), we get,

$$f'(z) = u_x\left(\frac{z+\bar{z}}{2}, \frac{z-\bar{z}}{2i}\right) - i.u_y\left(\frac{z+\bar{z}}{2}, \frac{z-\bar{z}}{2i}\right) \quad (2)$$

Equation (2) can be regarded as an identity in two independent variables z and $\bar{z}$
Putting $z = \bar{z}$ in (2), we get,
$f'(z) = u_x(z, 0) - i\, u_y(z, 0)$
Integrating we get, $f(z) = \int [u_x(z,0) - i.u_y(z,0)]dz + c$

Note:

(i) If the imaginary part v(x, y) is known, then the analytic function
$f(z) = \int [v_y(z,0) + i.v_x(z,0)]dz + c$

(ii) If f(z) = u(x, y) + i v(x, y), then f(z) can be obtained as a function of z by putting x = z and y = 0. i.e., f(z) = u(z, 0) + i v(z, 0).
This rule of computing f(z) as a function of z by replacing x by z and y by zero is called **Milne-Thomson rule.**

Example 3.2.9 Prove that the function $u = 3x^2y + 2x^2 - y^3 - 2y^2$ is harmonic and find its harmonic conjugate. Hence find f(z) as a function of z.

Solution:

$u = 3x^2y + 2x^2 - y^3 - 2y^2$

$\therefore u_x = 6xy + 4x$ and $u_y = 3x^2 - 3y^2 - 4y$

$u_{xx} = 6y + 4$ and $u_{yy} = -6y - 4$

$\therefore u_{xx} + u_{yy} = 6y + 4 - 6y - 4$
$= 0$

$\therefore$ u satisfies the Laplace equation.
Also the second order partial derivatives of u are continuous. $\therefore$ u is harmonic function.
v be the harmonic conjugate of u.

Then $dv = \frac{\partial v}{\partial x}.dx + \frac{\partial v}{\partial y}.dy$

$= -\frac{\partial u}{\partial y}.dx + \frac{\partial u}{\partial x}.dy$ (Using Cauchy-Riemann equations)

$= (-3x^2 + 3y^2 + 4y)dx + (6xy + 4x)\, dy$

$= 4(xdy + ydx) + (-3x^2)\, dx + (3y^2dx + 6xy\, dy)$

Integrating, we get,

$v = 4xy - x^3 + 3xy^2 + c$ where c is a constant.

$\therefore f(z) = u + iv$

$= (3x^2y + 2x^2 - y^3 - 2y^2) + i(4xy - x^3 + 3xy^2 + c)$

To obtain f(z) as a function of z, replace x by z and y by 0 (by Milne-Thomson's rule)

$\therefore \quad f(z) = 2z^2 + i(-z^3 + c)$

i.e., $f(z) = 2z^2 - iz^3 + c$

Example 3.2.10 Show that $u = \log\sqrt{x^2+y^2}$ is harmonic and determine its conjugate and hence find the corresponding analytic function f(z).

Solution:

$$u = \log\sqrt{x^2+y^2} = \frac{1}{2}\log(x^2+y^2)$$

$$\therefore u_x = \frac{x}{x^2+y^2} \text{ and } u_{xx} = \frac{(x^2+y^2)-2x^2}{(x^2+y^2)^2} = \frac{y^2-x^2}{(x^2+y^2)^2}$$

Similarly, $u_y = \dfrac{y}{x^2+y^2}$ and $u_{yy} = \dfrac{x^2-y^2}{(x^2+y^2)^2}$

Then $u_{xx} + u_{yy} = 0$. $\therefore$ u satisfies the Laplace equation.

Hence u is a harmonic function.

Let v be its harmonic conjugate.

$\therefore f(z) = u + iv$ is an analytic function and u and v satisfy the Cauchy-Riemann equations, $u_x = v_y$ and $u_y = -v_x$.

Now $v_y = u_x = \dfrac{x}{x^2+y^2}$

Integrating w.r.t y we get,

$v = \tan^{-1}(y/x) + \varphi(x)$ where $\varphi(x)$ is a function of x.

Now differentiating w.r.t x, we get,

$$v_x = \frac{1}{1+(y/x)^2}\left(\frac{-y}{x^2}\right) + \varphi'(x)$$

$$\text{i.e.,}\quad v_x = \frac{-y}{x^2+y^2} + \varphi'(x)$$

$$\text{i.e.,}\quad -u_y = \frac{-y}{x^2+y^2} + \varphi'(x)$$

$$\text{i.e.,}\quad \frac{-y}{x^2+y^2} = \frac{-y}{x^2+y^2} + \varphi'(x)$$

$\therefore \varphi'(x) = 0$ Hence $\varphi(x) = c$, a constant.

$\therefore v = \tan^{-1}(y/x) + c$

$\therefore f(z) = u + iv$

i.e., $f(z) = \log\sqrt{x^2 + y^2} + i[\tan^{-1}(y/x) + c]$

Example 3.2.11 If $u = \dfrac{\sin 2x}{\cosh 2y + \cos 2x}$, find the corresponding analytic function $f(z) = u + iv$.

Solution:

$$u = \frac{\sin 2x}{\cosh 2y + \cos 2x}$$

$$\therefore u_x = \frac{(\cosh 2y + \cos 2x)\, 2\cos 2x - \sin 2x\,(-2\sin 2x)}{(\cosh 2y + \cos 2x)^2}$$

$$= \frac{2 + 2\cos 2x \,.\cosh 2y}{(\cosh 2y + \cos 2x)^2}$$

$$\therefore u_y = \frac{(\cosh 2y + \cos 2x) .0 - \sin 2x\ .2\sinh 2y}{(\cosh 2y + \cos 2x)^2}$$

$$= \frac{-2\sin 2x \,.\sinh 2y}{(\cosh 2y + \cos 2x)^2}$$

Now by Milne-Thomson's method,

$$f(z) = \int [u_x(z,0) - iu_y(z,0)]\, dz + c$$

$$= \int \left[\frac{2 + 2\cos 2z}{(1 + \cos 2z)^2} - i.0\right] dz + c$$

$$= \int \frac{2}{1 + \cos 2z}\, dz + c$$

$$= \int \sec^2 z\, dz + c$$

$$= \tan z + c$$

Example 3.2.12 Find the analytic function $f(z) = u + iv$ such that $v = x^3 - 3xy^2 + 3x^2 - 3y^2 + 1$.

Solution:

$v = x^3 - 3xy^2 + 3x^2 - 3y^2 + 1.$

$\therefore v_x = 3x^2 - 3y^2 + 6x$ and $v_y = -6xy - 6y$

By Milne-Thomson's method,

$$f(z) = \int [v_y(z,0) + iv_x(z,0)]\, dz + c = \int [0 + i(3z^2 + 6z)]\, dz + c$$

i.e., $f(z) = i\,[z^3 + 3z^2] + c.$

Example 3.2.13 Find the analytic function $f(z) = u + iv$ if $u - v = e^x(\cos y - \sin y)$.

Solution:

We have $u - v = e^x(\cos y - \sin y)$ (1)

Method 1. Partial differentiating (1) w.r.t x, we get,

$u_x - v_x = e^x(\cos y - \sin y)$ (2)

Partial differentiating (1) w.r.t y, we get,

$u_y - v_y = -e^x(\sin y + \cos y)$

i.e., $-v_x - u_x = -e^x(\sin y + \cos y)$ (Using Cauchy-Riemann equations) (3)

Solving (2) and (3), we get

$u_x = e^x.\cos y$

$v_x = e^x.\sin y$

$f'(z) = u_x(x, y) + iv_x(x, y)$

By Milne Thomson's method,

$$f(z) = \int[u_x(z, 0) + iv_x(z, 0)]dz + c$$

$$= \int e^z dz + c$$

i.e., $f(z) = e^z + c$.

Method 2. $f(z) = u + iv$

$\therefore$ $i f(z) = iu - v$

Adding $(1 + i)f(z) = (u - v) + i(u + v)$

$f(z)$ is analytic implies $(1 + i)f(z)$ is also analytic.

$\therefore$ $P = u - v$ and $Q = u + v$ are harmonic conjugates.

Now $P = u - v = e^x(\cos y - \sin y)$

$\therefore$ $P_x = e^x(\cos y - \sin y)$ (1)

$P_y = -e^x(\sin y + \cos y)$ (2)

By Milne-Thomson's method,

$$\therefore (1+i)f(z) = \int[P_x(z, 0) - iP_y(z, 0)]dz + c$$

$$= \int[e^z - i(-e^z)]dz + c \quad \text{(Using (1) and (2))}$$

$$= \int(1 + i)e^z dz + c$$

$$= (1 + i)\, e^z + c$$

$\therefore$ $f(z) = e^z + c_1$ where c_1 is a constant.

Example 3.2.14 Find the analytic function $f(z) = u + iv$ given

$$u + v = \frac{2\sin 2x}{e^{2y} + e^{-2y} - 2\cos 2x}.$$

Solution:

$f(z) = u + iv$.

$\therefore$ i f(z) = iu – v .

Adding (1 + i)f(z) = (u - v) + i(u + v)

$= P + i\,Q.$

f(z) is analytic implies (1 + i) f(z) is analytic.

$\therefore$ P = u - v and Q = u + v are harmonic conjugates.

We have, $Q = \dfrac{2\sin 2x}{e^{2y} + e^{-2y} - 2\cos 2x}$

Partially differentiating Q w.r.t x, we get,

$$Q_x = \frac{\left(e^{2y} + e^{-2y} - 2\cos 2x\right)4\cos 2x - 2\sin 2x.4\sin 2x}{\left[e^{2y} + e^{-2y} - 2\cos 2x\right]^2}$$

$$= \frac{4\cos 2x\left(e^{2y} + e^{-2y}\right) - 8}{\left[e^{2y} + e^{-2y} - 2\cos 2x\right]^2}$$

$$\therefore Q_x(z, 0) = \frac{8\cos 2z - 8}{\left[2 - 2\cos 2z\right]^2}$$

$$= \frac{2}{\cos 2z - 1} \qquad (1)$$

Partially differentiating Q w.r.t y, we get,

$$Q_y = \frac{-2\sin 2x\left(2e^{2y} - 2e^{-2y}\right)}{\left[e^{2y} + e^{-2y} - 2\cos 2x\right]^2}$$

$\therefore Q_y(z, 0) = 0$ (2)

Now by Milne-Thomson's method, we get,

$$(1+i)f(z) = \int\left[Q_y(z,0) + iQ_x(z,0)\right]dz + c$$

i.e., $(1+i)f(z) = \int\left[0 + i\left(\dfrac{2}{\cos 2z - 1}\right)\right]dz + c$ (Using (1) and (2))

$$= 2i\int\frac{dz}{\cos 2z - 1} + c$$

$= -i\int \operatorname{cosec}^2 z.dz + c$ (Since $\dfrac{1 - \cos 2z}{2} = \sin^2 z$)

$$= i\cot z + c$$

$\therefore\ f(z) = \dfrac{i}{1+i}\cot z + c_1 = \dfrac{i}{2}(1-i)\cot z + c_1 = \dfrac{1}{2}(1+i)\cot z + c_1$, where c_1 is a constant

Example 3.2.15 If f(z) = u + iv is an analytic function and $u - v = \dfrac{e^y - \cos x + \sin x}{\cosh y - \cos x}$,

find f(z) subject to the condition $f\left(\pi/2\right) = \dfrac{3-i}{2}$.

Solution:

We have f(z) = u + iv

if(z) = iu – v

$\therefore$ (1 + i) f(z) = (u - v) + i(u + v) = P + i Q.

Where P = u - v and Q = u + v.

P and Q are conjugate harmonic functions.

$$P = \frac{e^y - \cos x + \sin x}{\cosh y - \cos x}$$

Partially differentiating w.r.t x, we get,

$$P_x = \frac{(\cosh y - \cos x)(\sin x + \cos x) - (e^y - \cos x + \sin x).\sin x}{(\cosh y - \cos x)^2}$$

$$\therefore P_x(z, 0) = \frac{(1 - \cos z)(\sin z + \cos z) - (1 - \cos z + \sin z).\sin z}{(1 - \cos z)^2}$$

$$= \frac{\cos z - 1}{(1 - \cos z)^2} = \frac{-1}{1 - \cos z} = -\frac{1}{2}\operatorname{cosec}^2 z/2$$

i.e., $P_x(z, 0) = -\frac{1}{2}\operatorname{cosec}^2 z/2$ (1)

Again differentiating P w.r.t y, we get,

$$P_y = \frac{(\cosh y - \cos x).e^y - (e^y - \cos x + \sin x).\sinh y}{(\cosh y - \cos x)^2}$$

$$\therefore P_y(z, 0) = \frac{(1 - \cos z) - (1 - \cos z + \sin z).0}{(1 - \cos z)^2} = \frac{1}{1 - \cos z}$$

i.e., $P_y(z, 0) = \frac{1}{2}\operatorname{cosec}^2 z/2$ (2)

By Milne-Thomson's method,

$$(1 + i)f(z) = \int [P_x(z, 0) - iP_y(z, 0)]dz + c$$

$$= \int \left[-\frac{1}{2}\operatorname{cosec}^2 z/2 - i\frac{1}{2}\operatorname{cosec}^2 z/2\right]dz + c \quad \text{(Using (1) and (2))}$$

$$= \int -\frac{1}{2}(1 + i)\operatorname{cosec}^2 z/2\, dz + c = (1 + i)\cot z/2 + c$$

$\therefore f(z) = \cot z/2 + c_1$, where c_1 is a constant

Given $f(\pi/2) = \frac{3 - i}{2}$

$$\therefore \frac{3 - i}{2} = \cot \pi/4 + c_1 = 1 + c_1$$

$\therefore c_1 = \frac{3 - i}{2} - 1 = \frac{1 - i}{2}$. Hence $f(z) = \cot z/2 + \frac{1 - i}{2}$

Example 3.2.16 Given $w = z^3$ where $w = u + iv$, show that the families of curves $u = c_1$ and $v = c_2$ (where c_1 and c_2 are constants) are orthogonal to each other.

Solution:

$w = z^3 = (x + iy)^3$

$= (x^3 - 3xy^2) + i\,(3x^2y - y^3)$

$\therefore\ u = x^3 - 3xy^2$ and $v = 3xy^2 - y^3$.

Now $u = c_1 \Rightarrow x^3 - 3xy^2 = c_1$

Differentiating w.r.t x, we get

$$3x^2 - 3\left(2xy.\frac{dy}{dx} + y^2\right) = 0$$

$$\therefore\ \frac{dy}{dx} = \frac{3(x^2 - y^2)}{6xy} = \frac{x^2 - y^2}{2xy}$$

The slope of any tangent to the curve $u = c_1$ is $m_1 = \dfrac{x^2 - y^2}{2xy}$ (1)

Now $v = c_2 \Rightarrow 3x^2y\text{-}y^3 = c_2$

Differentiating w.r.t x, we get

$$3\left(2xy + x^2.\frac{dy}{dx}\right) - 3y^2\frac{dy}{dx} = 0$$

$$\therefore\ \frac{dy}{dx} = \frac{-2xy}{x^2 - y^2}$$

The slope of any tangent to the curve $v = c_2$ is $m_2 = \dfrac{-2xy}{x^2 - y^2}$ (2)

Hence, at any point common to the two families of curves $u = c_1$ and $v = c_2$, we have $m_1m_2 = -1$ (using (1) and (2))

$\therefore$ The two families of curves are orthogonal.

Example 3.2.17 Show that the families of curves $r^n = a^n \cos n\theta$ and $r^n = b^n \sin n\theta$ cut orthogonally, where a and b are arbitrary constants.

Solution:

$r^n = a^n \cos n\theta$

Differentiating w.r.t x, we get $n.r^{n-1} = -a^n \sin n\theta.n.\dfrac{d\theta}{dr}$

$$\therefore\ r.\frac{d\theta}{dr} = \frac{-r^n}{a^n.\sin n\theta} = \frac{-a^n.\cos n\theta}{a^n.\sin n\theta}$$

i.e., $\left(r.\dfrac{d\theta}{dr}\right)_1 = -\dfrac{\cos n\theta}{\sin n\theta}$ (1)

Similarly $r^n = b^n \sin n\theta$

Differentiating w.r.t r, we get $n.r^{n-1} = b^n \cos n\theta . n . \dfrac{d\theta}{dr}$

$$\therefore \quad r.\frac{d\theta}{dr} = \frac{r^n}{b^n.\cos n\theta} = \frac{b^n.\sin n\theta}{b^n.\cos n\theta}$$

i.e., $$\left(r.\frac{d\theta}{dr}\right)_2 = \frac{\sin n\theta}{\cos n\theta} \qquad (2)$$

From (1) and (2) we get,

$$\left(r.\frac{d\theta}{dr}\right)_1 . \left(r.\frac{d\theta}{dr}\right)_2 = -1$$

$\therefore$ The two families of curves cut orthogonally.

EXERCISE 3.2

PART-A

1. State the Laplace equation for the function $\varphi(x, y)$.
2. State the Laplace equation for the function $\varphi(r, \theta)$ in polar form.
3. Define 'harmonic function' and give an example.
4. Define 'conjugate harmonic functions'.
5. Verify whether the function cos x. sinh y is harmonic.
6. Verify whether the function $e^y \cosh x$ is harmonic.
7. Find the harmonic conjugate of the function $u = x^2 - y^2$
8. Find the harmonic conjugate of the function $\log(x^2 + y^2)$.
9. Find the analytic function $f(z) = u + iv$, given $u = \dfrac{x}{x^2 + y^2}$.
10. Find the analytic function $f(z) = u + iv$, given $v = \text{amp}(z)$.
11. Prove that if f(z) and $\overline{f(z)}$ are simultaneously analytic in a region R, then f(z) is constant.
12. Prove that an analytic function in a region with constant real part is constant.
13. Prove that an analytic function with constant imaginary part is constant.
14. Prove that an analytic function $f(z) = u + iv$ with arg f(z) constant is itself a constant function.
15. Prove that an analytic function in a region with its derivative zero at every point of the region is a constant.
16. Prove that a function differentiable at a point, need not be analytic at that point.
17. If $f = u + iv$ is analytic and uv is a constant then prove that f reduces to a constant.
18. If $f = u + iv$ is analytic and $v = u^2$, then prove that f reduces to a constant.
19. If $f = u + iv$ is analytic and u/v is a constant then prove that f reduces to a constant.
20. If v is the harmonic conjugate of u, prove that u is the harmonic conjugate of $-v$.

21. If u and v are conjugate harmonic functions, prove that uv is a harmonic function.
22. Given $w = u + iv = z^3$ is analytic. Prove that the families of curves, $u = c_1$ and $v = c_2$ are orthogonal to each other.
23. If u = x, find the analytic function f(z) = u + iv.
24. State any two properties of analytic functions.
25. State Milne-Thomson's rule.
26. If a function f(z) is analytic in a domain D and f(z) is real valued for all z in D, prove that f(z) must be a constant.
27. Prove that two harmonic conjugates of a harmonic function u(x, y) in a domain D differ by an arbitrary additive constant.
28. If v is a harmonic conjugate of u in a domain D and u is a harmonic conjugate of v, prove that u(x, y) and v(x, y) must be constant throughout D.
29. For $f(z) = u + iv = z^2$, show that the families of curves $u(x, y) = c_1$ and $v(x, y) = c_2$ form an orthogonal system.
30. If f(z) is analytic prove $\left(\overline{f(z)}\right)' = \overline{f'(z)}$.

PART-B

31. If u(x, y) is a harmonic function in a domain D, prove that $f(z) = u_x - iu_y$ is analytic in D
32. Prove that the functions u(z) and $u(\overline{z})$ are simultaneously harmonic.

Prove that the following functions are harmonic. Also find their harmonic conjugates.

33. $u(x, y) = 2x(1- y)$.
34. $u(x, y) = (x -1)^3 - 3xy^2 + 3y^2$.
35. $u(x, y) = \sinh x \sin y$.

Find the function f(z) = u + iv such that f(z) is analytic; given that,

36. $u = e^x (x \cos y - y \sin y)$
37. $v = x^3 - 3xy^2 + 3x^2 - 3y^2 + 1$.
38. $u = \dfrac{2 \cos x . \cosh y}{\cos 2x + \cosh 2y}$
39. $u = \dfrac{2 \sin 2x}{e^{2y} + e^{-2y} - 2 \cos 2x}$.
40. Find the analytic function f(z) = u + iv, if $v = e^{-x}(x \cos y + y \sin y)$. Hence find u.
41. Find the analytic function f(z) = u + iv, if $u = e^{-2xy} \sin (x^2-y^2)$. Hence find v.
42. If $u-v = (x - y)(x^2 + 4xy + y^2)$ and f(z) = u + iv is an analytic function of z = x + iy, find f(z) interms of z.
43. If f(z) = u + iv is analytic and $u - v = \dfrac{\cos x + \sin x - e^{-y}}{2 \cos x - e^y - e^{-y}}$, find f(z) such that $f(\pi/2) = 0$.

44. Find the analytic function f(z) = u + iv if $u + v = \dfrac{\sin 2x}{\cosh 2y - \cos 2x}$.

45. Find the analytic function f(z) = u + iv if $u + v = \dfrac{x}{x^2 + y^2}$, given that f(1) = 1.

46. Show that the families of curves $u = c_1$ and $v = c_2$ cut orthogonally, where $u + iv = z^4$.

47. Show that the families of curves $r^n = \alpha \sec n\theta$ and $r^n = \beta \operatorname{cosec} n\theta$ intersect orthogonally, where α and β are arbitrary constants.

48. Find the orthogonal trajectories of the family of curves $2x\text{-}x^3+3xy^2 = a$.

49. If f(z) is analytic, prove that $\left(\dfrac{\partial^2}{\partial x^2} + \dfrac{\partial^2}{\partial y^2}\right)|\text{Re}(f(z))|^2 = 2|f'(z)|^2$.

50. If f(z) is analytic, prove that $\left(\dfrac{\partial}{\partial x}|f(z)|\right)^2 + \left(\dfrac{\partial}{\partial y}|f(z)|\right)^2 = |f'(z)|^2$.

3.3 CONFORMAL MAPPINGS

Consider the function w = f(z) where z = x + iy and w = u + iv. To each point (x, y) in the z- plane, w = f(z) determines a point (u, v) in the w- plane if f(z) is a single-valued function. When (x, y) moves along a curve C in the z-plane, the corresponding point (u, v) will, in general, move along a curve C′ in the w-plane. The correspondence thus defined by w = f(z) is called a **mapping** or **transformation** of elements or objects (points, curves, regions etc.) in the z–plane onto elements in the w-plane. These elements in the w-plane are called the images of those in the z-plane. To study the nature of a function w = f(z), we usually study the images of lines parallel to the coordinate axes, of concurrent lines passing through the origin, of concentric circles |z| = k and of regions enclosed by such curves in the z-plane. We also study the preimages in the z-plane of such elements in the w-plane. In this section we study a special class of mappings called conformal mappings.

Definition 3.3.1 A curve C in the z-plane given by z = z(t), $\alpha \le t \le \beta$ is said to be **differentiable** if z′(t) exists and is continuous. Also, if z′(t) ≠ 0, then the curve is said to be **regular** or **smooth** curve.

Let w = f(z) be a continuous function defined in a domain D in the z-plane. Let C be curve given by z = z(t), $\alpha \le t \le \beta$ and lying in D. Then the equation w = w(t) = f(z(t)), $\alpha \le t \le \beta$ defines a curve C′ in the w-plane called the **image of the curve C** under the mapping w = f(z).

Definition 3.3.2 Let w = f(z) be a continuous function defined in domain D. Let $z_0 \in D$. Let C_1 and C_2 be two smooth curves passing through z_0 and lying in D. Let C_1' and C_2'

be the images of C_1 and C_2 respectively under the mapping $w = f(z)$. If the angle between C_1 and C_2 at z_0 is equal to the angle between C_1' and C_2' at $f(z_0)$ both in magnitude and direction then $w = f(z)$ is said to be **conformal** at z_0.

Angle between two curves at a point of intersection is the angle between their tangents at the point of intersection. A conformal mapping preserves angle between curves both in magnitude and direction. If a mapping $w = f(z)$ preserves angles in magnitude only, and direction is reversed, then the mapping is said to be **isogonal** or **indirectly conformal**.

3.3.1 SUFFICIENT CONDITION FOR CONFORMALITY OF w = f(z)

Theorem 3.3.1 Let $w = f(z)$ be analytic in a domain D. Let $z_0 \in D$. If $f'(z_0) \neq 0$, then $w = f(z)$ is conformal at z_0.

Proof:

Let C be smooth curve passing through z_0 and lying in D, given by $z = z(t)$, $\alpha \le t \le \beta$

Let, $z_0 = z(t_0)$ for $\alpha < t_0 < \beta$.

The equation of the image C′ of C under the mapping

$w = f(z)$ is $w = w(t) = f(z(t))$, $\alpha \le t \le \beta$.

$\therefore w'(t) = f'(z(t)).\ z'(t)$

ie., $w'(t_0) = f'(z(t_0)).z'(t_0) = f'(z_0).z'(t_0)$

Given $f'(z_0) \neq 0$. Also, since C is a smooth curve passing through z_0, $z'(t_0) \neq 0$.

Hence $w'(t_0) \neq 0$.

Now, $\arg w'(t_0) = \arg f'(z_0) + \arg z'(t_0)$ (1)

Let $\arg w'(t_0) = \varphi$

$\arg f'(z_0) = \psi$ and $\arg z'(t_0) = \theta$

From (1), we have $\varphi = \theta + \psi$ (2)

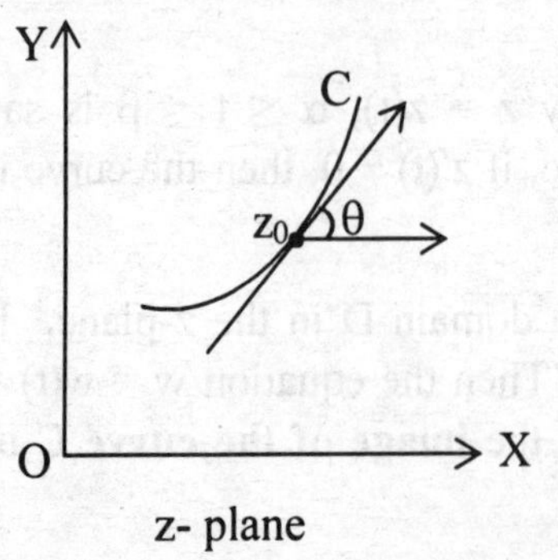

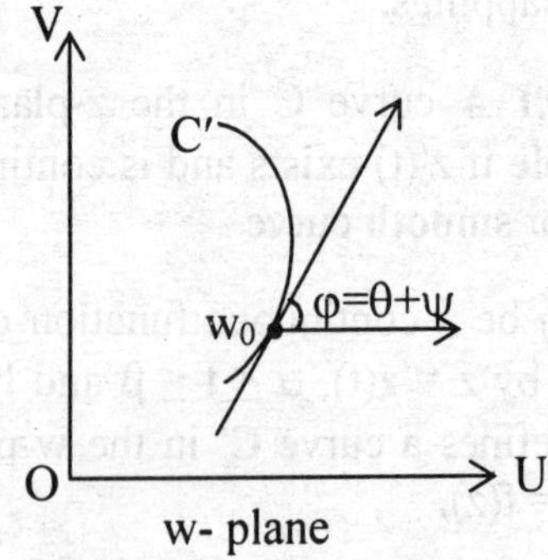

Fig. 3.4

θ is the angle made by the tangent to the curve C at z_0 with the positive direction of x-axis in the z-plane. φ is the angle made by the tangent to the curve C′ at $w_0 = f(z_0)$ with the positive direction of u-axis in the w-plane. Thus equation (2) implies that the tangent at z_0 is rotated through an angle ψ by the mapping w = f(z).

Let C_1 and C_2 be any two smooth curves passing through $z_0 \in D$ and lying in D. Let C_1' and C_2' be their images under the mapping w = f(z). C_1' and C_2' pass through $w_0 = f(z_0)$. Let θ_1 and θ_2 be the angles made by the tangents to the curves C_1 and C_2 respectively at z_0 with the positive direction of x-axis. Similarly, let φ_1 and φ_2 be the angles made by the tangents to the curves C_1' and C_2' at w_0 with the positive direction of u-axis in the w-plane.

Then $\varphi_1 = \theta_1 + \psi$ and $\varphi_2 = \theta_2 + \psi$ (Using (2))

$\therefore \varphi_2 - \varphi_1 = \theta_2 - \theta_1$ (3)

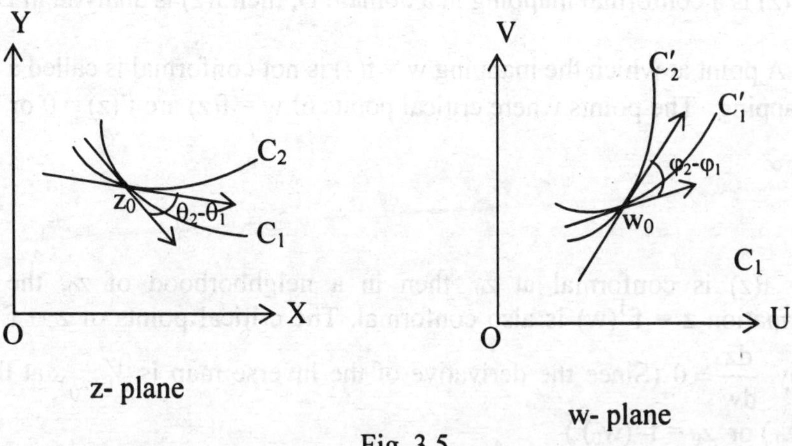

Fig. 3.5

$\theta_2 - \theta_1$ is the angle between C_1 and C_2. $\varphi_2 - \varphi_1$ is the angle between C_1' and C_2'. Equation (3) implies that the angle between C_1 and C_2 is equal to the angle between C_1' and C_2' both in magnitude and direction. Hence w = f(z) preserves angle between curves through z_0 both in magnitude and direction. $\therefore$ w = f(z) is conformal at z_0.

Note:

(i) From $w'(t_0) = f'(z_0).z'(t_0)$, we have

$\arg w'(t_0) = \arg f'(z_0) + \arg z'(t_0)$ (1)

$|w'(t_0)| = |f'(z_0)|.|z'(t_0)|$ (2)

$\arg f'(z_0)$ is called the **angle of rotation** at z_0 and $|f'(z_0)|$ is called the **scale factor** or **coefficient of magnification** at z_0.

Thus curves through z_0 are rotated through an angle $\arg f'(z_0)$ and infinitesimal lengths in the neighbourhood of z_0 are magnified by the factor $|f'(z_0)|$.

(ii) Let $w = f(z)$ be analytic at the point z_0.
Then the condition for conformality at the point z_0 is $f'(z_0) \neq 0$.
Now $f'(z_0) \neq 0 \Leftrightarrow |f'(z_0)| \neq 0$

$$\Leftrightarrow u_x^2 + v_x^2 \neq 0$$

$$\Leftrightarrow u_x v_y - v_x u_y \neq 0$$

$$\Leftrightarrow \frac{\partial(u, v)}{\partial(x, y)} \neq 0$$

Hence an equivalent condition for conformality of $w = f(z)$ at z_0 is the Jacobian of the transformation $u = u(x, y)$, $v = v(x, y)$ namely $\dfrac{\partial(u,v)}{\partial(x,y)} \neq 0$

(iii) (Necessary condition for $w = f(z)$ to be conformal)
If $w = f(z)$ is a conformal mapping in a domain D, then $f(z)$ is analytic in D.

Definition 3.3.3 A point at which the mapping $w = f(z)$ is not conformal is called a **critical point** of the mapping. The points where critical points of $w = f(z)$ are $f'(z) = 0$ or ∞.

i.e., $\dfrac{dw}{dz} = 0$ or ∞

Note:

If $w = f(z)$ is conformal at z_0, then in a neighborhood of z_0, the inverse transformation $z = f^{1}(w)$ is also conformal. The critical points of $z = f^{1}(w)$ are given by $\dfrac{dz}{dw} = 0$ (Since the derivative of the inverse map is $1/f'(z_0)$ at the point $w_0 = f(z_0)$ or $z_0 = f^{1}(w_0)$)

3.3.2 SOME SIMPLE MAPPINGS

1. Translation: $w = z + a$, where a is a complex constant.

Let $z = x + iy$, $w = u + iv$ and $a = a_1 + i\, a_2$.
Then $u + iv = x + iy + a_1 + i\, a_2$
$\therefore$ $u = x + a_1$ and $v = y + a_2$.
The image of the point (x, y) in the z-plane is the point $(x + a_1, y + a_2)$ in the w-plane. If we consider the z-plane and w-plane as the same plane, the transformation moves each point by a distance $|a| = \sqrt{a_1^2 + a_2^2}$ in the direction of a. i.e., the direction that makes an angle arg (a) with the positive direction of x-axis. Every point is moved by a distance a_1 in the direction of the x-axis and by a distance a_2 in the direction of the y-axis.

Thus the mapping $w = z + a$ is a **translation** by the vector representing a. Hence the mapping $w = z + a$ maps a straight line into a straight line, a circle into an equal circle and a region into another having the same shape, size and orientation.
When a is a real number (say) a_1, then $w = z + a_1$ is a translation parallel to the x-axis and when a is purely imaginary (say) ia_2, then $w = z + ia_2$ is a translation parallel to the y-axis. Since $\frac{dw}{dz} = 1$, a translation is conformal at all points in the complex plane.

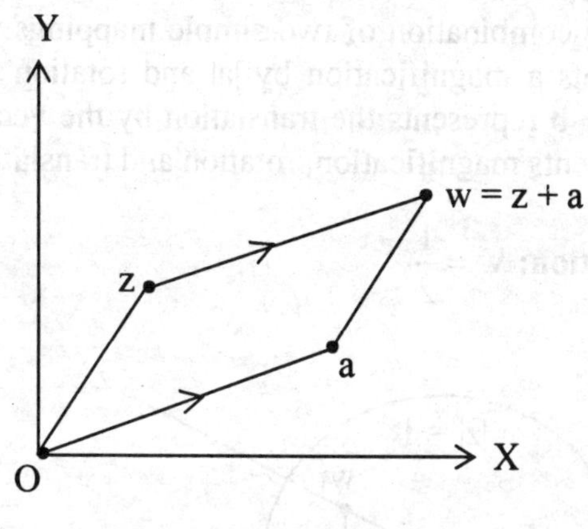

Fig. 3.6

The mapping $w = z + a$ takes the origin onto a and the points at ∞ onto the point at infinity. ∞ is the only fixed point under this mapping.

2. Magnification and rotation: $w = az$, where a is a complex constant $(a \neq 0)$

Let $z = re^{i\theta}$, $w = R.e^{i\varphi}$ and $a = \rho.e^{i\alpha}$.
Then $R.e^{i\varphi} = \rho.e^{i\alpha}.r.e^{i\theta}$.
$= \rho r.e^{i(\theta + \alpha)}$
$\therefore R = \rho r$ and $\varphi = \theta + \alpha$
The point (r, θ) is mapped onto $(\rho r, \theta + \alpha)$.
Thus the magnitude of z is magnified by $\rho = |a|$ and the direction is rotated through an angle $\alpha = \arg(a)$ about the origin in the anti-clockwise direction if $\alpha > 0$ and in the clockwise direction if $\alpha < 0$. Hence $w = az$ represents **magnification and rotation.** When $\rho > 1$ the mapping represents expansion and rotation, where as when $\rho < 1$ it is contraction and rotation. Note that 0 and ∞ are the two fixed points of the mapping $w = az$. The mapping takes straight lines onto straight lines and a circle with centre at origin and radius r is mapped onto a concentric circle with radius ρr. Since $\frac{dw}{dz} = a \neq 0$, $w = az$ is conformal at all points in the complex plane.

Note:

(i) $w = az$, where $|a| = 1$

In this case, the mapping represents a rotation about the origin, as the magnifying factor is $\rho = |a| = 1$

(ii) $w = az$, where a is real.

In this case, the mapping represents a magnification only, as the angle of rotation $\alpha = \arg(a) = 0$.

(iii) $w = az + b$, where a and b are complex constants. This mapping $w = az + b$ can be considered as the combination of two simple mappings $w_1 = az$ and $w = w_1 + b$. $w_1 = az$ represents a magnification by $|a|$ and rotation through arg(a) about the origin. $w = w_1 + b$ represents the translation by the vector representing b. Thus $w = az + b$ represents magnification, rotation and translation.

3. Inversion and Reflection: $w = \dfrac{1}{z}$

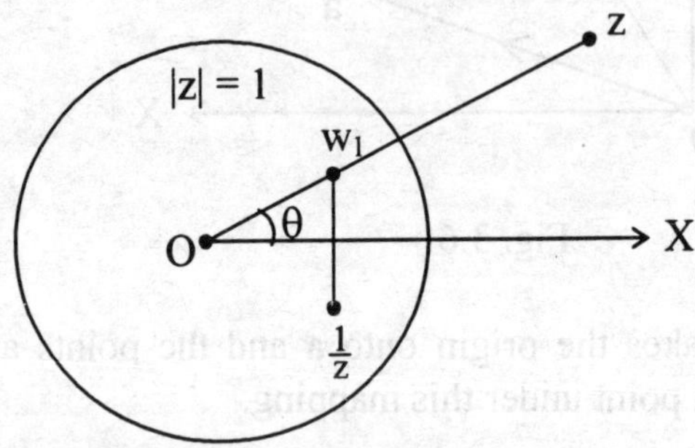

Fig. 3.7

Let $z = r.e^{i\theta}$. Then $w = \left(\dfrac{1}{r}\right)e^{-i\theta}$.

This mapping can be considered as the combination of two simple mappings $w_1 = \dfrac{1}{r}.e^{i\theta}$ and $w = \overline{w}_1$.

Now $|z| = r$ and $|w_1| = \dfrac{1}{r}$

$\therefore |z|.|w_1| = 1$.

Also $\arg(z) = \arg(w_1) = \theta$

Hence $w_1 = \dfrac{1}{r}e^{i\theta}$ represents **inversion** with respect to the unit circle $|z| = 1$. $w = \overline{w}_1$ represents **reflection** about the x-axis. Thus the mapping $w = \dfrac{1}{z}$ is the inversion with respect to the unit circle followed by reflection about the real axis. Points outside the unit circle are mapped into points inside the unit circle and vice versa. Points on the circle are reflected about the real axis and hence are mapped onto the unit circle itself. The origin is mapped onto the point at infinity and vice versa. The mapping has two fixed points, namely 1 and -1, the points of intersection of the unit circle with the real axis.

Example 3.3.1 Show that the transformation w = az maps straight lines onto straight lines and circles onto circles.

Solution:

Then equation $c.z\bar{z}+\bar{\alpha}z+\alpha\bar{z}+r=0$ (1)

Where c, r are real represents a circle if $c \neq 0$ and a straight line if c = 0. Substituting w = az or z = w/a in (1) we get the image of curve represented by (1).

Now equation (1) becomes $c.\frac{w}{a}\frac{\bar{w}}{\bar{a}}+\bar{\alpha}\frac{w}{a}+\alpha\frac{\bar{w}}{a}+r=0$

i.e., $\frac{c}{|a|^2}.w\bar{w}+\left(\frac{\bar{\alpha}}{a}\right).w+\left(\frac{\alpha}{\bar{a}}\right).\bar{w}+r=0$

i.e., $c_1.w\bar{w}+\bar{k}w+k.\bar{w}+r=0$ (2)

where $k=\frac{\alpha}{\bar{a}}$ and $c_1=\frac{c}{|a|^2}$

$c \neq 0 \Rightarrow c_1 \neq 0$ and $c = 0 \Rightarrow c_1 = 0$.

∴ When equation (2) represents a circle or a straight line, then equation (1) also represents a circle or a straight line.

Hence w = az maps a straight line onto a straight line and circle onto a circle.

Example 3.3.2 Show that under the mapping $w = 1/z$, circles and straight lines in the z-plane are transformed into circles and straight lines in the w-plane.

Solution:

The equation $a.z\bar{z}+\bar{\alpha}z+\alpha\bar{z}+r=0$ (1)

Represents a circle or straight line according as $a \neq 0$ or a = 0. Similarly the curve passes through the origin or not according as r = 0 or $r \neq 0$. The image of (1) in the w-plane is obtained by substituting $w = 1/z$. i.e., $z = 1/w$ ∴ equation (1) becomes,

$a.\frac{1}{w}\frac{1}{\bar{w}}+\bar{\alpha}\frac{1}{w}+\alpha\frac{1}{\bar{w}}+r=0$

i.e., $r.w\bar{w}+\alpha w+\bar{\alpha}.\bar{w}+a=0$ (2)

Equation (2) represents a circle or a straight line in the w-plane according as $r \neq 0$ or r = 0. This curve passes through the origin or not according as a = 0 or $a \neq 0$. Thus

(i) A circle ($a \neq 0$) not passing through the origin ($r \neq 0$) is mapped on to a circle not passing through the origin.

(ii) A circle ($a \neq 0$) passing through the origin ($r = 0$) is mapped onto a straight line not passing through the origin.

(iii) A straight line ($a = 0$) not passing through the origin ($r \neq 0$) is mapped onto a circle passing through the origin.

(iv) A straight line ($a = 0$) passing through the origin ($r = 0$) is mapped onto a straight line passing through the origin.

Example 3.3.3 Prove that $w = f(\bar{z})$ where f(z) is analytic is isogonal, but not conformal.

Solution:

The mapping $w = \bar{z}$ represents reflection about the real axis. Hence it preserves the angle between curves in magnitude, but reverses the direction. Hence it is a isogonal mapping. Since f(z) is analytic, $w = f(z)$ is conformal at all points where $f'(z) \neq 0$. Now $w = f(\bar{z})$ is a combination of the mappings

$$w_1 = \bar{z} \qquad (1)$$

$$w = f(w_1) \qquad (2)$$

In (1) the magnitude of the angles are preserved, but their signs are changed and in (2) both the magnitude and signs are preserved so that in the resulting mapping $w = f(\bar{z})$, magnitude of the angles are preserved but their signs are changed. i.e., $w = f(\bar{z})$ is isogonal but not conformal.

Example 3.3.4 Find the critical points of the transformation $w = 1/z$.

Solution:

$$w = 1/z$$

$$\therefore \frac{dw}{dz} = -\frac{1}{z^2} \to \infty \text{ as } z \to 0$$

$$\therefore w = 1/z \text{ is not analytic at } z = 0.$$

Hence the mapping $w = 1/z$ is not conformal at $z = 0$.

i.e., $z = 0$ is a critical point of the mapping.

Note:

In $w = 1/z$, put $z = re^{i\theta}$, $w = Re^{i\varphi}$.

Then $Re^{i\varphi} = -\frac{1}{r}e^{-i\theta}$

i.e., $R = 1/r$ and $\varphi = -\theta$.

If $\theta = \alpha_1$ and $\theta = \alpha_2$ ($\alpha_2 > \alpha_1$) are any two lines through the origin in the z-plane, angle between them is $\alpha_2 - \alpha_1$. But the angle between their images $\varphi = -\alpha_1$ and $\varphi = -\alpha_2$ is $\alpha_1 - \alpha_2$. Thus the angel between the lines through $z = 0$ is preserved in magnitude but not in direction. $\therefore$ w = $1/z$ is not conformal at z = 0. and hence z = 0 is a critical point of the mapping. Since the magnitude of the angle is preserved the mapping w = $1/z$ is isogonal at z = 0.

Example 3.3.5 Show that the mapping w = iz represents a rotation through angle $\pi/2$.

Solution:

Let $z = re^{i\theta}$ and $w = Re^{i\varphi}$.

Now $i = \cos \pi/2 + i \sin \pi/2 = e^{i\pi/2}$

$\therefore$ w = iz becomes $Re^{i\varphi} = e^{i\pi/2} . re^{i\theta} = re^{i(\theta + \pi/2)}$

$\therefore$ $R = r$ and $\varphi = \theta + \pi/2$

Hence the given mapping is a rotation through an angle $\pi/2$ about the origin, in the anti-clockwise direction.

Example 3.3.6 Find the image of the circle | z - 2i | = 2 under the mapping w = 1/z.

Solution:

| z - 2i | = 2 is a circle with centre at 2i and radius 2. Hence it passes through the origin. Under the mapping w = $1/z$, the image of the circle passing through the origin will be a straight line not passing through the origin. The equation of this line is given by $\left|\frac{1}{w} - 2i\right| = 2$.

i.e., $|1 - 2iw| = 2.|w|$

Substituting w = u + iv, we get,

$|1 - 2i(u + iv)| = 2.|(u + iv)|$

i.e., $|1 + 2v - 2iu| = 2.\sqrt{u^2 + v^2}$

i.e., $\sqrt{(1+2v)^2 + 4u^2} = 2.\sqrt{u^2 + v^2}$

i.e., $1 + 4v + 4v^2 + 4u^2 = 4u^2 + 4v^2$

i.e., $4v + 1 = 0$, which is a straight line parallel to real axis

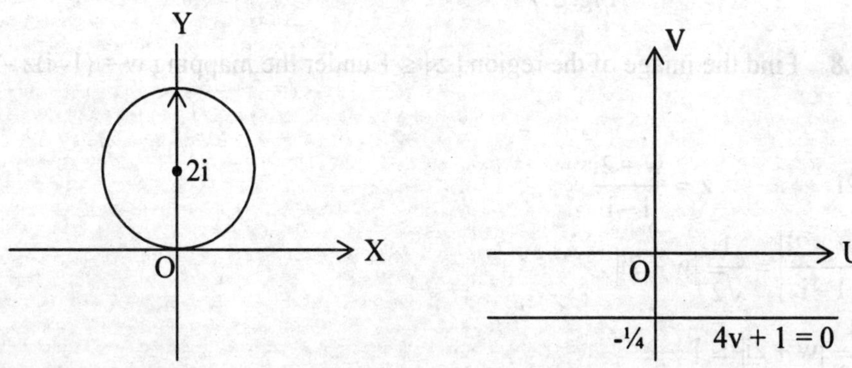

Fig. 3.8

Example 3.3.7 Find the mapping of the circle $|z - 2| = 3$ under the mapping $w = 1/z$.

Solution:

$|z - 2| = 3$ is a circle not passing through the origin. $\therefore$ the image of this circle will again a circle not passing through the origin. Its equation is given by

Under $\left|\frac{1}{w} - 2\right| = 3$

i.e., $|1 - 2w| = 3|w|$

Substituting $w = u + iv$, we get

$|1 - 2(u + iv)| = 3|u + iv|$

i.e., $(1 - 2u)^2 + 4v^2 = 9(u^2 + v^2)$

i.e., $5(u^2 + v^2) + 4u - 1 = 0$

i.e., $u^2 + v^2 + \frac{4}{5}u - \frac{1}{5} = 0$

This is a circle with center $\left(-2/5, 0\right)$ and radius $\sqrt{4/25 + 1/5} = 3/5$

The equation of the image circle in the w-plane can be written as $\left|w + \frac{2}{5}\right| = 3/5$

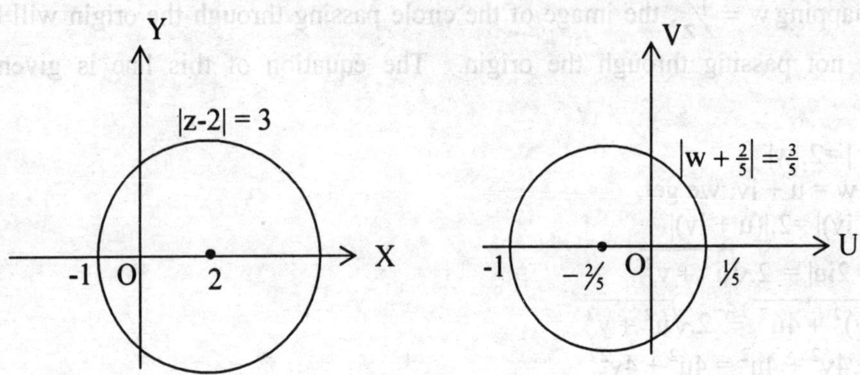

Fig. 3.9

Example 3.3.8 Find the image of the region $|z| \leq 1$ under the mapping $w = (1 - i)z - 2i$.

Solution:

$w = (1 - i)z - 2i \qquad \therefore z = \dfrac{w + 2i}{1 - i}$

Hence $|z| = \left|\dfrac{w + 2i}{1 - i}\right| = \dfrac{1}{\sqrt{2}}|w + 2i|$

$|z| \leq 1 \Rightarrow \dfrac{1}{\sqrt{2}}|w + 2i| \leq 1$

$\Rightarrow |w + 2i| \leq \sqrt{2}$

$\therefore$ The image of the unit disk $|z| \le 1$ is the region $|w+2i| \le \sqrt{2}$, which is again a circular disk with center $-2i$ and radius $\sqrt{2}$.

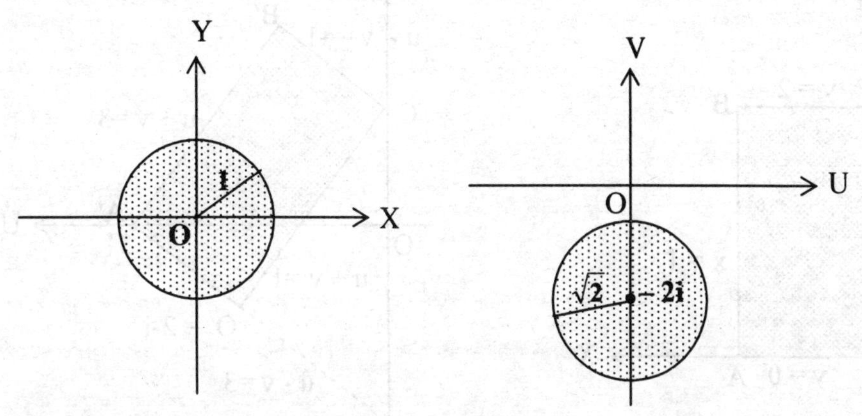

Fig. 3.10

Example 3.3.9 Determine the region in the w-plane into which the rectangular region bounded by the lines x = 0, y = 0, x = 1, y = 2 in the z-plane is mapped under the transformation $w = (1+i)z + 2 - i$.

Solution:

The rectangular region in the z-plane will be magnified by the factor $|1+i| = \sqrt{2}$, then rotated through an angel $\arg(1+i) = \pi/4$ and finally translated in the direction of $2 - i$ through a distance $|2 - i| = \sqrt{5}$.

The origin O is mapped onto the point $O' = 2 - i$.

Now, $w = (1+i)z + 2 - i$

i.e., $u + iv = (1+i)(x+iy) + 2 - i$

i.e., $u = (x - y + 2)$ and $v = x + y - 1$

The image of the line $x = 0$ is $u = -y + 2$, $v = y - 1$

i.e., $u + v = 1$

The image of the line $y = 0$ is $u = x + 2$, $v = x - 1$

i.e., $u - v = 3$

The image of the line $x = 1$ is $u = -y + 3$, $v = y$

i.e., $u + v = 3$

The image of the line $y = 2$ is $u = x$, $v = x + 1$

i.e., $u - v = -1$

Thus the rectangular region bounded by the lines x = 0, y = 0, x = 1 and x = 2 in the z-plane is mapped onto the rectangular region bounded by the lines u + v = 1, u - v = 3, u + v = 3 and u - v = -1 in the w-plane. Also note that the area of the new rectangle is double the area of the original rectangle (since the magnifying factor of length is $\sqrt{2}$).

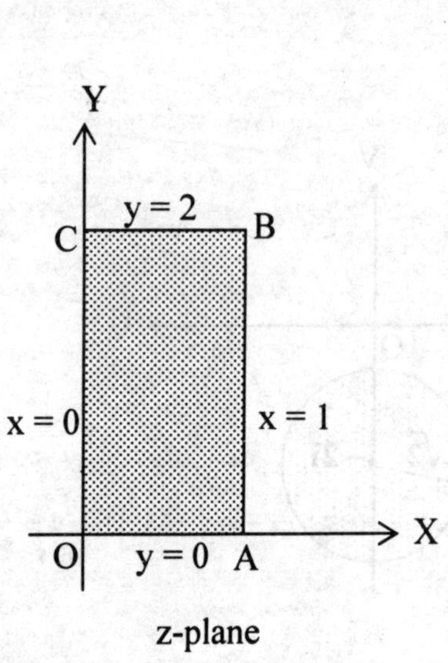

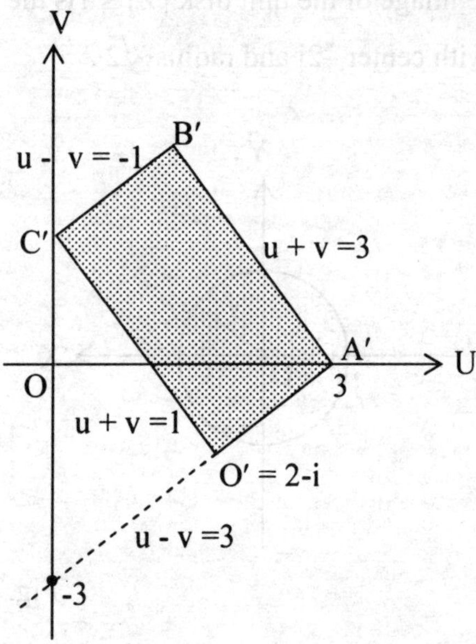

Fig. 3.11

Example 3.3.10 Find the image of the strip $1 < x < 2$ under the mapping $w = \frac{1}{z}$.

Solution:

Given $w = \frac{1}{z}$

i.e., $z = \frac{1}{w}$

i.e., $x + iy = \dfrac{1}{u + iv} = \dfrac{u - iv}{u^2 + v^2}$

$\therefore x = \dfrac{u}{u^2 + v^2}$ and $y = \dfrac{-v}{u^2 + v^2}$

Now $x > 1 \Rightarrow \dfrac{u}{u^2 + v^2} > 1$

$\Rightarrow u^2 + v^2 - u < 0$

$\therefore$ The region $x > 1$ is mapped onto the region $u^2 + v^2 - u < 0$.

The region $u^2 + v^2 - u < 0$ is the interior of the circle with center $\left(\frac{1}{2}, 0\right)$ and radius $\frac{1}{2}$.

Now $x < 2 \Rightarrow \dfrac{u}{u^2 + v^2} < 2$

$\Rightarrow u^2 + v^2 - \frac{1}{2}u > 0$

$\therefore$ The region $x < 2$ is mapped onto the region $u^2 + v^2 - \frac{1}{2}u > 0$.

The region $u^2 + v^2 - \frac{1}{2}u > 0$ is the exterior of the circle with center $\left(\frac{1}{4}, 0\right)$ and radius $\frac{1}{4}$.

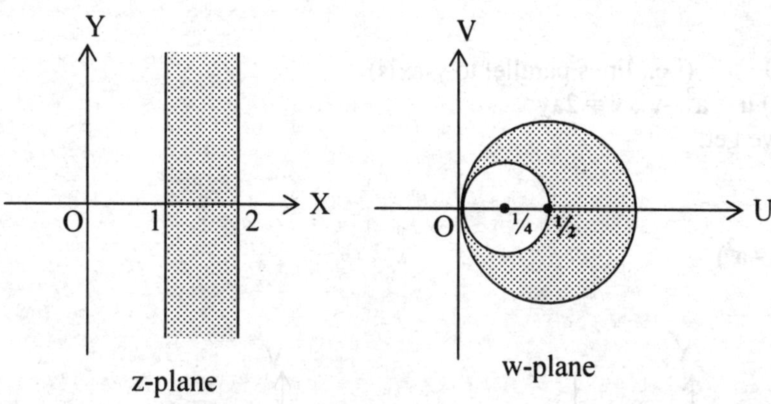

Fig. 3.12

$\therefore$ The strip $1 < x < 2$ is mapped onto the region bounded by the circles $u^2 + v^2 - u = 0$ and $u^2 + v^2 - \frac{1}{2}u = 0$ in the w –plane

3.3.3 SOME SPECIAL TRANSFORMATIONS

1. The transformation $w = z^2$

$$w = z^2 \tag{1}$$

Differentiating w.r.t z, $\dfrac{dw}{dz} = 2z = 0$ only at $z = 0$

$\therefore$ $w = z^2$ is conformal at all points except at the origin. $z = 0$ is the only critical point of the transformation. Substituting $z = re^{i\theta}$ and $w = R.e^{i\varphi}$, we get

$$R.e^{i\varphi} = r^2 e^{i2\theta}$$

$$\therefore R = r^2 \text{ and } \varphi = 2\theta$$

$r = a$ represents a family of concentric circles with center at the origin in the z-plane. Its image $R = a^2$ is again a family of concentric circles with center at the origin in the w-plane. The equation $\theta = \alpha$ represents a family of concurrent lines through the origin in the z-plane. Its image $\varphi = 2\alpha$ is again a family of concurrent lines through the origin in the w-plane. If $\theta = \alpha_1$ and $\theta = \alpha_2$ are two such lines and their images are $\varphi = 2\alpha_1$ and $\varphi = 2\alpha_2$, then the angle between the lines in the z-plane is $\alpha_2 - \alpha_1$. But the angle between the image line is $2\alpha_2 - 2\alpha_1 = 2(\alpha_2 - \alpha_1)$. Thus the mapping $w = z_2$ is not conformal at the origin.

Now substituting $w = u + iv$ and $z = x + iy$ in $w = z^2$, we get

$$u + iv = (x + iy)^2$$

$$= x^2 - y^2 + 2ixy$$

$\therefore u = x^2 - y^2$ (2)

and $v = 2xy$ (3)

(i) $x = a > 0$ (i.e., lines parallel to y-axis)

From (2) and (3) $u = a^2 - y^2$, $v = 2ay$

Eliminating y, we get,

$$u = a^2 - \frac{v^2}{4a^2}$$

i.e., $v^2 = -4a^2 (u - a^2)$ (4)

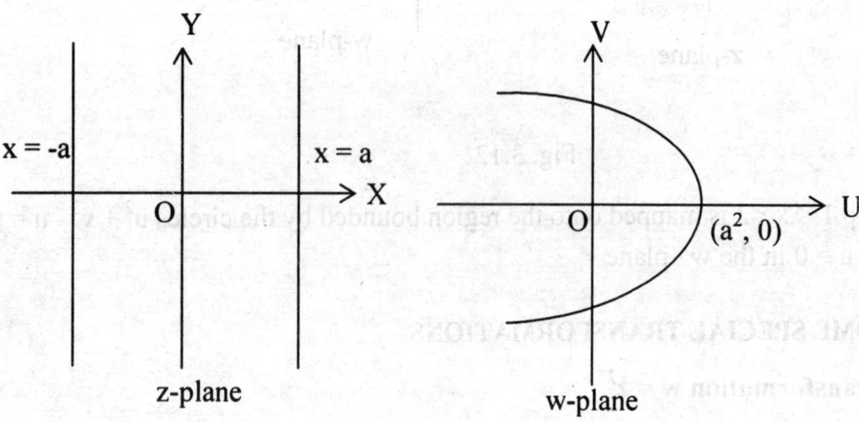

Fig. 3.13

which is a parabola in the w-plane with its vertex at $(a^2, 0)$, focus at the origin and axis along the real axis. The relation $v = 2ay$ implies that v is positive when y is positive and v is negative when y is negative. Hence the two parts of the lines $x = a$ lying above or below the real axis in the z-plane correspond to the parts of the parabola lying above and below the real axis in the w-plane. Replacing a by –a in (3), we get the same parabola. $\therefore$ the line $x = -a$ corresponds to the same parabola, but upper and lower parts of the line correspond to lower and upper parts of the parabola. Thus in general the line $x = \lambda$ $(\lambda \neq 0)$ in the z-plane corresponds to the parabola $v^2 = -4\lambda^2(u - \lambda^2)$ in the w-plane, where λ is a parameter.

(ii) When λ takes the values continuously from a to b, the line $x = \lambda$ describes the area lying between the lines $x = a$, $x = b$ in the z-plane. The parabola $v^2 = -4\lambda^2(u - \lambda^2)$ in the w-plane describes the area lying between the parabolas $v^2 = -4a^2(u - a^2)$ and $v^2 = -4b^2(u - b^2)$. Replacing a by –a and b by –b we get the same parabolas. Hence both the infinite strips defined by $a < x < b$ and $-b < x < -a$ are conformally mapped on the domain included between the parabolas $v^2 = -4a^2(u - a^2)$, $v^2 = -4b^2(u - b^2)$ (Fig. 3.14).

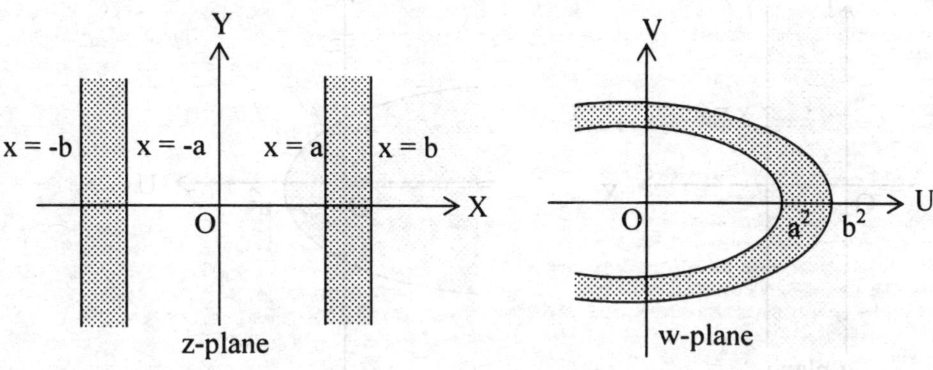

Fig. 3.14

(iii) As λ takes the values continuously from a to ∞ the line $x = \lambda$ describes the entire region of the z-plane lying to the right of the line x = a and the parabola $v^2 = -4\lambda^2(u - \lambda^2)$ describes the entire w-plane exterior to the parabola $v^2 = -4a^2(u - a^2)$. Hence to the half plane to the right of x = a is mapped conformally on the exterior of the parabola $v^2 = -4a^2(u - a^2)$ not containing the origin (Fig. 3.15). The half plane defined by $x < -a$ is also mapped onto the exterior of the same parabola.

(iv) When λ takes the values continuously from a to 0, the line $x = \lambda$ describes the area in the z-plane lying between the lines x = a and x = 0 and the parabola $v^2 = -4\lambda^2(u - \lambda^2)$ describes the region of the w-plane interior to the parabola $v^2 = -4a^2(u - a^2)$ with a cut along the negative real axis from $-\infty$ to 0 as shown below (Fig. 3.16).

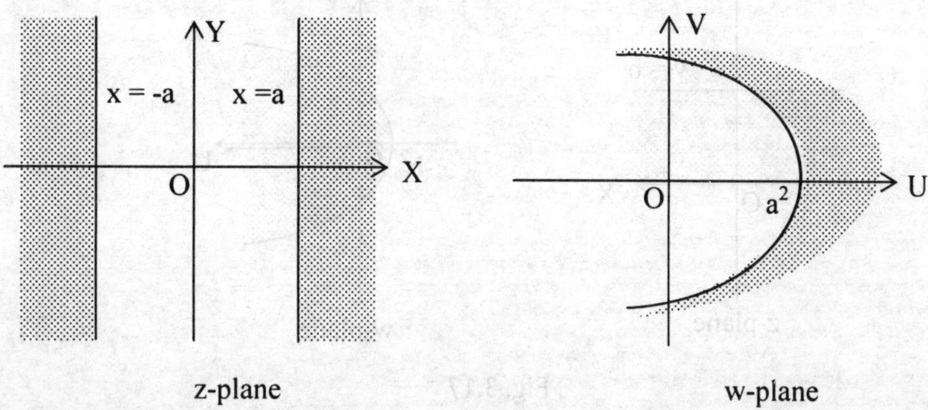

Fig. 3.15

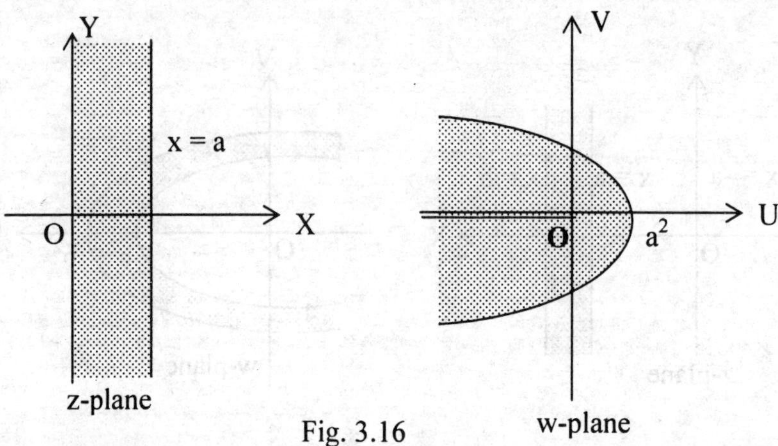

Fig. 3.16

The slit in the w-plane corresponds to the line $x = 0$ in the z-plane. Hence the image of the region $0 < x < a$ is the interior of the parabola $v^2 = -4a^2(u - a^2)$ with a slit along the negative real axis from $-\infty$ to 0. If $a \to \infty$, then the region $0 < x < \infty$ corresponds to the entire w-plane with a slit along the negative real axis from $-\infty$ to 0. The region $x < 0$ is also mapped onto the same whole plane with a slit.

(v) $y = b$ (i.e., lines parallel to x-axis)

From (2) and (3)

$$u = x^2 - b^2,\ v = 2bx. \quad \therefore\ u = \frac{v^2}{4b^2} - b^2 \qquad \text{i.e., } v^2 = 4b^2(u + b^2)$$

i.e., the image of the line $y = b$ is the parabola $v^2 = 4b^2(u + b^2)$ whose vertex is $(-b^2, 0)$, focus is the origin and axis is the real axis. The image of the half plane $y > 0$ is mapped on the whole w – plane with a slit along the positive real axis from 0 to ∞. The half plane

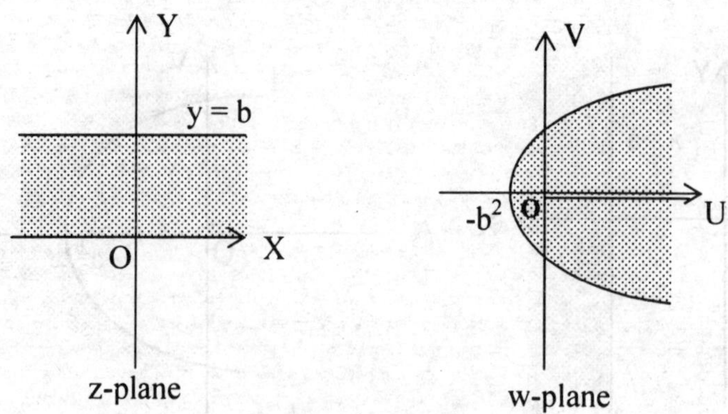

Fig. 3.17

$y < 0$ is also mapped onto the same whole plane with a slit.

(vi) The two families of parabolas $u^2 = -4a^2(u - a^2)$ and $v^2 = 4b^2(u + b^2)$ correspond to two orthogonal families of straight lines $x = \pm a$ and $y = \pm b$.

2. The transformation $w = e^z$ (1)

Differentiating w.r.t z, $\dfrac{dw}{dz} = e^z$ and $e^z \neq 0$ for all values of z

$\therefore$ The transformation $w = e^z$ is conformal at all points in the z-plane.

Substituting $w = \rho.e^{i\varphi}$ and $z = x + iy$, we get, $\rho.e^{i\varphi} = e^{x + iy}$

i.e., $\rho.e^{i\varphi} = e^x.e^{iy}$ $\therefore \rho = e^x$ and $\varphi = y$ (2).

(i) $y = c\ (0 < c < 2\pi)$

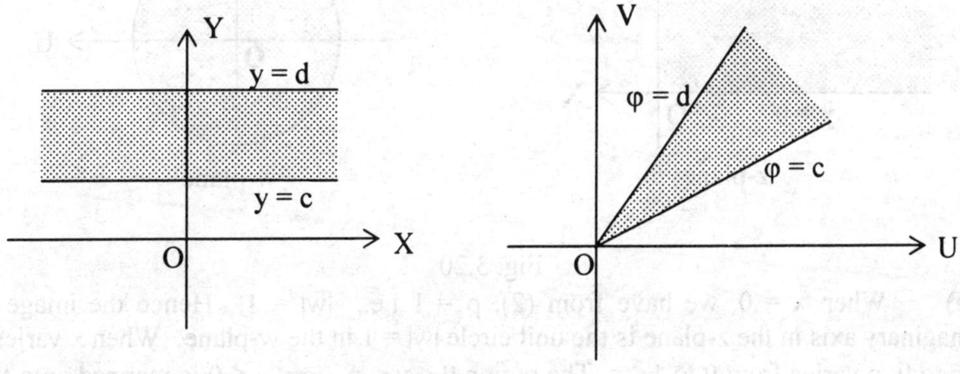

Fig. 3.18

When $y = c$, from (2) we have $\varphi = c$. Also $e^x = \rho$ increases from 0 to ∞ when x takes the values from $-\infty$ to ∞. Hence the image of the line $y = c$ is the ray $\varphi = c$. Hence lines parallel to the x-axis in the z-plane correspond to radial lines in the w-plane. Image of the infinite strip $c < y < d$ in the z-plane is the region $c < \varphi < d$ in the w-plane (Fig. 3.18) When $c = 0$, $d = \pi$ we get the infinite strip $0 < y < \pi$ and this infinite strip is mapped onto

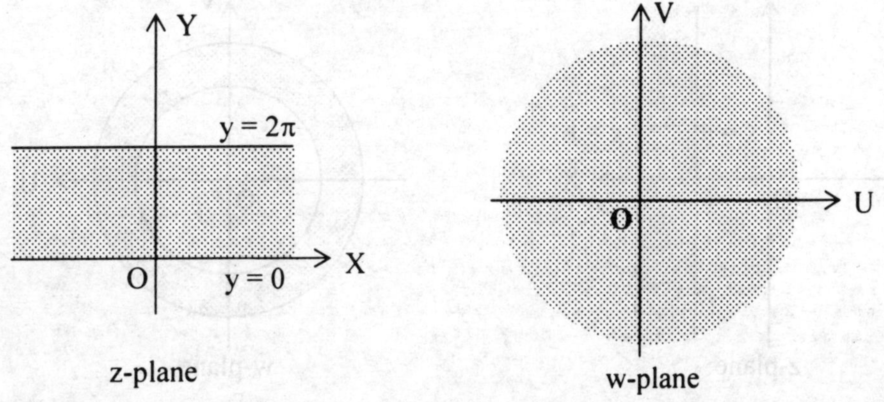

Fig. 3.19

the region $0 < \varphi < \pi$., the upper half of the w-plane. The strip $0 < y < 2\pi$ is mapped onto the whole w-plane with a slit along the positive real axis from 0 to ∞, the upper and lower edge of the slit corresponding to the lines $y = 0$ and $y = 2\pi$. The strip $0 \le y < 2\pi$ will be mapped onto the entire w-plane. As e^z is periodic with period $2\pi i$, we see that every parallel strip given by $n.2\pi \le y < (n+1)\,2\pi$, $n = 0, \pm1, \pm2, \ldots$ is mapped on to the entire w-plane (Fig. 3.19).

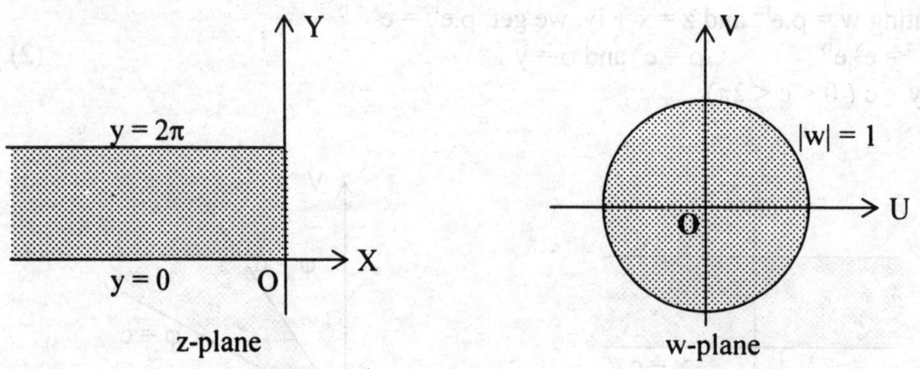

Fig. 3.20

(ii) When $x = 0$, we have from (2), $\rho = 1$ i.e., $|w| = 1$. Hence the image of the imaginary axis in the z-plane is the unit circle $|w| = 1$ in the w-plane. When x varies from $-\infty$ to 0, ρ varies from 0 to 1. $\therefore$ The region $0 \le y \le \pi$, $-\infty < x \le 0$ is mapped onto the unit semi circle $|w| \le 1$, $0 \le \varphi \le \pi$. The region $0 \le y < 2\pi$, $-\infty < x \le 0$ is mapped onto the region $|w| \le 1$ (Fig. 3.20)

(iii) $x = a$ (i.e, lines parallel to y-axis)

When $x = a$, from (1) we have $\rho = e^a$. If $w = u + iv$, then $\rho = \sqrt{u^2 + v^2}$.

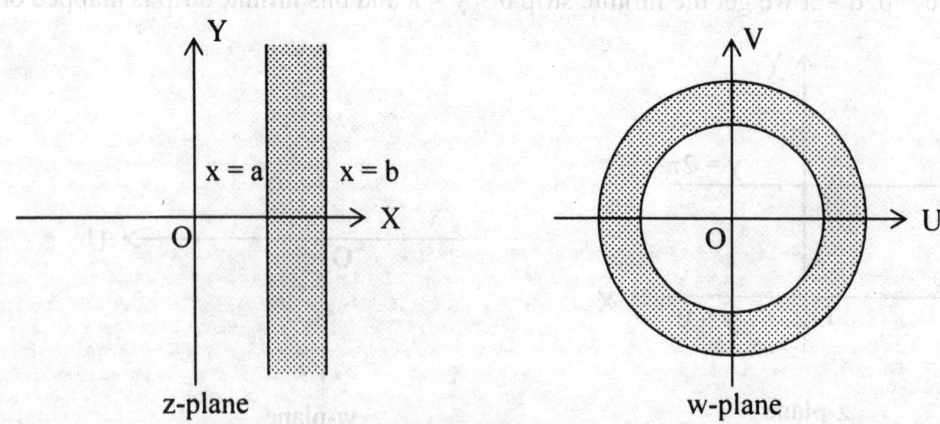

Fig. 3.21

i.e., $u^2 + v^2 = \rho^2 = e^{2a}$ = constant. ∴ The images of the lines parallel to y-axis are circles with center at the origin in the w-plane. Infinite strips bounded by the lines x = a and x = b (a < b) are transformed into the region between concentric circles given by $e^a < \rho < e^b$ (Fig. 3.21)

(iv) The image of the rectangular region $a \le x \le b$, $c \le y \le d$ is the region $e^a \le \rho \le e^b$, $c \le \varphi \le d$. (Fig. 3.22)

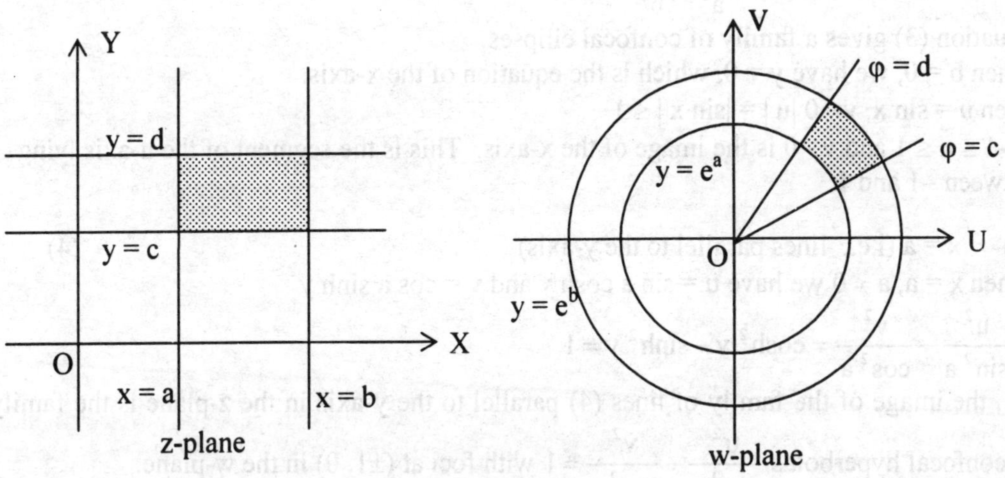

Fig. 3.22

3. The transformation w = sin z (1)

$$\frac{dw}{dz} = \cos z = 0 \text{ when } z = (2n+1)\frac{\pi}{2} \text{ for } n = 0, \pm 1, \pm 2, \ldots$$

∴ The mapping w = sin z is conformal at all points in the z-plane except at

$$z = (2n+1)\frac{\pi}{2},\ n = 0, \pm 1, \pm 2, \ldots$$

These points are the critical points of the mapping.

Substituting w = u + iv, z = x + iy in w = sin z, we get,

$$u + iv = \sin(x + iy)$$
$$= \sin x \cosh y + i \cos x \sinh y$$

i.e., $u = \sin x \cosh y$

$v = \cos x \sinh y$

(i) y = b (i.e., lines parallel to x-axis) (2)

When y = b, b ≠ 0 u = sin x cosh b and v = cos x sinh b

$$\therefore \frac{u^2}{\cosh^2 b} + \frac{v^2}{\sinh^2 b} = \sin^2 x + \cos^2 x = 1$$

Hence the image of the family of lines (2) parallel to the x-axis in the z-plane is the family of ellipses

$$\frac{u^2}{\cosh^2 b}+\frac{v^2}{\sinh^2 b}=1 \qquad (3)$$

These ellipse have their principal axes along the u and v axes, center at the origin and foci at the points (±1, 0)

[∵ The foci of the ellipse $\frac{x^2}{a^2}+\frac{y^2}{b^2}=1$ are at the points (±ae, 0). i.e., $(\pm\sqrt{a^2-b^2}, 0)$]

Equation (3) gives a family of confocal ellipses.

When b = 0, we have y = 0, which is the equation of the x-axis.

Then u = sin x, v= 0 |u | = |sin x | ≤ 1

∴ -1 ≤ u ≤ 1 and v =0 is the image of the x-axis. This is the segment of the u-axis lying between –1 and 1.

(ii) x = a (i.e., lines parallel to the y-axis) (4)

When x = a, a ≠ 0 we have u = sin a cosh y and v = cos a sinh y

$$\therefore \frac{u^2}{\sin^2 a}-\frac{v^2}{\cos^2 a}=\cosh^2 y-\sinh^2 y=1$$

i.e., the image of the family of lines (4) parallel to the y-axis in the z-plane is the family of confocal hyperbolas $\frac{u^2}{\sin^2 a}-\frac{v^2}{\cos^2 a}=1$ with foci at (±1, 0) in the w-plane.

[∵The foci of the hyperbola $\frac{x^2}{a^2}-\frac{y^2}{b^2}=1$ are at the points (±ae, 0). i.e., $(\pm\sqrt{a^2+b^2}, 0)$]

When a = 0, we have x = 0, which is the equation of y-axis. Then u = 0, v = sinh y. u = 0 is the v-axis in the w-plane. Thus the image of the y-axis is the v-axis.

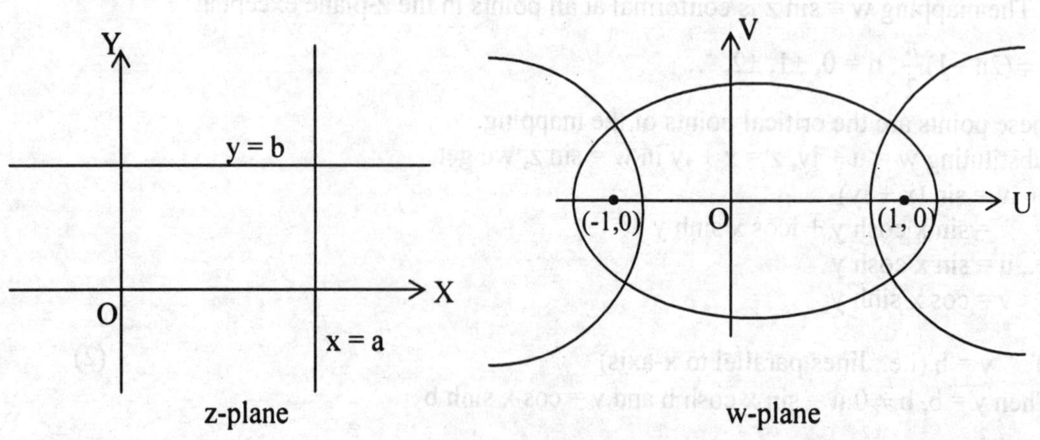

Fig. 3.23

Thus the transformation w = sin z transforms the family of lines parallel to the x-axis into the family of confocal ellipses with foci at (±1, 0) and the family of lines parallel to the y-axis into the family of confocal hyperbolas with foci at (±1, 0).

Example 3.3.11 Prove that the transformation $w = \frac{1}{2}(z + \frac{1}{z})$ maps the two ring shaped region $\lambda_1 < |z| < \lambda_2$ and $\frac{1}{\lambda_2} < |z| < \frac{1}{\lambda_1}$ conformally onto the region bounded by two confocal ellipses, where $\lambda_2 > \lambda_1 > 1$.

Solution:

$$w = \frac{1}{2}(z + \frac{1}{z}) \qquad (1)$$

$$\frac{dw}{dz} = \frac{1}{2}\left(1 - \frac{1}{z^2}\right) = 0 \text{ when } z = \pm 1.$$

∴ The transformation is conformal except at z = ±1.

Let w = u + iv and z = r (cos θ + i sin θ)

The equation (1) becomes,

$$u + iv = \frac{1}{2}\left[r(\cos\theta + i\sin\theta) + \frac{1}{r}(\cos\theta - i\sin\theta)\right]$$

$$\therefore u = \frac{1}{2}(r + \frac{1}{r})\cos\theta \qquad (2)$$

$$v = \frac{1}{2}(r - \frac{1}{r})\sin\theta \qquad (3)$$

When |z| = r = λ, we have from (2) and (3)

$$u = \frac{1}{2}(\lambda + \frac{1}{\lambda})\cos\theta \qquad v = \frac{1}{2}(\lambda - \frac{1}{\lambda})\sin\theta$$

$$\cos^2\theta = \frac{u^2}{\frac{1}{4}\left(\lambda + \frac{1}{\lambda}\right)^2}, \quad \sin^2\theta = \frac{v^2}{\frac{1}{4}\left(\lambda - \frac{1}{\lambda}\right)^2}$$

Eliminating θ, we get,

$$\frac{u^2}{\frac{1}{4}\left(\lambda + \frac{1}{\lambda}\right)^2} + \frac{v^2}{\frac{1}{4}\left(\lambda - \frac{1}{\lambda}\right)^2} = 1 \qquad (4)$$

Put $a = \frac{1}{2}(\lambda + \frac{1}{\lambda})$, $b = \frac{1}{2}(\lambda - \frac{1}{\lambda})$. Then $\frac{u^2}{a^2} + \frac{v^2}{b^2} = 1$, which is an ellipse with semi-major and semi-minor axes a and b and eccentricity $e = \sqrt{1 - b^2/a^2} = \frac{2}{\lambda + 1/\lambda}$.

The foci of the ellipse are at (±ae, 0) where $ae = \frac{1}{2}(\lambda + 1/\lambda).\frac{2}{\lambda + 1/\lambda} = 1$

$\therefore$ The foci of the ellipse are at $(\pm 1, 0)$, which is independent of λ. Thus the image of any circle $|z| = \lambda$ is an ellipse with foci at $(\pm 1, 0)$. The equation (4) of the ellipse remains unchanged, when λ is replaced by $1/\lambda$. Hence both the circles $z = \lambda$ and $z = 1/\lambda$ are mapped onto the same ellipse when $\lambda > 1$, $1/\lambda < 1$. $\therefore$ When $z = \lambda$ contains the unit circle, $z = 1/\lambda$ is contained in the unit circle. The region bounded by $\lambda_1 < |z| < \lambda_2$ is mapped onto the region bounded by two confocal ellipses. When $\lambda_1 < \lambda_2$, $1/\lambda_2 < 1/\lambda_1$. Further the region bounded by $1/\lambda_2 < |z| < 1/\lambda_1$ is also mapped onto the region bounded by the same confocal ellipses which are the images of $|z| = \lambda_1$ and $|z| = \lambda_2$. Thus the two ring shaped regions $\lambda_1 < |z| < \lambda_2$ and $1/\lambda_2 < |z| < 1/\lambda_1$ in the z-plane are mapped conformally on the region bounded by the two confocal ellipses.

Example 3.3.12 Find the image of the infinite strip $a \le x \le b$ $(0 < a < b < \pi/2)$ under the transformation $w = \cos z$.

Solution:

$w = \cos z$ (1)

Put $w = u + iv$, $z = x + iy$.

Then from (1), we have $u + iv = \cos x \cos iy - \sin x \sin iy$.

i.e., $u + iv = \cos x \cosh y - i \sin x \sinh y$.

$u = \cos x \cosh y$ (2)

$v = -\sin x \sinh y$ (3)

When $x = \lambda$, from (2) and (3) we get,

$$\cosh y = \frac{u}{\cos\lambda} \text{ and } \sinh y = \frac{-v}{\sin\lambda}$$

$\therefore$ Eliminating y we get, $\dfrac{u^2}{\cos^2\lambda} - \dfrac{v^2}{\sin^2\lambda} = 1 \quad (\because \cosh^2 y - \sinh^2 y = 1)$

This is a hyperbola with transverse and conjugate axes $\cos\lambda$ and $\sin\lambda$.

Eccentricity e is given by $e = \sqrt{\dfrac{b^2}{a^2} + 1} = \sqrt{\dfrac{\sin^2\lambda}{\cos^2\lambda} + 1} = \sec\lambda$.

Foci of the hyperbola is $(\pm ae, 0) = (\pm 1, 0)$

When $0 < x < \pi/2$ and $-\infty < y < \infty$, (2) implies that u is positive and (3) implies that v can be both positive and negative. Thus the image of the line $x = \lambda$ is the branch of the hyperbola $\dfrac{u^2}{\cos^2\lambda} - \dfrac{v^2}{\sin^2\lambda} = 1$ that lies to the right of the imaginary axis.

When λ varies from a to b $(0 < a < b < \pi/2)$, $\cos\lambda$ decreases and $\sin\lambda$ increases.

$\therefore$ The successive hyperbolas go on falling outside the preceding ones.

Thus the infinite strip $a \le x \le b$ in the z-plane transforms into the region between the hyperbola,

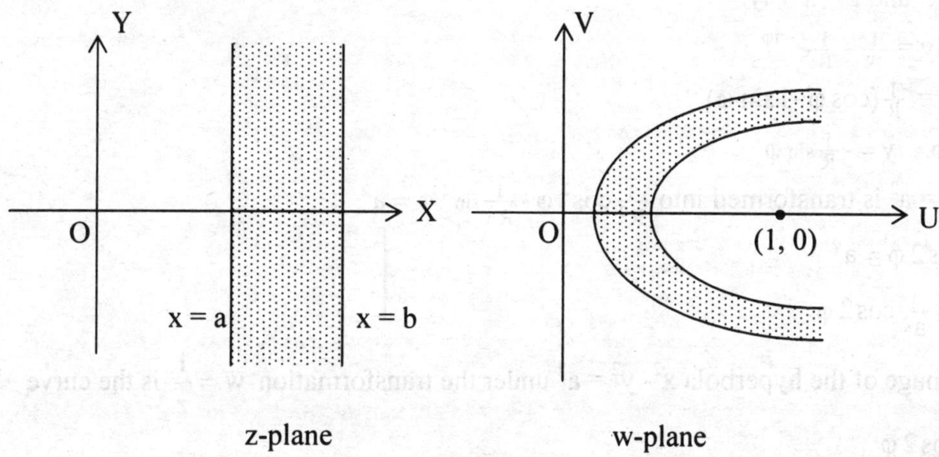

Fig. 3.24

$\frac{u^2}{\cos^2 a} - \frac{v^2}{\sin^2 a} = 1$ and the hyperbola, $\frac{u^2}{\cos^2 b} - \frac{v^2}{\sin^2 b} = 1$ lying to the right of the imaginary axis in the w-plane.

Example 3.3.13 Prove that the transformation w = log z transforms circles with centre at origin onto straight lines parallel to the imaginary axis in the w-plane and sectorial area bounded by radial lines $\theta = \alpha, \theta = \beta$ onto the infinite strip $\alpha \le v \le \beta$.

Solution:

$w = \log z$ (1)

Put $w = u + iv$ and $z = re^{i\theta}$ so that $w = u + iv = \log z = \log r + i\theta$.

$\therefore u = \log r$ and $v = \theta$.

The circle $r = \lambda$ in the z-plane is transformal into the straight line $u = \log \lambda$ in the w-plane. The circle $r = \lambda$ has centre at origin and line $u = \log \lambda$ is parallel to the imaginary axis in the w-plane.

The radial lines $\theta = \alpha$ and $\theta = \beta$ are transformal into straight lines $v = \alpha$ and $v = \beta$ that are parallel to the real axis in the w-plane. Hence the area bounded by radial lines $\theta = \alpha$ and $\theta = \beta$ is mapped onto the infinite strip $\alpha \le v \le \beta$ parallel to the real axis.

Example 3.3.14 Find the image of the hyperbola $x^2 - y^2 = a^2$ under the transformation $w = 1/z$ and $w = z^2$.

Solution:

$$w = \frac{1}{z} \quad (1)$$

Put $w = Re^{i\varphi}$ and $z = x + iy$.

$\therefore z = x + iy = \frac{1}{w} = \frac{1}{R} e^{-i\varphi}$

i.e., $x + iy = \frac{1}{R}(\cos\varphi - i\sin\varphi)$

$x = \frac{1}{R}\cos\varphi, \quad y = -\frac{1}{R}\sin\varphi$

Now $x^2 - y^2 = a^2$ is transformed into $\frac{1}{R^2}\cos^2\varphi - \frac{1}{R^2}\sin^2\varphi = a^2$

i.e., $\frac{1}{R^2}\cos 2\varphi = a^2$

i.e., $R^2 = \frac{1}{a^2}\cos 2\varphi$

Thus the image of the hyperbola $x^2 - y^2 = a^2$ under the transformation $w = \frac{1}{z}$ is the curve

$R^2 = \frac{1}{a^2}\cos 2\varphi$.

In $w = z^2$, put $w = u + iv$ and $z = x + iy$

$\therefore u + iv = (x + iy)^2$

$= x^2 - y^2 + 2ixy$

i.e., $u = x^2 - y^2$ (2)

$v = 2xy$ (3)

When $x^2 - y^2 = a^2$, from (1) we get, $u = a^2$

$\therefore$ The image of the hyperbola under the transformation $w = z^2$ is a straight line $u = a^2$, parallel to the v-axis in the w-plane.

Example 3.3.15 Under the mapping $w = f(z) = z^2$, find the image of the region bounded by the lines $x = 1$, $y = 1$ and $x + y = 1$.

Solution:

$w = z^2$

i.e., $u + iv = (x + iy)^2$

$= x^2 + 2ixy - y^2$

$\therefore u = x^2 - y^2$ (1)

$v = 2xy$ (2)

When $x = 1$, $u = 1 - y^2$ and $v = 2y$

$y^2 = 1 - u$ and $y = v/2$

$\therefore \left(\frac{v}{2}\right)^2 = 1 - u$

i.e., $v^2 = -4(u - 1)$ (3)

Hence image of the line $x = 1$ is the parabola $v^2 = -4(u - 1)$

When $y = 1$, from (1) and (2) we get

$u = x^2 - 1$, $v = 2x$

$x^2 = u + 1$, $x = v/2$

$$\therefore \left(\frac{v}{2}\right)^2 = u + 1$$

$$v^2 = 4(u+1) \qquad (4)$$

Hence the image of the line y = 1 is the parabola $v^2 = 4(u + 1)$

The image of x + y = 1 is obtained by eliminating x and y from (1), (2) and x + y = 1

Now y = 1 - x

$$\therefore (1) \Rightarrow u = x^2 - (1-x)^2 = -1 + 2x$$

$$\therefore x = \frac{u+1}{2}$$

$$(2) \Rightarrow v = 2x(1 - x)$$

$$= 2\left(\frac{u+1}{2}\right)\left(1 - \frac{u+1}{2}\right)$$

$$v = (u+1)\left(\frac{1-u}{2}\right)$$

$$\text{i.e., } v = \frac{1-u^2}{2}$$

$$u^2 = -2(v - \tfrac{1}{2}) \qquad (5)$$

(5) gives the image of x + y = 1, which is again a parabola. Hence the region bounded by the three lines x = 1, y = 1 and x + y = 1 is the region bounded by the three parabolas given by equations (3), (4) and (5), shown in fig. 3.25.

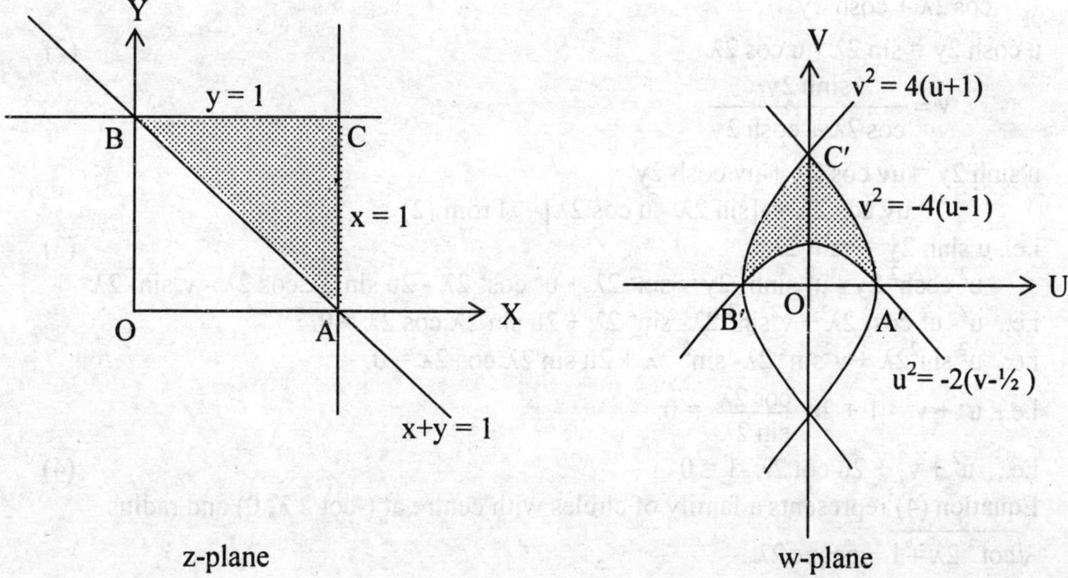

Fig. 3.25

The three lines in the z-plane meet at A(1, 0), B(0, 1) and C(1, 1). Their images under $w = z^2$ are A′(1, 0), B′(-1, 0) and C′(0, 2). Hence the required region is the one that contains the points A′, B′ and C′ in the w-plane.

Example 3.3.16 Show that the transformation w = tan z transforms the family of straight lines x = λ into the family of circles $u^2+v^2+2u^2 \cot 2\lambda - 1 = 0$ passing through a pair of fixed points in the w-plane.

Solution: $w = \tan z$ (1)

Let w = u + iv and z = x + iy.

Then we have

$$u + iv = \tan(x + iy)$$

$$= \frac{\sin(x+iy)}{\cos(x+iy)} = \frac{2\sin(x+iy)\cos(x-iy)}{2\cos(x+iy)\cos(x-iy)}$$

$$= \frac{\sin 2x + \sin 2iy}{\cos 2x + \cos 2iy}$$

$$= \frac{\sin 2x + i\sinh 2y}{\cos 2x + \cosh 2y}$$

$$\therefore u = \frac{\sin 2x}{\cos 2x + \cosh 2y}, \quad v = \frac{\sinh 2y}{\cos 2x + \cosh 2y}$$

Let x = λ, then

$$u = \frac{\sin 2\lambda}{\cos 2\lambda + \cosh 2y}$$

$$u \cosh 2y = \sin 2\lambda - u \cos 2\lambda \quad (2)$$

$$v = \frac{\sinh 2y}{\cos 2\lambda + \cosh 2y}$$

$$u \sinh 2y = uv \cos 2\lambda + uv \cosh 2y$$

$$= uv \cos 2\lambda + v[\sin 2\lambda - u \cos 2\lambda] \quad \text{From (2)}$$

i.e., $u \sinh 2y = v\sin 2\lambda$ (3)

$$u^2 \cosh^2 2y - u^2 \sinh^2 2y = \sin^2 2\lambda + u^2 \cos^2 2\lambda - 2u \sin 2\lambda.\cos 2\lambda - v^2\sin^2 2\lambda$$

i.e., $u^2 - u^2 \cos^2 2\lambda + v^2\sin^2 2\lambda - \sin^2 2\lambda + 2u \sin 2\lambda.\cos 2\lambda = 0.$

i.e., $u^2 \sin^2 2\lambda + v^2\sin^2 2\lambda - \sin^2 2\lambda + 2u \sin 2\lambda.\cos 2\lambda = 0.$

i.e., $u^2 + v^2 - 1 + 2u \dfrac{\cos 2\lambda}{\sin 2\lambda} = 0.$

i.e., $u^2 + v^2 + 2u \cot 2\lambda - 1 = 0$ (4)

Equation (4) represents a family of circles with centre at (-cot 2 λ, 0) and radius $\sqrt{\cot^2 2\lambda + 1} = \operatorname{cosec} 2\lambda$.

All these circles pass through the points (0, ±1).

i.e., w = i and w = -i.

EXERCISE 3.3

PART-A

1. Define 'conformal mapping'. Give an example.
2. Define 'isogonal mapping'. Give an example.
3. What is the necessary condition for $w = f(z)$ to be conformal?
4. What is the sufficient condition for $w = f(z)$ to be conformal?
5. Define 'critical points' of a transformation.
6. How will you find the critical points of $w = f(z)$ when $f(z)$ is an analytic function?
7. How will you find the critical points of the transformation $w = f(z)$?
8. Find the critical points of $w = \log z$.
9. Find the critical points of $w = \frac{1}{2}\left(ze^{-\alpha} + \frac{e^{\alpha}}{z}\right)$.
10. Find the critical points of $w = z + 1/z$.
11. Find the points where $w = \cosh z$ is not conformal.
12. Find the critical points of $w = \sinh z$.
13. Find the angle of rotation and scale factor at $z = 1 + i$ under the mapping $w = z^2$.
14. Find the coefficient of magnification and angle of rotation for $w = z^3$ at $1+i$.
15. Find the scale factor and angle of rotation of the transformation $w = 1/z$ at $z = 1$.
16. What is the image of the circle $|z| = 1$ under the transformation $w = 1/z$?
17. What are the fixed points of the transformation $w = 1/z$.
18. Find the image of $|z - 2i| = 2$ under the transformation $w = 1/z$.
19. Find the image of the strip $1 < x < 2$ under the mapping $w = 1/z$.
20. Find the image of the strip $0 < x < 2$ under the mapping $w = iz$.
21. Find the image of the region $y > 1$ under the mapping $w = iz + 1$.
22. Find the image of the semi-infinite strip $x > 0$, $0 < y < 2$ under $w = iz + 1$.
23. Find the image of $x^2 - y^2 = 1$ under the transformation $w = 1/z$.
24. Find the image of $|z| = a$ under $w = (3 + 4i)z$.
25. Find the image of the region $x > 0$ under $w = iz + i$.
26. Find the image of the region $x < 0$ under $w = -iz + i$.
27. Find the image of the hyperbola $xy = 1$ under $w = 2z + 1$.
28. Find the image of the region $|z| < r$ under $w = z^n$.
29. Find the image of the region $0 < \theta < 2\pi/n$ in the z-plane under $w = z^n$, where n is a positive integer.
30. Find the image of the region $0 < \theta < 2\pi/n$ in the z-plane under $w = z^n$.

31. Find the image of the region $0 \le \arg z \le \pi/2$ under $w = z^2$.
32. Find the region into which the sector $0 < \theta < \pi/4$, $r < 1$ is mapped by $w = z^2$.
33. What is the image of the line $y = c$ under $w = e^z$?
34. What is the image of the line $x = c$ under $w = e^z$?
35. Find the image of the region $1 < |z| < 3$ under $w = z^2$.
36. What is the image of the x-axis under the transformation $w = \sin z$.
37. Find the image of $y = \pi/2$ under $w = \cosh z$.
38. Find the image of the y-axis under $w = \cosh z$.

PART-B

39. Discuss the transformation $w = z^n$, where n is a positive integer. Prove that the sector $0 < \arg z < \pi/n$ is mapped conformally onto the entire upper half plane $I(w)>0$.
40. Show that the transformation $z = \sqrt{w}$ transforms the family of circles $|w - 1| = \lambda$ into the family of lemniscates $|z - 1|.|z + 1| = \lambda$ with focal points at $z = \pm 1$.
41. Find the image of the square region with vertices (0, 0), (2, 0), (2, 2), (0, 2) under the transformation $w = (1 + i)z + 2 + i$.
42. Find the image of the square region with vertices (0, 0), (1, 0), (1, 1), (0, 1) under the transformation $w = z + 3 - i$.
43. Find the image of the rectangular region bounded by $x = 0$, $y = 0$, $x = 2$, $y = 1$ under the transformation $w = z + 1 - 2i$.
44. Find the image of the region bounded by $x = 0$, $y = 0$, $x = 2$, $y = 1$ under the transformation $w = (1 + i)z + 2 - i$.
45. Prove that $w = 1/z$ transforms the circle $|z - 2| = 7$ into the circle $|w + 2/45| = 7/45$
46. Find the image of the circle $|z - 3i| = 3$ under the map $w = 1/z$.
47. Find the preimage in the z-plane, of the square region in the w-plane bounded by the lines $u = 1$, $u = 2$, $v = 1$ and $v = 2$ under the transformation $w = z^2$.
48. Find the image of the infinite strip $x \le 0$, $0 \le y \le \pi$ under the transformation $w = e^z$.
49. Show that the image of the unit circle $|z| = 1$ under the transformation $w = \frac{1}{2}\left(z + \frac{1}{z}\right)$ is the line segment $-1 \le u \le 1$ on the u-axis in the w-plane and this line segment is described twice as arg z varies from 0 to 2π.
50. Show that the transformation $w = z + \frac{1}{z}$ maps the circle $r = c$ into the ellipse

$$u = \left(c + \frac{1}{c}\right)\cos\theta,\ v = \left(c - \frac{1}{c}\right)\sin\theta.$$

51. Show that under the transformation $w = z + \frac{a^2}{z}$, when z describes a circle $x^2 + y^2 = a^2$, w describes a straight line of length 4a.

52. Show that under the transformation $w = z + \frac{4}{z}$, when z describes the circle $|z| = a$, w describes an ellipse with foci at the points $w = \pm 4$.
53. Show that under the transformation $w = z + \frac{4}{z}$, the area between $|z| = 4$ and $|z| = 6$ maps into the area between the ellipse whose semi-major axes are 5 and $20/3$.
54. Under the mapping $w = z^2$, find the image of the region bounded by the lines $x = 1$, $y = 1$ and $x + y = 1$. Is the mapping conformal in this region.

3.4 BILINEAR TRANSFORMATIONS

Definition 3.4.1 The transformation T defined by $w = T(z) = \frac{az+b}{cz+d}$, where a, b, c, d are complex constants such that ad - bc ≠ 0, is called a **bilinear transformation.** It is also called a **linear fractional transformation.**

The bilinear transformation is also called a **Mobius transformation,** after the mathematician **Mobius**. The determinant $\begin{vmatrix} a & b \\ c & d \end{vmatrix} = ad - bc$ is called the **determinant** of the bilinear transformation. If ad - bc = 1, the bilinear transformation is called a **normalized bilinear transformation**.

The transformation $w = \frac{az+b}{cz+d}$ (1)

can be written as: $cwz + dw - az - b = 0$ (2)

Equation (2) is linear in z as well as w. Hence T is called a **bilinear** transformation.

From (2) we have $z = \frac{-dw+b}{cw-a}$

The **inverse transformation** T^{-1} is defined by $z = T^{-1}(w) = \frac{-dw+b}{cw-a}$.

The determinant of T^{-1} is $= \begin{vmatrix} -d & b \\ c & -a \end{vmatrix} = (-d)(-a) - c.b = ad - bc \neq 0.$

The simple transformations w = z + a (translation), w = az (magnification and rotation) and $w = 1/z$ (inversion and reflection) are special types of bilinear transformations.

The transformation T can be written as $w = \frac{a}{c} - \frac{(ad-bc)}{c^2(z+d/c)}$ (3)

If ad - bc = 0, $w = a/c$, which is a constant function. Hence all the points of the z-plane are mapped onto a single point a/c in the w-plane.

3.4.1 CRITICAL POINTS OF THE BILINEAR TRANSFORMATION

$$w = \frac{az+b}{cz+d},\ ad - bc \neq 0$$

$$\frac{dw}{dz} = \frac{(cz+d).a-(az+b).c}{(cz+d)^2} = \frac{ad-bc}{(cz+d)^2}$$

If $z = -d/c$, $\frac{dw}{dz} = \infty$

If $z = \infty$, $\frac{dw}{dz} = 0$

Hence **critical points of the bilinear transformation** T are $z = -d/c$ and $z = \infty$. At these two points the mapping is not conformal.

For the inverse transformation, $z = T^{-1}(w) = \frac{-dw+b}{cw-a}$

$$\frac{dw}{dz} = \frac{(cw-a).(-d)-(-dw+b).c}{(cw-a)^2} = \frac{ad-bc}{(cw-a)^2}$$

If $w = a/c$, then $\frac{dz}{dw} = \infty$

If $w = \infty$, then $\frac{dz}{dw} = 0$

$\therefore$ The critical points of T^{-1} are $w = a/c$ and $w = \infty$.

Note:

The bilinear transformation $w = T(z) = \frac{az+b}{cz+d}$, $ad - bc \neq 0$ associates with each point in the z-plane a unique point in the w-plane except a point $z = -d/c$ $(c \neq 0)$. With $z = -d/c$ we can associate the point ∞ in the extended w-plane i.e., $\mathbb{C} \cup \{\infty\}$. Similarly the inverse transformation $z = T^{-1}(w) = \frac{-dw+b}{cw-a}$ associates with each point in the w-plane a unique point in the z-plane except the point $w = a/c\ (c \neq 0)$. We can associate with $w = a/c$, the point ∞ in the extended z-plane $\mathbb{C} \cup \{\infty\}$. Thus the bilinear transformation T is a bijection (one-one and onto map) from the extended z-plane to the extended w-plane. The extended planes are also known as closed z-plane and closed w-plane.

3.4.2 THE LINEAR GROUP OF BILINEAR TRANSFORMATION

Theorem 3.4.1 The product of two bilinear transformations is again a bilinear transformation.

Proof:

Let $T_1(z) = \dfrac{a_1 z + b_1}{c_1 z + d_1}$ $a_1d_1 - b_1c_1 \neq 0$ (1)

and $T_2(z) = \dfrac{a_2 z + b_2}{c_2 z + d_2}$ $a_2d_2 - b_2c_2 \neq 0$ (2)

Then the product of these two bilinear transformations is

$w = T_2(T_1(z))$

$$= T_2\left(\frac{a_1 z + b_1}{c_1 z + d_1}\right)$$

$$= \frac{a_2\left(\dfrac{a_1 z + b_1}{c_1 z + d_1}\right) + b_2}{c_2\left(\dfrac{a_1 z + b_1}{c_1 z + d_1}\right) + d_2}$$

$$= \frac{a_1a_2 z + a_2b_1 + c_1b_2 z + d_1b_2}{a_1c_2 z + c_2b_1 + c_1d_2 z + d_1d_2}$$

$$= \frac{(a_1a_2 + c_1b_2)z + (a_2b_1 + d_1b_2)}{(a_1c_2 + c_1d_2)z + (c_2b_1 + d_1d_2)}$$

i.e., $w = \dfrac{Az + B}{Cz + D}$ (3)

Where $AD - BC = (a_1a_2 + c_1b_2)(c_2b_1 + d_1d_2) - (a_1c_2 + c_1d_2)(a_2b_1 + d_1b_2)$

$= a_1a_2c_2b_1 + c_1b_2c_2b_1 + a_1a_2d_1d_2 + c_1b_2d_1d_2 - a_1c_2a_2b_1$
$- a_1c_2b_2d_1 - b_1c_1a_2d_2 - c_1d_2b_2d_1$

$= (a_1d_1 - b_1c_1)(a_2d_2 - b_2c_2) \neq 0$ by (1) and (2).

Hence equation (3) also represents a bilinear transformation. Thus the product of two bilinear transformations is again a bilinear transformation.

Theorem 3.4.2 The set of all bilinear transformations forms a non-abelian group under the operation of product of transformations.

Proof:

As the product of two bilinear transformations is again a bilinear transformation, the closure property is satisfied. In general, the product of transformations is associative.

$\therefore T_1(T_2T_3) = (T_1T_2)T_3$ for bilinear transformations T_1, T_2 and T_3.

Hence associative property is satisfied.

The identity mapping defined by $w = T(z) = z$ is a bilinear transformation as $ad - bc = 1.1 - 0.0 = 1 \neq 0$. Thus the set of bilinear transformations contains the identity transformation also.

Finally if $w = T(z) = \dfrac{az+b}{cz+d}$, ad - bc ≠ 0 then the inverse transformation,

$z = T^{-1}(w) = \dfrac{-dw+b}{cw-a}$ is also a bilinear transformation,

since $\begin{vmatrix} -d & b \\ c & -a \end{vmatrix} = (-d)(-a) - c.b = ad - bc \neq 0.$

$$T^{-1}T(z) = T^{-1}\left(\frac{az+b}{cz+d}\right) = \frac{-d\left(\dfrac{az+b}{cz+d}\right)+b}{c\left(\dfrac{az+b}{cz+d}\right)d-a}$$

$$= \frac{(bc-ad)z}{bc-ad}$$

$$= z \qquad \text{(Since ad - bc} \neq 0)$$

$$= I(z).$$

$\therefore T^{-1}T = I.$

Similarly we can prove that $TT^{-1}(w) = w$.

Thus $T^{-1}.T = T.T^{-1} = I$.

Hence the set of all bilinear transformations form a group under the binary operation of product of transformation.

As in general $T_1T_2(z) \neq T_2T_1(z)$, the group is a non-abelian group.

Example 3.4.1 Given the bilinear transformations $w = T_1(z) = \dfrac{z+i}{z-i}$ and $w = T_2(z) = \dfrac{z-i}{z+i}$, find the products T_1 T_2 (z) and T_2 T_1 (z). Do they commute?

Solution:

$$T_1\,T_2\,(z) = T_1\left(\frac{z-i}{z+i}\right)$$

$$= \frac{\dfrac{z-i}{z+i}+i}{\dfrac{z-i}{z+i}-i}$$

$$= \frac{z-i+iz-1}{z-i-iz+1} = \frac{(1+i)z-1-i}{(1-i)z+1-i}$$

$$T_2T_1(z) = T_2\left(\frac{z+i}{z-i}\right) = \frac{\dfrac{z+i}{z-i}-i}{\dfrac{z+i}{z-i}+i}$$

$$= \frac{z+i-iz-1}{z+i+iz+1}$$

$$= \frac{(1-i)z-1+i}{(1+i)z+1+i}$$

$T_1T_2(z) \neq T_2T_1(z)$ $\therefore$ T_1 and T_2 do not commute.

Example 3.4.2 Prove that a bilinear transformation is a product of the simple (elementary) transformations $w = z + a$, $w = az$ and $w = \frac{1}{z}$. Hence prove that every bilinear transformation transforms circles or straight lines onto circles or straight lines.

Solution:

Consider the bilinear transformation $w = T(z) = \frac{az+b}{cz+d}$, $ad - bc \neq 0$

Case 1. $c = 0$

Then $w = \frac{az}{d} + \frac{b}{d}$ (1)

$d \neq 0$ (since $d = 0 \Rightarrow ad - bc = 0$)

Put $w_1 = \frac{a}{d}z$ (2)

Then $w = w_1 + \frac{b}{d}$ (3)

Hence w is a product of the transformations of the type $w= z + a$ and $w = az$.

Case 2. $c \neq 0$

Then $w = \frac{az+b}{cz+d} = \frac{a}{c} + \frac{bc-ad}{c^2} \cdot \frac{1}{(z+d/c)}$ (4)

Put $w_1 = z + d/c$

$w_2 = \frac{1}{w_1}$

$w_3 = \frac{(bc-ad)}{c^2} w_2$

Then $w = w_3 + \frac{a}{c}$

Then w is a product of the transformation of the type $w = z + a$, $w= az$ and $w = \frac{1}{z}$. The transformation $w = z + a$ (translation) maps circles onto circles and straight lines onto straight lines. The transformation $w = az$ (magnification or rotation) also maps circles onto circles and straight lines onto straight lines. But the transformation $w = \frac{1}{z}$ (inversion and reflection) transforms circles or straight lines onto circles or straight lines. Hence the bilinear transformation which is a product of these transformations also transforms circles or straight lines onto circles or straight lines.

Example 3.4.3 Show that the transformation $w = \dfrac{iz+2}{4z+i}$ maps the real axis in the z-plane to a circle in the w-plane. Find the center and radius of the circle.

Solution: $w = \dfrac{iz+2}{4z+i}$

$w(4z + i) = iz + 2$

$z(4w - i) = -wi + 2$

$z = \dfrac{-iw+2}{4w-i}$

The equation of the real axis in the z-plane is $y = 0$

i.e., $z - \bar{z} = 0$

i.e., $\dfrac{-iw+2}{4w-i} - \dfrac{i\bar{w}+2}{4\bar{w}+i} = 0$

i.e., $-4iw\bar{w} + w + 8\bar{w} + 2i - 4iw\bar{w} - 8w - \bar{w} + 2i = 0$

i.e., $8iw\bar{w} + 7(w - \bar{w}) - 4i = 0$

i.e., $8i(u^2 + v^2) + 7.2iv - 4i = 0$

i.e., $u^2 + v^2 + \frac{7}{4}v - \frac{1}{2} = 0$, which is a circle in the w-plane with center $\left(0, -7/8\right)$ and radius $\sqrt{\left(\frac{7}{8}\right)^2 + \frac{1}{2}} = \frac{9}{8}$.

3.4.3 CROSS RATIO

Definition 3.4.2 Let z_1, z_2, z_3, z_4 be any four points in the z-plane. Then the quantity $\dfrac{(z_1 - z_2)(z_3 - z_4)}{(z_1 - z_4)(z_3 - z_2)}$ is called the **cross ratio** of the four points and is denoted by (z_1, z_2, z_3, z_4).

The cross ratio of four points is a complex number and it can be defined as the image of a complex number under a bilinear transformation.

Consider the bilinear transformation $w = T(z) = \dfrac{(z - z_2)(z_3 - z_4)}{(z - z_4)(z_3 - z_2)}$ (1)

where z_2, z_3, z_4 are any three distinct points in the extended complex plane and none of these points is ∞. Then,

$T(z_1)=\dfrac{(z_1-z_2)(z_3-z_4)}{(z_1-z_4)(z_3-z_2)}=(z_1z_2z_3z_4)$ Thus cross ratio of the four points z_1, z_2, z_3, z_4 is the image of the point z_1 under the bilinear transformation (1). We have $T(z_2) = 0$ $T(z_3) = 1$ and $T(z_4) = \infty$.

Note:

(i) Consider the bilinear transformation $w = T(z) = \dfrac{az+b}{cz+d}$

i.e., $w = T(z) = \dfrac{a/_d\, z + b/_d}{c/_d\, z + 1}$

i.e., $w = \dfrac{Az+B}{Cz+1}$. Since A, B, C are the unknown constants in the bilinear transformation, given any three points z_1, z_2, z_3 and their images w_1, w_2, w_3 then A, B, C can be determined from the three equations,

$w_1 = \dfrac{Az_1+B}{Cz_1+1}$, $w_2 = \dfrac{Az_2+B}{Cz_2+1}$, $w_3 = \dfrac{Az_3+B}{Cz_3+1}$. Thus we see that a bilinear transformation is uniquely determined by the images of any three points in the z-plane.

(ii) The bilinear transformation $w = \dfrac{(z-z_2)(z_3-z_4)}{(z-z_4)(z_3-z_2)}$ is uniquely determined as 0, 1, ∞ are the images of the points z_2, z_3, z_4 under this transformation.

(iii) In $w = T(z) = \dfrac{(z-z_2)(z_3-z_4)}{(z-z_4)(z_3-z_2)}$

$z_2 = \infty \Rightarrow T(z) = \dfrac{(z_3-z_4)}{(z-z_4)}$

$z_3 = \infty \Rightarrow T(z) = \dfrac{(z-z_2)}{(z-z_4)}$

$z_4 = \infty \Rightarrow T(z) = \dfrac{(z-z_2)}{(z_3-z_2)}$

Theorem 3.4.3 (Cross Ratio Property of bilinear transformation) The cross ratio of four points is invariant under a bilinear transformation.

Proof:

Let $w = T(z) = \dfrac{az+b}{cz+d}$ $ad - bc \neq 0$ be the given bilinear transformation. Let w_1, w_2, w_3, w_4 be the images of the four points z_1, z_2, z_3, z_4 under this bilinear transformation. Then, we have,

$$w_i = \frac{az_i + b}{cz_i + d} \quad i = 1, 2, 3, 4$$

$$w_1 - w_2 = \frac{az_1 + b}{cz_1 + d} - \frac{az_2 + b}{cz_2 + d}$$

$$= \frac{(ad - bc)(z_1 - z_2)}{(cz_1 + d)(cz_2 + d)}$$

$$w_3 - w_4 = \frac{(ad - bc)(z_3 - z_4)}{(cz_3 + d)(cz_4 + d)}$$

$$(w_1 - w_2)(w_3 - w_4) = \frac{(ad - bc)^2 (z_1 - z_2)(z_3 - z_4)}{(cz_1 + d)(cz_2 + d)(cz_3 + d)(cz_4 + d)}$$

Similarly $(w_1 - w_4)(w_3 - w_2) = \dfrac{(ad - bc)^2 (z_1 - z_4)(z_3 - z_2)}{(cz_1 + d)(cz_2 + d)(cz_3 + d)(cz_4 + d)}$

Hence $\dfrac{(w_1 - w_2)(w_3 - w_4)}{(w_1 - w_4)(w_3 - w_2)} = \dfrac{(z_1 - z_2)(z_3 - z_4)}{(z_1 - z_4)(z_3 - z_2)}$.

i.e., $(w_1, w_2, w_3, w_4) = (z_1, z_2, z_3, z_4)$

$\therefore$ **Cross ratio is invariant under a bilinear transformation**

Note:

We can find the bilinear transformation determined by three distinct points and their images in terms of cross ratios. Let w_2, w_3, w_4 be the images of three distinct points z_2, z_3, z_4 under a bilinear transformation. If w is the image of z under the transformation, then we have $(w, w_2, w_3, w_4) = (z, z_2, z_3, z_4)$

i.e., $$\frac{(w - w_2)(w_3 - w_4)}{(w - w_4)(w_3 - w_2)} = \frac{(z - z_2)(z_3 - z_4)}{(z - z_4)(z_3 - z_2)} \quad (1)$$

Solving for w, we get the equation (1) in the form $w = \dfrac{az + b}{cz + d}$. Thus the equation (1) gives the bilinear transformation that maps three distinct points z_2, z_3, z_4 onto The points w_2, w_3, w_4 respectively.

Example 3.4.4 Find the bilinear transformation that maps the points $z_1 = 2$, $z_2 = i$ and $z_3 = -2$ into the points $w_1 = 1$, $w_2 = i$ and $w_3 = -1$ respectively.

Solution:

The bilinear transformation that maps z_1, z_2, z_3 into w_1, w_2, w_3 respectively is given by

$$\frac{(w - w_1)(w_2 - w_3)}{(w - w_3)(w_2 - w_1)} = \frac{(z - z_1)(z_2 - z_3)}{(z - z_3)(z_2 - z_1)}$$

Substituting $z_1 = 2$, $z_2 = i$, $z_3 = -2$, $w_1 = 2$, $w_2 = i$ and $w_3 = -1$, we get,

$$\frac{(w-1)(i+1)}{(w+1)(i-1)}=\frac{(z-2)(i+2)}{(z+2)(i-2)}$$

$$\frac{(w-1)}{(w+1)}=\frac{(z-2)(i+2)(i-1)}{(z+2)(i-2)(i+1)}$$

$$\frac{(w-1)}{(w+1)}=\frac{(z-2)(-3+i)}{(z+2)(-3-i)}$$

$$\frac{(w-1)}{(w+1)}=\frac{(z-2)(4-3i)}{(z+2).5}$$

$$\frac{2w}{-2}=\frac{(z-2)(4-3i)+5(z+2i)}{(z-2)(4-3i)-5.(z+2)} \quad \left(\text{Since } \frac{a}{b}=\frac{c}{d} \Rightarrow \frac{a+b}{a-b}=\frac{c+d}{c-d}\right)$$

$$-w=\frac{(4-3i)z-8+6i+5z+10}{(4-3i)z-8+6i-5z-10}$$

$$=\frac{(9-3i)z+2+6i}{(-1-3i)z-18+6i}$$

$$=\frac{3(3-i)z+2i(3-i)}{-i(3-i)z-6(3-i)}$$

$$=\frac{3z+2i}{-iz-6}$$

$$w=\frac{3z+2i}{iz+6}$$

This is the required bilinear transformation that maps the points $z_1 = 2$, $z_2 = i$ and $z_3 = -2$ into the points $w_1 = 1$, $w_2 = i$ and $w_3 = -1$ respectively.

Example 3.4.5 Find the bilinear transformation which transforms the points z = 2, 1, 0 into w = 1, 0, i respectively.

Solution:

The bilinear transformation which transforms the points $z = z_1, z_2, z_3$ into $w = w_1, w_2, w_3$ respectively is given by

$$\frac{(w-w_1)(w_2-w_3)}{(w-w_3)(w_2-w_1)}=\frac{(z-z_1)(z_2-z_3)}{(z-z_3)(z_2-z_1)}$$

Substituting $z_1 = 2$, $z_2 = 1$, $z_3 = 0$, $w_1 = 1$, $w_2 = 0$ and $w_3 = i$, we get,

$$\frac{(w-1)(0-i)}{(w-i)(0-1)}=\frac{(z-2)(1-0)}{(z-0)(1-2)}$$

$$\frac{(w-1)}{(w-i)}.i=\frac{(z-2)}{-z}$$

$$-wiz+iz=wz-2w-iz+2i$$

$$w(-iz-z+2)=-2iz+2i$$

$$w = \frac{-2iz + 2i}{-(i+1)z + 2}$$

$$w = \frac{2iz - 2i}{(1+i)z - 2}.$$

This is the required bilinear transformation.

Example 3.4.6 Find the bilinear transformation which transforms the unit circle $|z| = 1$ into the real axis in such a way that the point $z = 1$, i and -1 are mapped into the points $w = 0$, 1 and ∞ respectively. Find the regions into which the interior and exterior of the circle are mapped.

Solution:

The required bilinear transformation is that which transforms the points $z = 1, i, -1$ into the points $w = 0, 1, \infty$.

$\therefore$ The required bilinear transformation is

$$\frac{(w - w_1)(w_2 - w_3)}{(w - w_3)(w_2 - w_1)} = \frac{(z - z_1)(z_2 - z_3)}{(z - z_3)(z_2 - z_1)}$$

where $z_1 = 1$, $z_2 = i$, $z_3 = -1$, $w_1 = 0$, $w_2 = 1$ and $w_3 = \infty$, Substituting these values, we get

$$\frac{(w - 0)(1 - \infty)}{(w - \infty)(1 - 0)} = \frac{(z-1)(i+1)}{(z+1)(i-1)}$$

i.e., $$\frac{w}{1} = \frac{z-1}{z+1} \cdot \frac{2i}{(-2)}$$

i.e., $$w = \frac{i(1-z)}{1+z} \qquad (1)$$

The inverse of the transformation (1) is obtained by computing z as a function of w.

$w(1 + z) = i(1 - z)$

$(w + i)z = -w + i$

$$z = -\frac{(w - i)}{w + i}$$

The circle $|z| = 1$ is transformed into $\left| -\frac{(w-i)}{w+i} \right| = 1$

i.e., $|w - i| = |w + i|$ which is the real axis in the w-plane as it is the locus of the points equidistant from $w = i$ and $w = -i$. Hence (1) is the required bilinear transformation.

The region $|z| < 1$ is mapped into $\left| -\frac{(w-i)}{w+i} \right| < 1$

i.e., $|w - i| < |w + i|$ which represents the upper half of the w-plane (i.e., $\text{Im}(w) > 0$) as the distance of any point above the real axis in the w-plane from $w = i$ is less than the distance from the point $w = -i$. Similarly the region exterior to the unit circle given by $|z| > 1$ is mapped into the lower half of the w-plane (i.e., $\text{Im}(w) < 0$).

Example 3.4.7 Determine the bilinear transformation which maps 0, 1, ∞ into i, -1, -i respectively. Show that this transformation maps the interior of the unit circle in the z-plane into the half plane Im(w) > 0.

Solution:

The bilinear transformation that maps the points z_1, z_2, z_3 into w_1, w_2, w_3 respectively is

$$\frac{(w-w_1)(w_2-w_3)}{(w-w_3)(w_2-w_1)} = \frac{(z-z_1)(z_2-z_3)}{(z-z_3)(z_2-z_1)}$$

Substituting $z_1 = 0$, $z_2 = 1$, $z_3 = \infty$, $w_1 = i$, $w_2 = -1$ and $w_3 = -i$, we get,

$$\frac{(w-i)(-1+i)}{(w+i)(-1-i)} = \frac{(z-0)(1-\infty)}{(z-\infty)(1-0)}$$

$$\frac{(w-i)(1-i)}{(w+i)(1+i)} = z$$

$$\frac{(w-i)}{(w+i)} = \frac{2iz}{2} \qquad (1)$$

$$\frac{(w-i)+(w+i)}{(w-i)-(w+i)} = \frac{iz+1}{iz-1}$$

$$\frac{2w}{-2i} = \frac{iz+1}{iz-1}$$

$$\frac{w}{-i} = \frac{z-i}{z+i}$$

$$w = -i\frac{(z-i)}{z+i} \qquad (2)$$

(2) is the equation of the required bilinear transformation. The interior of the unit circle in the z-plane is $|z| < 1$.

$$(1) \Rightarrow z = (-i)\frac{w-i}{w+i}$$

$$|z| < 1 \Rightarrow \left|\frac{w-i}{w+i}\right| < 1$$

$\Rightarrow |w - i| < |w + i|$

$\Rightarrow$ Distance of w from i is less than the distance of w from –i.

$\Rightarrow$ Half plane above the real axis in the w-plane.

$\Rightarrow$ Half plane Im(w) > 0.

3.4.4 FIXED POINTS OF BILINEAR TRANSFORMATIONS

While studying conformal mappings or bilinear transformations one can conveniently assume that the z-plane and w-plane are one and the same. Thus under a transformation an object (point, curve or region) and its image lie on the same plane. Hence the transformation will move points to new positions and at the same time leave some points of the plane undisturbed. For example, a translation will move all finite points of the plane, while a rotation by an angle α about the origin will move all finite points except the origin, the center of rotation.

Definition 3.4.3 If a transformation $w = f(z)$ maps a point z onto itself, then the point z is called a **fixed point** or **an invariant point** of the transformation.

The fixed points of the transformation $w = f(z)$ are the solutions of the equation $z = f(z)$.

The fixed points of the bilinear transformation are given by $z = \frac{az+b}{cz+d}$, $ad - bc \neq 0$.

i.e., $cz^2 + (d - a)z - b = 0$.

Case 1. $\mathbf{c \neq 0}$.

Then the fixed points are $z = \frac{(a-d) \pm \sqrt{(d-a)^2 + 4bc}}{2c}$

When $(d\text{-}a)^2 + 4bc \neq 0$, the bilinear transformation has two finite fixed points.

When $(d\text{-}a)^2 + 4bc = 0$, it has only one fixed point and that is the finite point $\frac{a-d}{2c}$.

Case 2. $\mathbf{c = 0}$.

Then the bilinear transformation becomes $w = \left(\frac{a}{d}\right)z + \frac{b}{d}$.

When $z \to \infty$, $w \to \infty$. $\therefore$ ∞ is a fixed point.

A finite fixed point is given by $z = \left(\frac{a}{d}\right)z + \frac{b}{d}$.

i.e., $(d - a)z = b$

$z = \frac{b}{d-a}$, when $d - a \neq 0$

When $d - a = 0$, we have ∞ is the only fixed point.

Thus we have,

$c \neq 0$, $(d - a)^2 + 4bc \neq 0 \Rightarrow$ two finite fixed point.

$c \neq 0$, $(d - a)^2 + 4bc = 0 \Rightarrow$ only one fixed point and that is a finite point.

$c = 0$, $a \neq d \Rightarrow$ one finite fixed point and ∞ is another fixed point.

$c = 0$, $a = d \Rightarrow \infty$ is the only fixed point.

3.4.5 NORMAL FORMS OF BILINEAR TRANSFORMATIONS

The bilinear transformations can be put in some standard form in terms of their fixed points. These are called **normal forms** of bilinear transformations. Further bilinear transformations can be classified as **hyperbolic, elliptic** or **parabolic.**

Theorem 3.4.4 Every bilinear transformation with two finite fixed points α, β can be put in the form $\dfrac{w-\alpha}{w-\beta} = k\left(\dfrac{z-\alpha}{z-\beta}\right)$

Proof:

Let T be any bilinear transformation having fixed points α and β. Let the image of a point γ under T be δ. Then T is given by

$(w, \alpha, \delta, \beta) = (z, \alpha, \gamma, \beta)$

i.e., $\dfrac{(w-\alpha)(\delta-\beta)}{(w-\beta)(\delta-\alpha)} = \dfrac{(z-\alpha)(\gamma-\beta)}{(z-\beta)(\gamma-\alpha)}$

i.e., $\dfrac{(w-\alpha)}{(w-\beta)} = k.\dfrac{(z-\alpha)}{(z-\beta)}$ where $k = \left(\dfrac{\gamma-\beta}{\gamma-\alpha}\right)\left(\dfrac{\delta-\alpha}{\delta-\alpha}\right)$

Note:

If the bilinear transformation T having two finite fixed points α and β is $w = \dfrac{az+b}{cz+d}$, then α and β are the roots of the equation

$$cz^2 + (d - a)z - b = 0 \tag{1}$$

By Theorem 3.4.4, T can be put in the form

$$\frac{w-\alpha}{w-\beta} = k\frac{z-\alpha}{z-\beta} \tag{2}$$

$$\text{where } k = \frac{\gamma-\beta}{\gamma-\alpha}.\frac{\delta-\alpha}{\delta-\alpha} \tag{3}$$

Also T transforms $z = -d/c$ into $w = \infty$. Substituting $\gamma = -d/c$ and $\delta = \infty$ in (3)

we get, $k = \dfrac{\left(-d/c\right)-\beta}{\left(-d/c\right)-\alpha}$

$= \dfrac{d+\beta c}{d+\alpha c}$, where α and β are given by equation (1)

Definition 3.4.4 Let T be a bilinear transformation $w = \dfrac{az+b}{cz+d}$ with two fixed points α and β. If the constant $k = \dfrac{d+\beta z}{d+\alpha z}$ is real and positive, then T is called **hyperbolic**. If $|k| = 1$, then T is called **elliptic.**

Theorem 3.4.5 Every bilinear transformation that has only one fixed point α (≠ ∞) can be put in the form $\dfrac{1}{w-\alpha} = \dfrac{1}{z-\alpha} + k.$

Proof:

Let $w = \dfrac{az+b}{cz+d}$, $ad - bc \neq 0$ be the bilinear transformation. Since α is the only fixed point, the equation $z = \dfrac{az+b}{cz+d}$ or $cz^2 + (d - a)z - b = 0$ has only one root $\alpha = \dfrac{a-d}{2c}$ and

$cz^2 + (d - a)z - b \equiv c(z - \alpha)^2$ (1)

Equating the corresponding coefficients in (1), we get $d - a = -2\alpha c$ and $-b = c\alpha^2$

$\therefore a = d + 2\alpha c$ and $b = -c\alpha^2$

Then $w = \dfrac{(d+2\alpha c)z + (-c\alpha^2)}{cz+d}$

$cwz + dw = dz + 2\alpha cz - c\alpha^2$

$c(w - \alpha)(z - \alpha) + c\alpha w + c\alpha z - c\alpha^2 + dw = dz + 2\alpha cz - c\alpha^2$

$c(w - \alpha)(z - \alpha) + (d + c\alpha)(w - \alpha) = dz + \alpha cz - c\alpha^2 - d\alpha$

$c(w - \alpha)(z - \alpha) + (d + c\alpha)(w - \alpha) = (d + c\alpha)(z - \alpha)$

$$\frac{c}{d+c\alpha} + \frac{1}{z-\alpha} = \frac{1}{w-\alpha}$$

$$\frac{1}{w-\alpha} = \frac{1}{z-\alpha} + k, \quad \text{where } k = \frac{c}{d+c\alpha} \text{ and } \alpha = \frac{a-d}{2c}$$

Substituting for α in k, we get, $k = \dfrac{c}{d + c.\left(\frac{a-d}{2c}\right)} = \dfrac{2c}{a+d}$

Note:

(i) If the bilinear transformation $w = \dfrac{az+b}{cz+d}$ has one finite fixed point α and the other fixed point is ∞, then $c = 0$ and $a \neq d$

$\therefore w = \dfrac{a}{d}z + \dfrac{b}{d}$

Since α is fixed, $\alpha = \dfrac{a}{d}\alpha + \dfrac{b}{d}$

$\therefore w - \alpha = \dfrac{a}{d}(z - \alpha)$

i.e., $w - \alpha = k(z - \alpha)$ where $k = \dfrac{a}{d}$

(ii) The bilinear transformation having ∞ as the only fixed point is a translation.

If ∞ is the only fixed point of $w = \dfrac{az+b}{cz+d}$, then $c = 0$ and $a = d$.

$\therefore w = z + \dfrac{b}{d}$, which is a translation.

(iii) The bilinear transformation having one finite fixed point and the other fixed point ∞ is elliptic if $|k| = \left|\frac{a}{d}\right| = 1$ i.e., if $|a| = |d|$

Definition 3.4.5 A bilinear transformation $w = \frac{az+b}{cz+d}$ is called **parabolic** if $c \neq 0$ and $(a - d)^2 + 4bc = 0$. i.e., the bilinear transformation has only one fixed point and that is finite.

Theorem 3.4.6 A bilinear transformation preserves inverse points with respect to circles. (or symmetric points w.r.t straight lines)

Proof:

The equation $\left|\frac{z-z_1}{z-z_2}\right| = \lambda \quad (\lambda > 0)$ (1)

represents a circle with z_1 and z_2 as inverse points. When $\lambda = 1$, (1) represents a straight line with z_1 and z_2 as symmetric points.

Let $w = \frac{az+b}{cz+d}$, $ad - bc \neq 0$ be any bilinear transformation. Let w_1 and w_2 be the inverses of z_1 and z_2. Then $w_1 = \frac{az_1+b}{cz_1+d}$, $w_2 = \frac{az_2+b}{cz_2+d}$

$$w - w_1 = \frac{az+b}{cz+d} - \frac{az_1+b}{cz_1+d} = \frac{(ad-bc)(z-z_1)}{(cz+d)(cz_1+d)}$$

$$w - w_2 = \frac{az+b}{cz+d} - \frac{az_2+b}{cz_2+d} = \frac{(ad-bc)(z-z_2)}{(cz+d)(cz_2+d)}$$

$$\frac{w-w_1}{w-w_2} = \left(\frac{cz_2+d}{cz_1+d}\right)\frac{z-z_1}{z-z_2}$$

$$\frac{w-w_1}{w-w_2} = k.\frac{z-z_1}{z-z_2} \quad (2)$$

Now $\left|\frac{w-w_1}{w-w_2}\right| = |k|.\left|\frac{z-z_1}{z-z_2}\right|$

i.e., $\left|\frac{w-w_1}{w-w_2}\right| = |k|.\lambda > 0$ (3)

Equation (3) represents a circle (or straight line when $|k|.\lambda = 1$) with w_1 and w_2 as inverse points (or symmetric points). Thus the image of a circle or straight line is again a circle or straight line and inverse points w.r.t circle (or symmetric points w.r.t straight line) are mapped into inverse points w.r.t circle (or symmetric points w.r.t straight line).

Example 3.4.8 Find the fixed points and the normal form of the bilinear transformation $w = \dfrac{z}{2-z}$.

Solution:

The fixed points of the bilinear transformation are given by $z = \dfrac{z}{2-z}$

i.e., $z^2 - z = 0$

i.e., $z(z-1) = 0$

i.e., $z = 0, 1$ are two finite fixed points.

$$\text{Now } w = \frac{z}{2-z} \Rightarrow w - 1 = \frac{z}{2-z} - 1 = \frac{2z-2}{2-z}$$

$$\therefore \frac{w}{w-1} = \frac{z}{2z-2}$$

i.e., $\dfrac{w-0}{w-1} = \dfrac{1}{2} \cdot \dfrac{z-0}{z-1}$ is the required normal form of the bilinear transformation.

Here $k = \dfrac{1}{2}$, a real number.

Hence the transformation is hyperbolic.

Example 3.4.9 Find the fixed points of the bilinear transformation $w = \dfrac{z-1}{z+1}$. Also determine whether it is elliptic, hyperbolic or parabolic.

Solution:

The fixed points are given by $z = \dfrac{z-1}{z+1}$

i.e., $z^2 + 1 = 0$

$z = \pm i$ are the two finite fixed points of the bilinear transformation. To determine whether it is elliptic or hyperbolic, put the bilinear transformation in its normal form.

$$\frac{w-\alpha}{w-\beta} = k\left(\frac{z-\alpha}{z-\beta}\right), \text{ where } \alpha = i \text{ and } \beta = -i.$$

$$w - i = \frac{z-1}{z+1} - i = \frac{(z-1)-i(z+1)}{z+1}$$

$$w + i = \frac{z-1}{z+1} + i = \frac{(z-1)+i(z+1)}{z+1}$$

$$\frac{w-i}{w+i} = \frac{(z-1)-i(z+1)}{(z-1)+i(z+1)}$$

$$= \frac{(1-i)(z-i)}{(1+i)(z+i)}$$

$$= \frac{(1-i)^2}{(1+i)(1-i)} \cdot \frac{z-i}{z+i}$$

i.e., $\frac{w-i}{w+i} = (-i).\frac{z-i}{z+i}$ is the normal form.

Here k = -i, so that |k| = 1.

Hence the bilinear transformation is elliptic.

Example 3.4.10 Find the fixed points and the normal form of the bilinear transformation $w = \frac{3z-4}{z-1}$.

Solution:

The fixed points are given by $z = \frac{3z-4}{z-1}$.

i.e., $z^2 - z = 3z - 4$

i.e., $z^2 - 4z + 4 = 0$

i.e., $(z-2)^2 = 0$

z =2 is the only fixed point. Hence the bilinear transformation is parabolic.

The normal form is $\frac{1}{w-2} = \frac{1}{z-2} + k$

where $k = \frac{1}{w-2} - \frac{1}{z-2}$

$$k = \frac{1}{\left(\frac{3z-4}{z-1}\right) - 2} - \frac{1}{z-2}$$

$$= \frac{z-1}{z-2} - \frac{1}{z-2}$$

$$= \frac{z-2}{z-2}$$

$$= 1$$

$\therefore$ The normal form is $\frac{1}{w-2} = \frac{1}{z-2} + 1$

Example 3.4.11 Prove that a bilinear transformation $w = \frac{az+b}{cz+d}$, ad - bc ≠ 0 maps the real axis into itself if and only if a, b, c, d are real. Also prove that this transformation maps the half plane Im(z) ≥ 0 into the halfplane Im(w) ≥ 0 if and only if ad - bc > 0.

Solution:

$w = \dfrac{az+b}{cz+d}$, ad - bc $\neq 0$.

If a, b, c, d are real, then z is real $\Rightarrow$ w is real.

$\therefore$ Real axis is mapped into itself.

Conversely suppose a bilinear transformation T maps the real axis into itself. Then there exists real numbers x_1, x_2, x_3 such that $T(x_1) = 0$, $T(x_2) = 1$ and $T(x_3) = \infty$. Then T is given by $(z, x_1, x_2, x_3) = (w, 0, 1, \infty)$

i.e., $$\frac{(z-x_1)(x_2-x_3)}{(z-x_3)(x_2-x_1)} = \frac{w-0}{1-0}$$

i.e., $$w = \frac{(x_2-x_3)z - x_1(x_2-x_3)}{(x_2-x_1)z - x_3(x_2-x_1)}$$

i.e., $w = \dfrac{az+b}{cz+d}$ where $a = x_2 - x_3$, $b = -x_1(x_2 - x_3)$, $c = x_2 - x_1$, $d = -x_3(x_2 - x_1)$

x_1, x_2, x_3 are real $\Rightarrow$ a, b, c, d are real.

Now $2i\,\text{Im}(w) = w - \overline{w}$

$$= \frac{az+b}{cz+d} - \frac{a\overline{z}+b}{c\overline{z}+d}$$

$$= \frac{(ad-bc)(z-\overline{z})}{|cz+d|^2}$$

$$= \frac{2i.(ad-bc)}{|cz+d|^2}.\text{Im}(z)$$

$$\therefore\ \text{Im}(w) = \frac{(ad-bc)}{|cz+d|^2}.\text{Im}(z)$$

Hence the transformation maps $\text{Im}(z) \geq 0$ into $\text{Im}(w) \geq 0$ if and only if ad - bc > 0.

Example 3.4.12 Find all the bilinear transformations which transform the half-plane $\text{Im}(z) \geq 0$ onto the unit circular disc $|w| \leq 1$.

Solution:

Let $w = \dfrac{az+b}{cz+d}$, ad - bc $\neq 0$ (1)

be the required bilinear transformation. $c \neq 0$ as otherwise ∞ will be a fixed point. The points $w = 0$ and $w = \infty$ in the w-plane correspond to the points $z = -b/a$ and $z = -d/c$ in the z-plane. Further the transformation maps the real axis onto the unit circle $|w| = 1$. As 0 and ∞ are inverse points with respect to the circle $|w| = 1$, $-b/a$ and $-d/c$ are inverse points are symmetric points w.r.t the real axis $\text{Im}(z) = 0$. Thus $-b/a = \alpha$ and $-d/c = \overline{\alpha}$

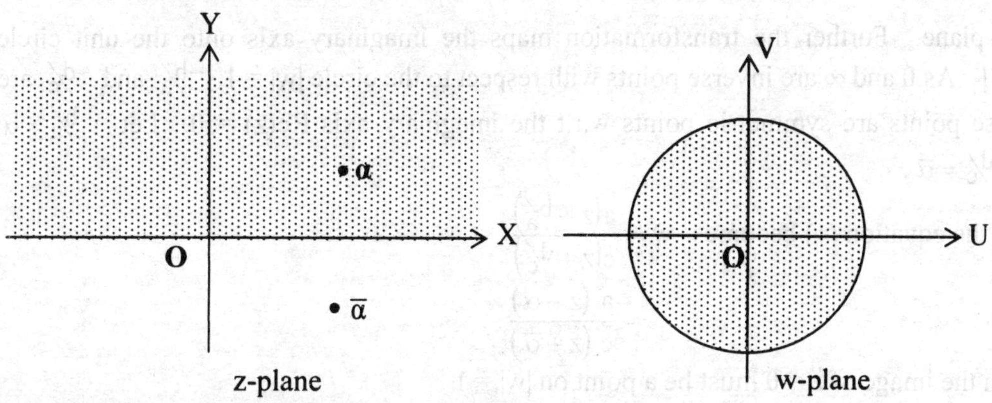

Fig. 3.26

Then the equation (1) becomes $w = \dfrac{a\left(z + \frac{b}{a}\right)}{c\left(z + \frac{d}{c}\right)}$

$$w = \frac{a(z+\alpha)}{c(z+\bar{\alpha})}$$

The image of $z = 0$ must be a point on $|w| = 1$

$$\therefore \left|\frac{a}{c}\right|\left|\frac{0-\alpha}{0-\bar{\alpha}}\right| = |w| = 1$$

i.e., $\left|\frac{a}{c}\right| = 1$ $\left(\text{Since}|\alpha| = |\bar{\alpha}|\right)$ Let $\frac{a}{c} = e^{i\lambda}$ where λ is real.

Then the bilinear transformation becomes $w = e^{i\lambda}.\left(\dfrac{z-\alpha}{z-\bar{\alpha}}\right)$

When $z = \alpha$, $w = 0$ which is the center of the circle $|w| = 1$ $\therefore$ $z = \alpha$ must be a point of the upper half-plane i.e., $\text{Im}(\alpha) > 0$

Thus the bilinear transformation which maps $\text{Im}(z) \geq 0$ onto the unit circular disc $|w| \leq 1$ is

$w = e^{i\lambda}.\left(\dfrac{z-\alpha}{z-\bar{\alpha}}\right)$ where λ is real and $\text{Im}(\alpha) > 0$

Example 3.4.13 Find all the bilinear transformations which transform the half-plane $\text{Re}(z) \geq 0$ onto the unit circular disc $|w| \leq 1$.

Solution:

Let $w = \dfrac{az+b}{cz+d}$, $ad - bc \neq 0$ (1)

be the required bilinear transformation. $c \neq 0$ as otherwise ∞ will be a fixed point. The points $w = 0$ and $w = \infty$ in the w-plane correspond to the points $z = -\frac{b}{a}$ and $z = -\frac{d}{c}$ in the z-plane. Further the transformation maps the imaginary axis onto the unit circle $|w| = 1$. As 0 and ∞ are inverse points with respect to the circle $|w| = 1$, $-b/a$ and $-d/c$ are inverse points are symmetric points w.r.t the imaginary axis $\text{Re}(z) = 0$. Let $-b/a = \alpha$ and $-d/c = \bar{\alpha}$.

Then the equation (1) becomes $w = \dfrac{a\left(z + b/a\right)}{c\left(z + d/c\right)}$

$$= \frac{a}{c}\frac{(z-\alpha)}{(z+\bar{\alpha})}$$

Again the image of $z = 0$ must be a point on $|w| = 1$

$$\therefore \left|\frac{a}{c}\right|\left|\frac{0-\alpha}{0+\bar{\alpha}}\right| = |w| = 1$$

i.e., $\left|a/c\right| = 1 \quad \left(\text{Since } |\alpha| = |\bar{\alpha}|\right)$

Let $a/c = e^{i\lambda}$ where λ is real. Then the bilinear transformation becomes $w = e^{i\lambda}.\left(\dfrac{z-\alpha}{z+\bar{\alpha}}\right)$

When $z = \alpha$, $w = 0$, which is the center of $|w| = 1$ $\therefore$ $z = \alpha$ must be a point of the half-plane $\text{Re}(z) > 0$. Hence $\text{Re}(\alpha) > 0$. Thus the bilinear transformation that maps the half plane $\text{Re}(z) \geq 0$ onto the unit circular disc $|w| \leq 1$ is

$w = e^{i\lambda}.\left(\dfrac{z-\alpha}{z+\bar{\alpha}}\right)$ where λ is real and $\text{Re}(\alpha) > 0$.

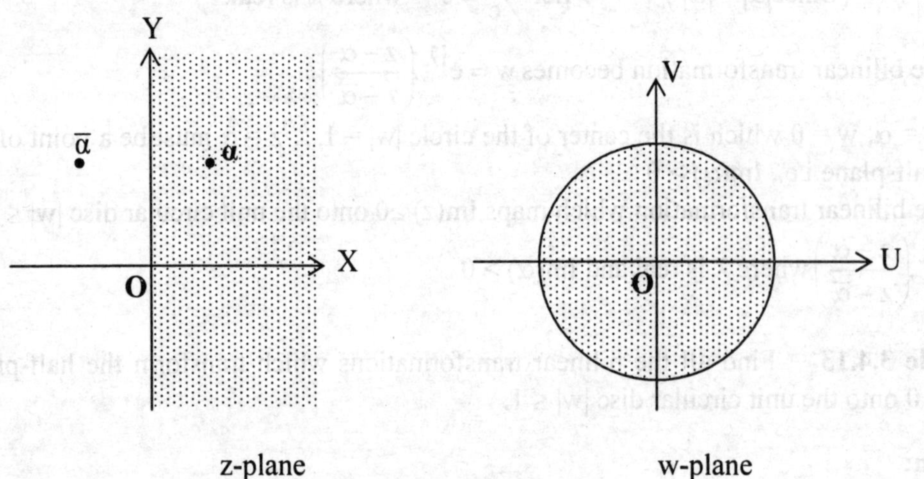

Fig. 3.27

Example 3.4.14 Prove that any bilinear transformation which maps the unit circle $|z| = 1$ onto the unit circle $|w| = 1$ can be written in the form $w = e^{i\lambda}.\left(\dfrac{z-\alpha}{\bar{\alpha}z-1}\right)$ where λ is real. Also prove that this transformation maps the circular disc $|z| \le 1$ onto the circular disc $|w| \le 1$ iff $|\alpha| < 1$

Solution:

Let $w = \dfrac{az+b}{cz+d}$, $ad - bc \neq 0$ (1)

be the required bilinear transformation which maps $|z| = 1$ onto $|w| = 1$. 0 and ∞ are inverse points with respect to the circle $|w| = 1$. These points correspond to the points $-b/a$ and $-d/c$ in the z-plane.

$\therefore$ $-b/a$ and $-d/c$ are inverse points with respect to the circle $|z| = 1$

Let $\alpha = -b/a$ so that $\dfrac{1}{\bar{\alpha}} = -d/c$.

Then the transformation (1) becomes $w = \dfrac{az+b}{cz+d}$

$$w = \frac{a}{c}\frac{\left(z + b/a\right)}{\left(z + d/c\right)}$$

$$= \frac{a}{c}\frac{(z-\alpha)}{\left(z - (1/\bar{\alpha})\right)}$$

$$= \frac{a\bar{\alpha}}{c}\frac{(z-\alpha)}{(\bar{\alpha}z - 1)} \qquad (2)$$

Now the image of $z = 1$ is a point on $|w| = 1$

$$\therefore \left|\frac{a\bar{\alpha}}{c}\right|\left|\frac{1-\alpha}{\bar{\alpha}-1}\right| = |w| = 1$$

i.e., $\left|a\bar{\alpha}/c\right| = 1$ $\left(\text{Since } |1-\alpha| = |\bar{\alpha}-1|\right)$

Let $a\bar{\alpha}/c = e^{i\lambda}$, where λ is real.

$\therefore$ Equation (2) becomes $w = e^{i\lambda}\left(\dfrac{z-\alpha}{\bar{\alpha}z-1}\right)$, where λ is real. (3)

$$\text{Now } w\bar{w} - 1 = e^{i\lambda}\left(\frac{z-\alpha}{\bar{\alpha}z-1}\right).e^{-i\lambda}\left(\frac{\bar{z}-\bar{\alpha}}{\alpha\bar{z}-1}\right) - 1$$

$$= \frac{(z-\alpha)(\bar{z}-\bar{\alpha}) - (\bar{\alpha}z-1)(\alpha\bar{z}-1)}{|\alpha\bar{z}-1|^2}$$

$$= \frac{1}{|\alpha\bar{z}-1|^2}(z\bar{z} + \alpha\bar{\alpha} - \alpha\bar{z} - \bar{\alpha}z - \alpha\bar{\alpha}z\bar{z} + \bar{\alpha}z + \alpha\bar{z} - 1)$$

$$= \frac{1}{|\alpha\bar{z}-1|^2}(z\bar{z} + \alpha\bar{\alpha} - \alpha\bar{\alpha}z\bar{z} - 1)$$

$$= \frac{1}{|\alpha\bar{z}-1|^2}(1-\alpha\bar{\alpha})(z\bar{z}-1)$$

i.e., $|w|^2 - 1 == \dfrac{(1-|\alpha|^2)(|z|^2-1)}{|\alpha\bar{z}-1|^2}$

The transformation (3) maps $|z| \le 1$ onto $|w| \le 1$

$\Leftrightarrow 1-|\alpha|^2 > 0$

$\Leftrightarrow |\alpha| < 1$

Example 3.4.15 Show that both the transformations $w = \dfrac{z-i}{z+i}$, and $w = \dfrac{i-z}{i+z}$ transform the upper half-plane $\text{Im}(z) \ge 0$ into $|w| \le 1$

Solution:

When $w = \dfrac{z-i}{z+i}$, we have,

$$w\bar{w} - 1 = \frac{z-i}{z+i}\cdot\frac{\bar{z}+i}{\bar{z}-i} - 1$$

$$= \frac{(z-i)(\bar{z}+i)-(z+i)(\bar{z}-i)}{|z+i|^2}$$

$$= \frac{2i.(z-\bar{z})}{|z+i|^2}$$

i.e., $$|w|^2 - 1 = \frac{-4\,\text{Im}(z)}{|z+i|^2} \quad (1)$$

When $w = \dfrac{i-z}{i+z}$, we have,

$$w\bar{w} - 1 = \frac{i-z}{i+z}\cdot\frac{-i-\bar{z}}{-i+z} - 1$$

$$= \frac{z-i}{z+i}\cdot\frac{\bar{z}+i}{\bar{z}-i} - 1$$

$$= \frac{2i.(z-\bar{z})}{|z+i|^2}$$

i.e., $$|w|^2 - 1 = \frac{-4\,\text{Im}(z)}{|z+i|^2} \quad (2)$$

(1) and (2) imply that for both the transformations $|w|^2 - 1 \le 0$ iff $\text{Im}(z) \ge 0$. Thus both the transformations transform the upper half-plane $\text{Im}(z) \ge 0$ into the unit circular disc $|w| \le 1$.

EXERCISE 3.4

PART-A

1. Define a bilinear transformation.
2. Define a Mobius transformation.
3. What is the determinant of the bilinear transformation $w = \dfrac{z+i}{z-i}$?
4. What is a normalized bilinear transformation?
5. Define 'fixed points' or 'invariant points' of a transformation.
6. What are the fixed points of $w = \dfrac{az+b}{cz+d}$, $ad - bc \neq 0$.
7. What are the critical points of $w = \dfrac{az+b}{cz+d}$, $ad - bc \neq 0$.
8. Prove that the inverse of a bilinear transformation is again a bilinear transformation.
9. Write $w = \dfrac{az+b}{cz+d}$ as a product of simple transformations.
10. Define cross ratio of four points in the complex plane.
11. Prove that cross ratio of four points is real iff the four points lie on a circle.
12. State the cross ratio property of a bilinear transformation.
13. What is the condition for a bilinear transformation to be hyperbolic? When is it elliptic?
14. When a bilinear transformation is called parabolic?
15. State the conditions for a bilinear transformation to have two finite fixed points.
16. Find the conditions for the fixed points of a bilinear transformation to be equal and finite.
17. Find the bilinear transformations whose only fixed point is the point at infinity.
18. Find the bilinear transformations whose only fixed points are 1 and –1.
19. Find the bilinear transformations whose only fixed points are i and –i.
20. Write $w = \dfrac{z-1}{z+1}$ as a product of elementary transformations.
21. Find the bilinear transformation which maps the points ∞, i, 0 to 0, i, ∞ respectively.
22. Find the fixed points of $w = \dfrac{1}{z-2i}$.
23. Prove that $w = \bar{z}$ is not a bilinear transformation.
24. Prove that the bilinear transformation having $z = 0$ as a fixed point is $w = \dfrac{z}{cz+d}$.
25. Prove that a bilinear transformation having 0 and ∞ as fixed points is of the form $w = az$.
26. Find the invariant points of $w = \dfrac{(2+i)z-2}{z+i}$.

27. Find the Mobius transformation which maps 2, 1 + i, 0 to 0, 1, ∞ respectively.
28. Show that the mapping w = az + b (a ≠ 0) is conformal in the entire complex plane.
29. Find the bilinear transformation which maps the points $z_1 = 1$, $z_2 = i$, $z_3 = -1$ into the points $w_1 = i$, $w_2 = -1$, $w_3 = -i$.
30. Find the bilinear transformation which maps the points $z_1 = 2$, $z_2 = i$, $z_3 = -2$ into the points $w_1 = 1$, $w_2 = i$, $w_3 = -1$.

PART-B

31. Prove that a bilinear transformation is a product of elementary (simple) transformations.
32. Prove that any bilinear transformation can be expressed as a product of translation, rotation, magnification or contraction and inversion.
33. Express the bilinear transformation $w = \frac{1+i-3z}{2-i+iz}$ as a product of elementary transformations.
34. Show that the transformation $w = \frac{iz+2}{4z+i}$ maps the real axis in the z-plane to a circle in the w-plane. Find the centre and radius of the circle.
35. Show that the transformation $w = \frac{2z+3}{z-4}$ maps the circle $z\bar{z} - 2(z+\bar{z}) = 0$ into a straight line $2(w+\bar{w})+3=0$.
36. Show that the relation $w = \frac{5-4z}{4z-2}$ transforms the circle |z| = 1 into a circle of radius unity in the w-plane and find the centre of this circle.
37. Find the bilinear transformation which maps z = 1, i, -1 onto w = i, 0, -i respectively. For this transformation find the images of (i) $|z| \le 1$ and (ii) $|z| = r$ $(r > 1)$.
38. Show that both the transformations $w = \frac{z+i}{z-i}$ and $w = \frac{i+z}{i-z}$ transform the lower half-plane $\text{Im}(z) \le 0$ into unit circular disc $|w| \le 1$.
39. Find the condition that the transformation $w = \frac{az+b}{cz+d}$ transforms a straight line of z-plane into the unit circle of w-plane.
40. Find the bilinear transformation which transforms the unit circle |z| = 1 into the real axis in such a way that the points $z_1 = 1$, $z_2 = i$ and $z_3 = -1$ are mapped into $w_1 = 0$, $w_2 = 1$ and $w_3 = \infty$ respectively. Into what regions the interior and exterior of the circle are mapped?
41. Find the bilinear transformation which maps the vertices 1 + i, -i, 2 - i of a triangle of the z-plane into the points 0, 1, i of the w-plane.

42. Find the bilinear transformation which maps the points $z_1 = -1$, $z_2 = 1$, $z_3 = \infty$ into the points $w_1 = -i$, $w_2 = -1$, $w_3 = i$ respectively.
43. Find the bilinear transformation which maps the points $z = 1, i, 2 + i$ in the z-plane onto the points $w = i, 1, \infty$ in the w-plane.
44. Find the image of the half-plane $x + y > 0$ under the bilinear transformation, $w = \dfrac{z-1}{z+i}$
45. Find the bilinear transformation which has 1 and i as fixed points and maps the point 0 into -1.
46. Find the bilinear transformation for which $z = i$ is the only fixed point and the point 1 is mapped into ∞.
47. Prove that a bilinear transformation having origin as a fixed point is in the form $w = \dfrac{z}{cz+d}$ and that having 0 and ∞ as fixed points is in the form $w = az$.
48. Find the fixed points of the bilinear transformations (i) $w = \dfrac{2z-1+3i}{z+3i}$ (ii) $w = \dfrac{z(i+1)-3}{z+i-3}$.
49. Find the fixed points and normal form of the bilinear transformation, $w = \dfrac{(2+i)z-2}{z+i}$.
50. Find the image of the segment of the real axis from $z = -1$ to $z = +1$, under the transformation, $w = \dfrac{1-iz}{z-i}$.
51. Prove that the general bilinear transformation which maps the circle $|z| = r$ onto the circle $|w| = \rho$ can be put in the form $w = r\rho\, e^{i\lambda}\left[\dfrac{z-\alpha}{\overline{\alpha}z-r^2}\right]$.

    Also prove that the transformation maps the disc $|z| \le r$ onto the disc $|w| \le \rho$ If and only if $|\alpha| < r$.
52. Find the bilinear transformation which maps the circle $|w| = 1$ onto the circle $|z - 1| = 1$ and maps $w = 0$, $w = 1$ respectively into $z = ½$, $z = 0$.
53. Find the bilinear transformation which maps the circle $|z| = 1$ onto the circle $|w + 1| = 1$ and maps the points $z = 0$ and $z = 1$ into $w = -3/2$ and $w = -2$ respectively.
54. Prove that z_1 and z_2 are inverse points with respect to a circle (line) passing through the points α, β, γ if and only if $(z_1, \alpha, \beta, \gamma) = \overline{(z_2, \alpha, \beta, \gamma)}$

ANSWERS

EXERCISE 3.1

(23) All points on the line $y = x$. (24) All points except $z = -1$.
(25) $a = b = c = -1$. (30) $a = d = 2, b = c = -1$.
(31) $p = -1$ (32) $a = b$.
(34) $a = -1, b = 1, f'(z) = 2(1+i)z$ (35) yes, analytic
(36) $z = \pm 1$.

EXERCISE 3.2

(5) Yes (6) No
(7) $2xy$ (8) $2\tan^{-1}(y/x)$
(9) $1/z$ (10) $\log z$
(16) $f(z) = |z|^2$ is differentiable at $z = 0$, but not analytic at $z = 0$
(33) $v(x, y) = x^2 - y^2 + 2y + c$ (34) $v(x, y) = 3x^2y - 6xy + 3y - y^3 + c$
(35) $v(x, y) = -\cosh x.\cos y + c$ (36) $z.e^z + c$.
(37) $i(z^3 + 3z) + c$ (38) $\sec z + c$
(39) $\cot z + c$ (40) $f(z) = i.ze^{-z} + c$; $u = e^{-x}(x \sin y - y \cos y) + c$
(41) $f(z) = -ie^{iz^2} + ic$; $v = e^{-2xy}\cos(x^2 - y^2) + c$
(42) $-iz^3 + c$ (43) $f(z) = \frac{1}{2}(1 - \cot z/2)$
(44) $\left(\frac{1+i}{2}\right)\cot z + c$ (45) $\frac{1+i}{2z} + \frac{1-i}{2}$
(48) $2y + y^3 - 3x^2y = b$

EXERCISE 3.3

(1) $w = z^2\ z \neq 0$ (2) $w = \bar{z}$
(6) Points where $\frac{dw}{dz} = 0$ (7) Points where $\frac{dw}{dz} = 0$ or $\frac{dw}{dz} = \infty$
(8) $z = 0$ (9) $z = \pm e^{\alpha}$ (10) $z = \pm 1$
(11) $z = 0, \pm\pi i, \pm 2\pi i, \ldots$ (12) $z = \pm\frac{\pi i}{2}, \pm\frac{3\pi i}{2}, \ldots$
(13) $\frac{\pi}{4}, 2\sqrt{2}$ (14) $6, \frac{\pi}{2}$
(15) $1, \pi$ (16) $|w| = 1$
(17) $1, -1$ (18) $4v + 1 = 0$
(19) The region between the circles $u^2 + v^2 - u = 0$ and $u^2 + v^2 - \frac{u}{2} = 0$
(20) The strip $0 < v < 2$ (21) The region $u < 0$

(22) $-1 < u < 1, v > 0$

(23) The lemniscate $R^2 = \cos 2\varphi$

(24) $| w | = 5a$

(25) $v > 1$

(26) $v > 1$

(27) $(u - 1)v = 4$

(28) Region $| w | < r^n$

(29) The entire w-plane

(30) The region $v > 0$

(31) Upper half of the w-plane

(32) The sector $0 < \theta < \frac{\pi}{2}$, $w < 1$

(33) The ray $\varphi = c$

(34) The circle $\rho = e^c$

(35) $1 < |w| < 9$

(36) The segment of the u-axis $-1 \leq u \leq 1$

(37) The v-axis

(38) The segment of the u-axis $-1 \leq u \leq 1$

(41) Square region with vertices (2, 1), (4, 3), (2, 5) and (0, 3)

(42) Square region with vertices (3, -1), (4, -1), (4, 0) and (3, 0)

(43) The region bounded by $u = 1$, $u = 3$, $v = -1$, $v = -2$

(44) The region bounded by $u + v = 1$, $u + v = 5$, $u - v = 3$, $u - v = 1$

(46) $6v + 1 = 0$

(47) The two regions bounded by the rectangular hyperbolas $x^2 - y^2 = 1$, $x^2 - y^2 = 2$, $xy = \frac{1}{2}$, $xy = 1$

(48) The semi-circular region given by $0 < \rho \leq 1, 0 \leq \varphi \leq \pi$

(54) The region bounded by the parabolas $v^2 = -4(u - 1)$, $v^2 = 4(u + 1)$, and $u^2 = -2(v - \frac{1}{2})$

Since $\frac{dw}{dz} \neq 0$ inside the region, the mapping is conformal.

EXERCISE 3.4

(3) $\begin{vmatrix} 1 & i \\ 1 & -i \end{vmatrix} = -2i$

(17) Translations $w = z + a$

(18) $w = \dfrac{az + 1}{z + a}$

(19) $w = \dfrac{az + 1}{a - z}$

(20) $w = 1 + \dfrac{(-2)}{z + 1}$, $T_1(z) = z + 1$, $T_2(z) = \frac{1}{2}$, $T_3(z) = -2z$, $T_4(z) = 1 + z$

(21) $w = -\frac{1}{z}$

(22) i

(26) 1+i, 1-i.

(27) $w = i\left(\dfrac{2 - z}{z}\right)$

(29) $w = iz$

(30) $w = \dfrac{3z + 2i}{iz + 6}$

(33) $w_1 = z - (2i + 1)$, $w_2 = 1/w_1$, $w_3 = -(5 - 2i)\, w_2$, $w = w_3 + 3i$

(36) $u^2 + v^2 + u - \frac{3}{4} = 0$, $(-\frac{1}{2}, 0)$

(37) $w = \frac{i-z}{i+z}$ (i) $Re(w) \geq 0$ (ii) $u^2 + v^2 - \left(\frac{1+r^2}{1-r^2}\right)2u + 1 = 0$

(39) $|a| = |c|$

(40) $w = \frac{i(1-z)}{1+z}$ $|z| < 1$ is mapped into the half plane $Im(z) > 0$

(41) $w = \frac{z-(2+2i)}{(i-1)z-(3+5i)}$ (42) $w = i\left(\frac{z+(1+2i)}{z+(1-2i)}\right)$

(43) $w = \frac{(2+i)z-(2i+1)}{z-(2+i)}$ (44) $|w| < 1$

(45) $w = \frac{(1+2i)z-i}{z+i}$ (46) $w = \frac{(2i-1)z+1}{z-1}$

(48) (i) 1, 1-3i (ii) 1, 3

(49) $1+i, 1-i$ $\frac{w-(1+i)}{w-(1-i)} = \left(\frac{(1-2i)}{5}\right)\left(\frac{z-(1+i)}{z-(1-i)}\right)$

(50) Upper half of the circle $|w| = 1$

(52) $w = \frac{1-2z}{1+z}$ (53) $w = \frac{-3(z+1)}{z+2}$

CHAPTER 4

COMPLEX INTEGRATION

4.0 INTRODUCTION

In the case of a real valued function of a real variable, integration is defined as the inverse process of differentiation or as a limit of a sum. The concept of integration as the inverse process of differentiation is also applicable in the case of a function of a complex variable provided that the function is analytic in the region where it is to be integrated. If f(z) is analytic in a region R, $\int f(z)\,dz = F(z) + c$ where $F'(z) = f(z)$ and c is an arbitrary constant.

The concept of Riemann integral, i.e., integral as a limit of a sum can be extended to a function of a complex variable. The definite integral $\int_a^b f(z)\,dz$ depends on the path from a to b. On the other hand, if f(z) is analytic in a simply-connected region R, then $\int_a^b f(z)\,dz$ will be independent of the path from a to b. We develop the theory of integration for complex functions in the following sections.

4.0.1 LINE INTEGRALS OR CONTOUR INTEGRALS

Definition 4.0.1 A continuous arc or path $\{ z(t) = x(t) + iy(t) \ / \ a \le t \le b \}$ in the complex plane is said to be **smooth** if the derivatives $x'(t)$ and $y'(t)$ exist and are continuous for $a \le t \le b$.

A path consisting of a finite number of smooth pieces is called a **contour**.

Definition 4.0.2 A contour is a piecewise smooth curve.

Let $C = \{z(t) = x(t) + i\,y(t) \ / \ a \le t \le b \}$ be a piecewise smooth curve (contour) in the complex plane. Let f(z) be a continuous function defined on C. Let $a = t_0 < t_1 < t_2 \ldots < t_n = b$. Then the points $z_k = z(t_k)$, k = 0, 1, 2, …, n lie on the curve C. Let $\Delta z_k = z_k - z_{k-1}$. Then $|\Delta z_k| = |z_k - z_{k-1}|$ is the length of the chord joining the points z_k and z_{k-1} on the curve. Let s_k be any point on C such that it lies between z_{k-1} and z_k.

Let $S_n = \sum_{k=1}^{n} f(s_k)\,\Delta z_k$

Let n, the number of subdivisions tend to ∞ such that max $|\Delta z_k| \to 0$. If the sum S_n approaches a limit as $n \to \infty$, regardless of the choice of the points z_k and s_k, then f(z) is said to be integrable along the curve C and the limit of S_n denoted by $\int_C f(z)\,dz$ is called the **line integral** or **contour integral** of f(z) along the curve C. C is called the **path of integration**. When C is a closed curve and is traversed in the anti-clockwise direction, we can also write $\int_C f(z)\,dz$ as $\oint_C f(z)\,dz$.

Thus $$\int_C f(z)\,dz = \lim_{\substack{n\to\infty \\ \max|\Delta z_k|\to 0}} \sum_{k=1}^{n} f(s_k)(z_k - z_{k-1}) \qquad (1)$$

Properties of Line Integrals

(i) Let f(z) be a continuous function defined on a piecewise smooth curve C. Let the curve traversed in the direction opposite to that of C be denoted by -C. Then $\int_{-C} f(z)\,dz = -\int_C f(z)\,dz$.

(ii) Let f(z) and g(z) be two continuous functions defined on a piecewise smooth curve C. α and β be complex constants. Then we have,
$$\int_C [\alpha f(z) + \beta g(z)]\,dz = \alpha \int_C f(z)\,dz + \beta \int_C g(z)\,dz$$

(iii) Let C be the join of two piecewise smooth curves C_1 and C_2 with the terminal point of C_1 coinciding with the initial point of C_2. Then $C = C_1 + C_2$ and we have
$$\int_C f(z)\,dz = \int_{C_1} f(z)\,dz + \int_{C_2} f(z)\,dz$$
In general, if $C = C_1 + C_2 + \ldots + C_n$
$$\int_C f(z)\,dz = \int_{C_1} f(z)\,dz + \int_{C_2} f(z)\,dz + \cdots + \int_{C_n} f(z)\,dz.$$

Note:

(i) If $f(z) = u(x, y) + i\,v(x, y)$
$$\int_C f(z)\,dz = \int_C (u + iv)(dx + i\,dy)$$
i.e., $$\int_C f(z)\,dz = \int_C (u\,dx - v\,dy) + i\int_C (v\,dx + u\,dy) \qquad (2)$$

(ii) When $z(t) = x(t) + iy(t)$, $a \le t \le b$ is a piecewise smooth curve, $z'(t) = x'(t) + iy'(t)$ exists. Then we have,

$$\int_C f(z)\,dz = \int_C (u(t)+iv(t))\,dz$$

$$= \int_a^b (u(t)+iv(t))\frac{dz}{dt}.dt$$

$$= \int_a^b f(z(t)).z'(t).dt \qquad (3)$$

(iii) Let L be the length of the piecewise smooth curve C given by $z(t) = x(t) + iy(t)$, $a \le t \le b$ and $M > 0$ be such that $|f(z)| \le M$ everywhere on C. Then we have,

$$|S_n| = \left|\sum_{k=1}^{n} f(s_k).\Delta z_k\right|$$

$$\le \sum_{k=1}^{n} |f(s_k)|.|\Delta z_k|$$

$$\le M\sum_{k=1}^{n} |z_k - z_{k-1}|$$

When $n \to \infty$ such that $|\Delta z_k| \to 0$, $\sum_{k=1}^{n} |z_k - z_{k-1}|$ tends to the length of the curve C, denoted by L.

Hence we get, $\left|\int_C f(z)\,dz\right| = \lim_{n\to\infty} |S_n| \le M.L$

i.e., $$\left|\int_C f(z)\,dz\right| \le M.L \qquad (4)$$

The inequality (4) is called the **ML-inequality**.

(iv) $$\left|\int_C f(z)\,dz\right| \le \int_C |f(z)|.|dz| \qquad \text{(Prove this!)} \qquad (5)$$

(v) For the contour C given by $z(t) = x(t) + iy(t)$, $a \le t \le b$, the length L is given by

$$L = \lim_{n\to\infty} \sum_{k=1}^{n} |\Delta z_k|$$

$$= \int_C |dz|$$

$$= \int_a^b |z'(t)|\,dt$$

i.e., $$L = \int_a^b \sqrt{(x'(t))^2 + (y'(t))^2}\,dt \qquad (6)$$

(Since $z'(t) = x'(t) + iy'(t)$)

Example 4.0.1 Find the length of the circumference of the circle with centre at a and radius r.

Solution:

The parametric equation of the circle with centre at a and radius r is $z(t) = a + re^{it}$ where $0 \le t \le 2\pi$.

Then $z'(t) = ire^{it}$

$$L = \int_0^{2\pi} | z'(t) | \, dt = \int_0^{2\pi} | ire^{it} | \, dt = \int_0^{2\pi} r \, dt = 2\pi r$$

Example 4.0.2 Prove that $\int_C \frac{dz}{z-a} = 2\pi i$, where C is the circle with centre a and radius r.

Solution:

The equation of the circle C is $z(t) = a + re^{it}$, $0 \le t \le 2\pi$. Then $\frac{dz}{dt} = z'(t) = ire^{it}$.

i.e., $dz = ire^{it}. \, dt$

$$\therefore \int_C \frac{dz}{z-a} = \int_0^{2\pi} \frac{ire^{it} dt}{re^{it}}$$

$$= i \int_0^{2\pi} dt = 2\pi i$$

Example 4.0.3 Prove that $\int_C \frac{dz}{(z-a)^n} = \begin{cases} 0 & \text{if } n \ne 1 \\ 2\pi i & \text{if } n = 1 \end{cases}$ where C is the circle with centre a and radius r, and $n \in Z$.

Solution: The equation of the circle C is given by $z(t) = a + re^{it}$, $0 \le t \le 2\pi$ Then $dz = ire^{it}. \, dt$

$$\therefore \int_C \frac{dz}{(z-a)^n} = \int_0^{2\pi} \frac{ire^{it} dt}{(re^{it})^n}$$

$$= \frac{i}{r^{n-1}} \int_0^{2\pi} e^{i(1-n)t} dt$$

$$= \frac{i}{r^{n-1}} \left[\frac{e^{i(1-n)t}}{i(1-n)} \right]_0^{2\pi} \quad \text{when } n \ne 1$$

$$= \frac{1}{(1-n)r^{n-1}} \left[e^{i(1-n)2\pi} - 1 \right] = 0 \qquad (\text{Since } e^{i(1-n)2\pi} = 1)$$

When n = 1, we have $\int_C \frac{dz}{z-a} = i \int_0^{2\pi} dt = 2\pi i$.

Example 4.0.4 Find the value of the integral $\int_0^{1+i}(x-y+ix^2)dz$

(i) Along the straight line from z = 0 to z = 1 + i.

(ii) Along the real axis from z = 0 to z = 1 and then along a line parallel to the imaginary axis from z = 1 to z = 1 + i.

Solution:

(i) $z = x + iy$

$\therefore \quad dz = dx + idy$

Let B denote the point z = 1+ i in the Argand plane.

Along OB, we have y = x

$\therefore$ dy = dx and dz = dx + idx = (1+ i) dx.

$\therefore \int_0^{1+i}(x-y+ix^2)dz = \int_{x=0}^{1}(x-x+ix^2)(1+i)dx$, when path of integration is along OB

$$= \int_{x=0}^{1} ix^2(1+i)dx$$

$$= i(1+i).\frac{x^3}{3}\Big]_0^1 = \frac{i(1+i)}{3} = \frac{1}{3}(i-1)$$

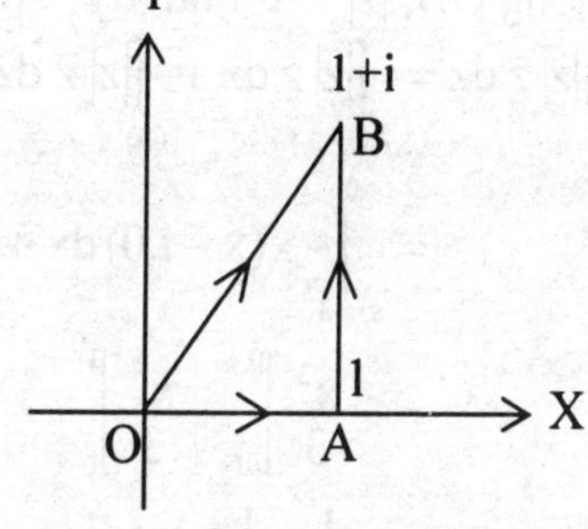

Fig. 4.1

(ii) Let A denote the point z =1. Along OA we have y = 0.

$\therefore$ dy = 0 and dz = dx.

$$\int_{OA}(x-y+ix^2)dz = \int_{x=0}^{1}(x-0+ix^2)dx = \frac{x^2}{2}+\frac{ix^3}{3}\Big]_0^1$$

$$= \frac{1}{2}+\frac{i}{3}$$

Along the line AB, we have x = 1 and y varies from 0 to 1.

$\therefore$ dx = 0 and dz = idy.

$$\int_{AB}(x-y+ix^2)dz = \int_{y=0}^{1}(1-y+i)idy = i\left[y-\frac{y^2}{2}+iy\right]_0^1$$

$$= i\left[1-\frac{1}{2}+i\right] = \frac{i}{2}-1$$

Hence $\int_0^{1+i}(x-y+ix^2)dz$ along the contour OAB

$$= \int_{OA}(x-y+ix^2)dz + \int_{AB}(x-y+ix^2)dz$$

$$= \frac{1}{2}+\frac{i}{3}+\frac{i}{2}-1 = \frac{-1}{2}+\frac{5i}{6}$$

Example 4.0.5 Evaluate $\int_C |z|\,\bar{z}\,dz$ where C is a closed curve consisting of the upper semi circle $|z| = 1$ and the segment $-1 \le x \le 1,\ y = 0$.

Solution:

$z = x + iy$

$dz = dx + i\,dy$

Let AB be the line segment $-1 \le x \le 1,\ y = 0$

Along AO, $|z| = -x$ and $dy = 0$

Along OB, $|z| = x$ and $dy = 0$

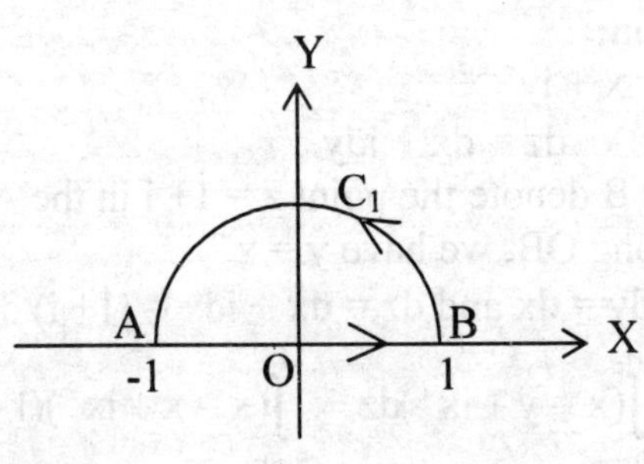

Fig. 4.2

$$\int_{AB} |z|\,\bar{z}\,dz = \int_{AO} |z|\,\bar{z}\,dz + \int_{OB} |z|\,\bar{z}\,dz$$

$$= \int_{x=-1}^{0} -x(x - i.0)\,dx + \int_{x=0}^{1} x(x - i.0)\,dx$$

$$= -\left[\frac{x^3}{3}\right]_{-1}^{0} + \left[\frac{x^3}{3}\right]_{0}^{1}$$

$$= -\frac{1}{3} + \frac{1}{3} = 0$$

Let C_1 be the upper semi circle $|z| = 1$.

Along C_1, $z = e^{i\theta},\ 0 \le \theta \le \pi$

$\therefore$ Along C_1, $\bar{z} = e^{-i\theta}$, $|z| = 1$ and $dz = ie^{i\theta}d\theta$

$$\therefore \int_{C_1} |z|\,\bar{z}\,dz = \int_{\theta=0}^{\pi} e^{-i\theta} ie^{i\theta} d\theta = i.\int_{\theta=0}^{\pi} d\theta = \pi\, i$$

Hence $\int_C |z|\,\bar{z}\,dz = \int_{AB} |z|\,\bar{z}\,dz + \int_{C_1} |z|\,\bar{z}\,dz$

$= 0 + \pi i = \pi i.$

Example 4.0.6 Evaluate $\int_{(0,3)}^{(2,4)} (2y + x^2)dx + (3x - y)dy$ along the parabola $x = 2t,\ y = t^2 + 3$.

Solution:

$x = 2t \quad \Rightarrow dx = 2dt$

$y = t^2 + 3 \quad \Rightarrow dy = 2tdt$

At (0, 3) we have $t = 0$

At (2, 4) we have $t = 1$.

$\therefore$ The given integral $= \int_{t=0}^{1} \left(2(t^2 + 3) + 4t^2\right)2dt + \left(3.2t - t^2 - 3\right)2tdt$

$$= \int_0^1 \left(4t^2 + 12 + 8t^2 + 12t^2 - 2t^3 - 6t\right)dt$$

$$= \int_0^1 \left(-2t^3 + 24t^2 - 6t + 12\right)dt$$

$$= \left.(-2)\frac{t^4}{4} + 24\frac{t^3}{3} - 6\frac{t^2}{2} + 12t\right]_0^1$$

$$= -\frac{2}{4} + \frac{24}{3} - \frac{6}{2} + 12$$

$$= -\frac{1}{2} + 8 - 3 + 12 = \frac{33}{2}$$

Example 4.0.7 Evaluate $\int_C \frac{z+2}{z} dz$ where C is

(i) the semicircle $z = 2e^{i\theta}$, $\pi \le \theta \le 2\pi$

(ii) the circle $z = 2e^{i\theta}$, $-\pi \le \theta \le \pi$

Solution:

(i) $z = 2e^{i\theta}$, $\pi \le \theta \le 2\pi$

$\therefore dz = 2ie^{i\theta}\, d\theta$

$$\int_C \frac{z+2}{z} dz = \int_\pi^{2\pi} \frac{2e^{i\theta} + 2}{2e^{i\theta}} .2ie^{i\theta} d\theta$$

$$= 2i \int_\pi^{2\pi} \left(e^{i\theta} + 1\right) d\theta$$

$$= 2i\left[\frac{e^{i\theta}}{i} + \theta\right]_\pi^{2\pi}$$

$$= 2i\left[\frac{1}{i} + 2\pi - \frac{(-1)}{i} - \pi\right]$$

$$= 4 + 2\pi i$$

(ii) $z = 2e^{i\theta}$, $-\pi \le \theta \le \pi$

$\therefore dz = 2ie^{i\theta}\, d\theta$

$$\int_C \frac{z+2}{z} dz = \int_{-\pi}^{\pi} \frac{2e^{i\theta} + 2}{2e^{i\theta}} .2ie^{i\theta} d\theta$$

$$= 2i \int_{-\pi}^{\pi} \left(e^{i\theta} + 1\right) d\theta$$

$$= 2i\left[\frac{e^{i\theta}}{i}+\theta\right]_{-\pi}^{\pi}$$

$$= 2i\left[\left(\frac{-1}{i}+\pi\right)-\left(\frac{-1}{i}-\pi\right)\right] = 4\pi i$$

Example 4.0.8 Find the upper bound for the value of the integral $\int_C e^z dz$ where C is the line segment joining the points (0, 0) and $(1, 2\sqrt{2})$.

Solution:

Let $f(z) = e^z = e^{x+iy}, 0 \le x \le 1, 0 \le y \le 2\sqrt{2}$

$|f(z)| = |e^x.e^{iy}| = |e^x| \le e$ since $0 \le x \le 1$

$\therefore M = \max|f(z)| = e$

L = length of C

= distance between (0, 0) and $(1, 2\sqrt{2})$.

$= \sqrt{1^2+(2\sqrt{2})^2} = \sqrt{9} = 3$

$\therefore \left|\int_C f(z)\,dz\right| \le ML = e.3 = 3e$

$\therefore$ The upper bound for the value of $\int_C e^z dz$ is 3e.

Example 4.0.9 Show that $\left|\int_C \frac{dz}{z^2+1}\right| \le \frac{\pi}{3}$ where C is the arc of the circle $|z| = 2$ from $z = 2$ to $z = 2i$ that lies in the first quadrant.

Solution:

On C, $|z^2+1| \ge |z|^2-1 = 2^2-1 = 3$

$$\therefore \frac{1}{|z^2+1|} \le \frac{1}{3}$$

$$\therefore |f(z)| = \left|\frac{1}{z^2+1}\right| \le \frac{1}{3}$$

i.e., $M = \frac{1}{3}$

Length of C is $L = \frac{1}{4}(2\pi r) = \frac{1}{4}2\pi.2 = \pi$

$$\therefore \left|\int_C \frac{dz}{z^2+1}\right| \le M.L = \frac{1}{3}.\pi = \frac{\pi}{3}$$

Example 4.0.10 Find an upper bound for $\int_C e^{z^2} dz$ where C is the broken lines from z = 0 to z = 1 and then from z = 1 to z = 1+i.

Solution:

Length of the contour C is L = 1+1 = 2.

On the line segment from z = 0 to z = 1, we have y = 0, z = x and $0 \le x \le 1$.

$$\therefore |f(z)| = \left|e^{z^2}\right|$$

$$= \left|e^{x^2}\right| \le e$$

On the line segment from z = 1 to z = 1+i, we have x = 1, z = 1+iy and $0 \le y \le 1$.

$$\therefore |f(z)| = = \left|e^{z^2}\right| = \left|e^{(1+iy)^2}\right|$$

$$= \left|e^{1-y^2+2iy}\right|$$

$$= \left|e^{1-y^2} e^{2iy}\right|$$

$$= \left|e^{1-y^2}\right| \le e \text{ as } 0 \le y \le 1 \text{ and hence } 0 \le 1 - y^2 \le 1$$

Thus on C, $|f(z)| \le e$

i.e., M = e

Hence $\int_C e^{z^2} dz \le M.L = e.2 = 2e$

Hence an upper bound for $\int_C e^{z^2} dz$ is 2e.

4.1 CAUCHY'S THEOREM AND APPLICATIONS

The theorem known as Cauchy's integral theorem or Cauchy's fundamental theorem has been proved by Cauchy using an important theorem of integral calculus, viz. Green's theorem. Cauchy proved the theorem for a function f(z) that is analytic and its derivative f'(z) continuous at all points on and inside a simple closed curve C. Later a French Mathematician E. Goursat (1858-1936) proved Cauchy's theorem without assuming the continuity of f'(z). Also it has been proved that if a function is analytic at a point, its derivatives of all orders exist and are also analytic at that point. In this section we give the proof of Cauchy's theorem using Green's theorem. Cauchy's integral theorem is perhaps the most important theorem in the theory complex analysis. Its application to the theory of integration – in the evaluation of contour integrals – will be discussed in the last section of this chapter.

Theorem 4.1.1 (Cauchy's Integral Theorem or Cauchy's Fundamental Theorem)
If f(z) is analytic and its derivative f′(z) is continuous at all points on and inside a simple closed curve C, then $\int_C f(z)dz = 0$.

Proof:
Let R be the closed region in the z-plane consisting of all points inside and on the simple closed curve C and let the curve C be described in the anti-clockwise direction. We prove Cauchy's integral theorem using **Green's Theorem**.

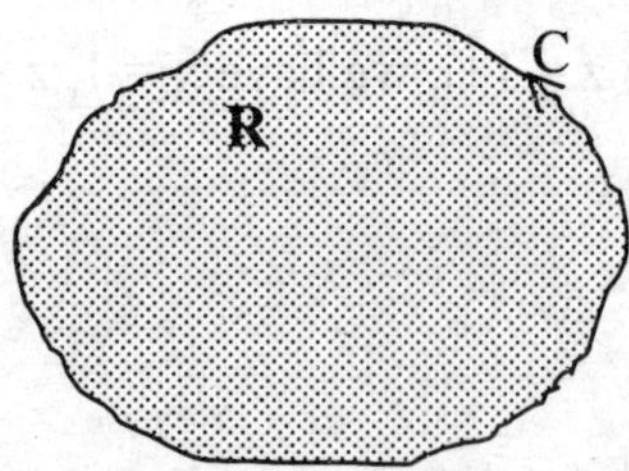

Fig. 4.3

Green's Theorem: If P(x, y), Q(x, y) and the partial derivatives $\frac{\partial P}{\partial y}$ and $\frac{\partial Q}{\partial x}$ are continuous single-valued functions over the closed region R bounded by the curve C, then

$$\int_C (Pdx + Qdy) = \iint_R \left(\frac{\partial Q}{\partial x} - \frac{\partial P}{\partial y}\right) dx.dy \qquad (1)$$

Let f(z) = u(x, y) + iv(x, y)
Since f(z) is analytic in R, u and v are continuous in R and satisfy the Cauchy-Riemann equations.

$$\therefore \frac{\partial u}{\partial x} = \frac{\partial v}{\partial y} \text{ and } \frac{\partial u}{\partial y} = -\frac{\partial v}{\partial x} \qquad (2)$$

Since f′(z) is continuous, the four partial derivatives $\frac{\partial u}{\partial x}, \frac{\partial u}{\partial y}, \frac{\partial v}{\partial x}$ and $\frac{\partial v}{\partial y}$ are also continuous on R.

$$\text{Now } \int_C f(z)dz = \int_C (u + iv)(dx + idy)$$

$$= \int_C (udx - vdy) + i\int_C (vdx + udy)$$

$$= \iint_R \left(-\frac{\partial v}{\partial x} - \frac{\partial u}{\partial y}\right) dx.dy + i\iint_R \left(\frac{\partial u}{\partial x} - \frac{\partial v}{\partial y}\right) dx.dy \quad \text{(By Green's theorem)}$$

$$= \iint_R \left(\frac{\partial u}{\partial y} - \frac{\partial u}{\partial y} \right) dx.dy + i \iint_R \left(\frac{\partial v}{\partial y} - \frac{\partial v}{\partial y} \right) dx.dy \quad = 0 \qquad \text{(Using (2))}$$

Note:

(i) Cauchy's integral theorem can be stated as follows: If f(z) is analytic and f'(z) is continuous in a simply connected region R, then $\int_C f(z)dz = 0$ along every simple closed curve C contained in R.

(ii) **Cauchy-Goursat's theorem** : If f(z) is analytic, in a simply connected region R, then $\int_C f(z)dz = 0$ along every simple closed curve C contained in R.

(iii) If f(z) is analytic, in a simply connected region R, then the line integral $\int_{z_0}^{z_1} f(z)dz$ is independent of the path joining the points z_0 and z_1 in R.

4.1.1 EXTENSION OF CAUCHY'S INTEGRAL THEOREM FOR MULTIPLY CONNECTED REGIONS

In the last section we proved Cauchy's integral theorem for a simply connected region. The theorem can be extended to doubly and multiply connected regions. A multiply connected region can be converted into a simply connected region by introducing sufficient number of cuts in the region. For example, consider a doubly connected region whose outer boundary is C and inner boundary is C_1. Introduce a cut in the region along PQ.

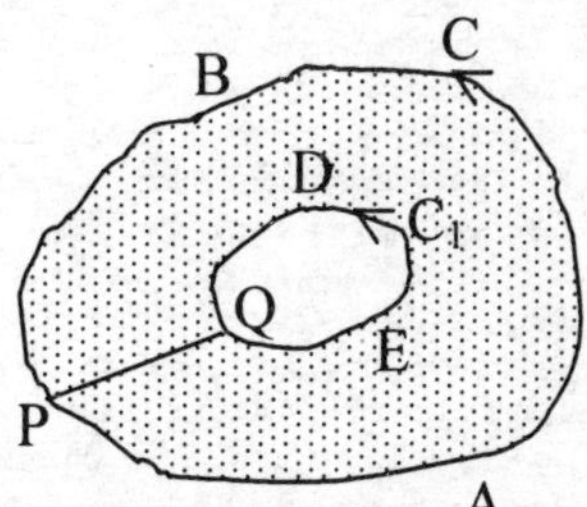

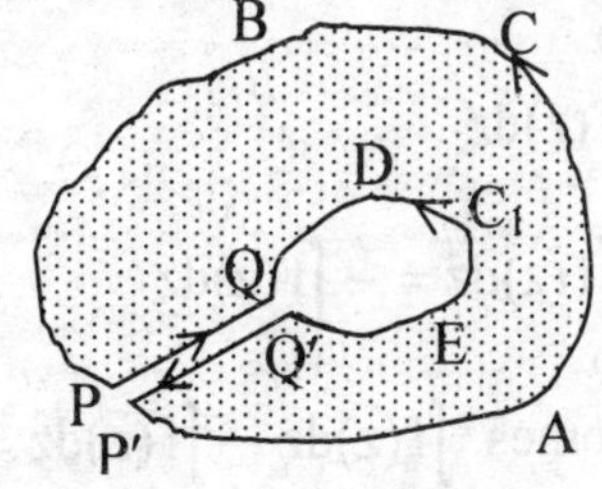

Fig. 4.4

The region bounded by the closed curve P'ABPQDEQ'P' is simply connected. Note that the boundary consists of the closed curves C, $-C_1$ and the line segments PQ and Q'P'. In a similar process, a multiply connected region with n inner boundaries can be converted into a simply connected region by introducing n cuts in the region joining the inner boundaries with the outer boundary.

Theorem 4.1.2 Let R be a multiply connected region bounded by non-intersecting simple closed curves C, C_1, C_2, ..., C_n where C_1, C_2, ..., C_n be inside C. Let the function f(z) be defined and analytic inside and on the boundaries of the region R. Then

$$\int_C f(z)dz = \int_{C_1} f(z)dz + \int_{C_2} f(z)dz + \cdots + \int_{C_n} f(z)dz$$

Proof:

Consider the case when n = 1. The region R has outer boundary C and inner boundary C_1. Convert the region R into a simply connected region by introducing a cut PQ. Then R is bounded by the simple closed curve PABPQDEQP.

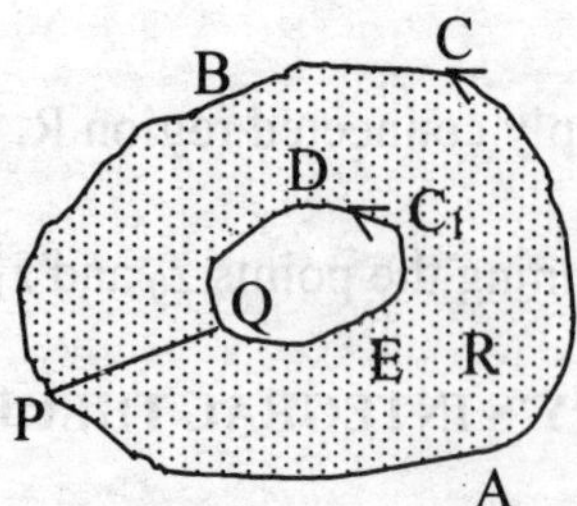

Fig. 4.5

By Cauchy's integral theorem for a simply connected region, we get $\int_{PABPQDEQP} f(z)dz = 0$

i.e., $$\int_{PABP} f(z)dz + \int_{PQ} f(z)dz + \int_{QDEQ} f(z)dz + \int_{QP} f(z)dz = 0 \quad (1)$$

$$\int_{QP} f(z)dz = -\int_{PQ} f(z)dz$$

$$\int_{PABP} f(z)dz = \int_{C} f(z)dz$$

$$\int_{QDEQ} f(z)dz = \int_{-C_1} f(z)dz = -\int_{C_1} f(z)dz$$

Hence (1) becomes $\int_C f(z)dz - \int_{C_1} f(z)dz = 0$

i.e., $\int_C f(z)dz = \int_{C_1} f(z)dz$

In general, let the region R be multiply connected with n inner boundaries C_1, C_2, ..., C_n. and outer boundary C. R can be converted into a simply connected region by introducing n cuts P_1Q_1, P_2Q_2, ..., P_nQ_n in the region. Note that while applying Cauchy's integral theorem for the resulting simply connected region, the integrals evaluated along the cuts in the opposite directions get cancelled and hence we get, (Fig. 4.6)

$$\int_C f(z)dz = \int_{C_1} f(z)dz + \int_{C_2} f(z)dz + \cdots + \int_{C_n} f(z)dz$$

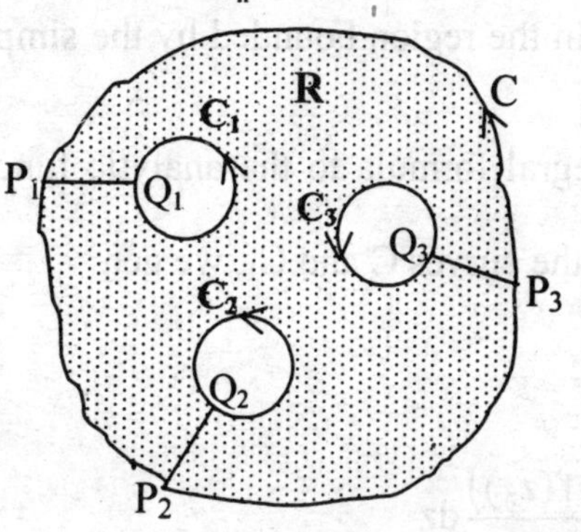

Fig. 4.6

4.1.2 CAUCHY'S INTEGRAL FORMULA

Cauchy's integral formula is a consequence of cauchy's integral theorem. It is a representation of an analytic function f(z) at any interior point z_0 of a simply connected region R as an integral along the boundary of a simple closed curve C that lies inside R and encloses the point z_0.

Theorem 4.1.3 (Cauchy's Integral Formula)
Let f(z) be analytic in a simply connected region R. Let z_0 be any interior point of R and C be any simple closed curve in R enclosing z_0.

Then $$f(z_0) = \frac{1}{2\pi i}\int_C \frac{f(z)}{z - z_0}dz \qquad (1)$$

where C is traversed in the anti-clockwise direction.

Proof: Let C_1 be a circle $|z - z_0| = \rho$ such that it lies entirely inside C.

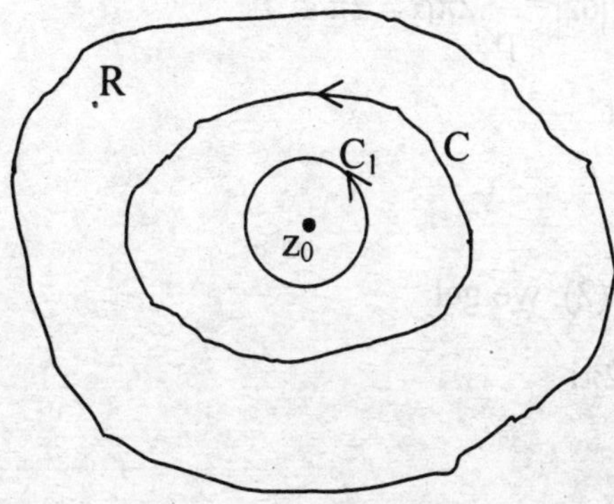

Fig. 4.7

f(z) is analytic on and inside C. $\therefore$ $\dfrac{f(z)}{z-z_0}$ is also analytic on and inside C except at the point z_0. Hence it is analytic in the region bounded by the simple closed curve C and the circle C_1.

Hence applying Cauchy's integral formula to the analytic function $\dfrac{f(z)}{z-z_0}$ in the doubly connected region bounded by the curves C and C_1, we get,

$$\int_C \frac{f(z)}{z-z_0}\,dz = \int_{C_1} \frac{f(z)}{z-z_0}\,dz$$

$$= \int_{C_1} \frac{f(z_0)+\left(f(z)-f(z_0)\right)}{z-z_0}\,dz$$

$$= f(z_0)\int_{C_1} \frac{dz}{z-z_0} + \int_{C_1} \frac{f(z)-f(z_0)}{z-z_0}\,dz \qquad (2)$$

On the circle C_1, $z = z_0 + \rho . e^{i\theta}$ $0 \le \theta \le 2\pi$

$$dz = \rho\, e^{i\theta} i\, d\theta$$

$$\therefore \quad f(z_0)\int_{C_1} \frac{dz}{z-z_0} = f(z_0)\int_0^{2\pi} \frac{\rho\, e^{i\theta} i\, d\theta}{\rho\, e^{i\theta}}$$

$$= f(z_0).\,2\pi i \qquad (3)$$

f(z) is analytic and hence is continuous in R.

$\therefore$ Given a real number $\in > 0$ there exists a real number $\delta > 0$ such that $|f(z) - f(z_0)| < \in$, whenever $|z-z_0| < \delta$. Choosing the radius ρ of the circle C_1 such that $\rho < \delta$, we get,

$$\left|\int_{C_1} \frac{f(z)-f(z_0)}{z-z_0}\,dz\right| \le \int_{C_1} \frac{|f(z)-f(z_0)|}{|z-z_0|}|dz|$$

$$< \int_{C_1} \frac{\in}{\rho}|dz| = \frac{\in}{\rho}.2\pi\rho = 2\pi \in$$

Since $\in$ is arbitrary, we get

$$\int_{C_1} \frac{f(z)-f(z_0)}{z-z_0}\,dz = 0 \qquad (4)$$

Hence using (3) and (4) in (2), we get

$$\int_C \frac{f(z)-f(z_0)}{z-z_0}\,dz = f(z_0).2\pi i$$

$$\text{i.e., } f(z_0) = \frac{1}{2\pi i}\int_C \frac{f(z)}{z-z_0}\,dz$$

Note: (Extension of Cauchy's Integral Formula to Multiply Connected Regions)

Let R be a multiply connected region bounded by non-intersecting simple closed curves C, $C_1, C_2, \ldots, C_n$, where $C_1, C_2, \ldots, C_n$ lie inside C. Let f(z) be analytic inside and on the boundaries of R and z_0 be any interior point of R. Then we have,

$$f(z_0) = \frac{1}{2\pi i}\int_C \frac{f(z)}{z-z_0}dz - \frac{1}{2\pi i}\int_{C_1} \frac{f(z)}{z-z_0}dz - \frac{1}{2\pi i}\int_{C_2} \frac{f(z)}{z-z_0}dz - \cdots - \frac{1}{2\pi i}\int_{C_n} \frac{f(z)}{z-z_0}dz$$

4.1.3 CAUCHY'S INTEGRAL FORMULA FOR THE DERIVATIVES OF AN ANALYTICAL FUNCTION

As a consequence of Cauchy's integral formula, we can now prove that if f(z) is analytic in a region R, then its derivatives of all orders exist and are also analytic in R. Note that this result is available only for functions of a complex variable. Such a result does not exist for functions of real variables.

Theorem 4.1.4 Let f(z) be analytic in a simply connected region R. z_0 be any point in R and C be a simple closed curve in R enclosing z_0. Then,

$$f'(z_0) = \frac{1}{2\pi i}\int_C \frac{f(z)}{(z-z_0)^2}dz \tag{1}$$

where C is traversed in the anti-clockwise direction.

Proof:

By the definition of derivative, we have

$$f'(z_0) = \lim_{h\to 0}\frac{f(z_0+h)-f(z_0)}{h} \tag{2}$$

Since z_0 lies inside C, $z_0 + h$ also lies inside C when |h| is sufficiently small.

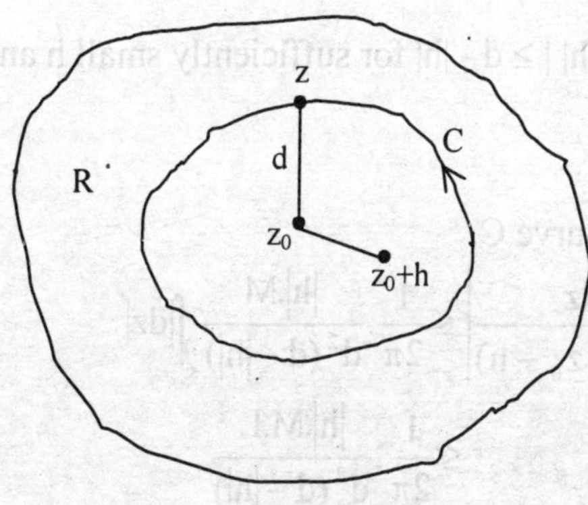

Fig. 4.8

By Cauchy's integral formula, we have $f(z_0) = \frac{1}{2\pi i}\int_C \frac{f(z)}{(z-z_0)}dz$

$$f(z_0+h) = \frac{1}{2\pi i}\int_C \frac{f(z)}{(z-(z_0+h))}dz$$

Hence, $\frac{f(z_0+h)-f(z_0)}{h} = \frac{1}{2\pi i.h}\int_C \left[\frac{1}{z-(z_0+h)} - \frac{1}{z-z_0}\right] f(z).dz$

$$= \frac{1}{2\pi i.}\int_C \frac{f(z).dz}{(z-z_0-h).(z-z_0)} \tag{3}$$

$$\frac{1}{(z-z_0-h)(z-z_0)} = \frac{1}{(z-z_0)^2}\left[\frac{z-z_0-h+h}{z-z_0-h}\right]$$

$$= \frac{1}{(z-z_0)^2} + \frac{h}{(z-z_0)^2(z-z_0-h)}$$

Therefore (2) becomes,

$$f'(z_0) = \lim_{h\to 0}\frac{1}{2\pi i}\int_C \frac{f(z).dz}{(z-z_0-h)(z-z_0)} \quad \text{(Using (3))}$$

$$= \lim_{h\to 0}\frac{1}{2\pi i}\int_C \frac{f(z).dz}{.(z-z_0)^2} + \lim_{h\to 0}\frac{1}{2\pi i}\int_C \frac{h.f(z).dz}{.(z-z_0)^2.(z-z_0-h)}$$

$$= \frac{1}{2\pi i}\int_C \frac{f(z).dz}{.(z-z_0)^2} + \frac{1}{2\pi i}\lim_{h\to 0}\int_C \frac{h.f(z).dz}{.(z-z_0)^2.(z-z_0-h)} \tag{4}$$

Since f(z) is analytic and hence is continuous on C, it is bounded on C. Hence there exists a positive real number M such that $|f(z)| \le M$ for all z on C. If d is the distance of z_0 from the point on C nearest to it, $|z-z_0| \ge d$ for any point z on C.

i.e., $\frac{1}{|z-z_0|^2} \le \frac{1}{d^2}$

Also $|z - z_0 - h| \ge |\,|z - z_0| - |h|\,| \ge d - |h|$ for sufficiently small h and z any point on C.

$$\left|\frac{1}{z-z_0-h}\right| \le \frac{1}{d-|h|}$$

Let L be the length of the curve C.

Hence, $\left|\frac{1}{2\pi i}\int_C \frac{h.f(z).dz}{(z-z_0)^2(z-z_0-h)}\right| \le \frac{1}{2\pi}.\frac{|h|.M}{d^2(d-|h|)}\int_C |dz|$

$$\le \frac{1}{2\pi}.\frac{|h|.M.L}{d^2(d-|h|)}$$

$$\to 0 \text{ as } h \to 0$$

Thus from (4), we get $f'(z_0) = \dfrac{1}{2\pi i}\displaystyle\int_C \dfrac{f(z)}{(z-z_0)^2}\,dz$

Note:

(i) In general one can prove by the method of mathematical induction on n that $f^{(n)}(z_0) = \dfrac{n!}{2\pi i}\displaystyle\int_C \dfrac{f(z)}{(z-z_0)^{n+1}}\,dz, \quad n = 1, 2, 3, \ldots$

(ii) When f(z) is analytic in a region R and C is any simple closed curve in R then the derivatives $f^{(n)}(z)$ for n = 1, 2, … exist at all points inside C. Now $f''(z)$ exist implies $f'(z)$ is analytic inside C, $f'''(z)$ exists implies $f''(z)$ is analytic inside C, etc. Thus f(z) and all its derivatives are analytic inside C.

(iii) If f(z) is analytic inside and on a simple closed curve C and C encloses a point z_0 then $\displaystyle\int_C \dfrac{f(z)}{(z-z_0)^{n+1}}\,dz = \dfrac{2\pi i}{n!} f^{(n)}(z_0)$ (1)

The integral in the L.H.S of (1) can be computed by computing $f^{(n)}(z_0)$.

Theorem 4.1.5 (Morera's theorem) If f(z) is continuous in a simply connected region R and $\displaystyle\int_C f(z)dz = 0$ for every simple closed curve C in R, then f(z) is analytic in R.

Proof:

Let z be any arbitrary point in R and z_0 be a fixed point in R. Define a function F(z) in R by $F(z) = \displaystyle\int_{z_0}^{z} f(s)ds$ (1)

Since $\displaystyle\int_C f(z)dz = 0$ for every simple closed curve C in R, the integral in equation (1) is independent of the path joining z_0 and z. For, if C_1 and C_2 are any two paths joining z_0 and z, $\displaystyle\int_{C_1 - C_2} f(z)dz = 0$

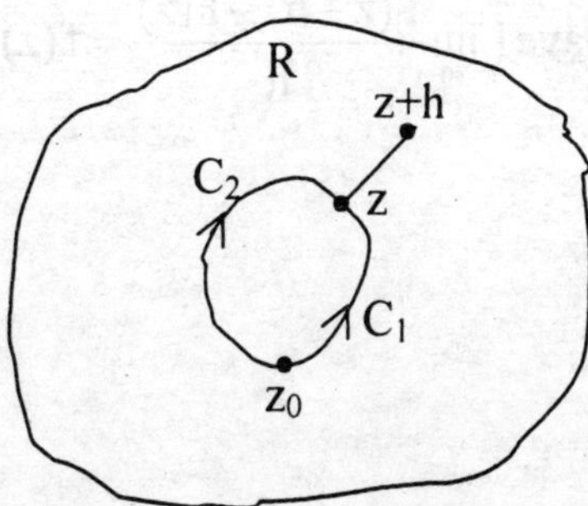

Fig. 4.9

i.e., $\int_{C_1} f(z)dz - \int_{C_2} f(z)dz = 0$

i.e., $\int_{C_1} f(z)dz = \int_{C_2} f(z)dz$

Hence F(z) is well-defined.

Now take a point z + h in R so that the line segment joining z and z + h is in R. Using this line segment as the path of integration between z and z + h, we can write

$$\frac{F(z+h)-F(z)}{h} = \frac{1}{h}\left[\int_{z_0}^{z+h} f(w)dw - \int_{z_0}^{z} f(w)dw\right]$$

$$= \frac{1}{h}\left[\int_{z_0}^{z} f(w)dw + \int_{z}^{z+h} f(w)dw - \int_{z_0}^{z} f(w)dw\right]$$

$$= \frac{1}{h}\int_{z}^{z+h} f(w)dw$$

$$\therefore \frac{F(z+h)-F(z)}{h} - f(z) = \frac{1}{h}\int_{z}^{z+h} f(w)dw - \frac{1}{h}\int_{z}^{z+h} f(z)dw$$

$$= \frac{1}{h}\int_{z}^{z+h} (f(w)-f(z))dw$$

$$\left|\frac{F(z+h)-F(z)}{h} - f(z)\right| \le \frac{1}{|h|}\int_{z}^{z+h} |f(w)-f(z)||dw|$$

Since f(z) is continuous in R, it is continuous on the line segment joining z and z + h.

$\therefore$ Given a real number $\in > 0$, there exist a real number $\delta > 0$ such that $|f(w) - f(z)| < \in$ whenever $|w - z| < \delta$. Now choosing $|h| < \delta$, we get,

$$\left|\frac{F(z+h)-F(z)}{h} - f(z)\right| \le \frac{1}{|h|}.\in.|h| = \in$$

Since $\in > 0$ is arbitrary, we have $\lim_{h\to 0}\frac{F(z+h)-F(z)}{h} = f(z)$

i.e., $F'(z) = f(z)$

Hence F(z) is analytic in R.

$\therefore$ The derivative of F(z).

i.e., F′(z) is analytic in R.

i.e., f(z) is analytic in R.

Note:

Theorem 4.1.5 is the converse of Cauchy's integral theorem. In the proof of this theorem we have used the result **'derivative of an analytic function is analytic'**.

Theorem 4.1.6 (Cauchy's Inequalities)
Let f(z) be analytic in a region R. Let C be any circle in R with centre z_0 and radius r.

Let M denote the maximum of |f(z)| on C. Then $\left|f^{(n)}(z_0)\right| \le \dfrac{n!M}{r^n}$ $\quad n = 1, 2, 3, \ldots$

Proof:
We have $f^{(n)}(z_0) = \dfrac{n!}{2\pi i}\displaystyle\int_C \dfrac{f(z)}{(z-z_0)^{n+1}}\,dz, \quad n = 1, 2, 3, \ldots$

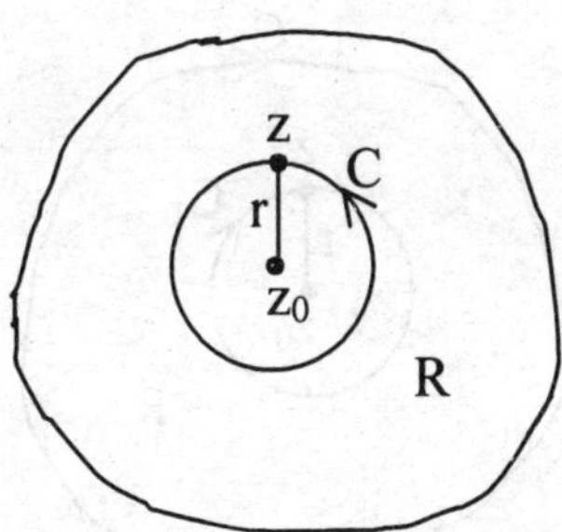

Fig. 4.10

$$\therefore \left|f^{(n)}(z_0)\right| \le \frac{n!}{2\pi}.\frac{M}{r^{n+1}}\int_C |dz| = \frac{n!}{2\pi}.\frac{M}{r^{n+1}}.2\pi r$$

$$= \frac{n!M}{r^n}$$

Hence $\left|f^{(n)}z_0\right| \le \dfrac{n!M}{r^n}$.

Theorem 4.1.7 (Liouville's Theorem)
A bounded entire function in the complex plane is a constant.

Proof:
Let f(z) be a bounded entire function. Then f(z) is analytic in the entire complex plane and then exists a real number M > 0 such that $|f(z)| \le M$ for all z.
Let z_0 be any point in the complex plane. Let C be any circle with centre z_0 and radius r.

Since f(z) is analytic, by Cauchy's inequality we have $\left|f'(z_0)\right| \le \dfrac{M}{r}$. Further r can be chosen as large as possible.

$\therefore$ Taking limit as $r \to \infty$, we get $|f'(z_0)| \to 0$
i.e., $f'(z_0) = 0$
Since z_0 is arbitrary, $f'(z) = 0$ for all z.
$\therefore$ f(z) is a constant.

Theorem 4.1.8 (Maximum Modulus Theorem)
If f(z) is non-constant and analytic within and on a simple closed curve C, then |f(z)| assumes its maximum value on C.

Proof:
Let f(z) be analytic in the region R bounded by the simple closed curve C. Since R is closed and bounded and f(z) is continuous, |f(z)| attains its maximum at some point in R. Let the maximum of |f(z)| be attained at a point z_0 in the interior of R and $|f(z_0)| = M$. Let C_1 be any circle with centre at z_0 and radius r such that C_1 is contained in R.

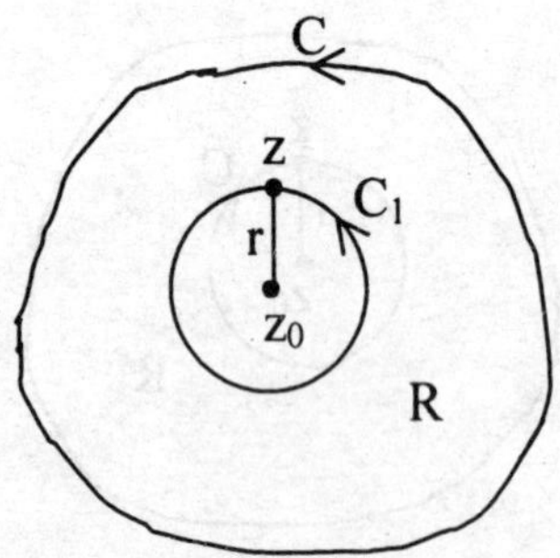

Fig. 4.11

By Cauchy's integral formula, we have $f(z_0) = \dfrac{1}{2\pi i}\displaystyle\int_{C_1}\frac{f(z)}{z-z_0}dz$

If $|f(z)| < |f(z_0)| = M$ at some point of C_1, then

$$\therefore\ |f(z_0)| = \left|\frac{1}{2\pi i}\int_{C_1}\frac{f(z)}{z-z_0}dz\right|$$

$$\le \frac{1}{2\pi}\int_{C_1}\frac{|f(z)|}{|z-z_0|}|dz|$$

$$< \frac{1}{2\pi}.\frac{M}{r}.2\pi r = M$$

i.e., $|f(z_0)| < M$, which is a contradiction.
$\therefore$ $|f(z)| = |f(z_0)| = M$ at every point inside C_1
i.e., |f(z)| is constant inside C_1.
i.e., f(z) is constant inside C_1.
Hence f(z) is constant in the entire region R, which is a contradiction.
$\therefore$ |f(z)| attains its maximum value on the boundary C of the region R.

Theorem 4.1.9 (Fundamental Theorem of Algebra)
If f(z) is a non-constant polynomial in z with complex coefficients, then there exists at least one complex number z_0 such that $f(z_0) = 0$

Solution:

Suppose f(z) is a non-constant polynomial and $f(z) \neq 0$ for all z.
Being a polynomial in z, f(z) is analytic in the entire complex plane.

$\therefore \dfrac{1}{f(z)}$ is also analytic in the entire complex plane. When $|z| \to \infty$ $|f(z)| \to \infty$.

Hence, there exists an $r > 0$ such that $|f(z)| > 1$ for $|z| > r$

$\therefore \dfrac{1}{|f(z)|} < 1$ for $|z| > r$

i.e., $\dfrac{1}{f(z)}$ is bounded in the region $|z| > r$ (1)

Since $\dfrac{1}{f(z)}$ is analytic for all z, $\dfrac{1}{f(z)}$ is continuous in $|z| \leq r$.

Hence $\dfrac{1}{f(z)}$ is bounded in the region $|z| \leq r$ (2)

(1) and (2) $\Rightarrow \dfrac{1}{f(z)}$ is bounded in the entire complex plane..

i.e., $\dfrac{1}{f(z)}$ is a bounded entire function. By Liouville's theorem, $\dfrac{1}{f(z)}$ is constant.

i.e., f(z) is a constant function, which is a contradiction.
$\therefore f(z) = 0$ for at least one complex number $z = z_0$
i.e., there exists at least one complex number z_0 such that $f(z_0) = 0$.

Example 4.1.1 Evaluate using Cauchy's integral formula $\dfrac{1}{2\pi i}\int_C \dfrac{z^2+7}{z-2}dz$ where C is $|z| = 5$.

Solution:

$f(z) = z^2 + 7$ is analytic inside and on the circle $|z| = 5$ and $z = 2$ lies inside it.
$\therefore$ By Cauchy's integral formula,

$$\frac{1}{2\pi i}\int_C \frac{z^2+7}{z-2}dz = f(2) = 2^2 + 7 = 11$$

Example 4.1.2 Evaluate $\int_C \dfrac{\sin 3z}{z + \pi/2}dz$ where C is the circle $|z| = 5$

Solution:

$f(z) = \sin 3z$ is analytic inside and on C and $z = -\pi/2$ lies inside C.

$\therefore$ By Cauchy's integral formula, $\dfrac{1}{2\pi i}\int_C \dfrac{\sin 3z}{z+\pi/2}dz = f(-\pi/2) = \sin(-3\pi/2) = 1$

$$\therefore \int_C \frac{\sin 3z}{z+\pi/2}dz = 2\pi i$$

Example 4.1.3 Show that $\frac{1}{2\pi i}\int_C \frac{e^{zt}dz}{z^2+1} = \sin t$ if $t > 0$ and C is the circle $|z| = 3$.

Solution:

$$\frac{1}{z^2+1} = \frac{1}{(z+i)(z-i)} = \frac{1}{2i}\left[\frac{1}{z-i} - \frac{1}{z+i}\right]$$

$$\therefore \frac{1}{2\pi i}\int_C \frac{e^{zt}}{z^2+1} = \frac{1}{2\pi i.2i}\int_C \left[\frac{e^{zt}}{z-i} - \frac{e^{zt}}{z+i}\right]dz$$

$$= -\frac{1}{4\pi.}\int_C \frac{e^{zt}}{z-i}dz + \frac{1}{4\pi.}\int_C \frac{e^{zt}}{z+i}dz \quad (1)$$

The points z = i and z = -i lie inside C

$$\therefore \int_C \frac{e^{zt}}{z-i}dz = 2\pi i e^{it}$$

and $\int_C \frac{e^{zt}}{z+i}dz = 2\pi i e^{-it}$

Substituting in (1) we get,

$$\frac{1}{2\pi i}\int_C \frac{e^{zt}}{z^2+1} = -\frac{1}{4\pi}.2\pi i.e^{it} + \frac{1}{4\pi}.2\pi i.e^{-it}$$

$$= -\frac{i}{2}\left(e^{it} - e^{-it}\right)$$

$$= \frac{\left(e^{it} - e^{-it}\right)}{2i} = \sin t$$

Example 4.1.4 Evaluate $\int_C \frac{zdz}{(9-z^2)(z+i)}$ where C is the circle $|z| = 2$.

Solution:

Let $f(z) = \frac{z}{9-z^2}$. Then f(z) is analytic inside and on C(f(z) is not analytic at z = 3 and z = -3 and these points lie outside C).

The point z = -i lies inside C.

∴ By Cauchy's integral formula,

$$\int_C \frac{zdz}{(9-z^2)(z+i)} = 2\pi i.f(-i)$$

$$= 2\pi i.\frac{(-i)}{9-(-i)^2} = \frac{2\pi}{10} = \pi/5$$

Example 4.1.5 Evaluate $\int_C \frac{dz}{z(z^2+4)}$ where C is

(i) $|z| = 1$
(ii) $|z - 2i| = 3$
(iii) $|z| = 3$

Solution:

(i) C is $|z| = 1$

Let $f(z) = \frac{1}{z^2+4}$

f(z) is analytic inside and on $|z| = 1$ and $z = 0$ lies inside it.

$\therefore$ By Cauchy's integral formula,

$$\int_C \frac{dz}{z(z^2+4)} = 2\pi i\, f(0)$$

$$= 2\pi i.\frac{1}{4} = \frac{\pi i}{2}$$

(ii) C is $|z - 2i| = 3$

$$\frac{1}{z(z^2+4)} = \frac{1/4}{z} - \frac{1/8}{z+2i} - \frac{1/8}{z-2i}$$

$$\int_C \frac{dz}{z(z^2+4)} = \frac{1}{4}\int_C \frac{dz}{z} - \frac{1}{8}\int_C \frac{dz}{z+2i} - \frac{1}{8}\int_C \frac{dz}{z-2i}$$

The points $z = 0$ and $z = 2i$ lie inside $|z-2i| = 3$ and $z = -2i$ lies outside it.

$$\therefore \frac{1}{4}\int_C \frac{dz}{z} = \frac{1}{4} 2\pi i = \frac{\pi i}{2}$$

$$\frac{1}{8}\int_C \frac{dz}{z+2i} = \frac{1}{8}.0 = 0$$

$$\frac{1}{8}\int_C \frac{dz}{z-2i} = \frac{1}{8}.2\pi i = \frac{\pi i}{4}$$

Hence $\int_C \frac{dz}{z(z^2+4)} = \frac{\pi i}{2} - 0 - \frac{\pi i}{4} = \frac{\pi i}{4}$

(iii) C is $|z| = 3$

The points $z = 0$, $z = 2i$, $z = -2i$ lie inside $|z| = 3$.

$$\int_C \frac{dz}{z(z^2+4)} = \frac{1}{4}\int_C \frac{dz}{z} - \frac{1}{8}\int_C \frac{dz}{z+2i} - \frac{1}{8}\int_C \frac{dz}{z-2i}$$

$$= \frac{1}{4} 2\pi i - \frac{1}{8} 2\pi i - \frac{1}{8} 2\pi i = 0$$

Example 4.1.6 Evaluate $\int_C \frac{dz}{z^2+4}$ where C is $|z| = 4$.

Solution:

$$\frac{1}{z^2+4} = \frac{1}{(z+2i)(z-2i)}$$

$\frac{1}{z^2+4}$ is not analytic at z = 2i and z = -2i. Both the points lie inside $|z| = 4$.

Enclose the points z = 2i and z = -2i by circles C_1: $|z - 2i| = 1$ and C_2: $|z + 2i| = 1$.

Then $f(z) = \frac{1}{z^2+4}$ is analytic in the region bounded by the outer curve C and inner curves C_1 and C_2

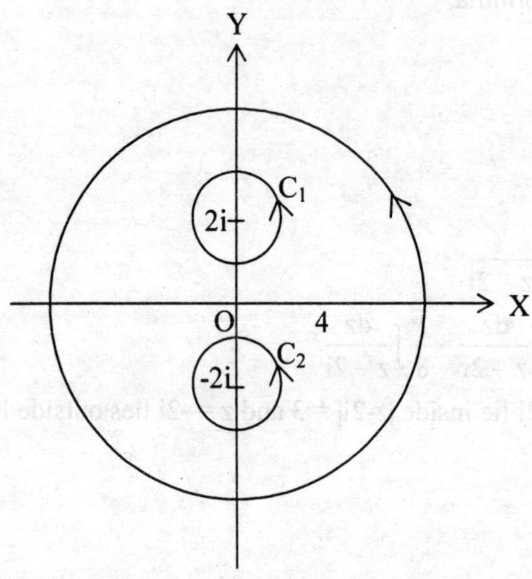

Fig. 4.12

$$\int_C \frac{dz}{z^2+4} = \int_{C_1} \frac{dz}{z^2+4} + \int_{C_2} \frac{dz}{z^2+4}$$

(Using Cauchy's integral theorem for multiply connected region)

$$= \int_{C_1} \frac{1/(Z+2i)}{z-2i} dz + \int_{C_2} \frac{1/(Z-2i)}{z+2i} dz$$

$$= 2\pi i.\frac{1}{(2i+2i)} + 2\pi i.\frac{1}{(-2i-2i)} = \frac{\pi}{2} - \frac{\pi}{2} = 0$$

Example 4.1.7 Evaluate $\int_C \frac{dz}{z^3(z-1)}$ where C is (i) $|z| = 2$ (ii) $|z - 1| = ½$

Solution:

$\frac{1}{z^3(z-1)}$ is analytic at all points except z = 0 and z = 1.

(i) When C is |z| = 2, both z = 0 and z = 1 lie inside C. Enclose z = 0 and z = 1 by two non-intersecting circles C_1 and C_2 such that they lie inside C.

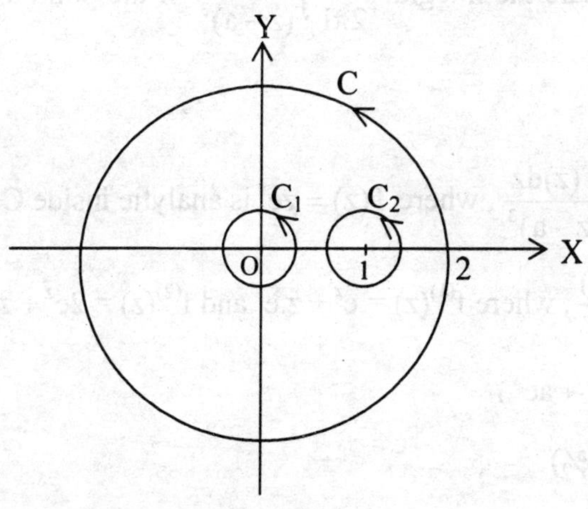

Fig. 4.13

Then

$$\int_C \frac{dz}{z^3(z-1)} = \int_{C_1} \frac{dz}{z^3(z-1)} + \int_{C_2} \frac{dz}{z^3(z-1)} \qquad (1)$$

$$\int_{C_1} \frac{dz}{z^3(z-1)} = \int_{C_1} \frac{f(z)}{z^3} dz \quad \text{where } f(z) = \frac{1}{z-1} \text{ is analytic inside } C_1$$

$$= \frac{2\pi i}{2!} f^{(2)}(0)$$

$$= \frac{2\pi i}{2!} \cdot \frac{2}{(-1)^3}$$

$$= -2\pi i \qquad (2)$$

$$\int_{C_2} \frac{dz}{z^3(z-1)} = \int_{C_2} \frac{f(z)}{z-1} dz \quad \text{where } f(z) = \frac{1}{z^3} \text{ is analytic inside } C_2$$

$$= 2\pi i.\ f(1)$$

$$= 2\pi i \qquad (3)$$

Using (2) and (3) in (1) we get

$$\int_C \frac{dz}{z^3(z-1)} = -2\pi i + 2\pi i = 0$$

(ii) When C is |z -1| = ½ , z = 0 lies outside C and z = 1 lies inside it.

Hence $\int_C \frac{dz}{z^3(z-1)} = \int_C \frac{f(z)}{z-1} dz$ where $f(z) = \frac{1}{z^3}$ is analytic inside C.

$= 2\pi i.\ f(1)$

$= 2\pi i$

Example 4.1.8 Evaluate the integral $\frac{1}{2\pi i}\int_C \frac{ze^z dz}{(z-a)^3}$ if the point a lies inside the simple closed curve C.

Solution:

$\frac{1}{2\pi i}\int_C \frac{ze^z dz}{(z-a)^3} = \frac{1}{2\pi i}\int_C \frac{f(z)dz}{(z-a)^3}$, where $f(z) = ze^z$ is analytic inside C.

$= \frac{f^{(2)}(a)}{2!}$, where $f^{(1)}(z) = e^z + z.e^z$ and $f^{(2)}(z) = 2e^z + z.e^z$

$= \frac{1}{2}(2e^a + ae^a)$

$= e^a(1 + a/2)$

Example 4.1.9 Evaluate $\int_C \frac{\tan z\, dz}{(z - \pi/4)^2}$ where C is |z| = 1

Solution:

Given C is |z| = 1 the point $z = \pi/4$ lies inside C

$\int_C \frac{\tan z\, dz}{(z-\pi/4)^2} = \int_C \frac{f(z)\, dz}{(z-\pi/4)^2}$, where f(z) = tan z is analytic inside C

$= \frac{2\pi i}{1!} f^{(1)}(z)$ where $f^{(1)}(z) = \sec^2 z$

$= 2\pi i(\sqrt{2})^2 = 4\pi i$

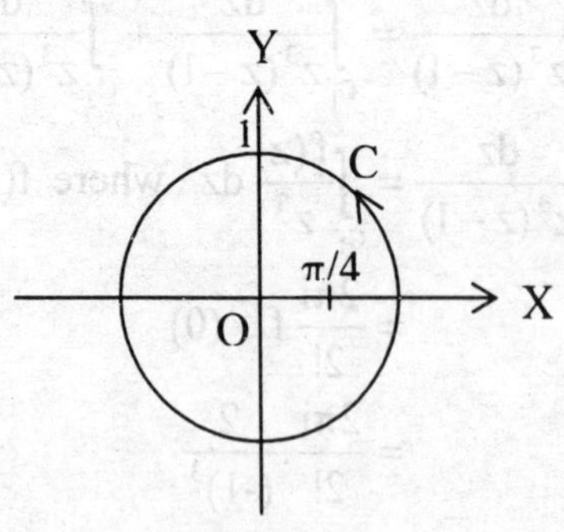

Fig. 4.14

Example 4.1.10 Evaluate $\int_C \frac{\cos z\, dz}{z^{2n+1}}$ where C is |z| =1

Solution:

The point z = 0 lie inside C

$\int_C \frac{\cos z\, dz}{z^{2n+1}} = \int_C \frac{f(z)\, dz}{z^{2n+1}}$, where f(z) = cos z is analytic inside C

$$= \frac{2\pi i}{(2n)!} f^{(2n)}(0) \qquad (1)$$

$f(z) = \cos z$

$f^{(1)}(z) = -\sin z$

$f^{(2)}(z) = -\cos z$

$f^{(4)}(z) = (-1)^2 \cos z$

$f^{(6)}(z) = (-1)^3 \cos z$

$\vdots$

$f^{(2n)}(z) = (-1)^n \cos z \quad \therefore f^{(2n)}(0) = (-1)^n$

Hence from (1), we get,

$$\int_C \frac{\cos z\, dz}{z^{2n+1}} = \frac{2\pi i}{(2n)!}(-1)^n$$

Example 4.1.11 Evaluate $\int_C \frac{z^3 + z + 1}{z^2 - 7z + 6} dz$, where C is the ellipse $4x^2 + 9y^2 = 1$

Solution:

C is the ellipse $4x^2 + 9y^2 = 1$

i.e., $\frac{x^2}{(1/4)} + \frac{y^2}{(1/9)} = 1$

$f(z) = \frac{z^3 + z + 1}{z^2 - 7z + 6} = \frac{z^3 + z + 1}{(z-1)(z-6)}$ is analytic except at z = 1 and z = 6. Both the points lie outside C

$\therefore$ f(z) is analytic inside and on C. By Cauchy's integral theorem $\int_C f(z)\, dz = 0$

i.e., $\int_C \frac{z^3 + z + 1}{z^2 - 7z + 6} dz = 0$

Example 4.1.12 Evaluate $\int_C \frac{\sin \pi z^2 + \cos \pi z^2}{(z+1)(z+2)} dz$, where C is the circle |z| = 3.

Solution:

The function $f(z) = \frac{\sin \pi z^2 + \cos \pi z^2}{(z+1)(z+2)}$ is not analytic at z = -1 and z = -2.

Both the points lie inside the circle C : |z| = 3.

Enclose –1 and –2 by nonintersecting circles C_1 and C_2 with centers at –1 and –2 such that both C_1 and C_2 lie inside C. Then f(z) is analytic in the multiply connected region bounded by C, C_1 and C_2

$$\therefore \int_C f(z)\,dz = \int_{C_1} f(z)\,dz + \int_{C_2} f(z)\,dz \qquad (1)$$

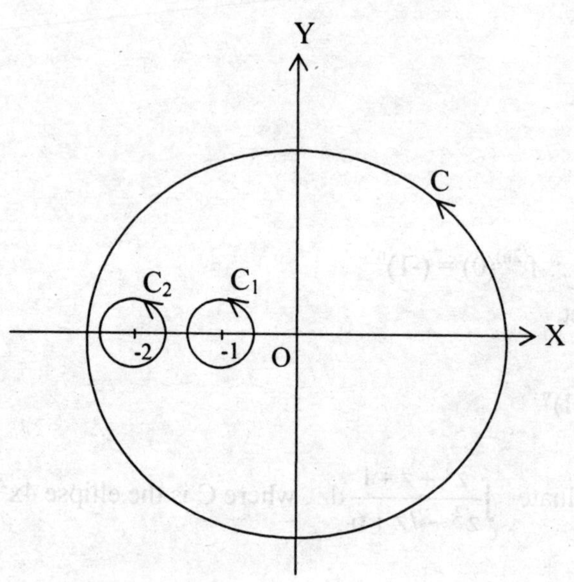

Fig. 4.15

$$\int_{C_1} f(z)\,dz = \int_{C_1} \frac{\varphi(z)}{z+1}\,dz \text{, where } \varphi(z) = \frac{\sin\pi z^2 + \cos\pi z^2}{z+2} \text{ is analytic inside and on } C_1$$

$= 2\pi i\ \varphi(-1)$ (By Cauchy's integral formula)

$= 2\pi i.\ \cos\pi$

$= -2\pi i$ (2)

$$\int_{C_2} f(z)\,dz = \int_{C_2} \frac{\chi(z)}{z+2}\,dz \text{, where } \chi(z) = \frac{\sin\pi z^2 + \cos\pi z^2}{z+1} \text{ is analytic inside and on } C_1$$

$= 2\pi i\ \chi(-2)$ (By Cauchy's integral formula)

$= 2\pi i.\ (-\cos 4\pi)$

$= -2\pi i$ (3)

Using (2) and (3) in (1) we get $\int_C f(z)\,dz = -2\pi i - 2\pi i = -4\pi i$

Example 4.1.13 Evaluate $\int_C \frac{z\,dz}{(z^3 - 6z + 25)^2}$, where C is the circle $|z - (3 + 4i)| = 4$

Solution:

$z^2 - 6z + 25 = (z - (3 + 4i))(z - (3 - 4i)) = (z - \alpha)(z - \beta)$, where $\alpha = 3 + 4i$ and $\beta = 3 - 4i$

$$\therefore f(z) = \frac{z}{(z^3 - 6z + 25)^2} = \frac{z}{(Z-\alpha)^2(Z-\beta)^2}$$

$\alpha = 3 + 4i$ is the centre of c and $\beta = 3 - 4i$ lies outside C.

$$\int_C f(z)\,dz = \int_C \frac{\varphi(z)dz}{(Z-\alpha)^2} \text{ where } \varphi(z) = \frac{z}{(Z-\beta)^2} \text{ is analytic inside and on C}$$

$$= \frac{2\pi i}{1!}\varphi'(\alpha) \quad \text{where } \varphi'(z) = \frac{1}{(z-\beta)^2} + \frac{z(-2)}{(z-\beta)^3} = \frac{-(z+\beta)}{(z-\beta)^3}$$

$$= 2\pi i \frac{-(\alpha+\beta)}{(\alpha-\beta)^3}$$

$$= \frac{-2\pi i.6}{(8i)^3} = \frac{3\pi}{128}$$

Example 4.1.14 Evalute $\int_C \frac{e^{z^2}dz}{z^2(z-i)^2}$ where C is the square with vertices at $\pm 3 \pm 3i$

Solution:

$f(z) = \frac{e^{z^2}}{z^2(z-i)^2}$ is not analytic at z= 0 and z = i.

Both the points lie inside C. Enclose z = 0 and z = i by non-intersecting circles C_1 and C_2 with centeres at z = 0 and z = i. Then f(z) is analytic in the multiply connected region bounded by C, C_1, C_2.

$$\therefore \int_C f(z)\,dz = \int_{C_1} f(z)\,dz + \int_{C_2} f(z)\,dz \qquad (1)$$

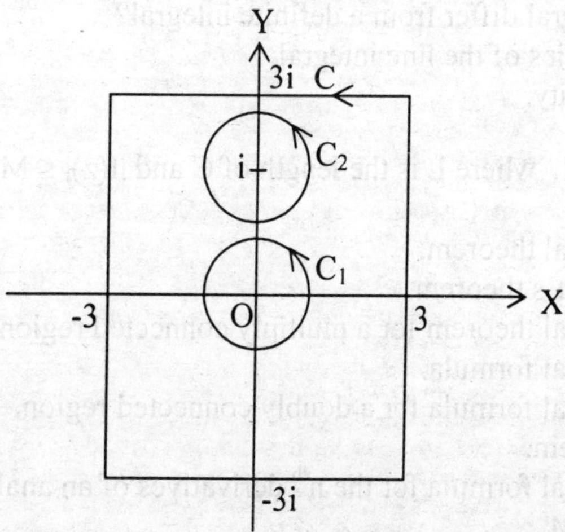

Fig. 4.16

$$\int_{C_1} f(z)\,dz = \int_{C_1} \frac{\varphi(z)}{z^2}\,dz, \text{ where } \quad \varphi(z) = \frac{e^{z^2}}{(z-i)^2} \text{ is analytic inside and on } C_1$$

$$= 2\pi i \varphi^{(1)}(0), \text{ where } \varphi^{(1)}(z) = \frac{(z-i)^2 e^{z^2}.2z - e^{z^2}.2(z-i)}{(z-i)^4}$$

$$= 2\pi i.\frac{2i}{(-i)^4}$$

$$= -4\pi \qquad (2)$$

$$\int_{C_2} f(z)\,dz = \int_{C_2} \frac{\chi(z)}{(z-i)^2}\,dz, \text{ where } \quad \chi(z) = \frac{e^{z^2}}{z^2} \text{ is analytic inside and on } C_2$$

$$= 2\pi i\, \chi^{(1)}(i), \text{ where } \chi^{(1)}(z) = \frac{z^2 e^{z^2}.2z - e^{z^2}.2z}{z^4} = \frac{2.e^{z^2}(z^2-1)}{z^3}$$

$$= 2\pi i.\,2.\frac{e^{-1}(-2)}{(i)^3}$$

$$= 8\pi e^{-1} \qquad (3)$$

Using (2) and (3) in (1), we get,

$$\int_C \frac{e^{z^2}\,dz}{z^2(z-i)^2} = -4\pi + 8\pi e^{-1} = 4\pi(2e^{-1} - 1)$$

EXERCISE 4.1

PART-A

1. Define ‘a line integral’.
2. How does a line integral differ from a definite integral?
3. State any two properties of the line integral.
4. State the ML-inequality.
5. Prove $\left|\int_C f(z)dz\right| \le M.L$ where L is the length of C and $|f(z)| \le M$ on C.
6. State Cauchy's integral theorem.
7. State Cauchy-Goursat’s theorem.
8. State Cauchy's integral theorem for a multiply connected region.
9. State Cauchy's integral formula.
10. State Cauchy's integral formula for a doubly connected region.
11. State Morera’s theorem.
12. State Cauchy's integral formula for the nth derivatives of an analytic function.
13. Prove $\left|f^{(n)}(z_0)\right| \le \frac{n!.M}{r^n}$, stating the necessary conditions.

14. State Liouville's theorem.
15. Evaluate $\int_C (y - x - i3x^2)dz$, where C is the curve x = t, y = t from (0, 0) to (1, 1).
16. Evaluate $\int_C (x^2 - iy^2)dz$, where C is the parabola $y = 2x^2$ from (1, 2) to (2, 8).
17. Evaluate $\int_C x dz$, where C is the directed line segment from 0 to 1+i.
18. Evaluate $\int_C x dz$, where C is the circle |z| = r.
19. Evaluate $\int_C \bar{z} dz$ along the curve $z = t^2 + it$ from 0 to 4 + 2i.
20. Evaluate $\int_C \bar{z}^2 dz$, where C is the circle |z-1| = 1.
21. Prove that $\int_C x dz = i\frac{\pi}{2}$, where C is the semicircle |z| = 1, 0 ≤ arg z ≤ π with initial point z = 1.
22. Prove that $\int_C \frac{1}{z} dz = \pi i$, where C is the semicircle |z| = 1 above the real axis.
23. Show that $\int_C \log z \, dz = 4\pi i$, where C is the circle |z| = 2.
24. Evaluate $\int_C (x^2 - 2ixy)dz$, where C is the curve x = t, $y = 2t^2$ from A(1, 2) to B(2, 8).
25. Prove $\int_C |z| \, dz = -2$, where C is |z| = 1, 0 ≤ arg z ≤ π.
26. Prove $\int_C |z| \, dz = \frac{8i}{3}$, where C is the circle |z-1| = 1.
27. Show that $\int_C z^2 \, dz = 0$, where C is |z-2| = 1.
28. Show that $\int_C e^{-z} dz = -1 - e^{-1}$, where C is the straight line path from z = πi to z = 1.
29. Show that $\int_C z^i dz = (1 + i) \cosh (\pi/2)$, where C is $z = e^{it}$, $-\pi/2 \le t \le \pi/2$.
30. Show that $\int \frac{e^{2z}}{(z+1)^4} dz = \frac{8\pi i}{3e^2}$, where C is |z| = 2.
31. Evaluate $\int_C \frac{\sinh (2z)}{z^4} dz$, where C is the boundary of the square whose sides lie along x = ± 2 and y = ± 2, described in the positive sense.

32. Verify Cauchy's integral theorem for $\int_C (2z^2 - iz)dz$, where C is the circle $|z| = 1$.

33. What is the value of $\int_C \frac{e^z}{z^2 - 5iz - 6} dz$, where C is $|z| = 1$?

34. What is the value of $\int_C z.e^z \, dz$, where C is $|z + 2i| = 4$?

35. What is the value of $\int_C z^3 \, dz$, where C is the square with vertices at $1 \pm i$, $-1 \pm i$?

36. What is the value of $\int_C \sinh z \, dz$, where C is the boundary of the rectangle with vertices at $2 \pm i$, $-2 \pm i$?

37. Show that $\int_C \frac{e^z + 1}{z} dz = 4\pi i$, when C is $|z| = 1$.

38. Show that $\int_C \frac{\sin z}{z - \pi/4} dz = \frac{2\pi i}{\sqrt{2}}$, when C is $|z| = 1$.

39. Show that $\int_C \frac{e^{2z}}{z + \pi i} dz = 2\pi i$, when C is $|z-1| = 5$.

40. Show that $\int_C \frac{\cos^2 z}{(z + \pi/2)^2} dz = 0$, when C is $|z| = 2$.

41. Show that $\int_C \frac{\cos(z^4)}{(z+2)^2} dz = 64\pi i \sin 16$, when C is $|z| = 3$.

PART-B

42. Show that $\int_C |z|^2 \, dz = -1 + i$, where C is the square with vertices (0, 0), (1, 0), (1, 1) and (0, 1).

43. If C is the curve $y = x^3 - 3x^2 + 4x - 1$ joining the points (1, 1) and (2, 3), evaluate $\int_C (12z^2 - 4iz)dz$.

44. Evaluate $\int_0^{1+i} (z^2 + z)dz$ by choosing two different paths of integration and show that the results are the same.

45. IF C is the boundary of a square with vertices at the points $z = 0$, $z = 1$, $z = 1+i$ and $z = i$ and the orientation of C is in the anti-clockwise direction, show that $\int_C (3z + 1)dz = 0$.

46. Evaluate $\int_C (x+2y)dx+(y-2x)dy$ where C is the ellipse defined by $x = 4\cos\theta$, $y = 3\sin\theta$ and C is described in the anti-clockwise direction.

47. Evaluate $\int_C (x^2 - iy^2)\,dz$ along
    (i) The parabola $y = 2x^2$ from (1, 1) to (2, 8).
    (ii) The straight lines from (1, 1) to (1, 8) and then from (1, 8) to (2, 8).
    (iii) The straight line from (1, 1) to (2, 8).

48. Evaluate $\int_C (z^2 + 2iz)\,dz$ when C is
    (i) The straight line path from $z = 2 + 3i$ to $z = 3 - 2i$.
    (ii) The straight line path from $z = 2 + 3i$ to $z = 3 + 3i$ and then along the straight line path form $z = 3 + 3i$ to $z = 3 - 2i$.

49. Evaluate $\int_C \frac{z+2}{3z}dz$ where C is the sector of the circle $|z| = 3$,
    (i) $0 \le \theta \le \pi/2$ (ii) $\pi \le \theta \le 2\pi$ (iii) $-\pi \le \theta \le \pi$

50 Find the upper bound of $\left|\int_C \frac{z}{z+1}dz\right|$, where C is the upper half of the circle $|z| = 2$.

51 Prove $\left|\int_C \frac{e^{iz}}{1+z^2}dz\right| \le \frac{\pi r}{r^2 - 1}$, where C is the semicircle $z = re^{it}$, $0 \le t \le \pi$, $r > 1$

52. Prove $\left|\int_C \frac{dz}{z^4}\right| \le \frac{1}{2\sqrt{2}}$, when C is the line segment from $z = 1 + i$ to $z = 2$.

53. Evaluate $\int_C \frac{dz}{\sqrt{z}}$, where C is
    (i) The semi circle $|z| = 1$, $y \ge 0$ in the anti-clockwise direction.
    (ii) The semi circle $|z| = 1$, $y \le 0$in the clockwise direction.

54 Evaluate $\int_C \frac{3z+5}{z^2+2z}dz$, where C is $|z| = 1$.

55 Evaluate $\int \frac{(z-1)dz}{z(z+i)(z+3i)}$, where C is $|z + i| = ½$.

56 Evaluate $\int_C \frac{(2z-3)}{z^2-3z-18}dz$, where C is $|z| = 8$.

57 Prove $\int_C \frac{7z+12}{z^2+z-2}dz = \frac{38\pi i}{3}$, where C is $|z - 2| = 2$.

58 If $0 < r < R$, prove $\int_C \frac{R+z}{z(R-z)}dz = 2\pi i$, where C is $|z| = r$. Hence deduce that

(i) $\int_0^{2\pi} \frac{d\theta}{R^2 - 2rR\cos\theta + r^2} = \frac{2\pi}{R^2 - r^2}$ (ii) $\int_0^{2\pi} \frac{\sin\theta}{R^2 - 2rR\cos\theta + r^2}d\theta = 0$.

59 Prove $\int_C \frac{dz}{(z^2+1)(z-1)} = 0$, when C is $|z| = 2$.

60 Evaluate $\int_C \frac{zdz}{z^4 - 1}$, when C is $|z-2| = 2$.

61 Evaluate $\int_C \frac{zdz}{(9-z^2)(z+1)}$, where C is $|z| = 2$.

62 Evaluate $\int_C \frac{dz}{z^2(z^2+4)}$, where C is (i) $|z| = 3/2$ (ii) $|z-2i| = 3$ (iii) $|z| = 3$.

63 Evaluate $\int_C \frac{dz}{z^2(z-3)}$, where C is $|z| = 2$.

64 Prove $\int_C \frac{e^{az}}{z^{n+1}}dz = \frac{a^n 2\pi i}{n!}$,where C is $|z| = 1/2$.

65 Prove $\int_C \frac{\cos z}{z^{2n+1}}dz = \frac{(-1)^n 2\pi i}{(2n)!}$, where C is $|z| = 1$.

66 Prove $\int_C \frac{\sin^6 z}{(z - \pi/6)^3}dz = \frac{21\pi i}{16}$, where C is $|z| = 1$.

67 Prove $\int_C \frac{z^3+1}{(3z+1)^3}dz = -\frac{4\pi i}{27}$, where C is $|z| = 1$.

68 Prove that $\frac{1}{2\pi i}\int_C \frac{e^{zt}dt}{(z^2+1)^2} = \frac{t\sin t}{2}$ if $t > 0$ and C is the circle $|z| = 3$.

69 Evaluate $\int_C \frac{\cos 2z}{z^2 - 4z + 5}dz$, where C is $|z-1-2i| = 2$.

70 Show that $\int_C \frac{3z^4 + 5z^2 + 2}{(z+1)^4}dz = -24\pi i$, where C is the circle $|z| = 2$.

71 Show that $\int_C \frac{dz}{(z^2+4)^2} = \frac{\pi}{16}$, where C is $|z-i| = 2$.

72 Show that $\int_C \frac{e^z}{z^2(z+1)^3}dz = (11e^{-1} - 4)\pi i$, where C is $|z| = 2$.

73 Show that $\int_C \frac{ze^z dz}{(z+a)^3} = \pi i(2-a)e^{-a}$; C is any simple closed curve enclosing z = -a.

74 Evaluate $\int_C \frac{z^2 - z - 1}{z(z-i)^2} dz$, where C is $|z - \frac{1}{2}| = 1$.

75 Evaluate $\int_C \frac{e^z}{z(1+z)^3} dz$, where C is the square with vertices at ±2 ± 2i.

76 Evaluate $\int_C \frac{(\cos z - \sin z)}{(z+i)^3} dz$, where C is $|z| = 2$.

4.2 TAYLOR AND LAURENT EXPANSIONS

In this section we prove one of the important results in the theory of complex variables, that a complex function f(z) which is analytic in a domain D can be expanded into a power series. If f(z) is analytic in a domain D containing a neighbourhood of z_0, then f(z) can be represented in this neighbourhood by a power series consisting of non-negative powers of $z-z_0$ called **Taylor's series.** If f(z) is analytic in domain D containing a neighbourhood of z_0, but not analytic at z_0, we can still expand f(z) into an infinite series having positive and negative powers of $z - z_0$. This series is called the **Laurent's series.**

4.2.1 POWER SERIES

Definition 4.2.1 Let $\sum_{n=1}^{\infty} f_n(z)$ be a series of complex functions defined in a domain D. If the series of positive real numbers $\sum_{n=1}^{\infty} |f_n(z_0)|$ is convergent for a given $z_0 \in D$, then the series $\sum_{n=1}^{\infty} f_n(z)$ is said to be **absolutely convergent** at z_0.

Note: An absolutely convergent series is convergent.

Definition 4.2.2 Let $\sum_{n=1}^{\infty} f_n(z)$ be a series of complex functions defined in a domain D. Let $S_k(z) = f_1(z) + f_2(z) + \ldots + f_k(z)$ be the kth partial sum. If at a point $z_0 \in D$, the sequence of partial sums $\{S_k(z_0)\}$ converges to $f(z_0)$, then we say that the series $\sum_{n=1}^{\infty} f_n(z_0)$ converges to $f(z_0)$. If $\sum_{n=1}^{\infty} f_n(z_0)$ converges to $f(z_0)$ for each point $z_0 \in D$, we say that the series $\sum_{n=1}^{\infty} f_n(z)$ **converges pointwise** to the function f(z) in D.

Definition 4.2.3 The series $\sum_{n=1}^{\infty} f_n(z)$ defined in a domain D is said to **converge uniformly** to the function f(z) if given a real number $\in > 0$, there exists a positive integer N such that $|S_n(z) - f(z)| < \in$ for all n > N and for all $z \in D$.

Note:

(i) A series which is uniformly convergent is also pointwise convergent.

(ii) **(Weierstrass's M-test)** Let $\sum_{n=1}^{\infty} f_n(z)$ be an infinite series defined in some domain D of the complex plane and Let $\{M_n\}$ be a sequence of positive numbers such that $|f_n(z)| \le M_n$ for all n and for all $z \in D$. If the series $\sum_{n=1}^{\infty} M_n$ is convergent, then the series $\sum_{n=1}^{\infty} f_n(z)$ is uniformly and absolutely convergent.

Definition 4.2.4 An infinite series of the form

$\sum_{n=0}^{\infty} a_n (z - z_0)^n = a_0 + a_1(z - z_0) + a_2(z - z_0)^2 + \cdots + a_n(z - z_0)^n + \cdots$ where a_n are complex constants, is called a **power series** about $z = z_0$. The point z_0 is called the centre of the power series.

If the power series converges at a point $z_1 \neq z_0$, then it converges absolutely for all z in the disk $|z - z_0| < |z_1 - z_0|$. If the power series diverges at a point z_1, then it diverges for all z, where $|z - z_0| > |z_1 - z_0|$. It can be proved that there exists a real number R, $0 \le R < \infty$, such that the power series is convergent for all z for which $|z - z_0| < R$ and divergent for all z for which $|z - z_0| > R$.

Definition 4.2.5 Let R be the radius of the circle with centre at z_0 such that the power series $\sum_{n=0}^{\infty} a_n (z - z_0)^n$ converges for $|z - z_0| < R$ and diverges for $|z - z_0| > R$, then R is called the **radius of convergence** and the circle $|z - z_0| = R$ is called the **circle of convergence** of the power series.

Note:

(i) For the power series $\sum_{n=0}^{\infty} a_n (z - z_0)^n$, compute $\lim_{n\to\infty} \left|\frac{a_{n+1}}{a_n}\right| = \rho$. Then the radius of convergence of the power series is $R = 1/\rho$.

(ii) **(Cauchy's Root Test).** The radius of convergence R can be computed using Cauchy's root test. The formula $\frac{1}{R} = \lim_{n\to\infty} |a_n|^{1/n}$, gives R, provided the limit exists.

(iii) If $R > 0$ is the radius of convergence of a power series $\sum_{n=0}^{\infty} a_n (z - z_0)^n$ and $r < R$, then the power series is uniformly and absolutely convergent within the circle $|z - z_0| < r$.

(iv) A power series is convergent with in the circle of convergence and represents an analytic function. i.e., if $\sum_{n=0}^{\infty} a_n (z - z_0)^n$ converges to f(z), then f(z) is analytic within the circle of convergence of the series.

Example 4.2.1 Find the radius of convergence and circle of convergence of the power series $\sum_{n=1}^{\infty} a_n (z - 1)^n$ where $a_n = \left(\frac{3n+4}{5n+6}\right)^n$.

Solution:

$$\lim_{n\to\infty} |a_n|^{1/n} = \lim_{n\to\infty} \left|\frac{3n+4}{5n+6}\right|$$

$$= \lim_{n\to\infty} \left|\frac{3 + 4/n}{5 + 6/n}\right| = \frac{3}{5}$$

$\therefore$ Radius of convergence $R = \frac{5}{3}$

Circle of convergence is $|z-1| = \frac{5}{3}$.

Example 4.2.2 Find the radius of convergence and circle of convergence of $\sum \frac{i^{n+1}}{3^{n+1}} (z - i)^n$.

Solution:

$$\lim_{n\to\infty} \left|\frac{a_{n+1}}{a_n}\right| = \lim_{n\to\infty} \left|\frac{i^{n+1}}{3^{n+1}} \cdot \frac{3^n}{i^n}\right| = \lim_{n\to\infty} \left|\frac{i}{3}\right| = \frac{1}{3}$$

$\therefore$ Radius of convergence = 3

Circle of convergence is $|z - i| = 3$.

Example 4.2.3 Find the radius of convergence of the power series $\sum \frac{(1+i)^n}{n+2} \cdot (z + 3i)^n$.

Solution:

$$a_n = \frac{(1+i)^n}{n+2}$$

$$\lim_{n\to\infty}\left|\frac{a_{n+1}}{a_n}\right| = \lim_{n\to\infty}\left|\frac{(1+i)^{n+1}}{n+3}\cdot\frac{n+2}{(1+i)^n}\right|$$

$$= \lim_{n\to\infty}\left|(1+i).\frac{n+2}{n+3}\right| = |1+i| = \sqrt{2} \text{ i.e., Radius of convergence } = 1/\sqrt{2}$$

4.2.2 TAYLOR'S SERIES

Every power series with non-zero radius of convergence defines an analytic function with in the circle of convergence. The converse of this result is also true. i.e., every function f(z) which is analytic inside a circle $|z - z_0| < R$ can be expanded into a power series $\sum_{n=0}^{\infty} a_n (z - z_0)^n$ that converges absolutely to f(z) inside the circle of convergence.

Theorem 4.2.1 (Taylor's Theorem)
If f(z) is analytic inside a circle $|z - z_0| = R$, then inside this circle, f(z) can be represented by a convergent power series as $f(z) = \sum_{n=0}^{\infty} a_n (z - z_0)^n$ where $a_n = \frac{f^{(n)}(z_0)}{n!}$

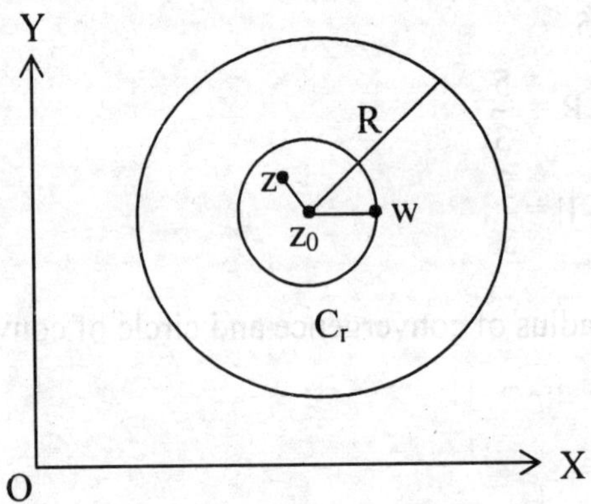

Fig. 4.17

Proof:
Let z be any point inside the circle with centre at z_0 and radius R. Choose r < R such that the circle C_r with centre at z_0 and radius r contains the point z. Since f(z) is analytic inside the circle C_r, by Cauchy's integral formula, we get,

$$f(z) = \frac{1}{2\pi i}\int_{C_r}\frac{f(w)}{w-z}dw \text{ , where w lies on } C_r \tag{1}$$

$$\text{Also } f^{(n)}(z) = \frac{n!}{2\pi i}\int_{C_r}\frac{f(w)}{(w-z)^{n+1}}dw \tag{2}$$

Now $\dfrac{1}{w-z} = \dfrac{1}{(w-z_0)-(z-z_0)}$

$$= \frac{1}{(w-z_0)\left[1-\frac{(z-z_0)}{(w-z_0)}\right]}$$

$$= \frac{1}{(w-z_0)}\left[1+\frac{z-z_0}{w-z_0}+\left(\frac{z-z_0}{w-z_0}\right)^2+\ldots+\left(\frac{z-z_0}{w-z_0}\right)^{n-1}+\frac{\left(\frac{z-z_0}{w-z_0}\right)^n}{1-\frac{z-z_0}{w-z_0}}\right]$$

$$\left(\text{Using the identity} \frac{1}{1-\alpha} = 1+\alpha+\alpha^2+\cdots+\alpha^{n-1}+\frac{\alpha^n}{1-\alpha}\right)$$

$$= \frac{1}{(w-z_0)}+\frac{z-z_0}{(w-z_0)^2}+\frac{(z-z_0)^2}{(w-z_0)^3}+\ldots+\frac{(z-z_0)^n}{(w-z_0)^{n+1}}+\frac{(z-z_0)^n}{(w-z_0)^n(w-z)}$$

Substituting in (1), we get,

$$f(z) = \frac{1}{2\pi i}\int_{C_r}\frac{f(w)}{(w-z_0)}dw + (z-z_0)\frac{1}{2\pi i}\int_{C_r}\frac{f(w)}{(w-z_0)^2}dw + (z-z_0)^2\frac{1}{2\pi i}\int_{C_r}\frac{f(w)}{(w-z_0)^3}dw + \ldots$$

$$+ (z-z_0)^{n-1}\frac{1}{2\pi i}\int_{C_r}\frac{f(w)}{(w-z_0)^n}dw + R_n \quad (3)$$

$$\text{where } R_n = (z-z_0)^n\frac{1}{2\pi i}\int_{C_r}\frac{f(w)}{(w-z_0)^n(w-z)}dw \quad (4)$$

using (2) in (3), we get,

$$f(z) = f(z_0)+\frac{f'(z_0)}{1!}(z-z_0)+\frac{f''(z_0)}{2!}(z-z_0)^2+\cdots++\frac{f^{(n-1)}(z_0)}{(n-1)!}(z-z_0)^{n-1}+R_n$$

$$\text{i.e., } f(z) = a_0+a_1(z-z_0)+a_2(z-z_0)^2+\cdots+a_{n-1}(z-z_0)^{n-1}+R_n \quad (5)$$

where $a_m = \dfrac{f^{(m)}(z_0)}{m!}$, m = 0, 1, 2,……, n-1.

When w lies on C_r and z lies inside C_r, we have $|w - z_0| = r$ and $|z - z_0| < r$

$\therefore |w - z| = |(w - z_0) - (z - z_0)| \geq |(w - z_0)| - |(z - z_0)| = r - |z - z_0|$

$\therefore \dfrac{1}{|w-z|} \leq \dfrac{1}{r-|z-z_0|}$

Let M be the maximum of $|f(z)|$ on C_r.

Then from (4) we get,

$$|R_n| = \left| \frac{(z-z_0)^n}{2\pi i} \int_{C_r} \frac{f(w)}{(w-z_0)^n (w-z)} dw \right|$$

$$\leq \frac{|z-z_0|^n}{2\pi} \int_{C_r} \frac{|f(w)|}{|w-z_0|^n |w-z|} |dw|$$

$$\leq \frac{|z-z_0|^n}{2\pi} \frac{M.2\pi r}{r^n (r - |z-z_0|)}$$

$$\leq \frac{M.|z-z_0|}{(r-|z-z_0|)} \left(\frac{|z-z_0|}{r^n} \right)^{n-1} \to 0 \text{ as } n \to \infty \qquad \left(\text{Since } \frac{|z-z_0|}{r} < 1 \right)$$

$$\lim_{n\to\infty} R_n = 0$$

Thus taking limit as $n \to \infty$, (5) becomes,

$$f(z) = a_0 + a_1(z-z_0) + a_2(z-z_0)^2 + \cdots + a_{n-1}(z-z_0)^{n-1} + \cdots$$

i.e., $f(z) = \sum_{n=0}^{\infty} a_n (z-z_0)^n$ where $a_n = \frac{f^{(n)}(z_0)}{n!}$

Note:

(i) The Taylor's series expansion of f(z) about the point z_0 is unique.

(ii) If f(z) is analytic in a domain D containing the origin, then f(z) can be expanded as a power series about the origin. Then we have,

$f(z) = \sum_{n=0}^{\infty} a_n z^n$ where $a_n = \frac{f^{(n)}(0)}{n!}$ This series is called the **Maclaurin's Seires.**

(iii) The Maclaurin's Seires expansions of some of the frequently used functions are given below.

1. $e^z = 1 + \frac{z}{1!} + \frac{z^2}{2!} + \frac{z^3}{3!} + \cdots + \frac{z^n}{n!} + \cdots, \quad |z| < \infty$

2. $\sin z = \frac{z}{1!} - \frac{z^3}{3!} + \frac{z^5}{5!} - \cdots + (-1)^{n+1} \frac{z^{2n-1}}{(2n-1)!} + \cdots, \quad |z| < \infty$

3. $\cos z = 1 - \frac{z^2}{2!} + \frac{z^4}{4!} - \cdots + (-1)^n \frac{z^{2n}}{(2n)!} + \cdots, \quad |z| < \infty$

4. $\log(1+z) = z - \frac{z^2}{2} + \frac{z^3}{3} - \cdots + (-1)^{n+1} \frac{z^n}{n} + \cdots, \quad |z| < 1$

5. $\frac{1}{1-z} = 1 + z + z^2 + \cdots + z^n + \cdots, \quad |z| < 1$

6. $\frac{1}{(1-z)^2} = 1 + 2z + 3z^2 + \cdots + (n+1)z^n + \cdots, \quad |z| < 1$

7. $\sinh z = z + \frac{z^3}{3!} + \frac{z^5}{5!} + \cdots + \frac{z^{2n-1}}{(2n-1)!} + \cdots, \qquad |z| < \infty$

8. $\cosh z = z + \frac{z^2}{2!} + \frac{z^4}{4!} - \cdots + \frac{z^{2n}}{(2n)!} + \cdots, \qquad |z| < \infty$

9. $\tan z = z + \frac{z^3}{3!} + \frac{2z^5}{15} + \cdots, \qquad |z| < \pi/2$

Example 4.2.4 Obtain the Taylor's series expansion of $f(z) = e^z$ about (i) $z = 0$ (ii) $z = 2$. Find also the radius of convergence.

Solution:

$f(z) = e^z$

$f^{(n)}(z) = e^z$ for $n = 1, 2, ,3 ,\ldots$

(i) $a_n = \frac{f^{(n)}(0)}{n!} = \frac{e^0}{n!} = \frac{1}{n!}$

Now $f(z) = \sum_{n=0}^{\infty} a_n z^n$

i.e, $e^z = \sum_{n=0}^{\infty} \frac{z^n}{n!}$

$$\lim_{n\to\infty}\left|\frac{a_{n+1}}{a_n}\right| = \lim_{n\to\infty}\left|\frac{n!}{(n+1)!}\right|$$

$$= \lim_{n\to\infty}\left|\frac{1}{n+1}\right| = 0 \qquad \therefore \text{ Radius of convergence } R = \infty.$$

(ii) $a_n = \frac{f^{(n)}(2)}{n!} = \frac{e^2}{n!}$

$f(z) = \sum_{n=0}^{\infty} a_n (z-2)^n$

i.e, $e^z = \sum_{n=0}^{\infty} \frac{e^2}{n!}(z-2)^n$

$$= e^2\left[1 + \frac{(z-2)}{1!} + \frac{(z-2)^2}{2!} + \cdots + \frac{(z-2)^n}{n!} + \cdots\right]$$

$$\lim_{n\to\infty}\left|\frac{a_{n+1}}{a_n}\right| = \lim_{n\to\infty}\left|\frac{n!}{(n+1)!}\right|$$

$$= \lim_{n\to\infty}\left|\frac{1}{n+1}\right| = 0 \qquad \therefore \text{ Radius of convergence } R = \infty.$$

Example 4.2.5 Obtain the Taylor's series expansion of tan z about the points z = 0.

Solution:

$$f(z) = f(0) + \frac{f'(0)}{1!}z + \frac{f''(0)}{2!}z^2 + \frac{f'''(0)}{3!}z^3 + \cdots$$

$f(z) = \tan z$ $\qquad \therefore f(0) = 0$

$f'(z) = \sec^2 z$ $\qquad \therefore f'(0) = 1$

$f''(z) = 2\sec^2 z \tan z$ $\qquad \therefore f''(0) = 0$

$f'''(z) = 2\sec^2 z.\sec^2 z + 2.2\sec^2 z.\tan^2 z = 2\sec^4 z + 4\sec^2 z.\tan^2 z$ $\qquad \therefore f'''(0) = 2.$

$f^{(iv)}(z) = 2.4.\sec^3 z.\sec z.\tan z + 4.\sec^2 z.2\tan z.\sec^2 z + 4.2.\sec^2 z.\tan z.\tan^2 z$

$= 16.\sec^4 z.\tan z + 8\sec^2 z.\tan^3 z$ $\qquad \therefore f^{(iv)}(0) = 0$

$f^{(v)}(z) = 16.4.\sec^4 z.\tan^2 z + 16\sec^4 z.\sec^2 z + 8.2.\sec^2 z.\tan^4 z + 8\sec^2 z.3.\tan^2 z.\sec^2 z$

$= 88\sec^4 z\tan^2 z + 16\sec^6 z + 16\sec^2 z\tan^4 z$ $\qquad \therefore f^{(v)}(0) = 16$

Hence $f(z) = \frac{1}{1!}z + \frac{2}{3!}z^3 + \frac{16}{5!}z^5 + \cdots$

i.e., $\tan z = z + \frac{z^3}{3} + \frac{2z^5}{15} + \cdots$

Example 4.2.6 Expand $\frac{z}{z^2 - 4z + 13}$ about z = 0.

Solution:

$$\frac{z}{z^2 - 4z + 13} = \frac{z}{(z-a)(z-b)} \quad \text{where } a = 2 + 3i \text{ and } b = 2 - 3i$$

$$= \frac{A}{(z-a)} + \frac{B}{(z-b)} \quad \text{where } A = \frac{a}{a-b},\ B = \frac{b}{b-a}$$

$$= \frac{A}{-a}(1 - z/a)^{-1} + \frac{B}{-b}(1 - z/b)^{-1}$$

$$= \frac{A}{-a}\left(1 + \frac{z}{a} + \frac{z^2}{a^2} + \cdots\right) - \frac{B}{b}\left(1 + \frac{z}{b} + \frac{z^2}{b^2} + \cdots\right)$$

$$\left(\text{Using } \frac{1}{1-z} = 1 + z + z^2 + z^3 + \cdots\right)$$

$$= \frac{A}{-a}\sum_{n=0}^{\infty}\frac{z^n}{a^n} - \frac{B}{b}\sum_{n=0}^{\infty}\frac{z^n}{b^n}$$

$$= \frac{-1}{6i}\sum_{n=0}^{\infty}\frac{z^n}{(2+3i)^n} + \frac{1}{6i}\sum_{n=0}^{\infty}\frac{z^n}{(2-3i)^n} \qquad (1)$$

This expansion is valid for $\left|\frac{z}{a}\right| < 1$ and $\left|\frac{z}{b}\right| < 1$

i.e., for $|z| < |a| = \sqrt{13}$ and $|z| < |b| = \sqrt{13}$

i.e., the Taylor's expansion (1) of $\frac{z}{z^2 - 4z + 13}$ is valid in the neighbourhood of origin given by $|z| < \sqrt{13}$

Example 4.2.7 For the function $f(z) = \frac{2z^3 + 1}{z^2 + z}$, find Taylor's series valid in the neighbourhood of the point $z = i$.

Solution:

$$f(z) = \frac{2z^3 + 1}{z^2 + z} = Az + B + \frac{C}{z} + \frac{D}{z+1}$$

$$= 2(z-1) + \frac{1}{z} + \frac{1}{z+1}$$

Expanding 2(z -1) in Taylor's series about z = i, we get $2(z-1) = \sum_{n=0}^{\infty} a_n (z-i)^n$ where

$a_n = \frac{f_1^{(n)}(i)}{n!}$ and $f_1(z) = 2(z - 1)$

$a_0 = f_1(i) = 2(i - 1)$

$a_1 = \frac{f_1'(i)}{1!} = 2$

$a_n = \frac{f_1^{(n)}(i)}{n!} = 0$ for $n \geq 2$

$\therefore\ 2(z - 1) = 2(i - 1) + 2(z - i)$ (1)

Expanding $\frac{1}{z}$ in Taylor's series about z = i, we get $\frac{1}{z} = \sum_{n=0}^{\infty} a_n (z-i)^n$

where $a_n = \frac{f_2^{(n)}(i)}{n!}$ and $f_2(z) = \frac{1}{z}$

$a_0 = f_2(i) = \frac{1}{i}$. $\qquad f_2^{(n)}(z) = \frac{(-1)^n . n!}{z^{n+1}}$ for $n \geq 1$

$\therefore a_n = \frac{f_2^{(n)}(i)}{n!} = \frac{(-1)^n}{i^{n+1}}$ for $n \geq 0$

$\therefore \frac{1}{z} = \sum_{n=0}^{\infty} (-1)^n \frac{(z-i)^n}{i^{n+1}}$ (2)

Expanding $\frac{1}{z+1}$ in Taylor's series about z = i, we get $\frac{1}{z+1} = \sum_{n=0}^{\infty} a_n (z-i)^n$, where

$a_n = \frac{f_3^{(n)}(i)}{n!}$ and $f_3(z) = \frac{1}{z+1}$

$a_0 = f_3(i) = \frac{1}{i+1}$. $f_3^{(n)}(z) = \frac{(-1)^n . n!}{(z+1)^n}$ for $n \geq 1$

$\therefore a_n = \frac{f_3^{(n)}(i)}{n!} = \frac{(-1)^n}{(i+1)^n}$ for $n \geq 0$

$$\therefore \frac{1}{z+1} = \sum_{n=0}^{\infty} (-1)^n \frac{(z-i)^n}{(i+1)^n} \quad (3)$$

From (1), (2) and (3), we get

$$f(z) = \frac{2z^3+1}{z^2+z} = 2(i-1) + 2(z-i) + \sum_{n=0}^{\infty} (-1)^n \frac{(z-i)^n}{i^{n+1}} + \sum_{n=0}^{\infty} (-1)^n \frac{(z-i)^n}{(i+1)^n}$$

Theorem 4.2.2 (Laurent's Theorem)

If f(z) is analytic between and on two concentric circles C_1 and C_2 having centre at z_0 and radii r_1, r_2 ($r_2 < r_1$), then at each point z in the annulus $r_2 < |z - z_0| < r_1$, f(z) can be expanded into a convergent series of positive and negative powers of $z - z_0$ as

$$f(z) = a_0 + a_1(z - z_0) + a_2(z - z_0)^2 + \ldots + a_n(z - z_0)^n + \ldots + \frac{b_1}{z-z_0} + \frac{b_2}{(z-z_0)^2} + \ldots + \frac{b_n}{(z-z_0)^n} + \ldots$$

where $a_n = \frac{1}{2\pi i} \int_{C_1} \frac{f(w)dw}{(w-z_0)^{n+1}}$, $n = 0, 1, 2, \ldots$

$b_n = \frac{1}{2\pi i} \int_{C_2} \frac{f(w)dw}{(w-z_0)^{-n+1}}$, $n = 1, 2, 3, \ldots$

Proof:

Let z be any point in the annulus $r_2 < |z - z_0| < r_1$ and $|z - z_0| = r$.
By applying Cauchy's Integral Formula, we get

$$f(z) = \frac{1}{2\pi i} \int_{C_1 - C_2} \frac{f(w)dw}{w-z}$$

$$= \frac{1}{2\pi i} \int_{C_1} \frac{f(w)dw}{w-z} - \frac{1}{2\pi i} \int_{C_2} \frac{f(w)dw}{w-z} \quad (1)$$

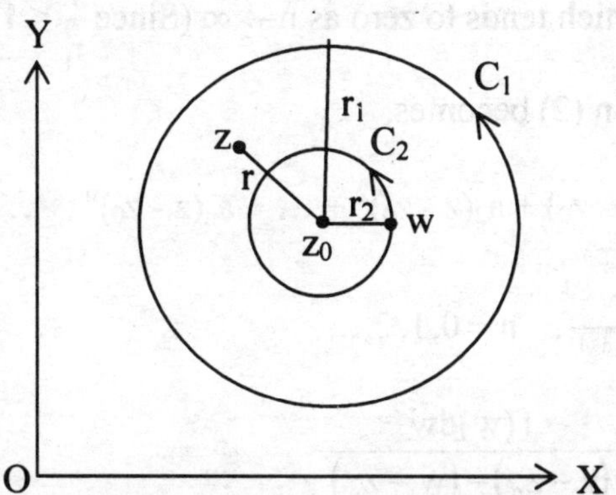

Fig. 4.18

$$\frac{1}{2\pi i}\int_{C'_1}\frac{f(w)dw}{w-z}=\frac{1}{2\pi i}\int_{C'_1}\frac{f(w)dw}{(w-z_0)-(z-z_0)}$$

$$=\frac{1}{2\pi i}\int_{C'_1}\frac{f(w)dw}{(w-z_0)\left(1-\frac{z-z_0}{w-z_0}\right)}$$

$$=\frac{1}{2\pi i}\int_{C'_1}\frac{f(w)dw}{(w-z_0)}\left[1+\frac{z-z_0}{w-z_0}+\cdots+\left(\frac{z-z_0}{w-z_0}\right)^{n-1}+\frac{\left(\frac{z-z_0}{w-z_0}\right)^n}{1-\frac{z-z_0}{w-z_0}}\right]ds$$

$$=\frac{1}{2\pi i}\int_{C'_1}\frac{f(w)dw}{(w-z_0)}+\frac{(z-z_0)}{2\pi i}\int_{C'_1}\frac{f(w)dw}{(w-z_0)^2}+\cdots+\frac{(z-z_0)^{n-1}}{2\pi i}\int_{C'_1}\frac{f(w)dw}{(w-z_0)^n}+R_n \qquad (2)$$

where $R_n=\frac{1}{2\pi i}\int_{C'_1}\frac{f(w)}{(w-z_0)}\frac{\left(\frac{z-z_0}{w-z_0}\right)^n}{1-\frac{z-z_0}{w-z_0}}dw=\frac{(z-z_0)^n}{2\pi i}\int_{C'_1}\frac{f(w)dw}{(w-z_0)^n(w-z)}$

As f(z) is analytic on C_1, it is bounded. Let $|f(z)| \le M$ for z on C_1. When w lies on C_1, we have $|f(w)| \le M$, $|w - z_0| = r_1$ and $|w - z| = |(w - z_0) - (z - z_0)| \ge |w - z_0| - |z - z_0| = r_1 - r$.

$$\therefore |R_n| \le \frac{|z-z_0|^n|f(w)|2\pi r_1}{2\pi|w-z_0|^n.|w-z|}$$

$$\le \frac{r^n.M.2\pi r_1}{2\pi r_1^n.(r_1-r)}$$

$$\le \left(\frac{M.r_1}{r_1 - r}\right)\left(\frac{r}{r_1}\right)^n$$, which tends to zero as $n \to \infty$ (Since $\frac{r}{r_1} < 1$)

Hence $\lim_{n\to\infty} R_n = 0$ equation (2) becomes,

$$\frac{1}{2\pi i}\int_{C_1}\frac{f(w)dw}{w-z} = a_0 + a_1(z - z_0) + a_2(z - z_0)^2 + \ldots + a_n(z - z_0)^n + \ldots \quad (3)$$

where $a_n = \frac{1}{2\pi i}\int_{C_1}\frac{f(w)dw}{(w-z_0)^{n+1}}, \quad n = 0, 1, 2, \ldots$

$$-\frac{1}{2\pi i}\int_{C_2}\frac{f(w)dw}{w-z} = \frac{1}{2\pi i}\int_{C_2}\frac{f(w)dw}{(z-z_0)-(w-z_0)}$$

$$= \frac{1}{2\pi i}\int_{C_2}\frac{f(w)dw}{(z-z_0)\left(1-\frac{w-z_0}{z-z_0}\right)}$$

$$= \frac{1}{2\pi i}\int_{C_2}\frac{f(w)dw}{(z-z_0)}\left[1+\frac{w-z_0}{z-z_0}+\cdots+\left(\frac{w-z_0}{z-z_0}\right)^{n-1}+\frac{\left(\frac{w-z_0}{z-z_0}\right)^n}{1-\frac{w-z_0}{z-z_0}}\right]dw$$

$$= \frac{1}{2\pi i}\int_{C_2}\frac{f(w)dw}{(z-z_0)} + \frac{1}{2\pi i}\int_{C_2}\frac{(w-z_0)f(w)dw}{(z-z_0)^2} + \frac{1}{2\pi i}\int_{C_2}\frac{(w-z_0)^2 f(w)dw}{(z-z_0)^3} + \cdots$$

$$+ \frac{1}{2\pi i}\int_{C_2}\frac{(w-z_0)^{n-1} f(w)dw}{(z-z_0)^n} + \frac{1}{2\pi i}\int_{C_2}\frac{(w-z_0)^n f(w)dw}{(z-z_0)^n(z-w)}$$

$$= \frac{(z-z_0)^{-1}}{2\pi i}\int_{C_2} f(w)dw + \frac{(z-z_0)^{-2}}{2\pi i}\int_{C_2}\frac{f(w)dw}{(w-z_0)^{-1}} + \frac{(z-z_0)^{-3}}{2\pi i}\int_{C_2}\frac{f(w)dw}{(w-z_0)^{-2}} + \cdots$$

$$+ \frac{(z-z_0)^{-n}}{2\pi i}\int_{C_2}\frac{f(w)dw}{(w-z_0)^{-n+1}} + S_n \quad (4)$$

where $S_n = \frac{1}{2\pi i}\int_{C_2}\frac{(w-z_0)^n f(w)dw}{(z-z_0)^n(z-w)}$

As f(z) is analytic on C_2, it is bounded. Let $|f(z)| \le M$ for z on C_2. When w lies on C_2, we have $|f(w)| \le M$, $|w - z_0| = r_2$ and $|z - w| = |(z - z_0) - (w - z_0)| \ge |z - z_0| - |w - z_0| = r - r_2$.

$$|S_n| \le \frac{1}{2\pi}\frac{|w-z_0|^n |f(w)| 2\pi r_2}{|z-z_0|^n . |z-w|}$$

$$\le \frac{r_2^n . M . r_2}{r^n . (r - r_2)}$$

$$\le \left(\frac{M.r_2}{r - r_2}\right)\left(\frac{r_2}{r}\right)^n, \text{ which tends to zero as } n \to \infty, \text{ since } \frac{r_2}{r} < 1$$

Hence $\lim\limits_{n\to\infty} S_n = 0$ and equation (4) becomes,

$$-\frac{1}{2\pi i}\int_{C_2}\frac{f(w)dw}{w - z} = b_1(z - z_0)^{-1} + b_2(z - z_0)^{-2} + \ldots + b_n(z - z_0)^{-n} + \ldots \qquad (5)$$

where $b_n = \dfrac{1}{2\pi i}\displaystyle\int_{C_2}\frac{f(w)dw}{(w - z_0)^{-n+1}}, \quad n = 1, 2, 3, \ldots$

Thus using (3) and (5) in (1) we get,

$$f(z) = a_0 + a_1(z - z_0) + a_2(z - z_0)^2 + \ldots + a_n(z - z_0)^n + \ldots + b_1(z - z_0)^{-1} + b_2(z - z_0)^{-2} + \ldots + b_n(z - z_0)^{-n} + \ldots$$

Note:

(i) If $f(z) = \sum\limits_{n=0}^{\infty} a_n (z - z_0)^n + \sum\limits_{n=1}^{\infty}\dfrac{b_n}{(z - z_0)^n}$ is the Laurent's expansion of f(z) then $\sum\limits_{n=1}^{\infty}\dfrac{b_n}{(z - z_0)^n}$ is called the **principal part** or **singular part** of the Laurent's expansion. $\sum\limits_{n=0}^{\infty} a_n (z - z_0)^n$ is called the **analytic part** or **regular part** of the Laurent's series.

(ii) The Laurent's expansion of f(z) in the annulus $r_2 < |z - z_0| < r_1$ can be written as $f(z) = \sum\limits_{n=-\infty}^{\infty} a_n (z - z_0)^n$ where $a_n = \dfrac{1}{2\pi i}\displaystyle\int_C \frac{f(w)dw}{(w - z_0)^{n+1}}$ and C is any circle $|z - z_0| = r$, $r_2 < r < r_1$. Also as f(z) is analytic in the annulus, C can be any simple closed curve in the annulus, containing C_2, the inner circle of the annulus.

Example 4.2.8 Expand $f(z) = \dfrac{1}{z^2(1 - z)}$ in the Laurent's series for the region $0 < |z| < 1$

Solution:

$f(z) = \dfrac{1}{z^2(1 - z)}$ is analytic in the annulus $0 < |z| < 1$.

Hence it can be expanded into a Laurent's series in powers of z.

$$f(z) = \frac{1}{z^2(1 - z)} = \frac{1}{z^2}(1 - z)^{-1}$$

$$= \frac{1}{z^2}\left(1 + z + z^2 + z^3 + \cdots\right) \qquad \left(\text{Since } (1-z)^{-1} = 1 + z + z^2 + z^3 + \cdots \quad \text{for } |z| < 1\right)$$

$$= \frac{1}{z^2} + \frac{1}{z} + 1 + z + z^2 + z^3 + \cdots$$

Example 4.2.9 Find the Laurent's series expansion of $f(z) = z^2.e^{1/z}$ about $z = 0$.

Solution:

f(z) is analytic in the annulus $0 < |z| < \infty$. Hence it can be expanded into a Laurent's series in powers of z.

$$f(z) = z^2.e^{1/z}$$

$$= z^2\left(1 + \frac{1}{z} + \frac{1}{2!}\frac{1}{z^2} + \frac{1}{3!}\frac{1}{z^3} + \cdots\right)$$

$$= z^2 + z + \frac{1}{2!} + \frac{1}{3!}\frac{1}{z} + \frac{1}{4!}\frac{1}{z^2} + \cdots$$

Example 4.2.10 Expand the function $f(z) = \dfrac{1}{(z+1)(z+2)^2}$ about the point $z = 1$.

Solution:

$f(z) = \dfrac{1}{(z+1)(z+2)^2}$ is not analytic at the point $z = -1$ and $z = -2$.

It is analytic in the regions (i) $|z - 1| < 2$ (ii) $2 < |z - 1| < 3$ and (iii) $3 < |z - 1| < \infty$.

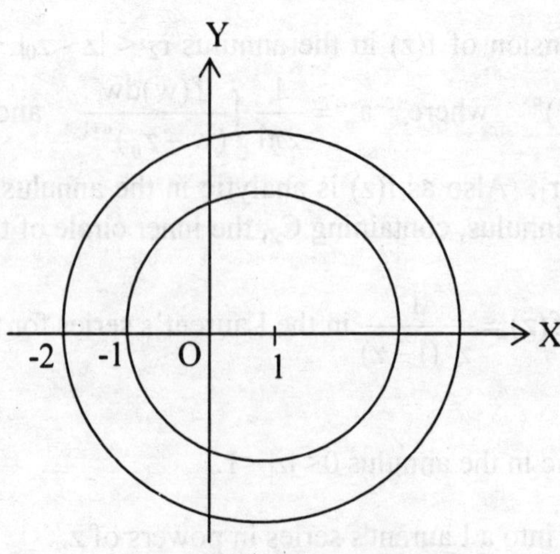

Fig. 4.19

In the region (i) f(z) can be expanded into a Taylor's series, whereas in the region (ii) and (iii) it can be expanded into Laurent's series in powers of z -1.
Now

$$f(z)=\frac{1}{(z+1)(z+2)^2}=\frac{1}{z+1}-\frac{1}{z+2}-\frac{1}{(z+2)^2}$$

$$f(z)=\frac{1}{w+2}-\frac{1}{w+3}-\frac{1}{(w+3)^2} \qquad \text{(Substituting } z-1=w\text{)}$$

(i) $|z-1|<2$ i.e., $|w|<2$

$$f(z)=\frac{1}{w+2}-\frac{1}{w+3}-\frac{1}{(w+3)^2}$$

$$=\frac{1}{2}\cdot\frac{1}{(1+w/2)}-\frac{1}{3}\cdot\frac{1}{(1+w/3)}-\frac{1}{9}\cdot\frac{1}{(1+w/3)^2}$$

$$=\frac{1}{2}\left[1-w/2+\left(w/2\right)^2-\left(w/2\right)^3+\cdots\right]-\frac{1}{3}\left[1-w/3+\left(w/3\right)^2-\left(w/3\right)^3+\cdots\right]$$

$$-\frac{1}{9}\left[1-2(w/3)+3\left(w/3\right)^2-4\left(w/3\right)^3+\cdots\right] \quad \left(\text{Since } |w/2|<1 \text{ and } |w/3|<1\right)$$

$$=\frac{1}{2}\sum_{n=0}^{\infty}(-1)^n(w/2)^n-\frac{1}{3}\sum_{n=0}^{\infty}(-1)^n(w/3)^n-\frac{1}{9}\sum_{n=0}^{\infty}(-1)^n(n+1)(w/3)^n$$

$$\text{i.e., } f(z)=\sum_{n=0}^{\infty}(-1)^n\left[\frac{1}{2^{n+1}}-\frac{1}{3^{n+1}}-(n+1)\frac{1}{3^{n+2}}\right](z-1)^n$$

(ii) $2<|z-1|<3$ i.e., $2<|w|<3$

$$f(z)=\frac{1}{w+2}-\frac{1}{w+3}-\frac{1}{(w+3)^2}$$

$$=\frac{1}{w(1+2/w)}-\frac{1}{3}\cdot\frac{1}{(1+w/3)}-\frac{1}{9}\cdot\frac{1}{(1+w/9)^2}$$

$$=\frac{1}{w}\left[1-2/w+\left(2/w\right)^2-\left(2/w\right)^3+\cdots\right]-\frac{1}{3}\left[1-w/3+\left(w/3\right)^2-\left(w/3\right)^3+\cdots\right]$$

$$-\frac{1}{9}\left[1-2(w/3)+3\left(w/3\right)^2-4\left(w/3\right)^3+\cdots\right] \quad \left(\text{since } |2/w|<1 \text{ and } |w/3|<1\right)$$

$$=\frac{1}{w}\sum_{n=0}^{\infty}(-1)^n(2/w)^n-\frac{1}{3}\sum_{n=0}^{\infty}(-1)^n(w/3)^n-\frac{1}{9}\sum_{n=0}^{\infty}(-1)^n(n+1)(w/3)^n$$

$$\text{i.e., } f(z)=\sum_{n=0}^{\infty}(-1)^n\frac{2^n}{(z-1)^{n+1}}-\sum_{n=0}^{\infty}(-1)^n\left[\frac{1}{3^{n+1}}+\frac{(n+1)}{3^{n+2}}\right](z-1)^n$$

(iii) $3<|z-1|<\infty$ i.e., $3<|w|<\infty$

$$f(z) = \frac{1}{w+2} - \frac{1}{w+3} - \frac{1}{(w+3)^2}$$

$$= \frac{1}{w(1+2/w)} - \frac{1}{w(1+3/w)} - \frac{1}{w^2(1+3/w)^2}$$

$$= \frac{1}{w}\left[1 - 2/w + \left(2/w\right)^2 - \left(2/w\right)^3 + \cdots\right] - \frac{1}{w}\left[1 - 3/w + \left(3/w\right)^2 - \left(3/w\right)^3 + \cdots\right]$$

$$- \frac{1}{w^2}\left[1 - 2(3/w) + 3(3/w)^2 - 4(3/w)^3 + \cdots\right] \left(\text{since } |2/w| < 1 \text{ and } |3/w| < 1\right)$$

$$= \frac{1}{w}\sum_{n=0}^{\infty}(-1)^n (2/w)^n - \frac{1}{w}\sum_{n=0}^{\infty}(-1)^n (3/w)^n - \frac{1}{w^2}\sum_{n=0}^{\infty}(-1)^n (n+1)(3/w)^n$$

$$= \sum_{n=0}^{\infty}(-1)^n\left[2^n - 3^n\right]\frac{1}{(z-1)^{n+1}} - \sum_{n=0}^{\infty}(-1)^n 3^n \frac{(n+1)}{(z-1)^{n+2}}$$

Example 4.2.11 Expand $\frac{1}{z(z^2-3z+2)}$ in the region (i) $0 < |z| < 1$ (ii) $1 < |z| < 2$ (iii) $|z| > 2$.

Solution:

Let $f(z) = \frac{1}{z(z^2-3z+2)} = \frac{1}{z(z-1)(z-2)}$

Resolving into partial fractions, we get $f(z) = \frac{1}{2z} - \frac{1}{z-1} + \frac{1}{2(z-2)}$

(i) $0 < |z| < 1$

f(z) is analytic in this annulus and $|z| < 1$, $\left|\frac{z}{2}\right| < 1$

$$\therefore f(z) = \frac{1}{2z} + \frac{1}{1-z} - \frac{1}{4}\frac{1}{1-z/2}$$

$$= \frac{1}{2z} + \left[1 + z + z^2 + \cdots\right] - \frac{1}{4}\left[1 + (z/2) + (z/2)^2 + (z/2)^3 + \cdots\right]$$

$$= \frac{1}{2z} + \sum_{n=0}^{\infty} z^n - \sum_{n=0}^{\infty}\frac{z^n}{2^{n+2}}$$

(ii) $1 < |z| < 2$

f(z) is analytic in this annulus. Also $\left|\frac{z}{2}\right| < 1$ and $\left|\frac{1}{z}\right| < 1$

$$\therefore f(z) = \frac{1}{2z} - \frac{1}{z(1-1/z)} - \frac{1}{4}\frac{1}{1-z/2}$$

$$= \frac{1}{2z} - \frac{1}{z}\left[1 + \tfrac{1}{z} + \left(\tfrac{1}{z}\right)^2 + \left(\tfrac{1}{z}\right)^3 + \cdots\right] - \frac{1}{4}\left[1 + (z/2) + (z/2)^2 + (z/2)^3 + \cdots\right]$$

$$= -\frac{1}{2z} - \frac{1}{z^2} - \frac{1}{z^3} - \frac{1}{z^4} - \frac{1}{4}\left[1 + \frac{z}{2} + \frac{z^2}{2^2} + \frac{z^3}{2^3} + \cdots\right]$$

$$= -\frac{1}{2z} + \sum_{n=0}^{\infty} \frac{1}{z^{n+2}} - \sum_{n=0}^{\infty} \frac{z^n}{2^{n+2}}$$

(iii) $|z| > 2$.

Then $\left|\frac{1}{z}\right| < 1$ and $\left|\frac{2}{z}\right| < 1$

$$\therefore f(z) = \frac{1}{2z} - \frac{1}{z(1 - 1/z)} - \frac{1}{2z(1 - 2/z)}$$

$$= \frac{1}{2z} - \frac{1}{z}\left[1 + \tfrac{1}{z} + \left(\tfrac{1}{z}\right)^2 + \left(\tfrac{1}{z}\right)^3 + \cdots\right] - \frac{1}{2z}\left[1 + \tfrac{2}{z} + \left(\tfrac{2}{z}\right)^2 + \left(\tfrac{2}{z}\right)^3 + \cdots\right]$$

$$= -\frac{1}{2z} - \frac{1}{z^2} - \frac{1}{z^3} - \frac{1}{z^4} - \frac{1}{2}\left[\frac{1}{z} + \frac{2}{z^2} + \frac{2^2}{z^3} + \cdots\right]$$

$$= -\frac{1}{2z} + \sum_{n=0}^{\infty} \frac{1}{z^{n+2}} - \frac{1}{2}\sum_{n=0}^{\infty} \frac{2^n}{z^{n+1}}$$

Example 4.2.12 Expand $\dfrac{1}{(1+z^2)(2+z^2)}$ in powers of z, when

(i) $|z| < 1$

(ii) $1 < |z| < \sqrt{2}$

(iii) $|z| > \sqrt{2}$

Solution:

Let $f(z) = \dfrac{1}{(1+z^2)(2+z^2)}$

Resolving into partial fractions we get, $f(z) = \dfrac{1}{1+z^2} - \dfrac{1}{2+z^2}$

f(z) is not analytic at i, -i, $\sqrt{2}$ i and $-\sqrt{2}$ i

(i) $|z| < 1$.

f(z) is analytic in this region and we have $|z|^2 < 1$, $\dfrac{|z|^2}{2} < 1$.

$$\therefore f(z) = \frac{1}{1+z^2} - \frac{1}{2}\frac{1}{\left(1 + z^2/2\right)}$$

$$= \left(1 - z^2 + z^4 - z^6 + \cdots\right) - \frac{1}{2}\left[1 - \frac{z^2}{2} + \left(\frac{z^2}{2}\right)^2 - \left(\frac{z^2}{2}\right)^3 + \cdots\right]$$

$$= \sum_{n=0}^{\infty}(-1)^n . z^{2n} - \frac{1}{2}\sum_{n=0}^{\infty}(-1)^n \frac{z^{2n}}{2^n}$$

(ii) $1 < |z| < \sqrt{2}$.

f(z) is analytic in this region and we have $\left|\frac{1}{z^2}\right| < 1,\ \frac{|z|^2}{2} < 1.$

$$\therefore\ f(z) = \frac{1}{z^2} . \frac{1}{\left(1 + 1/z^2\right)} - \frac{1}{2}\frac{1}{\left(1 + z^2/2\right)}$$

$$= \frac{1}{z^2}\left[1 - \frac{1}{z^2} + \left(\frac{1}{z^2}\right)^2 - \left(\frac{1}{z^2}\right)^3 + \cdots\right] - \frac{1}{2}\left[1 - \frac{z^2}{2} + \left(\frac{z^2}{2}\right)^2 - \left(\frac{z^2}{2}\right)^3 + \cdots\right]$$

$$= \left[\frac{1}{z^2} - \frac{1}{z^4} + \frac{1}{z^6} - \cdots\right] - \frac{1}{2}\left[1 - \frac{z^2}{2} + \left(\frac{z^2}{2}\right)^2 - \left(\frac{z^2}{2}\right)^3 + \cdots\right]$$

$$= \sum_{n=0}^{\infty}(-1)^n . \frac{1}{z^{2n+2}} - \frac{1}{2}\sum_{n=0}^{\infty}(-1)^n \frac{z^{2n}}{2^n}$$

(iii) $|z| > \sqrt{2}$.

f(z) is analytic in this region. Also $\left|\frac{2}{z^2}\right| < 1,\ \left|\frac{1}{z^2}\right| < 1.$

$$\therefore\ f(z) = \frac{1}{z^2} . \frac{1}{\left(1 + 1/z^2\right)} - \frac{1}{z^2}\frac{1}{\left(1 + 2/z^2\right)}$$

$$= \frac{1}{z^2}\left[1 - \frac{1}{z^2} + \left(\frac{1}{z^2}\right)^2 - \left(\frac{1}{z^2}\right)^3 + \cdots\right] - \frac{1}{z^2}\left[1 - \frac{2}{z^2} + \left(\frac{2}{z^2}\right)^2 - \left(\frac{2}{z^2}\right)^3 + \cdots\right]$$

$$= \left[\frac{1}{z^2} - \frac{1}{z^4} + \frac{1}{z^6} - \frac{1}{z^8} + \cdots\right] - \left[\frac{1}{z^2} - \frac{2}{z^4} + \frac{2^2}{z^6} - \frac{2^3}{z^8} + \cdots\right]$$

$$= (2 - 1)\frac{1}{z^4} - (2^2 - 1)\frac{1}{z^6} + (2^3 - 1)\frac{1}{z^8} \cdots$$

$$= \sum_{n=0}^{\infty}(-1)^n . (2^{n+1} - 1)\frac{1}{z^{2n+4}}$$

Example 4.2.13 Find the Laurent's series of $f(z) = \dfrac{z}{(z-1)(z-2)}$ valid in the region

(i) $|z + 2| < 3$ (ii) $3 < |z + 2| < 4$ (iii) $|z + 2| > 4$.

Solution:

$$f(z) = \frac{z}{(z-1)(z-2)}$$

Put $z + 2 = u$ i.e., $z = u - 2$

Then $f(z) = \dfrac{u-2}{(u-3)(u-4)} = \dfrac{-1}{u-3} + \dfrac{2}{u-4}$

(i) $|z+2|<3$

Then $\left|\dfrac{u}{3}\right| < 1$ and $\left|\dfrac{u}{4}\right| < 1$

$$\therefore\ f(z) = \frac{-1}{u-3} + \frac{2}{u-4}$$

$$= \frac{1}{3(1-u/3)} - \frac{1}{2(1-u/4)}$$

$$= \frac{1}{3}\sum_{n=0}^{\infty}(u/3)^n - \frac{1}{2}\sum_{n=0}^{\infty}(u/4)^n$$

$$= \sum_{n=0}^{\infty}\frac{(z+2)^n}{3^{n+1}} - \frac{1}{2}\sum_{n=0}^{\infty}\frac{(z+2)^n}{4^n}$$

(ii) $3 < |z+2| < 4$ i.e., $3 < |u| < 4$

Then $\left|\dfrac{u}{4}\right| < 1$ and $\left|\dfrac{3}{u}\right| < 1$

$$\therefore\ f(z) = \frac{-1}{u-3} + \frac{2}{u-4}$$

$$= \frac{-1}{u(1-3/u)} - \frac{1}{2(1-u/4)}$$

$$= -\frac{1}{u}\sum_{n=0}^{\infty}(3/u)^n - \frac{1}{2}\sum_{n=0}^{\infty}(u/4)^n$$

$$= -\sum_{n=0}^{\infty}\frac{3^n}{(z+2)^{n+1}} - \frac{1}{2}\sum_{n=0}^{\infty}\frac{(z+2)^n}{4^n}$$

(iii) $|z+2| > 4$ i.e., $|u| > 4$

Then $\left|\dfrac{4}{u}\right| < 1$ and $\left|\dfrac{3}{u}\right| < 1$

$$\therefore\ f(z) = \frac{-1}{u-3} + \frac{2}{u-4}$$

$$= \frac{-1}{u(1-3/u)} + \frac{2}{u(1-4/u)}$$

$$= -\frac{1}{u}\sum_{n=0}^{\infty}\left(\tfrac{3}{u}\right)^n + \frac{2}{u}\sum_{n=0}^{\infty}\left(\tfrac{4}{u}\right)^n$$

$$= -\sum_{n=0}^{\infty}\frac{3^n}{(z+2)^{n+1}} - 2.\sum_{n=0}^{\infty}\frac{4^n}{(z+2)^{n+1}}$$

$$= \sum_{n=0}^{\infty}\left[2.4^n - 3^n\right]\frac{1}{(z+2)^{n+1}}$$

Example 4.2.14 Represent $f(z) = \dfrac{z}{(z-1)(z-2)}$ by a series of powers of z -1 in the region $0 < |z-1| < 2$.

Solution:

$$f(z) = \frac{z}{(z-1)(z-2)}$$

Put $z - 1 = u$. Let f(z) becomes $\varphi(u)$.

Then $\varphi(u) = \dfrac{u+1}{u(u-2)}$

$0 < |z-1| < 2 \Rightarrow 0 < |u| < 2$

$\varphi(u)$ is not analytic at $u = 0$ and $u = 2$. Hence $\varphi(u)$ is analytic in the region $0 < |u| < 2$.

$$\therefore \varphi(u) = \frac{u+1}{u(u-2)}$$

$$= -\frac{1}{2u} + \frac{3}{2(u-2)}$$

$$= -\frac{1}{2u} + \frac{3}{(-4)(1-u/2)}$$

$$= -\frac{1}{2u} - \frac{3}{4}\left[1 + \left(\frac{u}{2}\right) + \left(\frac{u}{2}\right)^2 + \cdots\right] \text{ as } \left|\frac{u}{2}\right| < 1$$

$$= -\frac{1}{2u} - \frac{3}{4}\sum_{n=0}^{\infty}\left(\frac{u}{2}\right)^n$$

$$\therefore \quad f(z) = -\frac{1}{2(z-1)} - \frac{3}{4}\sum_{n=0}^{\infty}\frac{(z-1)^n}{2^n}$$

EXERCISE 4.2

PART-A

1. Define power series.
2. Define radius of convergence and circle of convergence of a power series.
3. State Taylor's theorem.

4. State Laurent's theorem.
5. What do you mean by principal part and analytic part of the Laurent's expansion of a function?
6. Give a formula for finding radius of convergence of a power series.
7. Find the radius of convergence of $\sum_{n=0}^{\infty} \frac{(n!)^2 z^n}{(2n)!}$.
8. Find the radius of convergence of $\sum_{n=0}^{\infty} \frac{(n!)(z-i)^n}{2^n}$.
9. Find the radius of convergence of the power series $\sum_{n=0}^{\infty} \left(\frac{2n+1}{3n+4}\right)^n z^n$.
10. What is the radius of convergence of the power series expansion $e^z = \sum_{n=0}^{\infty} \frac{z^n}{n!}$?

Write down the Maclaurin's expansions for the following functions (Ex: 11-20)

11. $\frac{1}{1-z}$
12. $\frac{z}{(1-z)^2}$
13. $\frac{z-1}{z+1}$
14. $z.e^{2z}$
15. $\frac{1}{2-z}$
16. $z^3.e^z$
17. $\frac{1}{1+z^2}$
18. $\log(1+z)$
19. $z.\sin z$
20. $(1-z)^{-3}$

21. Give the Taylor's series for $f(z) = \frac{1}{z}$ about $z = 1$.
22. Show that $\frac{1}{z^2} = 1 + \sum_{n=1}^{\infty} (n+1)(1+z)^n$, when $|1 + z| < 1$.
23. Show that $\frac{1}{z^2} = \sum_{n=0}^{\infty} (-1)^n (n+1)(z-1)^n$, when $|z - 1| < 1$.
24. Find the Taylor's series for e^z about $z = -i$.
25. Find the Taylor's series for e^{-z} about $z = 1$.
26. Find the Laurent's series expansion of $\frac{1}{z(z-1)}$ valid in $0 < |z - 1| < 1$.
27. Expand $\frac{z-1}{z^2}$ in the region $|z - 1| > 1$.
28. If $z \neq 0$, Show that $\frac{\sin(z^2)}{z^4} = \frac{1}{z^2} - \frac{z^2}{3!} + \frac{z^6}{5!} - \frac{z^{10}}{7!} + \ldots$.
29. Prove $e^{\frac{-1}{z-2}} = \sum_{n=0}^{\infty} \frac{(-1)^n}{n!(z-2)^n}$ for $|z - 2| > 0$.

PART-B

30. Expand ze^{2z} in a Taylor's series about the point z = -1.

31. Obtain the Taylor's series to represent $\dfrac{1}{(z+1)(z+3)}$ in |z| < 1.

32. Expand $\dfrac{1}{(z-1)(z-2)}$ in powers of z in the regions
(i) |z| <1 (ii) 1 < |z| < 2 (iii) |z| > 2

33. Expand $f(z) = \dfrac{z+3}{z(z^2-z-2)}$ in powers of z in the regions
(i) |z| <1 (ii) 1 < |z| < 2 (iii) |z| > 2.

34. Expand $\dfrac{1}{z^2(z-3)^2}$ as a Laurent's series about z = 3 and give the region of validity.

35. Expand $f(z) = \dfrac{1}{(z+1)(z+3)}$ in Laurent's series valid for
(i) 1 < |z| < 3 (ii) |z| > 3 (iii) 0 < |z + 1| < 2 (iv) |z| <1.

36. Show that $\dfrac{1}{4z-z^2} = \sum_{n=0}^{\infty} \dfrac{z^{n-1}}{4^{n+1}}$, when 0 < |z| <4.

37. Prove that $\dfrac{1+2z}{z^2+z^3} = \dfrac{1}{z^2} + \dfrac{1}{z} - 1 + z - z^2 + z^3 - \cdots$, when 0 < |z| < 1.

38. Find the Taylor's and Laurent's series which represents the function
$f(z) = \dfrac{z^2-1}{(z+2)(z+3)}$ in the region (i) |z| <2 (ii) 2 < |z| < 3 (iii) |z| >3.

39. If 0 < |z-1| <2, prove $\dfrac{z}{(z-1)(z-3)} = \dfrac{-1}{2(z-1)} - \dfrac{3}{4}\sum_{n=0}^{\infty}\left(\dfrac{z-1}{2}\right)^n$.

40. Find the Laurent's expansions of $f(z) = \dfrac{7z-2}{z(z-2)(z+1)}$ about z = -1.

41. Find the Laurent's expansion of $f(z) = \dfrac{z^3-6z-1}{(z-)(z-3)(z+2)}$ about the point z = 3.
Given the region of convergence of the series in each case.

42. Find the Laurent's expansion of $f(z) = \dfrac{1}{(z^2+1)^2}$ about the point z = i.

43. Expand $\dfrac{1}{(z+1)^2(z+3)}$ as a Laurent's series about z = -1. State the region of convergence of the series.

44. Obtain the Laurent's series expansion of $f(z) = z^2 \cos\left(\frac{1}{z-1}\right)$ about the point z = 1.

45. Prove $e^{\frac{z}{z-2}} = e.\sum_{n=0}^{\infty}\frac{1}{n!}\left(\frac{2}{z-2}\right)^n$ for $|z-2| > 0$.

46. Find the Laurent's series expansion of $f(z) = \dfrac{1}{2z^2 - iz}$ about the origin, which converges at the point $z = i$.

47. Find the Laurent's series expansion of $f(z) = \dfrac{1}{z^2 + 2iz + 3}$ about $z = 0$, which converges at $z = 4i$.

48. If $f(z) = \dfrac{z+4}{(z+3)(z-1)^2}$, find the Laurent's series expansion in the regions (i) $0 < |z-1| < 4$ and (ii) $|z-1| > 4$

49. Expand $f(z) = \dfrac{e^{2z}}{(z-1)^3}$ about $z = 1$ as a Laurent's series.

4.3 SINGULARITIES AND RESIDUES

In this section we study the behaviour of functions in the neighbourhood of certain points where the function ceases to be analytic.

4.3.1 ZEROS OF AN ANALYTIC FUNCTION

Definition 4.3.1 Let f(z) be analytic in a domain D and $z_0 \in D$. z_0 is said to be a zero of f(z) if $f(z_0) = 0$.

When f(z) is analytic at z_0, there exists a neighbourhood of z_0 at which f(z) can be expanded into a Taylor's series as $f(z) = a_0 + a_1(z - z_0) + a_2(z - z_0)^2 + \ldots + a_n(z - z_0)^n + \ldots$

where $a_0 = f(z_0)$ and $a_n = \dfrac{f^{(n)}(z_0)}{n!}$; $n = 1, 2, 3, \ldots$

If z_0 is a zero of f(z), then $f(z_0) = a_0 = 0$. If $a_1 \neq 0$, z_0 is said to be a **simple zero** of f(z). If $a_0 = a_1 = \ldots = a_{m-1} = 0$ and $a_m \neq 0$, then z_0 is called a **zero of order m**. Then

$f(z) = (z - z_0)^m\,[a_m + a_{m+1}(z - z_0) + a_{m+2}(z - z_0)^2 + \ldots]$

$\quad = (z - z_0)^m\, g(z)$

where g(z) is analytic in D and $g(z_0) = a_m \neq 0$

Note:

(i) If z_0 is a zero of order m for f(z), then $a_0 = a_1 = a_2 = \ldots = a_{m-1} = 0$ and $a_m \neq 0$ i.e., $f(z_0) = f'(z_0) = f''(z_0) = \ldots = f^{(m-1)}(z_0) = 0$ and $f^{(m)}(z_0) \neq 0$.

(ii) Zeros of non-constant analytic functions are isolated. i.e., if z_0 is a zero of a non-constant analytic function f(z), then there exists a neighbourhood of z_0 in which f(z) has no other zeros.

Example 4.3.1 Find the zeros of the functions (i) sin z (ii) $\dfrac{(z-1)(z-3)^3}{z+2}$

(iii) $(z-1)\sin\left(\dfrac{1}{z-2}\right)$.

Solution:

(i) sin z = 0 when z = 0, ±π, ±2π, ...
Hence the zeros of sinz are z = nπ, n = 0, ±1, ±2, ...

(ii) $\dfrac{(z-1)(z-3)^3}{z+2} = 0$ when $(z-1)(z-3)^3 = 0$
i.e., z = 1, z = 3.
z = 1 is a simple zero and z = 3 is a zero of order 3.

(iii) $(z-1)\sin\left(\dfrac{1}{z-2}\right) = 0$ when z = 1 or $\sin\left(\dfrac{1}{z-2}\right) = 0$

$\sin\left(\dfrac{1}{z-2}\right) = 0$ when $\dfrac{1}{z-2} = n\pi$, n = 0, ±1, ±2, ...

i.e., $z = 2 + \dfrac{1}{n\pi}$, n = 0, ±1, ±2, ...

∴ The zeros of $(z-1)\sin\left(\dfrac{1}{z-2}\right)$ are z = 1 and $z = 2 + \dfrac{1}{n\pi}$, n = 0, ±1, ±2, ...

4.3.2 SINGULARITIES

Definition 4.3.2 If a function f(z) is analytic at all points of a domain D except at some points of D, then these exceptional points are called the **singular points** or **singularities** of f(z).

Thus a singularity of a function f(z) is a point where f(z) ceases to be analytic. For example, $f(z) = \dfrac{1}{z}$ is analytic everywhere except at z = 0 (Since $f'(z) = -\dfrac{1}{z^2}$ is not finite at z = 0). Hence z = 0 is a singularity of $f(z) = \dfrac{1}{z}$.

Definition 4.3.3 Let z_0 be a singularity of f(z). If there exists some neighbourhood of z_0 such that f(z) is analytic at all points in the neighbourhood (except at z_0) then z_0 is called an **isolated singularity** of f(z). Otherwise z_0 is called a **non-isolated singularity** of f(z).

Note:

(i) If z_0 is an isolated singularity of f(z), there exists a δ > 0 such that the circle $|z - z_0| = \delta$ encloses no singularity of f(z) other than z_0.

(ii) If z_0 is a non-isolated singularity of f(z), then every neighbourhood of the point z_0 contains singularities of f(z) other than z_0.

Let f(z) be analytic in a domain D except at the point z_0. The point z_0 is an isolated singularity of f(z) and there exists a real number r > 0 such that f(z) is analytic in the annulus $0 < |z - z_0| < r$ contained in D. Then f(z) can be expanded into a Laurent's series in the annulus as $f(z) = \sum_{n=0}^{\infty} a_n (z - z_0)^n + \sum_{n=1}^{\infty} \frac{b_n}{(z - z_0)^n}$

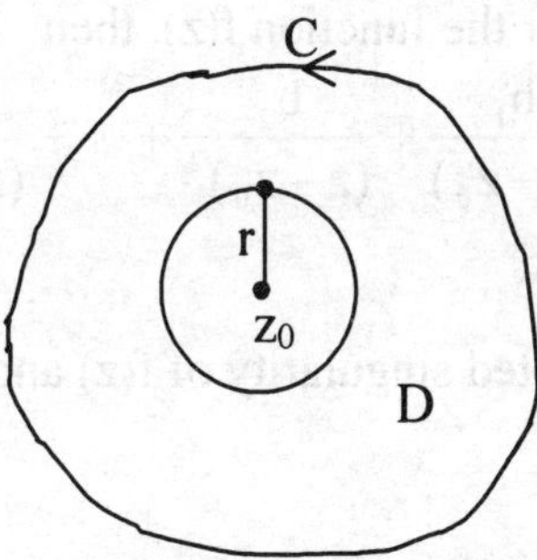

Fig. 4.20

Definition 4.3.4 An isolated singularity z_0 of f(z) is a **removable singularity** if the Laurent's expansion of f(z) about z_0 has no principal part. i.e., $b_n = 0$ for all n.

Definition 4.3.5 An isolated singularity z_0 of f(z) is called a **pole of order m** if the principal part of the Laurent's expansion of f(z) about z_0 contains only the first m terms. i.e., $b_n = 0$ for n > m and $b_m \neq 0$.

A pole of order 1 is called a **simple pole** and a pole of order 2 is called a double pole.

Note:

(i) If z_0 is a removable singularity of f(z), then $f(z) = \sum_{n=0}^{\infty} a_n (z - z_0)^n$.

Hence $\lim_{z \to z_0} f(z) = a_0$ (i.e., the limit exists). Conversly, if the $\lim_{z \to z_0} f(z) = a_0$ exists and is finite, then z_0 is a removable singularity of f(z).
By defining f(z) at z_0 as a_0. i.e., $f(z_0) = a_0$, the singularity can be removed.
Then f(z) becomes analytic at z_0 also, and

$$f'(z) = a_1 + a_2 2(z - z_0) + a_3 3(z - z_0)^2 + \ldots. = \sum_{n=1}^{\infty} a_n n(z - z_0)^{n-1} \text{ for } |z - z_0| < r.$$

For example, consider $f(z) = \frac{\sin z}{z}$ $z \neq 0$.

As f(z) is not defined at $z = 0$, it is not analytic there. But $z = 0$ is a removable singularity of f(z) since $\lim_{z\to 0} f(z) = \lim_{z\to 0} \frac{\sin z}{z} = 1$, which is finite.

Now define $$f(z) = \begin{cases} \frac{\sin z}{z} & z \neq 0 \\ = 1 & z = 0 \end{cases} \qquad (1)$$

The function f(z) defined by (1) is analytic in the entire complex plane.

(ii) If z_0 is a pole of order m for the function f(z), then

$$f(z) = \sum_{n=0}^{\infty} a_n (z - z_0)^n + \frac{b_1}{(z - z_0)} + \frac{b_2}{(z - z_0)^2} + \cdots + \frac{b_m}{(z - z_0)^m}, \; b_m \neq 0$$

Then $\lim_{z\to z_0} f(z) = \infty$

Conversely if, z_0 is an isolated singularity of f(z) and $\lim_{z\to z_0} f(z) = \infty$, then z_0 is a pole. (Prove this!)

(iii) If z_0 is a pole of order m for the function f(z), then

$$f(z) = \sum_{n=0}^{\infty} a_n (z - z_0)^n + \sum_{n=1}^{m} \frac{b_n}{(z - z_0)^n}, \quad b_m \neq 0$$

$$= \frac{1}{(z - z_0)^m} . \varphi(z)$$

where $\varphi(z) = b_m + b_{m-1}(z - z_0) + \ldots + b_1(z - z_0)^{m-1} + a_0(z - z_0)^m + \ldots$

$\lim_{z\to z_0} \varphi(z) = b_m \neq 0$. i.e., $\lim_{z\to z_0} (z - z_0)^m . f(z) = b_m \neq 0$.

Hence $\varphi(z)$ is analytic at z_0. Since $\varphi(z_0) = b_m \neq 0$, z_0 is not a zero of $\varphi(z)$.

(iv) If z_0 is a pole of order m for f(z), then z_0 is a zero of order m for $\frac{1}{f(z)}$.

(Prove this!).

(v) If z_0 is a pole of f(z), then f(z) is unbounded in every neighbourhood of z_0.

Definition 4.3.6 An isolated singularity z_0 of f(z) is called an **essential singularity** if the principal part of the Laurent's expansion of f(z) about z_0 contains infinite number of terms.

Note:

(i) If z_0 is an isolated essential singularity of f(z), then there exists no finite n such that $\lim_{z\to z_0} (z - z_0)^n . f(z)$ is a finite non-zero constant. (Otherwise z_0 is a pole).

(ii) **(Weiestrass's Theorem)**
If z_0 is an essential singularity of f(z), then f(z) comes arbitrarily close to any complex number in every neighbourhood of z_0. (Prove this!).
i.e., given any complex number b, and real numbers $\varepsilon>0$ and $\delta>0$, there exists z such that $|z-z_0|<\delta$ and $|f(z)-b|<\varepsilon$

Definition 4.3.7 A function f(z) which is analytic everywhere in the finite complex plane except at a finite number of poles is called a **meromorphic function.**

Note:

To check whether the point $z=\infty$ is a singular point of f(z) and to determine the type of singularity at $z=\infty$, we substitute $z = 1/w$ in f(z) and determine the nature of singularity of the resulting function $f(1/w)$ at $w = 0$. f(z) has a removable singularity, pole or isolated essential singularity at $z=\infty$ according as $f(1/w)$ has a removable singularity, pole or isolated essential singularity at $w = 0$.

Example 4.3.1 Find the singularities of $f(z)=\dfrac{z-\sin z}{z^3}$.

Solution:
f(z) is not defined at z = 0.
$\therefore$ z = 0 is an isolated singularity.

Also $f(z)=\dfrac{z-\sin z}{z^3}=\dfrac{1}{z^3}\left[z-\left(z-\dfrac{z^3}{3!}+\dfrac{z^5}{5!}-\cdots\right)\right]$

$$=\frac{1}{3!}-\frac{z^2}{5!}+\frac{z^4}{7!}-\cdots$$

$\lim\limits_{z\to 0} f(z)=\dfrac{1}{3!}$
$\therefore$ z = 0 is a removable singularity.
(If we define $f(0)=\dfrac{1}{3!}$. Then f(z) becomes analytic at the point z = 0).

Example 4.3.2 Find the singularities of $f(z)=\dfrac{\sinh z}{z^5}$.

Solution:
z = 0 is an isolated singularity.

$$f(z)=\frac{\sinh z}{z^5}=\frac{1}{z^5}\left[z+\frac{z^3}{3!}+\frac{z^5}{5!}+\cdots\right]$$

$=\dfrac{1}{z^4}+\dfrac{1}{z^2.3!}+\dfrac{1}{5!}+\dfrac{z^2}{7!}+\cdots$ is the Laurent's expansion of f(z) about z = 0.

Hence z = 0 is a pole of order 4.

Example 4.3.3 Find the singularities of $f(z) = ze^{1/z}$.

Solution:
$z = 0$ is an isolated singularity.

$$f(z) = ze^{1/z}$$

$$= z\left[1 + \frac{1}{z} + \frac{1}{z^2.2!} + \frac{1}{z^3.3!} + \cdots\right]$$

$$= z + 1 + \frac{1}{z.2!} + \frac{1}{z^2.3!} + \cdots$$ is the Laurent's expansion of $f(z)$ about $z = 0$.

Since the principal part contains infinite number of terms, $z = 0$ is an isolated essential singularity.

Example 4.3.4 Classify the singularity of $f(z) = ze^{1/z}$ at $z = \infty$.

Solution:
Put $z = 1/w$

Then $f(1/w) = \frac{1}{w}.e^w$

$$= \frac{1}{w}\left[1 + \frac{w}{1!} + \frac{w^2}{2!} + \cdots\right]$$

$$= \frac{1}{w} + \frac{1}{1!} + \frac{w}{2!} + \frac{w^2}{3!} + \cdots$$ is the Laurent's expansion of $f(1/w)$ about $w = 0$.

$\therefore$ $w = 0$ is a simple pole for $f(1/w)$. Hence $z = \infty$ is a simple pole for $f(z) = ze^{1/z}$.

4.3.3 RESIDUES AND COMPUTATION OF RESIDUES

In this section we introduce the concept of residue of a function f(z) at an isolated singularity. We also prove Cauchy's Residue Theorem. The theory of residues provides a simpler way for the evaluation of complex integrals. It can also be used for evaluating certain types of real definite integrals.

Definition 4.3.8 Let z_0 be an isolated singularity of the function f(z). Then the coefficient of $\frac{1}{z - z_0}$ in the Laurent's expansion of f(z) about z_0 is called the **residue of f(z) at z_0** and is denoted by $\text{Res}[f(z); z_0]$.

Thus $\text{Res}[f(z); z_0]$ = Coefficient of $\frac{1}{z - z_0}$ in the Laurent's expansion of f(z).

$$= b_1 = \frac{1}{2\pi i}\int_C f(z)dz$$

where C is any simple closed curve around z_0 such that f(z) is analytic on and inside C, except at the point z_0.

1. Residue at a removable singularity.

If z_0 is a removable singularity of f(z), then there are no terms in the principal part of the Laurent's expansion of f(z) about z_0. Hence Res[f(z); z_0] = $b_1 = 0$.
Also it follows that $\int_C f(z)dz = 0$.

2. Residue at a simple pole.

If z_0 is a simple pole, then the Laurent's expansion of f(z) about z_0 is

$$f(z) = \sum_{n=0}^{\infty} a_n (z - z_0)^n + \frac{b_1}{z - z_0}$$

$$\therefore \ (z - z_0)f(z) = \sum_{n=0}^{\infty} a_n (z - z_0)^{n+1} + b_1$$

$$\lim_{z \to z_0} (z - z_0)f(z) = b_1$$

Hence Res[f(z); z_0] = $\lim\limits_{z \to z_0} (z - z_0)f(z)$ (1)

If f(z) is in the form $f(z) = \dfrac{P(z)}{Q(z)}$, where P(z) and Q(z) have no common factors, $P(z_0) \neq 0$ and $Q(z_0) = 0$, then the Res[f(z); z_0] can be computed by the following method also. Now Q(z) is analytic at z_0 and hence has a Taylor's expansion about z_0.

i.e., $Q(z) = Q(z_0) + (z - z_0)Q'(z_0) + \frac{1}{2!}(z - z_0)^2 Q''(z_0) + ...$

$$= (z - z_0)Q'(z_0) + \frac{1}{2!}(z - z_0)^2 Q''(z_0) + ...$$

$$\therefore \text{Res}[f(z); z_0] = \lim_{z \to z_0} (z - z_0)f(z)$$

$$= \lim_{z \to z_0} (z - z_0)\frac{P(z)}{Q(z)}$$

$$= \lim_{z \to z_0} \frac{(z - z_0)P(z)}{(z - z_0)Q'(z_0) + \frac{1}{2!}(z - z_0)^2 Q''(z_0) + ...}$$

$$= \lim_{z \to z_0} \frac{P(z)}{Q'(z_0) + \frac{1}{2!}(z - z_0)Q''(z_0) + ...}$$

i.e., Res[f(z); z_0] = $\dfrac{P(z_0)}{Q'(z_0)}$ (2)

3. Residue at a pole of order m > 1.

If z_0 is a pole of order m for f(z), f(z) can be expanded about z_0 into a Laurent's series as

$$f(z) = \sum_{n=0}^{\infty} a_n (z - z_0)^n + \frac{b_1}{(z - z_0)} + \frac{b_2}{(z - z_0)^2} + \cdots + \frac{b_m}{(z - z_0)^m}$$

$$\therefore \; (z - z_0)^m f(z) = \sum_{n=0}^{\infty} a_n (z - z_0)^{m+n} + b_1 (z - z_0)^{m-1} + b_2 (z - z_0)^{m-2} + \cdots + b_m.$$

Now $\lim_{z \to z_0} \frac{d^{m-1}}{dz^{m-1}}\left((z - z_0)^m f(z)\right) = (m - 1)!.b_1$

$\therefore$ Res[f(z); z_0] = $b_1 = \frac{1}{(m-1)!} \lim_{z \to z_0} \frac{d^{m-1}}{dz^{m-1}}\left((z - z_0)^m f(z)\right)$

Putting $(z - z_0)^m f(z) = \varphi(z)$, we get Res[f(z); z_0] = $\frac{\varphi^{m-1}(z_0)}{(m-1)!}$ (3)

When z_0 is a pole of order 2,

Res[f(z); z_0] = = $\lim_{z \to z_0} \frac{d}{dz}\left((z - z_0)^2 f(z)\right) = \varphi'(z_0)$ (4)

4. Residue at an isolated essential singularity.

If z_0 is an isolated essential singularity of f(z), then there is only one way of computing the residue at z_0. Expand f(z) into a Laurent's series about z_0 and collect the coefficient b_1 of the term $\frac{1}{z - z_0}$. By definition, Res[f(z); z_0] = b_1.

5. Residue at z = ∞

Definition 4.3.9 Let z = ∞ be an isolated singularity of f(z). Then the residue of f(z) at z = ∞ is $-\frac{1}{2\pi i}\int_C f(z)dz$, where C is the circle |z| = r traversed in the anti-clockwise direction such that all the finite singularities of f(z) lie inside C.

Note:

(i) Res[f(z); ∞] = Negative of the coefficient of $\frac{1}{z}$ in the Laurent's expansion of f(z) in the neighbourhood of z = ∞.

(ii) By definition Res[f(z); ∞] = $-\frac{1}{2\pi i}\int_C f(z)dz$ (5)

Put $z = \frac{1}{w}$ so that $dz = -\frac{1}{w^2}dw$

The transformation $z = \frac{1}{w}$ transforms the circle C into a circle γ and changes the orientation.

$$\therefore \text{Res}[f(z); \infty] = -\frac{1}{2\pi i}\int_C f(z)dz$$

$$= \frac{1}{2\pi i}\int_\gamma f(1/w)\left(\frac{-1}{w^2}\right)dw$$

$$= -\frac{1}{2\pi i}\int_\gamma g(w)dw \text{ , where } g(w) = \frac{1}{w^2} f(1/w)$$

$$= -\text{Res}[g(w); 0] \qquad \text{(Since } z = \infty \text{ corresponds to } w = 0\text{)}$$

i.e., Res[f(z); ∞] = -(coefficient of $\frac{1}{w}$ in the Laurent's series expansion of $g(w) = \frac{1}{w^2} f(1/w)$ about $w = 0$)

(iii) A function that is analytic at $z = \infty$ may have a non-zero residue at infinity. For example $f(z) = \frac{1}{z-a}$ is analytic at $z = \infty$. Yet it has residue -1 at infinity.

Example 4.3.5 Find the residues at the poles of $f(z) = \frac{z+1}{z^2 - 2z}$.

Solution:

$$f(z) = \frac{z+1}{z(z-2)}$$

Poles of f(z) are given by $z(z - 2) = 0$ i.e., $z = 0, 2$.
Both the poles are simple poles.

$$\text{Res}[f(z); 0] = \lim_{z\to 0} z.\frac{(z+1)}{z(z-2)} \qquad \text{(Using (1))}$$

$$= \lim_{z\to 0}\frac{z+1}{z-2} = -\tfrac{1}{2}$$

$$\text{Res}[f(z); 2] = \lim_{z\to 2}(z-2).\frac{(z+1)}{z(z-2)} \qquad \text{(Using (1))}$$

$$= \lim_{z\to 2}\frac{z+1}{z} = \tfrac{3}{2}$$

Example 4.3.6 Find the residues at the poles of $f(z) = \frac{z^2}{z^2 + a^2}$.

Solution:

$f(z) = \frac{z^2}{z^2 + a^2} = \frac{z^2}{(z+ia)(z-ia)}$. Hence ia and –ia are simple poles.

$$\text{Res}[f(z); ia] = \lim_{z\to ia}(z-ia).\frac{z^2}{(z+ia)(z-ia)}$$

$$= \lim_{z\to ia}\frac{z^2}{(z+ia)}$$

$$= \frac{i^2a^2}{2ia} = \frac{ia}{2}$$

$$\text{Res}[f(z); -ia] = \lim_{z\to -ia}(z+ia).\frac{z^2}{(z+ia)(z-ia)}$$

$$= \lim_{z\to -ia}\frac{z^2}{(z-ia)}$$

$$= \frac{i^2a^2}{-2ia} = -\frac{ia}{2}$$

Example 4.3.7 Find the residue of $\dfrac{1+e^z}{\sin z + z.\cos z}$ at the pole $z = 0$.

Solution:

Let $f(z) = \dfrac{1+e^z}{\sin z + z.\cos z} = \dfrac{P(z)}{Q(z)}$

$z = 0$ is a simple pole.

$Q(z) = \sin z + z\cos z$

$Q'(z) = \cos z + \cos z + z.(-\sin z)$ $\quad\therefore Q'(0) = 1 + 1 = 2$

$P(z) = 1 + e^z$ $\quad\therefore P(0) = 1 + 1 = 2$

$$\text{Res}[f(z); 0] = \frac{P(0)}{Q'(0)} \qquad \text{(Using (2))}$$

$$= \frac{2}{2} = 1$$

Example 4.3.8 Find the residue of $\dfrac{1-e^{2z}}{z^3}$ at the pole $z = 0$.

Solution:

$$f(z) = \frac{1-e^{2z}}{z^3} = \frac{1}{z^3}\left[1-\left(1+\frac{2z}{1!}+\frac{4z^2}{2!}+\frac{8z^3}{3!}+\cdots\right)\right]$$

$$= \frac{-2}{z^2 1!} - \frac{4}{z.2!} - \frac{8}{3!} - \frac{16.z}{4!} - \cdots$$

$z = 0$ is a pole of order 2 for $f(z)$.

$\text{Res}[f(z); 0] =$ coefficient of $1/z$ (By definition)

$$= \frac{-4}{2!} = -2$$

Example 4.3.9 Find the residue of $f(z) = \frac{1}{(z+1)^3}$ at its pole.

Solution:

$z = -1$ is a pole of order 3.

$(z+1)^3 f(z) = 1$ i.e., $\varphi(z) = 1$

$\text{Res}[f(z); -1] = \frac{1}{2!}\varphi''(-1)$ (Using (3))

Now $\varphi''(z) = 0$ $\therefore \varphi''(-1) = 0$

Hence $\text{Res}[f(z); -1] = 0$

Example 4.3.10 Find the residue of $f(z) = \frac{(z+3)^3}{(z-1)^4}$ at the pole of $z = 1$.

Solution:

$$f(z) = \frac{(z+3)^3}{(z-1)^4}$$

$z = 1$ is a pole of order 4.

Let $f(z) = \frac{1}{(z-1)^4}\varphi(z)$ where $\varphi(z) = (z+3)^3$

Then $\text{Res}[f(z); 1] = \frac{1}{3!}\varphi^{(3)}(1)$ (Using (3))

$\varphi'(z) = 3(z+3)^2$ $\varphi''(z) = 3.2.(z+3)$ $\varphi'''(z) = 3.2.1$

i.e., $\varphi^{(3)}(1) = 6$

$$\therefore \text{Res}[f(z); 1] = \frac{1}{3!}.6 = 1$$

Example 4.3.11 Find the sum of residues at the pole of $\frac{z^2}{(z^2+a^2)^3}$.

Solution:

$$\text{Let } f(z) = \frac{z^2}{(z^2+a^2)^3} = \frac{z^2}{(z+ia)^3(z-ia)^3}$$

ia and –ia are poles of order 3.

$\text{Res}[f(z); ia] = \frac{1}{2!}\varphi''(ia)$ where $\varphi(z) = \frac{z^2}{(z+ia)^3}$

$$\varphi'(z) = \frac{(z+ia)^3.2z - z^2.3(z+ia)^2}{(z+ia)^6} = \frac{(z+ia).2z - 3z^2}{(z+ia)^4} = \frac{2iaz - z^2}{(z+ia)^4}$$

$$\varphi''(z) = \frac{(z+ia)^4(2ia-2z) - (2iaz - z^2)4.(z+ia)^3}{(z+ia)^8}$$

$$= \frac{(z+ia)(2ia-2z)-(2iaz-z^2).4}{(z+ia)^5}$$

$$\varphi''(ia) = \frac{-4\left[2ia.ia-(ia)^2\right]}{(ia+ia)^5} = \frac{-4(ia)^2}{(2ia)^5} = \frac{-4}{32.(ia)^3} = -\frac{1}{8a^3i^3} = \frac{-i}{8a^3}$$

$$\therefore \text{Res}[f(z);\ ia] = \frac{1}{2!}\left(\frac{-i}{8a^3}\right) = \frac{-i}{16a^3}$$

$$\text{Res}[f(z);\ \text{-}ia] = \frac{1}{2!}\varphi''(-ia) \quad \text{where } \varphi(z) = \frac{z^2}{(z-ia)^3}$$

$$\varphi'(z) = \frac{(z-ia)^3.2z - z^2.3(z-ia)^2}{(z-ia)^6} = \frac{-2iaz-z^2}{(z-ia)^4}$$

$$\varphi''(z) = \frac{(z-ia)^4(-2ia-2z)+(2iaz+z^2)\,4.(z-ia)^3}{(z-ia)^8} = \frac{(z-ia)(-2ia-2z)+4.(2iaz+z^2)}{(z-ia)^5}$$

$$\varphi''(\text{-}ia) = \frac{-4\left[-2ia.ia+(ia)^2\right]}{(-2ia)^5} = \frac{-4(ia)^2}{(-2ia)^5} = \frac{-4(ia)^2}{-32(ia)^5} = \frac{1}{8a^3i^3} = \frac{i}{8a^3}$$

$$\therefore \text{Res}[f(z);\ \text{-}ia] = \frac{1}{2!}\left(\frac{i}{8a^3}\right) = \frac{i}{16a^3}$$

Sum of the residues at the poles $= \dfrac{-i}{16a^3} + \dfrac{i}{16a^3} = 0$

Example 4.3.12 Find the residue of $\dfrac{z^3}{z^2-1}$ at $z = \infty$

Solution:

$$\text{Let } f(z) = \frac{z^3}{z^2-1}$$

$$= \frac{1}{1-\frac{1}{z^2}}$$

$$= z\left(1-\frac{1}{z^2}\right)$$

$$= z\left[1+\frac{1}{z^2}+\frac{1}{z^4}+\cdots\right] \text{ for } |z|>1$$

$$= z+\frac{1}{z}+\frac{1}{z^3}+\cdots \qquad (1)$$

(1) is the Laurent's expansion of f(z) about $z = \infty$.

$\therefore$ Res[f(z); ∞] = - coefficient of $\dfrac{1}{z}$ in (1) = -1

Example 4.3.13 Find the residue of $\dfrac{z^3}{(z-1)(z-2)(z-3)}$ at infinity.

Solution:

Let $f(z) = \dfrac{z^3}{(z-1)(z-2)(z-3)}$

$$g(w) = \frac{1}{w^2}.f(1/w) = \frac{1}{w^2}.\frac{1}{(1-w)(1-2w)(1-3w)}$$

$$= \frac{1}{w^2}.(1+w+w^2+\cdots)(1+2w+4w^2+\cdots)(1+3w+9w^2+\cdots) \text{ for } |w|<1$$

$$= \frac{1}{w^2}+\frac{6}{w}+\text{constant} + \text{terms containing powers of w.}$$

$\therefore$ Res[f(z); ∞] = -Res[g(w); 0]

= -coefficient of $\dfrac{1}{w}$ in the expansion of g(w) about w = 0.

= -6.

Example 4.3.14 Find the residue of $f(z) = z\sin(1/z)$ at z = 0.

Solution:

$f(z) = z\sin(1/z)$

$$= z\left[\frac{1}{z}-\frac{1}{3!}\frac{1}{z^3}+\frac{1}{5!}\frac{1}{z^5}-\cdots\right] \text{ for } |z|<1$$

$$= 1-\frac{1}{3!}\frac{1}{z^2}+\frac{1}{5!}\frac{1}{z^4}-\cdots$$

This is the Laplace expansion of f(z) about z = 0 and it contains infinite number of terms in the principal part.

$\therefore$ z = 0 is an essential singularity.

$\therefore$ Res[f(z); 0] = Coefficient of $\dfrac{1}{z}$ = 0

Example 4.3.15 Find the residues at the singularities of $\dfrac{\sin(1/z)}{z-1}$

Solution:

Let $f(z) = \dfrac{\sin(1/z)}{z-1}$

z = -1 is a simple pole.

$$\text{Res}[f(z); 1] = \lim_{z\to 1}(z-1).\frac{\sin(1/z)}{z-1} = \sin 1$$

z = 0 is an essential singularity.

$$f(z) = \frac{\sin(1/z)}{z-1} = -(1-z)^{-1}\sin(1/z)$$

$$= -(1+z+z^2+\ldots)\left(\frac{1}{z} - \frac{1}{3!}\frac{1}{z^3} + \frac{1}{5!}\frac{1}{z^5} + \cdots\right)$$

$$\therefore \text{Res}[f(z); 0] = \text{coefficient of } \frac{1}{z}$$

$$= -\left[1 - \frac{1}{3!} + \frac{1}{5!} + \frac{1}{7!} + \cdots\right]$$

$$= \sum_{n=0}^{\infty} \frac{(-1)^{n+1}}{(2n+1)!}$$

4.3.4 CAUCHY'S RESIDUE THEOREM

In this section we derive an important result on residues, known as Cauchy's residue theorem. This result will be used as a tool for the evaluation of several complex integrals of the form $\int_C f(z)dz$ where C is a simple closed curve in the complex plane. Integrals of the form $\int_C f(z)dz$ are known as contour integrals.

Theorem 4.3.1 (Cauchy's Residue Theorem)

Let f(z) be analytic inside and on a simple closed curve C except at a finite number of isolated singularities $z_1, z_2, \ldots, z_n$ lying inside C.

Then $\int_C f(z)dz = 2\pi i$ [Sum of the residues of f(z) at the isolated singularities]

$$= 2\pi i \sum_{k=1}^{n} \text{Res}[f(z); z_k]$$

Proof:

Since $z_1, z_2, \ldots, z_n$ are isolated singularities, we can find non-intersecting circles $C_1, C_2, \ldots, C_n$ with centres at $z_1, z_2, \ldots, z_n$ such that each circle C_k encloses only one singular point z_k, k = 1, 2, 3, …, n and lies inside C. By Cauchy's integral theorem on multiply connected domains,

$$\int_C f(z)dz = \int_{C_1} f(z)dz + \int_{C_2} f(z)dz + \cdots + \int_{C_n} f(z)dz \qquad (1)$$

Again, by the definition of residues, $\text{Res}[f(z); z_k] = \frac{1}{2\pi i}\int_{C_k} f(z)dz$

i.e., $\int_{C_k} f(z)dz = 2\pi i \,\text{Res}[f(z); z_k] \qquad k = 1, 2, 3, \ldots, n \qquad (2)$

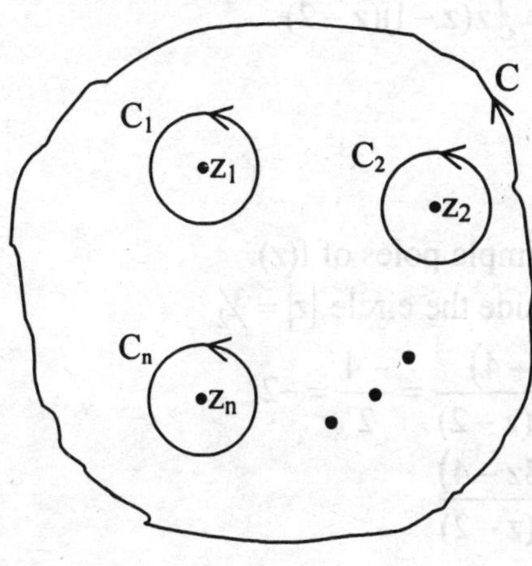

Fig. 4.21

From (1), we get,

$$\int_C f(z)dz = \sum_{k=1}^{n} \int_{C_k} f(z)dz$$

$$= 2\pi i \sum_{k=1}^{n} \text{Res}[f(z); z_k] \qquad \text{(Using (2))}$$

Hence the theorem

Note:

(i) Let C encloses all the isolated singularities $z_1, z_2, \ldots, z_n$ of f(z) in the finite complex plane. Then $\frac{1}{2\pi i}\int_C f(z)dz = \sum_{k=1}^{n} \text{Res}[f(z); z_k]$

Also $\text{Res}[f(z); \infty] = -\frac{1}{2\pi i}\int_C f(z)dz$ (by definition)

Hence $\sum_{k=1}^{n} \text{Res}[f(z); z_k] + \text{Res}[f(z); \infty] = 0$

(ii) If f(z) is analytic inside and on C and a is any point inside C, $\frac{f(z)}{z-a}$ has a simple pole at a and has residue f(a).

$\therefore \int_C \frac{f(z)}{z-a}dz = 2\pi i.f(a)$

i.e., $f(a) = \frac{1}{2\pi i.}\int_C \frac{f(z)}{z-a}dz$ (Cauchy's integral formula)

Example 4.3.16 Evaluate $\int_C \frac{3z-4}{z(z-1)(z-2)} dz$ where C is the circle $|z| = 3/2$

Solution:

Let $f(z) = \frac{3z-4}{z(z-1)(z-2)}$

$z = 0, z = 1$ and $z = 2$ are simple poles of f(z).

Only $z = 0$ and $z = 1$ lie inside the circle $|z| = 3/2$

$$Res[f(z); 0] = \lim_{z \to 0} \frac{z.(3z-4)}{z(z-1)(z-2)} = \frac{-4}{2} = -2.$$

$$Res[f(z); 1] = \lim_{z \to 1} \frac{(z-1).(3z-4)}{z(z-1)(z-2)}$$

$$= \lim_{z \to 1} \frac{.3z-4}{z(z-2)}$$

$$= \frac{3-4}{1-2} = 1$$

$\therefore \int_C f(z)dz = 2\pi i$ [Sum of the residues at the poles inside C]

$= 2\pi i\,[-2 + 1]$

$= -2\pi i.$

Example 4.3.17 Evaluate $\int_C \frac{1}{z^3(z+4)} dz$ where C is the circle $|z + 2| = 3$.

Solution:

Let $f(z) = \frac{1}{z^3(z+4)}$

$z = 0$ is a pole of order 3 and $z = -4$ is a simple pole.

Both the poles lie inside the circle $|z + 2| = 3$.

$Res[f(z); 0] = \frac{\varphi''(0)}{2!}$, where $\varphi(z) = \frac{1}{z+4}$

Now $\varphi'(z) = \frac{-1}{(z+4)^2}$

$$\varphi''(z) = \frac{2}{(z+4)^3} \qquad \varphi''(0) = \frac{2}{4^3} = \frac{1}{32}$$

Hence $Res[f(z); 0] = \frac{1}{2!}.\frac{1}{32} = \frac{1}{64}$

$$\text{Res}[f(z); -4] = \lim_{z \to -4}(z+4).\frac{1}{z^3(z+4)} = \frac{1}{(-4)^3} = \frac{-1}{64}$$

$$\therefore \int_C f(z)dz = 2\pi i \text{ [Sum of the residues at the poles inside C]} = 2\pi i\left[\frac{1}{64} - \frac{1}{64}\right] = 0$$

Example 4.3.18 Evaluate $\int_C ze^{1/z}dz$ where C is the circle $|z| = 5$.

Solution:

Let $f(z) = ze^{1/z}$

$$= z\left[1 + \frac{1}{1!}.\frac{1}{z} + \frac{1}{2!}.\frac{1}{z^2} + \cdots\right]$$

$$= z + \frac{1}{1!} + \frac{1}{2!}.\frac{1}{z} + \frac{1}{2!}.\frac{1}{z^3} + \cdots \qquad (1)$$

This is the Laurent's expansion of f(z) about $z = 0$. Hence $z = 0$ is an essential singularity of f(z) and lies inside $|z| = 5$.

$\text{Res}[f(z); 0]$ = coefficient of $\frac{1}{z}$ in (1) $= \frac{1}{2}$

$$\therefore \int_C ze^{1/z}dz = 2\pi i \text{ [Sum of the residues at the singularities of f(z) inside C]}$$

$$= 2\pi i.\frac{1}{2} = \pi i.$$

Example 4.3.19 Evaluate $\int_C \frac{e^{2z}}{(z+1)^n}dz$ where C is the circle $|z| = 2$.

Solution:

Let $f(z) = \frac{e^{2z}}{(z+1)^n}$.

$z = -1$ is a pole of order n for f(z) and lies inside $|z| = 2$.

$\text{Res}[f(z); -1] = \frac{\varphi^{(n-1)}(-1)}{(n-1)!}$, where $\varphi(z) = e^{2z}$.

$\varphi^{(n-1)}(z) = 2^{n-1}.e^{2z}$

$\therefore \varphi^{(n-1)}(-1) = 2^{n-1}.e^{-2}$.

$$\text{Res}[f(z); -1] = \frac{1}{(n-1)!}2^{n-1}.e^{-2}$$

Hence $\int_C f(z)dz = 2\pi i$ [Sum of the residues at the singularities inside C]

$$= 2\pi i.\frac{1}{(n-1)!}2^{n-1}.e^{-2}$$

$$= \frac{2^n.\pi i}{e^2.(n-1)!}$$

Example 4.3.20 Prove that $\int_C e^{1/z} \sin(1/z) dz = 2\pi i$ where C is the circle $|z| = 1$.

Solution:

Let $f(z) = e^{1/z} \sin(1/z)$

$z = 0$ is an essential singularity and lies inside C.

Now $f(z) = e^{1/z} \sin(1/z)$

$$= \left(1 + \frac{1}{1!}.\frac{1}{z} + \frac{1}{2!}.\frac{1}{z^2} + \cdots\right)\left(\frac{1}{z} - \frac{1}{3!}.\frac{1}{z^3} + \frac{1}{5!}.\frac{1}{z^5} + \cdots\right)$$

$$\text{Res}[f(z); 0] = \text{Coefficient of } \frac{1}{z}$$

$$= 1$$

$\therefore \int_C f(z)dz = 2\pi i$ [Sum of the residues at the singularities inside C]

$$= 2\pi i\,[1]$$

$$= 2\pi i.$$

EXERCISE 4.3

PART-A

1. Define 'singularity' and give an example.
2. Define 'isolated singularity'.
3. Define 'removable singularity' and give an example.
4. Define 'pole' and give an example.
5. Define 'essential singularity' and give an example.
6. Find the zeros and poles of $\frac{(z-1)^2(z-3)}{(z+1)^3}$.
7. Define 'meromorphic function'.
8. State Weiestrass's theorem on essential singularities.
9. Define 'residue' at a singularity.
10. What is the residue of a function at a removable singularity?
11. How will you find residue at a simple pole?
12. How will you find residue at a pole of order m?
13. Define 'residue at infinity'.
14. State Cauchy's residue theorem.
15. Derive Cauchy's integral formula from Cauchy's residue theorem.

Find the singularities of the following functions. (Ex: 16-25)

16. $\frac{\cos z}{z^2}$

17. $\frac{\sin z}{z}$

18. $ze^{1/z}$

19. $\frac{1}{z}.e^z$

20. $\frac{z}{e^z - 1}$

21. $(z-1)\sin\left(\frac{1}{z-1}\right)$

22. $\frac{1-e^{2z}}{z^4}$

23. $\frac{\sin z}{e^z - 1}$

24. $z^2 \cos(1/z)$

25. $\frac{e^{z^2}-1}{z^3}$

Find the residues at the singularities of the following functions. (Ex: 26-30)

26. $\cot z$ at $z = 0$

27. $\frac{ze^z}{(z-1)^3}$ at $z = 1$.

28. $\frac{e^{2z}}{(z-1)^2}$ at $z = 1$.

29. $\frac{\cos z}{z^3}$ at $z = 0$.

30. $\frac{e^z}{z^2 + \pi^2}$ at $-\pi i$.

Evaluate the following integrals using the Cauchy's residue theorem. (Ex: 31-45)

31. $\int_C \frac{2z}{z+1} dz$, where C is $|z| = 2$.

32. $\int_C \cot z\, dz$, where C is $|z| =1$.

33. $\int_C \tan z\, dz$, where C is $|z| =2$.

34. $\int_C \frac{dz}{2z+3}$, where C is $|z| = 2$.

35. $\int_C \frac{e^{-z}dz}{z^2}$, where C is $|z| = 2$.

36. $\int_C \frac{dz}{z^3(z-1)}$, where C is $|z| = 3$.

37. $\int_C \frac{(z+1)\,dz}{z^2 - 2z}$, where C is $|z| = 3$.

38. $\int_C \frac{(z+1)\,dz}{z^2}$, where C is $|z| = 1$.

39. $\int_C \frac{2z\,dz}{e^z - 1}$, where C is $|z| = 1$.

40. $\int_C \frac{z\,dz}{\sin z}$, where C is $|z| = 2$.

41. $\int_C \frac{e^{-z^2}}{\sin 2z} dz$, where C is $|z| = 1$.

42. $\int_C \frac{1-e^{z^2}}{z^3} dz$, where C is $|z-2| = 1$.

43. $\int_C \tanh z\, dz$, where C is $|z| = 4$.

44. $\int_C \tan \pi z\, dz$, where C is $|z| = 1$.

45. $\int_C \frac{dz}{z^2 + 4iz - 1}$, where C is $|z| = 1$.

PART-B

46. Find the residues at the singular points of (i) $\dfrac{z^2+4}{z^3+2z^2+2z}$ (ii) $\dfrac{e^z}{z^2(z^2+9)}$.

47. Find the residues at z = 0 of (i) $\dfrac{1+e^z}{2\cos z+\sin z}$ (ii) $\dfrac{1}{z-\sin z}$.

48. Find the residues at the poles of (i) $\dfrac{1-e^{2z}}{z^4}$ (ii) $\dfrac{2z}{(z+4)(z-1)^2}$.

49. Find the residues of (i) $\dfrac{z^{2n}}{(1+z)^n}$ at z = -1, given n∈ N (ii) $\dfrac{1}{(1+z^2)^n}$ at z = i.

50. Find the singularities and the corresponding residues for
(i) $\dfrac{1}{z^4+16}$ (ii) $\dfrac{z}{(z+1)(z^2-1)}$.

51. Find the singularities and the corresponding residues for
(i) $\dfrac{1}{z^4+a^4}$ (ii) $\dfrac{z^2-2z}{(z+1)^2(z^2+4)}$.

52. Find the residues at the singular points of (i) $\dfrac{1}{z^3+z^5}$ (ii) $\dfrac{z^2}{(z^2+1)^2}$.

53. Find the residue at z = ∞ for the functions (i) $\dfrac{z^2}{z^3+2}$ (ii) $\dfrac{e^z}{z^2(z^2+1)}$.

54. Find the reisdue at z = ∞ for the following functions (i) $\dfrac{z}{e^{-z^2}+1}$ (ii) $z^3\cos(1/z)$.

Evalute the following integrals using Cauchy's residue theorem (Ex: 55-60)

55. $\displaystyle\int_C \frac{dz}{z^3(z+4)}$, where C is |z| = 2.

56. $\displaystyle\int_C \frac{e^z-1}{z(z+1)}dz$, where C is |z| = 2.

57. $\displaystyle\int_C \frac{\sin z}{(z-1)^2(z^2+9)}dz$, where C is |z -3i| = 1.

58. $\displaystyle\int_C \frac{e^{\frac{1}{z-1}}}{(z-2)}dz$, where C is |z| = 4.

59. $\displaystyle\int_C \frac{\sin \pi z}{z^6}dz$, where C is |z| = 1.

60. $\int_C \frac{3z^3+2}{(z-1)(z^2+4)} dz$, where C is $|z - 2| = 2$.

61. $\int_C \frac{e^z dz}{z(z-1)^2}$, where C is $|z| = 2$.

62. $\int_C \frac{(3z^2+z)}{(z-1)(z^2+9)} dz$, where C is $|z - 2| = 2$.

63. $\int_C \frac{1}{(z^2+4)^3} dz$, where C is $|z - i| = 2$.

64. $\int_C \frac{e^z (z^2+4)dz}{(z-i)^3}$, where C is $|z| = 2$.

65. $\int_C \frac{(3+2z)dz}{1+\cos z}$, where C is $|z| = 4$.

4.4 CONTOUR INTEGRATION

A large number of real definite integrals can be evaluated using Cauchy's residue theorem. These real definite integrals, whose evaluation by usual methods is likely to be difficult, can be easily evaluated by converting them to the associated contour integrals. The contour integrals $\int_C f(z)dz$ are evaluated using Cauchy's residue theorem and then the values of the real definite integrals $\int_a^b f(x)dx$ are obtained. This is one of the important applications of Cauchy's residue theorem. In the following sections we discuss the evaluation of only certain types of definite integrals.

4.4.1 EVALUATION OF INTEGRALSOF THE FORM $\int_0^{2\pi} F(\cos\theta, \sin\theta)d\theta$

Consider integrals of the form $I = \int_0^{2\pi} F(\cos\theta, \sin\theta)d\theta$ where $F(\cos\theta, \sin\theta)$ is a rational function of $\cos\theta$ and $\sin\theta$. i.e., $F(\cos\theta, \sin\theta)$ is a quotient of two polynomials in $\cos\theta$ and $\sin\theta$. Substituting $z = e^{i\theta} = \cos\theta + i\sin\theta$, we get $dz = e^{i\theta}.id\theta$ i.e., $d\theta = \frac{dz}{i.e^{i\theta}} = \frac{dz}{iz}$

Then $\cos\theta = \frac{1}{2}(e^{i\theta} + e^{-i\theta}) = \frac{1}{2}\left(z + \frac{1}{z}\right)$ and $\sin\theta = \frac{1}{2}(e^{i\theta} - e^{-i\theta}) = \frac{1}{2i}\left(z - \frac{1}{z}\right)$

As θ varies from 0 to 2π, the variable z traverses the unit circle |z| = 1 once in the anti-clockwise direction.

$$\therefore\ I = \int_0^{2\pi} F(\cos\theta, \sin\theta)d\theta$$

$$= \int_C F\left(\frac{1}{2}\left(z+\frac{1}{z}\right), \frac{1}{2i}\left(z-\frac{1}{z}\right)\right)\frac{dz}{iz}$$

$$= \int_C f(z)dz \text{ where C is } |z| = 1.$$

Now $\int_C f(z)dz$ can be evaluated using Cauchy's residue theorem.

Thus I = 2πi [Sum of the residues at the singularities of f(z) inside C]

Example 4.4.1 Evaluate $\int_0^{2\pi} \frac{d\theta}{2+\cos\theta}$

Solution:

Let $I = \int_0^{2\pi} \frac{d\theta}{2+\cos\theta}$

Put $z = e^{i\theta}$

Then $d\theta = \frac{dz}{iz}$ and $\cos\theta = \frac{1}{2}\left(z+\frac{1}{z}\right)$

$$\therefore\ I = \int_C \frac{dz/iz}{2+\frac{1}{2}\left(z+\frac{1}{z}\right)} \text{ where C is } |z| = 1.$$

$$= \int_C \frac{2dz}{i(4z+z^2+1)}$$

$$= \int_C \frac{-2idz}{z^2+4z+1}$$

$$= \int_C f(z)dz, \qquad \text{where } f(z) = \frac{-2i}{z^2+4z+1}$$

$$= \frac{-2i}{(z+2)^2-3}$$

$$= \frac{-2i}{(z+2-\sqrt{3})(z+2+\sqrt{3})}$$

$-2+\sqrt{3}$ and $-2-\sqrt{3}$ are simple poles of f(z).

$-2+\sqrt{3}$ alone lies inside C.

$$\text{Res}\left[f(z); -2+\sqrt{3}\right] = \lim_{z\to -2+\sqrt{3}} \frac{\left(z+2-\sqrt{3}\right)(-2i)}{\left(z+2-\sqrt{3}\right)\left(z+2+\sqrt{3}\right)}$$

$$= \frac{-2i}{-2+\sqrt{3}+2+\sqrt{3}} = \frac{-i}{\sqrt{3}}$$

$\therefore$ I = $2\pi i$[sum of the residues at the singularities of f(z) inside C]

$$= 2\pi i.\frac{(-i)}{\sqrt{3}} = \frac{2\pi}{\sqrt{3}}$$

Example 4.4.2 Show that $\int_0^{2\pi} \frac{d\theta}{1+a\sin\theta} = \frac{2\pi}{\sqrt{1-a^2}}$ where $-1 < a < 1$

Solution:

Let $I = \int_0^{2\pi} \frac{d\theta}{1+a\sin\theta}$

Put $z = e^{i\theta}$

Then $d\theta = \frac{dz}{iz}$, $\sin\theta = \frac{1}{2i}\left(z - \frac{1}{z}\right)$

$\therefore\ I = \int_C \frac{dz/iz}{1+a.\frac{1}{2i}\left(z-\frac{1}{z}\right)}$ where C is $|z| = 1$.

$$= \int_C \frac{-2dz}{az^2+2iz-a}$$

$$= \int_C f(z)dz, \qquad \text{where } f(z) = \frac{-2}{az^2+2iz-a}$$

Pole of f(z) are $z = \frac{-2i \pm \sqrt{(2i)^2 - 4a.(-a)}}{2a}$

$$= \frac{-i \pm \sqrt{-1+a^2}}{a} = \frac{-i \pm i\sqrt{1-a^2}}{a} \qquad \text{(Since } -1 < a < 1\text{)}$$

Let $z_1 = \frac{-i + i\sqrt{1-a^2}}{a}$ and $z_2 = \frac{-i - i\sqrt{1-a^2}}{a}$

$|z_2| = \frac{1+\sqrt{1-a^2}}{|a|} > 1$ (Since $-1 < a < 1$)

Also $|z_1z_2| = |z_1||z_2| = \frac{1+\sqrt{1-a^2}}{|a|}.\frac{1-\sqrt{1-a^2}}{|a|} = \frac{1-\left(1-a^2\right)}{|a|^2} = \frac{|a|^2}{|a|^2} = 1$

$\therefore$ $|z_2| > 1$ implies $|z_1| < 1$

$\therefore$ z_1 is the only simple pole inside C.

$$\text{Res}[f(z); z_1] = \lim_{z \to z_1}(z - z_1).f(z)$$

$$= \lim_{z \to z_1}(z - z_1).\frac{2}{a(z - z_1)(z - z_2)}$$

$$= \frac{2}{a(z_1 - z_2)}$$

$$= \frac{2}{a\left(\frac{2i\sqrt{1-a^2}}{a}\right)} = \frac{1}{i\sqrt{1 - a^2}}$$

$$\therefore\ I = 2\pi i\, I = 2\pi i\left[\frac{1}{i\sqrt{1-a^2}}\right] = \frac{2\pi}{\sqrt{1-a^2}}$$

Example 4.4.3 Show that $\int_0^{2\pi} \frac{d\theta}{a + b\cos\theta} = \frac{2\pi}{\sqrt{a^2 - b^2}}$ given $a > b > 0$.

Hence deduce that $\int_0^{2\pi} \frac{d\theta}{(a + b\cos\theta)^2} = \frac{2\pi}{(a^2 - b^2)^{3/2}}$

Solution:

Put $z = e^{i\theta}$

Then $d\theta = \frac{dz}{iz}$ and $\cos\theta = \frac{1}{2}\left(z + \frac{1}{z}\right)$

$$I = \int_0^{2\pi} \frac{d\theta}{a + b\cos\theta}$$

$$= \int_C \frac{dz/iz}{a + b\frac{1}{2}\left(z + \frac{1}{z}\right)}, \text{ where C is } |z| = 1.$$

$$= \frac{-2i}{b}\int_C \frac{dz}{z^2 + \frac{2a}{b}z + 1}$$

$$= \int_C f(z)dz, \text{ where } f(z) = \frac{2i/b}{z^2 + \frac{2a}{b}z + 1}$$

Poles of f(z) are $z = \dfrac{-\frac{2a}{b} \pm \sqrt{\frac{4a^2}{b^2} - 4}}{2} = -\dfrac{a}{b} \pm \sqrt{\dfrac{a^2}{b^2} - 1}$

Let $z_1 = -\dfrac{a}{b} + \sqrt{\dfrac{a^2}{b^2} - 1}$ and $z_2 = -\dfrac{a}{b} - \sqrt{\dfrac{a^2}{b^2} - 1}$

Given $a > b > 0$ $\quad\therefore\ \dfrac{a}{b} > 1$ and hence $|z_2| > 1$

Also $|z_1z_2| = |z_1|.|z_2| = \left(\frac{a}{b} - \sqrt{\frac{a^2}{b^2}-1}\right)\left(\frac{a}{b} + \sqrt{\frac{a^2}{b^2}-1}\right) = \frac{a^2}{b^2} - \left(\frac{a^2}{b^2}-1\right) = 1$

$\therefore$ $|z_2| > 1$ implies $|z_1| < 1$. Hence z_1 is the only simple pole inside C.

$$\text{Res}[f(z); z_1] = \lim_{z\to z_1}(z - z_1).f(z)$$

$$= \lim_{z\to z_1}(z - z_1).\frac{(-2i/b)}{(z-z_1)(z-z_2)}$$

$$= \frac{-2i/b}{z_1 - z_2}$$

$$= \frac{-2i/b}{2\sqrt{\frac{a^2}{b^2}-1}} = \frac{-i}{\sqrt{a^2-b^2}}$$

$$\therefore I = 2\pi i\, I = 2\pi i\left[\frac{-i}{\sqrt{a^2-b^2}}\right] = \frac{2\pi}{\sqrt{a^2-b^2}}$$

i.e., $$\int_0^{2\pi}\frac{d\theta}{a + b\cos\theta} = \frac{2\pi}{\sqrt{a^2-b^2}}$$

Differentiating both sides partially w.r.t a, we get,

$$-\int_0^{2\pi}\frac{d\theta}{(a+b\cos\theta)^2} = \frac{-2\pi.2a}{2.(a^2-b^2)^{3/2}}$$

i.e., $$\int_0^{2\pi}\frac{d\theta}{(a+b\cos\theta)^2} = \frac{2\pi a}{(a^2-b^2)^{3/2}}$$

Exercise 4.4.4 Prove that $\int_0^{2\pi}\frac{d\theta}{a^2 - 2a\cos\theta + 1} = \frac{2\pi}{1-a^2}$, given $a^2 < 1$.

Solution:

Let $I = \int_0^{2\pi}\frac{d\theta}{a^2 - 2a\cos\theta + 1}$

Put $z = e^{i\theta}$

Then $d\theta = \frac{dz}{iz}$ and $\cos\theta = \frac{1}{2}\left(z + \frac{1}{z}\right)$

$$\therefore I = \int_C \frac{dz/iz}{a^2 - a\left(z+\frac{1}{z}\right)+1}, \text{ where C is } |z| = 1.$$

$$= \frac{1}{ai}\int_C \frac{dz}{\left(a+\frac{1}{a}\right)z - z^2 - 1}$$

$$= \frac{i}{a}\int_C \frac{dz}{z^2 - \left(a + \frac{1}{a}\right)z + 1}$$

$$= \int_C f(z)dz\text{, where } f(z) = \left(\frac{i}{a}\right)\frac{1}{z^2 - \left(a + \frac{1}{a}\right)z + 1}$$

$$= \left(\frac{i}{a}\right)\frac{1}{(z-a)\left(z - \frac{1}{a}\right)}$$

The singularities of f(z) are simple poles at a and $\frac{1}{a}$. $a^2 < 1$ implies $|a| < 1$ and $\frac{1}{|a|} > 1$

$\therefore$ The pole that lies inside C is z = a.

$$\text{Res}[f(z); a] = \lim_{z \to a}(z-a).\left(\frac{i}{a}\right)\frac{1}{(z-a)\left(z-\frac{1}{a}\right)}$$

$$= \left(\frac{i}{a}\right)\frac{1}{\left(a - \frac{1}{a}\right)} = \frac{i}{a^2 - 1}$$

$$\text{Hence } I = 2\pi i.\frac{i}{a^2-1} = \frac{2\pi}{1-a^2}$$

Example 4.4.5 Show that $\int_0^{2\pi} \frac{\cos 2\theta . d\theta}{5 + 4\cos\theta} = \frac{\pi}{6}$

Solution:

Let $I = \int_0^{2\pi} \frac{\cos 2\theta . d\theta}{5 + 4\cos\theta}$ Put $z = e^{i\theta}$

Then $d\theta = \frac{dz}{iz}$ and $\cos\theta = \frac{1}{2}\left(z + \frac{1}{z}\right)$

$$I = \text{Real Part of } \int_0^{2\pi} \frac{e^{i2\theta}.d\theta}{5 + 4\cos\theta}$$

$$= \text{Real Part of } \int_C \frac{z^2 . dz/iz}{5 + 2\left(z + \frac{1}{z}\right)}\text{, where C is } |z| = 1.$$

$$= \text{Real Part of } \frac{1}{2i}\int_C \frac{z^2 . dz}{z^2 + \frac{5}{2}z + 1}$$

$$= \text{Real Part of } \frac{1}{2i}\int_C \frac{z^2 . dz}{\left(z + \frac{1}{2}\right)(z+2)}$$

$$= \text{Real Part of } \int_C f(z)dz\text{, where } f(z) = \frac{1}{2i}.\frac{z^2}{\left(z + \frac{1}{2}\right)(z+2)}$$

$z = -\frac{1}{2}$ and $z = -2$ are simple poles of f(z).

$z = -\frac{1}{2}$ lies inside C.

$$\text{Res}[f(z); -\tfrac{1}{2}] = \lim_{z \to -\frac{1}{2}} (z + \tfrac{1}{2}).\frac{1}{2i}.\frac{z^2}{\left(z+\frac{1}{2}\right)(z+2)}$$

$$= \frac{1}{2i}.\frac{\frac{1}{4}}{\frac{3}{2}} = \frac{1}{12i}$$

$$\therefore \text{I} = \text{Real Part of } 2\pi i.\frac{1}{12i}$$

$$= \text{Real Part of } \pi/6$$

$$= \pi/6.$$

Example 4.4.6 Evaluate $\int_0^{2\pi} \frac{\sin n\theta}{1 + 2a\cos\theta + a^2} d\theta$ and $\int_0^{2\pi} \frac{\cos n\theta}{1 + 2a\cos\theta + a^2} d\theta$ given $a^2 < 1$ and n is a positive integer.

Solution:

Let $\text{I} = \int_0^{2\pi} \frac{e^{in\theta} d\theta}{1 + 2a\cos\theta + a^2}$ Put $z = e^{i\theta}$

Then $d\theta = \frac{dz}{iz}$, $\cos\theta = \frac{1}{2}\left(z + \frac{1}{z}\right)$

$$\therefore \text{I} = \int_C \frac{z^n . dz/iz}{1 + a\left(z + \frac{1}{z}\right) + a^2}, \text{ where C is } |z| = 1.$$

$$= \frac{1}{ia}\int_C \frac{z^n dz}{(z+a)\left(z+\frac{1}{a}\right)}$$

$$= \int_C f(z)dz, \text{ where } f(z) = \frac{1}{ia}.\frac{z^n}{(z+a)\left(z+\frac{1}{a}\right)}$$

Poles of f(z) are $z = -a$ and $z = -\frac{1}{a}$

$z = -a$ is the only pole inside C (since $|a| < 1$) and its is a simple pole.

$$\text{Res}[f(z); -a] = \lim_{z \to -a} (z + a).\frac{1}{ia}.\frac{z^n}{(z+a)\left(z+\frac{1}{a}\right)}$$

$$= \frac{(-a)^n}{ia.\left(-a+\frac{1}{a}\right)} = \frac{(-1)^n.a^n}{i.(1-a^2)}$$

Hence $I = 2\pi i.\dfrac{(-1)^n.a^n}{i.(1-a^2)} = \dfrac{2\pi(-1)^n.a^n}{1-a^2}$

$$\int_0^{2\pi} \frac{\cos n\theta}{1+2a\cos\theta+a^2}\,d\theta = \text{Real Part of } I = \frac{2\pi(-1)^n.a^n}{1-a^2}.$$

$$\int_0^{2\pi} \frac{\sin n\theta}{1+2a\cos\theta+a^2}\,d\theta = \text{Imaginary Part of } I = 0.$$

Example 4.4.7 Prove that $\displaystyle\int_0^{2\pi} \frac{\sin^2\theta}{a+b\cos\theta}\,d\theta = \frac{2\pi}{b^2}\left[a-\sqrt{a^2-b^2}\right]$ where $a > b > 0$.

Solution:

Let $I = \displaystyle\int_0^{2\pi} \frac{\sin^2\theta}{a+b\cos\theta}\,d\theta$ Put $z = e^{i\theta}$

Then $d\theta = \dfrac{dz}{iz}$, $\sin\theta = \dfrac{1}{2i}\left(z-\dfrac{1}{z}\right)$, $\cos\theta = \dfrac{1}{2}\left(z+\dfrac{1}{z}\right)$

$$\therefore\ I = \int_C \frac{\left[\frac{1}{2i}\left(z-\frac{1}{z}\right)\right]^2 \frac{dz}{iz}}{a+b.\frac{1}{2}\left(z+\frac{1}{z}\right)}, \text{ where C is } |z| = 1.$$

$$= \int_C \frac{-1}{2i}\frac{(z^2-1)^2\,dz}{z^2(2az+bz^2+b)}$$

$$= \frac{-1}{2bi}\int_C \frac{(z^2-1)^2\,dz}{z^2\left(z^2+\frac{2a}{b}z+1\right)}$$

$= \displaystyle\int_C f(z)dz$, where $f(z) = \dfrac{-1}{2bi}\dfrac{(z^2-1)^2}{z^2(z-\alpha)(z-\beta)}$ and α, β are roots of $z^2+\frac{2a}{b}z+1 = 0$

i.e., $\alpha = -\dfrac{a}{b}+\sqrt{\dfrac{a^2}{b^2}-1}$, $\beta = -\dfrac{a}{b}-\sqrt{\dfrac{a^2}{b^2}-1}$

$f(z)$ has a pole of order two at $z = 0$ and simple poles at $z = \alpha$ and $z = \beta$.

Since $a > b$, we have $\dfrac{a}{b} > 1$

$\therefore$ $|\beta| > 1$ and $|\alpha\beta| = 1$ implies $|\alpha| < 1$.
Hence the poles inside C are $z = 0$ and $z = \alpha$.

$$\text{Res}[f(z);\alpha] = \lim_{z\to\alpha}(z-\alpha).\left(\frac{-1}{2bi}\right).\frac{(z^2-1)^2}{z^2(z-\alpha)(z-\beta)}$$

$$= \left(\frac{-1}{2bi}\right).\frac{(\alpha^2-1)^2}{\alpha^2(\alpha-\beta)}$$

$$= \frac{-1}{2bi} \cdot \frac{\left(\alpha - \frac{1}{\alpha}\right)^2}{(\alpha - \beta)}$$

$$= \frac{-1}{2bi} \cdot \frac{(\alpha - \beta)^2}{(\alpha - \beta)} \qquad \text{(Since } \alpha\beta = 1)$$

$$= \frac{-1}{2bi} . (\alpha - \beta)$$

$$= \frac{-1}{2bi} . 2\sqrt{\frac{a^2}{b^2} - 1}$$

$$= \frac{-1}{b^2 i} . \sqrt{a^2 - b^2}$$

$$\text{Res}[f(z); 0] = \frac{\varphi'(0)}{1!} \quad \text{where } \varphi(z) = \frac{-1}{2bi} . \frac{(z^2 - 1)^2}{(z - \alpha)(z - \beta)}$$

$$\text{Now } \varphi'(z) = \left(\frac{-1}{2bi}\right) . \frac{(z - \alpha)(z - \beta).2(z^2 - 1)2z - (z^2 - 1)^2 [z - \alpha + z - \beta]}{(z - \alpha)^2 (z - \beta)^2}$$

$$\varphi'(0) = \frac{-1}{2bi} . \frac{\alpha + \beta}{\alpha^2 \beta^2} = \frac{-1}{2bi} . \frac{\left(-{}^{2a}/_{b}\right)}{1} = \frac{a}{ib^2} \qquad \text{(Since } \alpha + \beta = -{}^{2a}/_{b} \text{ and } \alpha\beta = 1)$$

$$\text{Hence } I = 2\pi i \left[\frac{-1}{ib^2} \sqrt{a^2 - b^2} + \frac{a}{ib^2}\right]$$

$$= \frac{2\pi}{b^2} \left[a - \sqrt{a^2 - b^2}\right]$$

Example 4.4.8 Prove $\int_0^{2\pi} \frac{\cos^3 3\theta}{5 - 4\cos 2\theta} d\theta = \frac{3\pi}{8}$

Solution:

$$\text{Let } I = \int_0^{2\pi} \frac{\cos^3 3\theta}{5 - 4\cos 2\theta} d\theta$$

$$= \frac{1}{2} \int_0^{2\pi} \frac{(1 + \cos 6\theta)}{5 - 4\cos 2\theta} d\theta$$

$$= \text{Real Part of } \frac{1}{2} \int_0^{2\pi} \frac{(1 + e^{i6\theta})}{5 - 4\cos 2\theta} d\theta$$

Put $z = e^{i\theta}$ Then $d\theta = \frac{dz}{iz}$, $e^{i6\theta} = z^6$, $\cos 2\theta = \frac{1}{2}\left(z^2 + \frac{1}{z^2}\right)$

$$\therefore \qquad I = \text{Real Part of } \int_C \frac{(1 + z^6).{}^{dz}/_{iz}}{5 - 4.\frac{1}{2}\left(z^2 + \frac{1}{z^2}\right)}$$

$$= \text{Real Part of } \frac{1}{2i} \int_C \frac{z(1+z^6).dz}{5z^2 - 2(z^4+1)}$$

$$= \text{Real Part of } \int_C f(z)dz \text{, where C is } |z| = 1.$$

$$f(z) = \frac{1}{2i} \cdot \frac{z(1+z^6)}{-2z^4 + 5z^2 - 2}$$

$$= \frac{1}{-4i} \cdot \frac{z(1+z^6)}{\left(z^4 - \frac{5}{2}z^2 + 1\right)}$$

$$= -\frac{1}{4i} \cdot \frac{z(1+z^6)}{(z^2-2)\left(z^2 - \frac{1}{2}\right)}$$

Poles of f(z) are $z = \pm\sqrt{2},\ z = \pm 1/\sqrt{2}$

The simple poles lying inside C are $z = 1/\sqrt{2}$ and $z = -1/\sqrt{2}$.

$$\text{Res}\left[f(z); 1/\sqrt{2}\right] = \lim_{z \to 1/\sqrt{2}} \left(-\frac{1}{4i}\right) \cdot \frac{\left(z - \frac{1}{\sqrt{2}}\right) z(1+z^6)}{(z^2-2)\left(z - \frac{1}{\sqrt{2}}\right)\left(z + \frac{1}{\sqrt{2}}\right)}$$

$$= \left(-\frac{1}{4i}\right) \cdot \frac{\frac{1}{\sqrt{2}} \cdot \left(1 + \frac{1}{8}\right)}{\left(\frac{1}{2} - 2\right)\left(\frac{1}{\sqrt{2}} + \frac{1}{\sqrt{2}}\right)}$$

$$= \left(-\frac{1}{4i}\right) \cdot \frac{\cdot\left(\frac{9}{8}\right)}{\left(-\frac{3}{2}\right)2} = \frac{3}{32i}$$

$$\text{Res}\left[f(z); -1/\sqrt{2}\right] = \lim_{z \to -1/\sqrt{2}} \left(-\frac{1}{2i}\right) \cdot \frac{\left(z + \frac{1}{\sqrt{2}}\right) z(1+z^6)}{(z^2-2)\left(z - \frac{1}{\sqrt{2}}\right)\left(z + \frac{1}{\sqrt{2}}\right)}$$

$$= \left(-\frac{1}{4i}\right) \cdot \frac{-\frac{1}{\sqrt{2}} \cdot \left(1 + \frac{1}{8}\right)}{\left(\frac{1}{2} - 2\right)\left(-\frac{1}{\sqrt{2}} - \frac{1}{\sqrt{2}}\right)}$$

$$= \left(-\frac{1}{4i}\right) \cdot \frac{(-1) \cdot \left(\frac{9}{8}\right)}{\left(-\frac{3}{2}\right)(-2)} = \frac{3}{32i}$$

$$\therefore I = \text{Real Part of } 2\pi i \left[\frac{3}{32i} + \frac{3}{32i}\right]$$

$$= \text{Real Part of } 2\pi i \frac{3}{16i}$$

$$= \text{Real Part of } \frac{3\pi}{8} = \frac{3\pi}{8}.$$

Example 4.4.9 Prove $\int_0^{\pi}\frac{a\,d\theta}{a^2+\sin^2\theta}=\frac{\pi}{\sqrt{1+a^2}}$, $a>0$

Solution:

Let $I=\int_0^{\pi}\frac{a\,d\theta}{a^2+\sin^2\theta}=\int_0^{\pi}\frac{a\,d\theta}{a^2+\frac{1}{2}(1-\cos 2\theta)}=\int_0^{\pi}\frac{2a\,d\theta}{2a^2+1-\cos 2\theta}$

Put $2\theta = t$, then $2d\theta = dt$

$\therefore\ I=\int_0^{2\pi}\frac{a\,dt}{2a^2+1-\cos t}$ Put $z=e^{it}$

Then $dt=\frac{dz}{iz}$, $\cos t=\frac{1}{2}\left(z+\frac{1}{z}\right)$

$\therefore\ I=\int_C\frac{a.{}^{dz}/_{iz}}{2a^2+1-\frac{1}{2}\left(z+\frac{1}{z}\right)}$, where C is $|z|=1$

$$=\frac{2a}{-i}\int_C\frac{dz}{z^2-2(2a^2+1)z+1}$$

$$=\int_C f(z)dz,\ \text{where}\ f(z)=-\frac{2a}{i}.\frac{1}{z^2-2(2a^2+1)z+1}$$

Poles of f(z) are given by

$z^2-2(2a^2+1)z+1=0$

i.e., $z=(2a^2+1)\pm\sqrt{(2a^2+1)^2-1}$

$=(2a^2+1)\pm 2a\sqrt{a^2+1}$

Let $\alpha=(2a^2+1)+2a\sqrt{a^2+1}$ $\quad\beta=(2a^2+1)-2a\sqrt{a^2+1}$

Then $a>0$ impiles $|\alpha|>1$ also $|\alpha\beta|=1$ implies $|\beta|<1$. Hence β is the only pole inside C and it is a simple pole.

$$\text{Res}[f(z);\beta]=\lim_{z\to\beta}(z-\beta)\left(-\frac{2a}{i}\right).\frac{1}{(z-\alpha)(z-\beta)}$$

$$=\left(-\frac{2a}{i}\right).\frac{1}{(\beta-\alpha)}$$

$$=\frac{2a}{i}\frac{1}{4a\sqrt{a^2+1}}=\frac{1}{2i\sqrt{a^2+1}}$$

$\therefore I=2\pi i$[sum of the residues at the poles inside C]

$$=2\pi i.\frac{1}{2i\sqrt{a^2+1}}=\frac{\pi}{\sqrt{a^2+1}}$$

Example 4.4.10 Evaluate $\int_0^{2\pi} e^{\cos\theta}\cos(n\theta - \sin\theta)\,d\theta$

Solution:

Let $I = \int_0^{2\pi} e^{\cos\theta}\cos(n\theta - \sin\theta)\,d\theta$

$$= \int_0^{2\pi} e^{\cos\theta}\frac{1}{2}\left[e^{i(n\theta-\sin\theta)} + e^{-i(n\theta-\sin\theta)}\right]d\theta$$

$$= \frac{1}{2}\int_0^{2\pi}\left[e^{in\theta}e^{\cos\theta - i\sin\theta} + e^{-in\theta}e^{\cos\theta + i\sin\theta}\right]d\theta$$

$$= \frac{1}{2}\int_0^{2\pi}\left[e^{in\theta}e^{e^{-i\theta}} + e^{-in\theta}e^{e^{i\theta}}\right]d\theta \qquad \text{Put } z = e^{i\theta}, \text{ then } d\theta = \frac{dz}{iz}$$

$$\therefore I = \frac{1}{2i}\int_C (z^n e^{1/z} + z^{-n}e^z)\frac{dz}{z}, \text{ where C is } |z| = 1$$

$$= \frac{1}{2i}\int_C z^{n-1}e^{1/z}dz + \frac{1}{2i}\int_C \frac{e^z}{z^{n+1}}dz \;= I_1 + I_2$$

$I_1 = \frac{1}{2i}\int_C z^{n-1}e^{1/z}dz = \int_C f(z)dz$, where $f(z) = \frac{1}{2i}z^{n-1}e^{1/z}$ has an essential singularity at $z = 0$.

$$\frac{1}{2i}z^{n-1}e^{1/z} = \frac{1}{2i}z^{n-1}\left[1 + \frac{1}{z} + \frac{1}{2!}\frac{1}{z^2} + \frac{1}{3!}\frac{1}{z^3} + \cdots + \frac{1}{n!}\frac{1}{z^n} + \cdots\right]$$

$\text{Res}[f(z); 0] = \text{Coefficent of } \frac{1}{z} = \frac{1}{2i\,n!}$

$\therefore\; I_1 = 2\pi i.\frac{1}{2i\,n!} = \frac{\pi}{n!}$

$I_2 = \int_C g(z)dz$, where $g(z) = \frac{1}{2i}\frac{e^z}{z^{n+1}}$ has a pole of order $n + 1$ at $z = 0$

$\text{Res}[g(z);0] = \frac{1}{n!}\varphi^{(n)}(0)$, where $\varphi(z) = \frac{e^z}{2i}$

$\varphi^{(n)}(z) = \frac{e^z}{2i} \qquad \therefore \varphi^{(n)}(0) = \frac{1}{2i}$

Hence $\text{Res}[g(z);0] = \frac{1}{2i.n!}$ And $I_2 = 2\pi i.\frac{1}{2i.n!} = \frac{\pi}{n!}$

$\therefore\; I = I_1 + I_2 = \frac{\pi}{n!} + \frac{\pi}{n!} = \frac{2\pi}{n!}$

4.4.2 EVALUATION OF INTEGRALS OF THE FORM $\int_{-\infty}^{\infty} f(x)dx$

In this section we consider the evaluation of only those improper integrals of the form $\int_{-\infty}^{\infty} f(x)dx$ satisfying the following conditions.

(i) f(x) is a rational function of x.

i.e., $f(x) = \dfrac{g(x)}{h(x)}$ where g(x) and h(x) are polynomials in x

(ii) g(x) and h(x) have no common factors and the degree of h(x) exceeds that of g(x) by atleast two.

(iii) h(x) does not vanish for any real value.

1. CAUCHY'S LEMMA

If f(z) is a continuous function such that $|zf(z)| \to 0$ uniformly as $|z| \to \infty$ on C_R, then $\int_{C_R} f(z)dz \to 0$ as $R \to \infty$, where C_R is the semicircle $|z| = R$, $Im(z) \geq 0$.

Proof:

On the semicircle $|z| = R$, $Im(z) \geq 0$ we have $z = Re^{i\theta}$, $0 \leq \theta \leq \pi$.
Then $dz = Re^{i\theta} i d\theta$ and $|dz| = Rd\theta$.

$$\therefore \left| \int_{C_R} f(z)dz \right| = \left| \int_{C_R} z.f(z).\frac{1}{z} dz \right|$$

$$\leq \int_{C_R} |z.f(z)|.\frac{1}{|z|}|dz| \qquad (1)$$

Since $|zf(z)| \to 0$ uniformly as $R \to \infty$, $0 \leq \theta \leq \pi$, given $\in > 0$ we can find a real number R^* sufficiently large such that $|zf(z)| < \in$ for $R > R^*$
Hence (1) becomes,

$$\left| \int_{C_R} f(z)dz \right| < \left| \int_0^{\pi} \in .\frac{1}{R}.Rd\theta \right| < \in\pi \quad \text{for } R > R^*$$

Since $\in$ is arbitrary, we get $\int_{C_R} f(z)dz \to 0$ as $R \to \infty$.

i.e., $\lim\limits_{R\to\infty} \int_{C_R} f(z)dz = 0$ Hence the lemma

2. To evaluate $\int_{-\infty}^{\infty} f(x)dx$, consider the contour integral $\int_C f(z)dz$, where $C = [-R, R] \cup C_R$ is traversed in the anti-clockwise direction as shown in Fig. 4.22

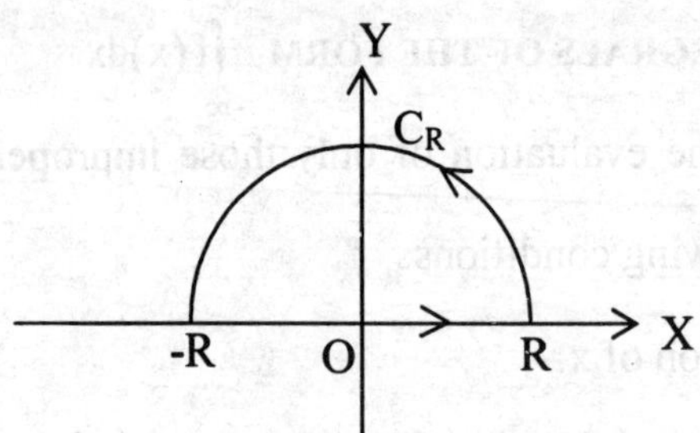

Fig. 4.22

By assumption $f(z) = \dfrac{g(z)}{h(z)}$ where

(i) g(z) and h(z) are polynomials in z.

(ii) g(z) and h(z) have no common factors and the degree of h(z) exceeds that of g(z) by atleast two.

(iii) h(z) does not vanish on the real axis.
i.e., f(z) has no poles on the real axis.

$$\int_C f(z)dz = \int_{-R}^{R} f(x)dx + \int_{C_R} f(z)dz \qquad \text{(Since } z = x \text{ on the real axis)}$$

When $R \to \infty$, we get,

$$\lim_{R\to\infty} \int_C f(z)dz = \int_{-\infty}^{\infty} f(x)dx + \lim_{R\to\infty} \int_{C_R} f(z)dz \tag{2}$$

Now $z.f(z) = \dfrac{z.g(z)}{h(z)}$ and degree of h(z) exceeds that of z.g(z) by atleast one.

$\therefore |z.f(z)| = \left|\dfrac{z.g(z)}{h(z)}\right| \to 0$ uniformly as $R \to \infty$

By Cauchy's Lemma, we have $\lim\limits_{R\to\infty} \int_{C_R} f(z)dz = 0$ (3)

Also by Cauchy's residue theorem,

$$\lim_{R\to\infty} \int_C f(z)dz = 2\pi i\{\text{Sum of the residues at the poles of } f(z) \text{ lying in the upper half plane}\} \tag{4}$$

Using (3) and (4) in (2) we get,

$$\int_{-\infty}^{\infty} f(x)dx = 2\pi i \text{ \{Sum of the residues at the poles of } f(z) \text{ lying in the upper half plane \}}$$

Example 4.4.11 Use the method of contour integration to

provethat $\int_{-\infty}^{\infty}\frac{x^2dx}{(x^2+1)(x^2+4)}=\frac{\pi}{3}$.

Solution:

Let $I=\int_{-\infty}^{\infty}\frac{z^2dz}{(z^2+1)(z^2+4)}=\int_C f(z)dz$, where the contour $C = [-R, R] \cup C_R$ and

$$f(z)=\frac{z^2}{(z^2+1)(z^2+4)}$$

$\int_{-\infty}^{\infty} f(x)dx+\lim_{R\to\infty}\int_{C_R} f(z)dz = 2\pi i$ {Sum of the residues at the poles of f(z) lying above the real axis} (1)

Now $z.f(z)=\frac{z^3}{(z^2+1)(z^2+4)}$

On $|z| = R$ we have $|z^2 + 1| \geq |z|^2 - 1 = R^2 - 1$ and $|z^2 + 4| \geq |z|^2 - 4 \geq R^2-4$.

$$\therefore |z.f(z)|=\frac{|z^3|}{|z^2+1||z^2+4|}$$

$$\leq\frac{R^3}{(R^2-1)(R^2-4)}$$

$$\leq\frac{1}{R\left(1-\frac{1}{R^2}\right)\left(1-\frac{4}{R^2}\right)} \to 0 \text{ as } R\to\infty$$

Hence by Cauchy's lemma we have $\lim_{R\to\infty}\int_{C_R} f(z)dz=0$.

The poles of f(z) are given by $(z^2 + 1)(z^2 + 4) = 0$

i.e., $z = \pm i, \pm 2i$.

$z = i$ and $z = 2i$ are the simple poles of f(z) in theupper half plane.

$$\text{Res}[f(z);i]=\lim_{z\to i}(z-i).\frac{z^2}{(z-i)(z+i).(z^2+4)}=\frac{i^2}{2i.(i^2+4)}=\frac{-1}{6i}$$

$$\text{Res}[f(z);2i]=\lim_{z\to 2i}(z-2i).\frac{z^2}{(z^2+1)(z-2i)(z+2i)}=\frac{(2i)^2}{((2i)^2+1)4i}=\frac{-4}{(-3).(4i)}=\frac{1}{3i}$$

Hence equation (1) becomes,

$$\int_{-\infty}^{\infty} f(x)dx=2\pi i\left[\frac{-1}{6i}+\frac{1}{3i}\right]=2\pi i\frac{1}{6i}=\frac{\pi}{3}$$

i.e., $\int_{-\infty}^{\infty}\frac{x^2dx}{(x^2+1)(x^2+4)}=\frac{\pi}{3}$.

Example 4.4.12 Prove $\int_0^{\infty}\frac{x^2dx}{(x^2+a^2)^3}=\frac{\pi}{16a^3}$, $a>0$.

Solution:

Let $I=\int_C f(z)dz$, where $C=[-R,R]\cup C_R$ and $f(z)=\frac{z^2}{(z^2+a^2)^3}$

The poles of f(z) are given by $(z^2+a^2)^3=0$. i.e., $z=\pm\, ia$ are poles of order 3.
$z=ia$ is the only pole of f(z) lying above the real axis (since $a>0$) and it is of order 3.

$$\text{Res}[f(z);ia]=\frac{\varphi''(ia)}{2!},\ \text{where}\ \varphi(z)=\frac{z^2}{(z+ia)^3}$$

$$\text{Then}\ \varphi'(z)=\frac{(z+ia)^3.2z-z^2.3(z+ia)^2}{(z+ia)^6}=\frac{(z+ia).2z-3z^2}{(z+ia)^4}=\frac{ia.2z-z^2}{(z+ia)^4}$$

$$\varphi''(z)=\frac{(z+ia)^4(2ia-2z)-(ia.2z-z^2).4(z+ia)^3}{(z+ia)^8}$$

$$\varphi''(ia)=\frac{[ia.2ia-(ia)^2].4(2ia)^3}{(2ia)^8}=\frac{-4.8.(ia)^5}{2^8.(ia)^8}=\frac{1}{8ia^3}$$

$$\therefore\ \text{Res}[f(z);ia]=\frac{1}{2!}.\frac{1}{8ia^3}=\frac{1}{16ia^3}$$

$$\text{Now}\ I=\int_C f(z)dz=\int_{-R}^{R}f(x)dx+\int_{C_R}f(z)dz \qquad (1)$$

$$\therefore\ |z.f(z)|=\left|z.\frac{z^2}{(z^2+a^2)^3}\right|\to 0\ \text{as}\ R\to\infty$$

$$\therefore\ \text{By Cauchy's lemma,}\ \lim_{R\to\infty}\int_{C_R}f(z)dz=0$$

When $R\to\infty$, (1) becomes, $\int_{-\infty}^{\infty}f(x)dx=2\pi i\left\{\frac{1}{16ia^3}\right\}$

$$\text{i.e.,}\ \int_{-\infty}^{\infty}\frac{x^2dx}{(x^2+a^2)^3}=\frac{\pi}{8a^3}$$

$$\text{i.e.,}\ 2\int_0^{\infty}\frac{x^2dx}{(x^2+a^2)^3}=\frac{\pi}{8a^3}\qquad\left(\text{Since}\ f(x)=\frac{x^2}{(x^2+a^2)^3}\ \text{is an even function}\right)$$

$$\text{i.e.,}\ \int_0^{\infty}\frac{x^2dx}{(x^2+a^2)^3}=\frac{\pi}{16a^3}$$

Example 4.4.13 Prove that $\int_{-\infty}^{\infty}\frac{x^2dx}{x^6+1}=\frac{\pi}{3}$.

Solution:

Let $I=\int_C f(z)dz$ where $C=[-R, R]\cup C_R$ and $f(z)=\frac{z^2}{z^6+1}$

The poles of f(z) are given by $z^6+1=0$

i.e., $z=(-1)^{1/6}$

$$=(\cos\pi+i\sin\pi)^{1/6}$$

$$=\cos\frac{2n\pi+\pi}{6}+i\sin\frac{2n\pi+\pi}{6}\text{, where } n = 0, 1, 2, 3, 4, 5.$$

$n=0,\ z_0=\cos\pi/6+i\sin\pi/6=e^{i\pi/6}$

$n=1,\ z_1=\cos 3\pi/6+i\sin 3\pi/6=e^{i3\pi/6}$

$n=2,\ z_2=\cos 5\pi/6+i\sin 5\pi/6=e^{i5\pi/6}$

$n=3,\ z_3=\cos 7\pi/6+i\sin 7\pi/6=\cos 5\pi/6-i\sin 5\pi/6=e^{-i5\pi/6}$

$n=4,\ z_4=\cos 9\pi/6+i\sin 9\pi/6=\cos 3\pi/6-i\sin 3\pi/6=e^{-i3\pi/6}$

$n=5,\ z_5=\cos 11\pi/6+i\sin 11\pi/6=\cos\pi/6-i\sin\pi/6=e^{-i\pi/6}$

The poles of f(z) lying above the real axis are $e^{i\pi/6}, e^{i3\pi/6}, e^{i5\pi/6}$ and these are simple poles.

$$\text{Res}\left[f(z);e^{i\pi/6}\right]=\frac{P\left(e^{i\pi/6}\right)}{Q'\left(e^{i\pi/6}\right)}$$, where $P(z) = z^2$, $Q(z) = z^6 + 1$ and $Q'(z) = 6z^5$.

$$=\frac{1}{6\left(e^{i\pi/6}\right)^3}=\frac{1}{6e^{i3\pi/6}}=\frac{1}{6i}$$

$$\text{Res}\left[f(z);e^{i3\pi/6}\right]=\frac{P\left(e^{i3\pi/6}\right)}{Q'\left(e^{i3\pi/6}\right)}=\frac{1}{6\left(e^{i3\pi/6}\right)^3}=\frac{1}{6i^3}=\frac{-1}{6i}$$

$$\text{Res}\left[f(z);e^{i5\pi/6}\right]=\frac{P\left(e^{i5\pi/6}\right)}{Q'\left(e^{i5\pi/6}\right)}$$

$$= \frac{1}{6\left(e^{i5\pi/6}\right)^3} = \frac{1}{6e^{i15\pi/6}} = \frac{1}{6e^{i3\pi/6}} = \frac{1}{6i}$$

$\therefore$ Sum of the residues at the poles of f(z) lying above the real axis $= \frac{1}{6i} - \frac{1}{6i} + \frac{1}{6i} = \frac{1}{6i}$

$$I = \int_C f(z)dz = \int_{-R}^{R} f(x)dx + \int_{C_R} f(z)dz \quad (1)$$

$$|z.f(z)| = \frac{|z^3|}{|z^6+1|} \to 0 \text{ as } R \to \infty$$

$\therefore$ By Cauchy's lemma, $\lim_{R\to\infty} \int_{C_R} f(z)dz = 0$

When $R \to \infty$, equation (1) becomes

$$\int_{-\infty}^{\infty} f(x)dx = 2\pi i\{\text{Sum of the residues at the poles of f(z) lying above the real axis}\} = 2\pi i.\frac{1}{6i} = \frac{\pi}{3}$$

Example 4.4.14 Using the method of contour integration evaluate $\int_{-\infty}^{\infty} \frac{x^2+1}{5x^4+26x^2+5}dx$

Solution:

Let $I = \int_C f(z)dz$ where $C = [-R, R] \cup C_R$ and $f(z) = \frac{z^2+1}{5z^4+26z^2+5}$

Poles of f(z) are given by $5z^4 + 26z^2 + 5 = 0$

i.e., $(5z^2 + 1)(z^2 + 5) = 0$

i.e., $z = \pm i/\sqrt{5}, \pm i\sqrt{5}$

The poles lying above the real axis are $z = i/\sqrt{5}$, and $z = i\sqrt{5}$ and these are simple poles.

$$\text{Res}\left[f(z); i/\sqrt{5}\right] = \lim_{z\to i/\sqrt{5}} .\frac{\left(z - i/\sqrt{5}\right)\left(z^2+1\right)}{\left(5z^2+1\right)\left(z^2+5\right)}$$

$$= \lim_{z\to i/\sqrt{5}} .\frac{\left(z - i/\sqrt{5}\right)\left(z^2+1\right)}{5\left(z - i/\sqrt{5}\right)\left(z + i/\sqrt{5}\right)\left(z^2+5\right)}$$

$$= \frac{\left(-1/5+1\right)}{5.2i/\sqrt{5}\left(-1/5+5\right)}$$

$$= \frac{4/5}{5.2i/\sqrt{5}\,24/5} = \frac{1}{12\sqrt{5}i}$$

$$\text{Res}\left[f(z);\ i\sqrt{5}\right] = \lim_{z\to i\sqrt{5}} \cdot \frac{\left(z-i\sqrt{5}\right)\left(z^2+1\right)}{\left(5z^2+1\right)\left(z-i\sqrt{5}\right)\left(z+i\sqrt{5}\right)}$$

$$= \frac{(-5+1)}{(-25+1)2i\sqrt{5}} = \frac{4}{24.2.i\sqrt{5}} = \frac{1}{12\sqrt{5}i}$$

$\therefore$ Sum of the residues at the poles above the real axis $= \dfrac{1}{12\sqrt{5}i} + \dfrac{1}{12\sqrt{5}i} = \dfrac{1}{6\sqrt{5}i}$

$$\text{Now } I = \int_C f(z)dz = \int_{-R}^{R} f(x)dx + \int_{C_R} f(z)dz \qquad (1)$$

$$\therefore \left|z.f(z)\right| = \left|\frac{z(z^2+1)}{\left(5z^2+1\right)\left(z^2+5\right)}\right| \to 0 \text{ as } R\to\infty$$

$\therefore$ By Cauchy's lemma, $\lim\limits_{R\to\infty} \int_{C_R} f(z)dz = 0$

When $R\to\infty$, equation (1) becomes

$$\int_{-\infty}^{\infty} f(x)dx = 2\pi i\{\text{Sum of the residues at the poles of } f(z) \text{ lying above the real axis}\} = 2\pi i.\frac{1}{6\sqrt{5}i} = \frac{\pi}{3\sqrt{5}}$$

$$\text{i.e., } \int_{-\infty}^{\infty} \frac{x^2+1}{5x^4+26x^2+5}dx = \frac{\pi}{3\sqrt{5}}$$

Example 4.4.15 Evaluate $\displaystyle\int_{-\infty}^{\infty} \frac{dx}{\left(1+x^2\right)^{n+1}}$

Solution:

Let $I = \int_C f(z)dz$ where $C = [-R, R] \cup C_R$ and $f(z) = \dfrac{1}{\left(1+z^2\right)^{n+1}}$

Poles of f(z) are given by $\left(1+z^2\right)^{n+1} = 0.$ i.e., $z = \pm i$

$z = i$ is a pole of order $n+1$ and lies above the real axis.

$$\text{Res}[f(z);i] = \frac{\varphi^{(n)}(i)}{n!}, \text{ where } \varphi(z) = \frac{1}{(z+i)^{n+1}}$$

$$\varphi^{(n)}(z) = \frac{(-1)^n.(n+1)(n+2)\cdots(n+n)}{(z+i)^{2n+1}}$$

$$\varphi^{(n)}(i) = \frac{(-1)^n.(2n)!}{n!(2i)^{2n+1}} = \frac{(-1)^n.(2n)!}{n!2^{2n+1}.i^{2n}.i} = \frac{(-1)^n.(2n)!}{n!2^{2n+1}.(-1)^n.i} = \frac{(2n)!}{n!\,2^{2n+1}.i}$$

$$\text{Res}[f(z);i] = \frac{(2n)!}{n!.n!.2^{2n+1}.i}$$

$\therefore\ |z.f(z)| = \left|\dfrac{z}{(1+z^2)^{n+1}}\right| \to 0 \text{ as } R \to \infty$

$\therefore$ By Cauchy's lemma, $\lim\limits_{R\to\infty} \int\limits_{C_R} f(z)dz = 0$

Now $I = \int\limits_C f(z)dz = \int\limits_{-R}^{R} f(x)dx + \int\limits_{C_R} f(z)dz$

When $R \to \infty$, we get,

$\int\limits_{-\infty}^{\infty} f(x)dx = 2\pi i$ {Sum of the residues at the poles of f(z) lying above the real axis}

$$= 2\pi i.\frac{(2n)!}{n!.n!.2^{2n+1}.i} = \frac{\pi(2n)!}{(n!)^2.2^{2n}}$$

i.e., $\displaystyle\int\limits_{-\infty}^{\infty} \frac{dx}{(1+x^2)^{n+1}} = \frac{\pi(2n)!}{(n!)^2.2^{2n}}$

4.4.3 EVALUATION OF INTEGRALS OF THE FORM $\int\limits_{-\infty}^{\infty} f(x).\cos ax.dx$ or $\int\limits_{-\infty}^{\infty} f(x).\sin ax.dx$

In this section we consider the evaluation of improper integrals of the form

$\int\limits_{-\infty}^{\infty} f(x).\cos ax.dx$ or $\int\limits_{-\infty}^{\infty} f(x).\sin ax.dx$ satisfying the following conditions.

(i) $a > 0$

(ii) $f(x) = \dfrac{g(x)}{h(x)}$ is a rational function of x.

(iii) g(x) and h(x) have no common factors and the degree of h(x) is greater than that of g(x) by atleast one.

(iv) h(x) does not vanish for any real values of x.

1. JORDAN'S LEMMA

If f(z) is a continuous function such that $|f(z)| \to 0$ uniformly as $|z| \to \infty$ on C_R, then $\int\limits_{C_R} e^{iaz}.f(z)dz \to 0$ as $R \to \infty$, where C_R is the semicircle $|z| = R$, $Im(z) > 0$ and $a > 0$.

Proof:

On the semicircle $|z| = R$, $Im(z) > 0$, we have $z = Re^{i\theta}$, $0 \le \theta \le \pi$

$\therefore\ dz = R.e^{i\theta}.i.d\theta \qquad |dz| = R.d\theta$

$$\therefore \left|\int_{C_R} e^{iaz}.f(z)dz\right| = \left|\int_{C_R} e^{iaR(\cos\theta + i\sin\theta)}.f(z).Re^{i\theta}\, id\theta\right|$$

$$\leq \int_0^{\pi} e^{-aR\sin\theta}|f(z)|.R.d\theta$$

$$\leq 2\int_0^{\pi/2} e^{-aR\sin\theta}|f(z)|.R.d\theta \qquad (1)$$

Since $|f(z)| \to 0$ uniformly as $R \to \infty$ on C_R, given $\in > 0$, we can find a real number R^* such that $|f(z)| < \in$ for $R > R^*$

Further $\frac{2}{\pi} \leq \frac{\sin\theta}{\theta} \leq 1$ **when** $0 \leq \theta \leq \pi/2$

$$\therefore -\sin\theta \leq -\frac{2}{\pi}\theta$$

i.e., $-aR\sin\theta \leq -\frac{2aR\theta}{\pi}$ **(as $a > 0$)**

i.e., $e^{-aR\sin\theta} \leq e^{-\frac{2aR\theta}{\pi}}$ **for** $0 \leq \theta \leq \pi/2$

∴ Equation (1) becomes,

$$\left|\int_{C_R} e^{iaz}.f(z)dz\right| < \int_0^{\pi/2} 2R \in e^{-\frac{2aR\theta}{\pi}} d\theta$$

$$< \left[\frac{2R \in e^{-\frac{2aR\theta}{\pi}}}{-\frac{2aR}{\pi}}\right]_0^{\pi/2} < \frac{\pi\in}{a}\left(1 - e^{-aR}\right) < \frac{\pi\in}{a}$$

Since $\in$ is arbitrary, we get $\left|\int_{C_R} e^{iaz}.f(z)dz\right| \to 0$ **as $R \to \infty$.**

i.e., $\lim_{R\to\infty} \int_{C_R} e^{iaz}.f(z)dz = 0$. **Hence the lemma**

2. Evaluation of $\int_{-\infty}^{\infty} f(x).\cos ax.dx$ or $\int_{-\infty}^{\infty} f(x).\sin ax.dx$

Consider the contour integral $I = \int_{C_R} e^{iaz}.f(z)dz$, **where $C = [-R, R] \cup C_R$ is traversed in the anti-clockwise direction. $f(z)$ has no zeros on the real axis.**

$$I = \int_C e^{iaz} f(z)dz = \int_{-R}^{R} e^{iax} f(x)dx + \int_{C_R} e^{iaz} f(z)dz \quad (1)$$

By Cauchy's residue theorem, we have,

$$\lim_{R\to\infty} \int_C e^{iaz} f(z)dz = 2\pi i\{\text{Sum of the residues at the poles of } e^{iaz} f(z) \text{ lying above the real axis}\} \quad (2)$$

$$I = \lim_{R\to\infty} \int_{-R}^{R} e^{iax} f(x)dx = \int_{-\infty}^{\infty} e^{iax} f(x)dx \quad (3)$$

By assumption $f(z) = \dfrac{g(z)}{h(z)}$ and degree of h(z) is greater than that of g(z) by atleast one.

$\therefore |f(z)| \to 0$ as $R \to \infty$.

Hence by Jordan's lemma, we have,

$$\lim_{R\to\infty} \int_{C_R} e^{iaz} f(z)dz = 0 \quad (4)$$

Using (2), (3) and (4) in (1) we get,

$$\int_{-\infty}^{\infty} e^{iax} f(x)dx = 2\pi i\{\text{Sum of the residues}\}$$

Equating the real parts and the imaginary parts on both sides we get,

$$\int_{-\infty}^{\infty} f(x).\cos ax.dx = \text{Re}\{2\pi i[\text{Sum of the residues}]\}$$

$$\int_{-\infty}^{\infty} f(x).\sin ax.dx = \text{Im}\{2\pi i[\text{Sum of the residues}]\}$$

Example 4.4.16 Use the method of contour integration to prove that

$$\int_{-\infty}^{\infty} \frac{\cos x.dx}{(x^2+a^2)(x^2+b^2)} = \frac{\pi}{a^2-b^2}\left(\frac{e^{-b}}{b} - \frac{e^{-a}}{a}\right) \text{ where } a > b > 0.$$

Solution:

Let $I = \displaystyle\int_C \frac{e^{iz}.dz}{(z^2+a^2)(z^2+b^2)}$ where $C = [-R, R] \cup C_R$.

$$= \int_C e^{iz}.f(z).dz \text{ where } f(z) = \frac{1}{(z^2+a^2)(z^2+b^2)}$$

$$= \int_{-R}^{R} e^{ix} f(x)dx + \int_{C_R} e^{iz} f(z)dz$$

When $R \to \infty$, we get,

$$\int_{-\infty}^{\infty} e^{ix} f(x)dx + \underset{R\to\infty}{\text{Lim}} \int_{C_R} e^{iz} f(z)dz = 2\pi i \{\text{Sum of the residues at the poles of } e^{iz} f(z) \text{ lying above the real axis}\} \quad (1)$$

$$|f(z)| = \frac{1}{|z^2 + a^2||z^2 + b^2|} \to 0 \text{ as } R \to \infty$$

$\therefore$ By Jordan's lemma, $\underset{R\to\infty}{\text{Lim}} \int_{C_R} e^{iz} f(z)dz = 0$ (2)

Poles of $e^{iz} f(z)$ are given by $(z^2 + a^2).(z^2 + b^2) = 0$ i.e., $z = \pm ia, \pm ib$.
$z = ia$ and $z = ib$ are simple poles and they lie above the real axis.

$$\text{Res}\left[e^{iz} f(z); ia\right] = \underset{z\to ia}{\text{Lim}} \frac{(z - ia)e^{iz}}{(z - ia)(z + ia).(z^2 + b^2)}$$

$$= \frac{e^{-a}}{2ia.(-a^2 + b^2)} = \frac{-e^{-a}}{2ia.(a^2 - b^2)}$$

Similarly $\text{Res}\left[e^{iz} f(z); ib\right] = \dfrac{e^{-b}}{2ib.(-b^2 + a^2)} = \dfrac{e^{-b}}{2ib.(a^2 - b^2)}$

Sum of the residues $= \dfrac{1}{2i(a^2 - b^2)}\left(\dfrac{e^{-b}}{b} - \dfrac{e^{-a}}{a}\right)$ (3)

Using (2) and (3) in (1), we get,

$$\int_{-\infty}^{\infty} \frac{e^{ix}.dx}{(x^2 + a^2)(x^2 + b^2)} = \frac{2\pi i}{2i(a^2 - b^2)}\left(\frac{e^{-b}}{b} - \frac{e^{-a}}{a}\right)$$

$$= \frac{\pi}{(a^2 - b^2)}\left(\frac{e^{-b}}{b} - \frac{e^{-a}}{a}\right)$$

Equating the real parts on both sides, we get,

$$\int_{-\infty}^{\infty} \frac{\cos x.dx}{(x^2 + a^2)(x^2 + b^2)} = \frac{\pi}{a^2 - b^2}\left(\frac{e^{-b}}{b} - \frac{e^{-a}}{a}\right)$$

Example 4.4.17 Prove that $\displaystyle\int_0^{\infty} \frac{\cos ax.dx}{(x^2 + b^2)^2} = \frac{\pi}{4b^3}(1 + ab)e^{-ab}$ where $a > 0, b > 0$.

Solution:

Let $I = \displaystyle\int_C \frac{e^{iaz}.dz}{(z^2 + b^2)^2}$ where $C = [-R, R] \cup C_R$.

$$= \int_C e^{iaz}.f(z).dz \text{ where } f(z) = \frac{1}{(z^2 + b^2)^2}$$

$$= \int_{-R}^{R} e^{iax} f(x)dx + \int_{C_R} e^{iaz} f(z)dz$$

When $R \to \infty$, we get,

$$\int_{-\infty}^{\infty} e^{iax} f(x)dx + \lim_{R\to\infty} \int_{C_R} e^{iaz} f(z)dz = 2\pi i \{\text{Sum of the residues at the poles of } e^{iaz} f(z) \text{ lying above the real axis}\} \quad (1)$$

$$|f(z)| = \left|\frac{1}{(z^2+b^2)^2}\right| \to 0 \text{ as } R \to \infty$$

$\therefore$ By Jordan's lemma, $\lim_{R\to\infty} \int_{C_R} e^{iaz} f(z)dz = 0$ (2)

Poles of $e^{iaz} f(z)$ are given by $(z^2 + b^2)^2 = 0$

i.e., $z = ib$ and $z = -ib$ are poles of order two and $z = ib$ lies in the half plane above the real axis.

$$\text{Res}[e^{iaz} f(z); ib] = \frac{\varphi'(ib)}{1!}, \text{ where } \varphi(z) = \frac{e^{iaz}}{(z+ib)^2}$$

and $\varphi'(z) = \frac{-2e^{iaz}}{(z+ib)^3} + \frac{e^{iaz}.ia}{(z+ib)^2}$ $\quad \therefore \varphi'(ib) = \frac{-2e^{-ab}}{(2ib)^3} + \frac{e^{-ab}.ia}{(2ib)^2}$

Hence $\text{Res}[e^{iaz} f(z); ib] = \frac{e^{-ab}}{4ib^3} + \frac{ae^{-ab}}{4ib^2}$

$\therefore$ Sum of the residues $= \frac{(1+ab)e^{-ab}}{4ib^3}$

Using (2) and (3) in (1) we get, $\int_{-\infty}^{\infty} \frac{e^{iax}.dx}{(x^2+b^2)^2} = 2\pi i.\frac{(1+ab)e^{-ab}}{4ib^3}$

i.e., $\int_{-\infty}^{\infty} \frac{e^{iax}.dx}{(x^2+b^2)^2} = \frac{\pi}{2b^3}(1+ab)e^{-ab}$

Equating the real parts on both sides, we get,

$$\int_{-\infty}^{\infty} \frac{\cos ax.dx}{(x^2+b^2)^2} = \frac{\pi}{2b^3}(1+ab)e^{-ab} \quad \text{i.e.,} \quad \int_{0}^{\infty} \frac{\cos ax.dx}{(x^2+b^2)^2} = \frac{\pi}{4b^3}(1+ab)e^{-ab}$$

Example 4.4.18 Evaluate by the method of contour integration $\int_{-\infty}^{\infty} \frac{x \sin x}{x^2+4x+5} dx$

Solution:

Let $I = \int_C \frac{ze^{iz}}{z^2+4z+5} dz$ where $C = [-R, R] \cup C_R$.

$$= \int_C e^{iz}.f(z).dz \text{ where } f(z) = \frac{z}{z^2+4z+5}$$

$$= \int_{-R}^{R} e^{ix} f(x)dx + \int_{C_R} e^{iz} f(z)dz$$

When R→ ∞, we get,

$$\int_{-\infty}^{\infty} e^{ix} f(x)dx + \lim_{R\to\infty} \int_{C_R} e^{iz} f(z)dz = 2\pi i \{\text{Sum of the residues at the poles of } e^{iz} f(z) \text{ lying above the real axis}\} \quad (1)$$

$$|f(z)| = \left|\frac{z}{z^2+4z+5}\right| \to 0 \text{ as } R\to\infty$$

∴ By Jordan's lemma, $\lim_{R\to\infty} \int_{C_R} e^{iz} f(z)dz = 0$ (2)

Poles of $e^{iz} f(z)$ are given by $z^2 + 4z + 5 = 0$

i.e., $z = -2 + i$ and $z = -2-i$ are simple poles and $z = -2 + i$ lies in the half plane above the real axis.

$$\text{Res}\left[e^{iz} f(z); -2+i\right] = \lim_{z\to -2+i} \frac{(z+2-i)e^{iz} z}{(z+2-i)(z+2+i)}$$

$$= \frac{e^{i(-2+i)}(-2+i)}{2i} = \frac{e^{-2i-1}}{2i}(-2+i)$$

∴ Sum of the residues $= \frac{e^{-2i-1}}{2i}(-2+i)$ (3)

Using (2) and (3) in (1) we get,

$$\int_{-\infty}^{\infty} e^{ix} f(x)dx = 2\pi i.\frac{e^{-2i-1}}{2i}(-2+i) = \pi e^{-1}.e^{-2i}(-2+i)$$

$$\int_{-\infty}^{\infty} \frac{xe^{ix}}{x^2+4x+5} dx = \frac{\pi}{e}(\cos 2 - i\sin 2)(-2+i)$$

Equating the imaginary parts on both sides, we get,

$$\int_{-\infty}^{\infty} \frac{x\sin x}{x^2+4x+5} dx = \frac{\pi}{e}(\cos 2 + 2\sin 2)$$

Example 4.4.19 Evaluate $\int_{-\infty}^{\infty} \frac{\cos^2 x}{(1+x^2)^2} dx$

Solution:

Let $I = \int_C \frac{1+e^{i2z}}{(1+z^2)^2} dz$ where $C = [-R, R] \cup C_R$.

$$= \int_C (1+e^{i2z}).f(z).dz \quad \text{where } f(z)=\frac{1}{(1+z^2)^2}$$

$$= \int_{-R}^{R} (1+e^{i2x})f(x)dx + \int_{C_R} (1+e^{i2z})f(z)dz$$

When R→ ∞, we get,

$$\int_{-\infty}^{\infty} (1+e^{i2x})f(x)dx + \lim_{R\to\infty}\int_{C_R} (1+e^{i2z})f(z)dz$$

= 2πi {Sum of the residues at the poles of

$(1+e^{i2z})f(z)$ lying in the half plane above the real axis} (1)

$$|z.f(z)| = \left|\frac{z}{(1+z^2)^2}\right| \to 0 \text{ as } R\to\infty$$

∴ By Cauchy's lemma, $\lim_{R\to\infty}\int_{C_R} f(z)dz = 0$

Also $|f(z)| = \left|\frac{1}{(1+z^2)^2}\right| \to 0$ as $R\to\infty$

∴ By Jordan's lemma we have, $\lim_{R\to\infty}\int_{C_R} e^{i2z}f(z)dz = 0$

Thus $\lim_{R\to\infty}\int_{C_R}(1+e^{i2z})f(z)dz = \lim_{R\to\infty}\int_{C_R} f(z)dz + \lim_{R\to\infty}\int_{C_R} e^{i2z}f(z)dz = 0+0=0$ (2)

Poles of $(1 + e^{i2z}) f(z)$ are given by $(1 + z^2)^2 = 0$

i.e., z = i and z = -i are poles of order 2 and z = i lies above the real axis.

$$\text{Res}[(1+e^{i2z})f(z); i] = \frac{\varphi'(i)}{1!}, \text{ where } \varphi(z) = \frac{(1+e^{i2z})}{(z+i)^2}$$

and $\varphi'(z) = \frac{(1+e^{2iz}).(-2)}{(z+i)^3} + \frac{e^{2iz}.2i}{(z+i)^2}$

$$\varphi'(i) = \frac{2(1+e^{-2})}{8i} + \frac{2e^{-2}}{4i}$$

$$= \frac{1+e^{-2}+2e^{-2}}{4i} = \frac{1+3e^{-2}}{4i}$$

$$\text{Res}[(1+e^{i2z})f(z); i] = \frac{1+3e^{-2}}{4i}$$

Sum of the residues $= \frac{1+3e^{-2}}{4i}$ (3)

Using (2) and (3) in (1) we get,

$$\int_{-\infty}^{\infty}(1+e^{i2x})f(x)dx = 2\pi i\frac{1+3e^{-2}}{4i}$$

$$\int_{-\infty}^{\infty}\frac{\left(1+e^{i2x}\right)}{\left(1+x^2\right)^2}dx = \frac{\pi}{2}\left(1+3e^{-2}\right)$$

Equating the real parts on both sides, we get,

$$\int_{-\infty}^{\infty}\frac{1+\cos 2x}{\left(1+x^2\right)^2}dx = \frac{\pi}{2}\left(1+3e^{-2}\right)$$

i.e., $$\int_{-\infty}^{\infty}\frac{2\cos^2 x}{\left(1+x^2\right)^2}dx = \frac{\pi}{2}\left(1+3e^{-2}\right)$$

i.e., $$\int_{-\infty}^{\infty}\frac{\cos^2 x}{\left(1+x^2\right)^2}dx = \frac{\pi}{4}\left(1+3e^{-2}\right)$$

Example 4.4.20 By integrating $\frac{e^{iz}}{z-ia}$, $(a>0)$ around a sutiable contour, prove that

$$\int_{-\infty}^{\infty}\frac{a\cos x + x\sin x}{x^2+b^2}dx = 2\pi e^{-a}$$

Solution:

Let $I = \int_C \frac{e^{iz}}{z-ia}dz$, where $C = [-\infty, \infty] \cup C_R$.

$= \int_C e^{iz}.f(z).dz$, where $f(z) = \frac{1}{z-ia}$

$= \int_{-R}^{R} e^{ix}f(x)dx + \int_{C_R} e^{iz}f(z)dz$

When $R \to \infty$, we get,

$\int_{-\infty}^{\infty} e^{ix}f(x)dx + \underset{R\to\infty}{\text{Lim}} \int_{C_R} e^{iz}f(z)dz = 2\pi i$ {Sum of the residues at the poles of $e^{iz}f(z)$ lying in the half plane above the real axis} (1)

$|f(z)| = \left|\frac{1}{z-ia}\right| \to 0$ as $R \to \infty$

$\therefore$ By Jordan's lemma, $\underset{R\to\infty}{\text{Lim}} \int_{C_R} e^{ia}f(z)dz = 0$ (2)

$e^{iz}f(z) = \frac{e^{iz}}{z-ia}$ has a simple pole at $z = ia$

$$\operatorname{Res}[e^{iz}f(z); ia] = \lim_{z\to ia}(z-ia)\frac{e^{iz}}{z-ia} = e^{-a}$$

Hence sum of the residues $= e^{-a}$ (3)

Using (2) and (3) in (1), we get, $\int_{-\infty}^{\infty} e^{ix} f(x)dx = 2\pi i e^{-a}$

i.e., $\int_{-\infty}^{\infty} \frac{e^{ix}}{x-ia}dx = 2\pi i e^{-a}$

i.e., $\int_{-\infty}^{\infty} \frac{e^{ix}(x-ia)}{x^2+a^2}dx = 2\pi i e^{-a}$

i.e., $\int_{-\infty}^{\infty} \frac{(\cos x + i\sin x)(x+ia)}{x^2+b^2}dx = 2\pi i e^{-a}$

Equating the imaginary parts on both sides, we get, $\int_{-\infty}^{\infty} \frac{a\cos x + x\sin x}{x^2+b^2}dx = 2\pi e^{-a}$

EXERCISE 4.4

PART-A

1. Convert $\int_0^{2\pi} F(\cos\theta, \sin\theta)\, d\theta$ into a contour integral.
2. How will you evaluate $\int_{-\infty}^{\infty} f(x)dx$ using method of contour integration?
3. To evaluate $\int_{-\infty}^{\infty} f(x)dx$ what contour you will choose?
4. State Cauchy's Lemma.
5. State Jordan's Lemma.
6. What contour integral you will choose for integrating $\int_{-\infty}^{\infty} \frac{\cos ax}{x^2+a^2}dx$?

PART-B

Evaluate the following integrals by the method of contour integration. (Qns: 7-16)

7. $\int_0^{2\pi} \frac{d\theta}{5+4\sin\theta}$
8. $\int_0^{2\pi} \frac{d\theta}{2+\cos\theta}$
9. $\int_0^{2\pi} \frac{d\theta}{13+5\sin\theta}$
10. $\int_0^{2\pi} \frac{d\theta}{\cos\theta+2\sin\theta+4}$.

11. $\int_0^{2\pi} \frac{a.d\theta}{a^2+\sin^2\theta}$ $(a>0)$

12. $\int_0^{2\pi} \frac{\cos 2\theta.d\theta}{5+4\cos\theta}$

13. $\int_0^{2\pi} \frac{\cos^2 3\theta.d\theta}{5-4\cos 2\theta}$

14. $\int_0^{\pi} \frac{d\theta}{a+\cos\theta}$ $(a>1)$

15. $\int_0^{\pi} \frac{(1+2\cos\theta)\,d\theta}{5+4\cos\theta}$

16. $\int_0^{2\pi} \frac{d\theta}{(a+\cos\theta)^2}$ $(a>1)$

17. Prove that $\int_0^{\pi} \frac{\cos 3\theta.d\theta}{5-4\cos\theta} = \frac{\pi}{12}$

18. Prove that $\int_0^{\pi} \frac{d\theta}{(2+\cos\theta)^2} = \frac{2\pi}{3\sqrt{3}}$.

19. Prove that $\int_0^{2\pi} \frac{d\theta}{a+b\sin\theta} = \frac{2\pi}{\sqrt{a^2-b^2}}$ $(a>b>0)$

20. Prove $\int_0^{2\pi} \frac{\sin^2\theta}{a+b\cos\theta}\,d\theta = \frac{2\pi}{b^2}\left[a - \sqrt{a^2-b^2}\right]$ $(a>b>0)$

21. Prove that $\int_0^{2\pi} \frac{d\theta}{(a+b\cos\theta)^2} = \frac{2a\pi}{(a^2-b^2)^{3/2}}$ $(a>b>0)$

22. Prove that $\int_0^{2\pi} e^{-\cos\theta}\cos(n\theta+\sin\theta)\,d\theta = (-1)^n \frac{2\pi}{n!}$

23. Prove that $\int_0^{\pi} \frac{\cos 2\theta.d\theta}{1-2a\cos\theta+a^2} = \frac{\pi a^2}{1-a^2}$ $(|a|<1)$

Evaluate the following integrals by the method of contour integration. (Ex: 24-31)

24. $\int_0^{\infty} \frac{dx}{1+x^4}$

25. $\int_{-\infty}^{\infty} \frac{x^2}{(x^2+a^2)(x^2+b^2)}\,dx$

26. $\int_{-\infty}^{\infty} \frac{x^2-x+2}{x^4+10x^2+9}\,dx$

27. $\int_0^{\infty} \frac{dx}{1+x^6}$

28. $\int_0^{\infty} \frac{dx}{x^4+x^2+1}$

29. $\int_0^\infty \frac{dx}{1+x^4}$

30. $\int_0^\infty \frac{dx}{(x^2+1)^3}$

31. $\int_0^\infty \frac{(2x^2-1)dx}{x^4+5x^2+4}$

32. Prove that $\int_{-\infty}^\infty \frac{dx}{(x^4+5x^2+4)^2} = \frac{11\pi}{432}$

33. Prove that $\int_{-\infty}^\infty \frac{x^2+a^2}{x^4+a^4}dx = \frac{\pi\sqrt{2}}{a}$ $(a>0)$

34. Prove that $\int_{-\infty}^\infty \frac{dx}{(x^2+1)(2x^4+5x^2+2)} = \frac{\pi}{6}(5\sqrt{2}-6)$

35. Prove that $\int_{-\infty}^\infty \frac{\cos 2x}{(x^2+a^2)(x^2+b^2)}dx = \frac{\pi(ae^{-2b}-be^{-2a})}{ab(a^2-b^2)}$ $(a>0, b>0)$

36. Prove that $\int_{-\infty}^\infty \frac{\cos ax}{(x^2+b^2)^2}dx = \frac{\pi(ab+1)e^{-ab}}{2b^3}$ $(a>0, b>0)$

37. Prove that $\int_{-\infty}^\infty \frac{x^3.\sin x}{(x^2+a^2)^2}dx = \frac{\pi}{2}(2-a)e^{-a}$

38. Prove that $\int_{-\infty}^\infty \frac{\cos^2 x}{(1+x^2)^2}dx = \frac{\pi}{4}(1+3e^{-2})$

39. Prove that $\int_{-\infty}^\infty \frac{\sin x}{x^2+4x+5}dx = \frac{-\pi\sin 2}{e}$

40. Prove that $\int_{-\infty}^\infty \frac{x.\sin ax}{x^4+4}dx = \frac{\pi e^{-a}\sin a}{2}$

41. Prove that $\int_0^\infty \frac{x.\sin x}{(x^2+a^2)}dx = \frac{\pi}{2}e^{-a}$

42. Prove $\int_0^\infty \frac{x.\sin x}{(x^2+a^2)^2}dx = \frac{\pi e^{-a}}{4a}$

43. Prove $\int_{-\infty}^\infty \frac{x.\cos x}{x^2-2x+2}dx = \frac{\pi}{e}(\cos 1 - \sin 1)$

44. Prove $\int_{-\infty}^{\infty}\frac{x^3.\sin mx}{(x^2+a^2)(x^2+b^2)}dx=\frac{\pi}{(a^2-b^2)}\left(a^2e^{-ma}-b^2e^{-mb}\right)$

45. Prove that $\int_0^{\infty}\frac{x.\sin x}{x^4+1}dx=\frac{\pi}{2}e^{-1/\sqrt{2}}\sin\left(1/\sqrt{2}\right)$

46. Prove $\int_{-\infty}^{\infty}\frac{x.\sin x}{(x^2+1)^2}dx=\frac{\pi}{2e}$

47. Prove that $\int_{-\infty}^{\infty}\frac{x.\sin x}{(x^2+1)(x^2+4)}dx=\frac{\pi}{3e^2}(e-1)$

48. Prove that $\int_{-\infty}^{\infty}\frac{x.\cos \pi x}{x^2+2x+5}dx=\frac{\pi}{2}e^{-2\pi}$

49. Prove $\int_0^{\infty}\frac{x^3.\sin x}{(x^2+a^2)(x^2+b^2)}dx=\frac{\pi}{2(a^2-b^2)}\left(a^2e^{-a}-b^2e^{-b}\right)$

ANSWERS

EXERCISE 4.1

(15) 1- i (16) $\frac{511}{3}-\frac{49}{5}i$ (17) $\frac{1+i}{2}$

(18) $i\pi r^2$ (19) $10-\frac{8}{3}i$ (20) $4\pi i$

(24) $\frac{1523}{15}$ (31) $\frac{8\pi i}{3}$ (33) 0

(34) 0 (35) 0 (36) 0

(43) -156 + 38i (44) $\frac{5}{3}i-\frac{2}{3}$ (46) -48π

(47) (i) $\frac{511}{3}-\frac{49}{5}i$ (ii) $\frac{518}{3}-57i$ (iii) $\frac{518}{3}-8i$

(48) (i) $\frac{109}{3}-\frac{25}{3}i$ (ii) $\frac{109}{3}-\frac{25}{3}i$

(49) (i) $\frac{-9}{18}+\frac{(4\pi+9\sqrt{3})}{18}i$ (ii) $\frac{2}{3}(3+\pi i)$ (iii) $\frac{4\pi}{3}i$

(50) 4π (53) (i) 2(i -1) (ii) -2(i + 1) (54) $5\pi i$

(55) π(1-i) (56) $4\pi i$ (60) $\frac{\pi i}{2}$

(61) $-\frac{\pi i}{2}$ (62) (i) 0 (ii) 2π (iii) 0 (63) $\frac{-2\pi i}{9}$

(69) $\pi \cos(4+2i)$ (74) $2\pi i$ (75) $\pi i(2-5e^{-1})$

(76) $-\pi i(\cosh 1 + i \sinh 1)$

EXERCISE 4.2

(7) 4 (8) 0

(9) $3/2$ (10) ∞

(11) $1+z+z^2+\ldots$ (12) $\sum_{n=0}^{\infty}(n+1)z^{n+1}$

(13) $-1+2z-2z^2+2z^3$ (14) $\sum_{n=0}^{\infty}\frac{2^n z^{n+1}}{n!}$ (15) $\sum_{n=0}^{\infty}\frac{z^n}{2^{n+1}}$

(16) $z^3+\frac{z^4}{1!}+\frac{z^5}{2!}+\frac{z^6}{3!}+\cdots$ (17) $1-z^2+z^4-z^6+\ldots$

(18) $z-\frac{z^2}{2}+\frac{z^3}{3}-\frac{z^4}{4}+\cdots$ (19) $z^2-\frac{z^4}{3!}+\frac{z^6}{5!}-\cdots$

(20) $1+3z+\frac{3.4}{1.2}z^2+\frac{4.5}{1.2}z^3+\cdots$ (21) $\frac{1}{z}=1-(z-1)+(z-1)^2-(z-1)^3+\cdots$

(24) $e^{-i}\sum_{n=0}^{\infty}\frac{(z+i)^n}{n!}$ (25) $e^{-1}\sum_{n=0}^{\infty}(-1)^n\frac{(z-1)^n}{n!}$

(26) $\frac{1}{z-1}-1+(z-1)-(z-1)^2+(z-1)^3+\cdots$

(27) $\frac{1}{z-1}-\frac{2}{(z-1)^2}+\frac{3}{(z-1)^3}-\cdots$

(30) $\frac{1}{e^2}\left[-1+\left(1-\frac{2}{1!}\right)(z+1)+\left(\frac{2}{1!}-\frac{2^2}{2!}\right)(z+1)^2+\left(\frac{2^2}{2!}-\frac{2^3}{3!}\right)(z+1)^3+\cdots\right]$

(31) $\frac{1}{2}\sum_{n=0}^{\infty}(-1)^n\left[1-\frac{1}{3^{n+1}}\right]z^n$

(32) (i) $\sum_{n=0}^{\infty}\left(\frac{1}{2^{n+1}}-1\right)z^n$ (ii) $\sum_{n=0}^{\infty}\frac{1}{z^{n+1}}+\sum_{n=0}^{\infty}\frac{z^n}{2^{n+1}}$ (iii) $\sum_{n=0}^{\infty}(1-2^n)\frac{1}{z^{n+1}}$

(33) (i) $\frac{-3}{2z}+\sum_{n=0}^{\infty}\left[\frac{2}{3}(-1)^n-\frac{5}{12}\left(\frac{1}{2}\right)^n\right]z^n$

(ii) $\frac{-3}{2z}+\frac{2}{3z}\sum_{n=0}^{\infty}\frac{(-1)^n}{z^n}-\frac{5}{12}\sum_{n=0}^{\infty}\left(\frac{z}{2}\right)^n$

(iii) $\frac{-3}{2z}+\frac{2}{3z}\sum_{n=0}^{\infty}\frac{(-1)^n}{2^n}-\frac{5}{6z}\sum_{n=0}^{\infty}\left(\frac{2}{z}\right)^n$

(34) $\frac{1}{9(z-3)^2}-\frac{2}{27(z-3)}+\frac{1}{27}-\frac{4(z-3)}{243}+\cdots$

(35) (i) $-\frac{1}{2z^4}+\frac{1}{2z^3}-\frac{1}{2z^2}+\frac{1}{2z}-\frac{1}{6}+\frac{z}{18}-\frac{z^2}{54}+\cdots$

(ii) $\frac{1}{z^2}-\frac{4}{z^3}+\frac{13}{z^4}-\frac{40}{z^5}+\cdots$

(iii) $\frac{1}{2(z+1)}-\frac{1}{4}+\frac{1}{8}(z+1)-\frac{1}{16}(z+1)^2+\cdots$

(iv) $\frac{1}{3}-\frac{4z}{9}+\frac{13}{27}z^2-\frac{40}{81}z^3+\cdots$

(38) (i) $1+\frac{3}{2}\sum_{n=0}^{\infty}(-1)^n\left(\frac{z}{2}\right)^n-\frac{8}{3}\sum_{n=0}^{\infty}(-1)^n\left(\frac{z}{3}\right)^n$

(ii) $1+\frac{3}{z}\sum_{n=0}^{\infty}(-1)^n\left(\frac{2}{z}\right)^n-\frac{8}{3}\sum_{n=0}^{\infty}(-1)^n\left(\frac{z}{3}\right)^n$

(iii) $1+\frac{3}{z}\sum_{n=0}^{\infty}(-1)^n\left(\frac{2}{z}\right)^n-\frac{8}{z}\sum_{n=0}^{\infty}(-1)^n\left(\frac{3}{z}\right)^n$

(40) (i) $\frac{-3}{z+1}-\sum_{n=0}^{\infty}\left[1+\frac{2}{3^{n+1}}\right](z+1)^n$ in $0<|z+1|<1$

(ii) $\frac{-3}{z+1}+\sum_{n=0}^{\infty}\frac{1}{(z+1)^{n+1}}-2\sum_{n=0}^{\infty}\frac{(z+1)^n}{3^{n+1}}$ in $1<|z+1|<3$

(iii) $\frac{-3}{z+1}+\sum_{n=0}^{\infty}\frac{(1+2.3^n)}{(z+1)^{n+1}}$ in $|z+1|>3$

(41) (i) $1+\frac{4/5}{z-3}+\sum_{n=0}^{\infty}(-1)^n\left(\frac{1}{2^{n+1}}+\frac{1}{5^{n+2}}\right)(z-3)^n$ in $0<|z-3|<2$

(ii) $1+\frac{4/5}{z-3}+\sum_{n=0}^{\infty}(-1)^n\left(\frac{2^n}{(z-3)^{n+1}}+\frac{1}{5^{n+2}}(z-3)^n\right)$ in $2<|z-3|<5$

(iii) $1+\frac{4/5}{z-3}+\sum_{n=0}^{\infty}(-1)\ (2^n+5^{n+1})\frac{1}{(z-3)^n}$ in $|z-3|>5$

(42) (i) $-\frac{1}{4(z-i)^2}\sum_{n=0}^{\infty}(-1)^n(n+1)\left(\frac{z-i}{2i}\right)^n,\ 0<|z-i|<2$

(ii) $\sum_{n=0}^{\infty}\frac{(-1)^n(n+1)(2i)^n}{(z-i)^{n+4}}, \quad |z-i|>2$

(43) (i) $\sum_{n=0}^{\infty}\frac{(-1)^n(z+1)^{n-2}}{2^{n+1}}, 0<|z+1|<2$

(ii) $\sum_{n=0}^{\infty}\frac{(-1)^n(2)^n}{(z+1)^{n+3}}, |z+1|>2$

(44) $\sum_{n=0}^{\infty}\frac{(-1)^n}{(2n)!}\left[\frac{(z-1)^2+2(z-1)+1}{(z-1)^{2n}}\right], |z-1|>0$

(46) $\sum_{n=0}^{\infty}\frac{(i)^n}{2^{n+1}z^{n+2}}, |z|>\frac{1}{2}$ convergentat $z=i$

(47) $\frac{1}{4}\sum_{n=0}^{\infty}(i)^{n-1}\left[\frac{1+(-1)^{n-1}3^n}{z^{n+1}}\right]$ in $|z|>3$ converges at $z=4i$

(48) (i) $\frac{5}{4(z-1)^2}-\frac{1}{16(z-1)}+\frac{1}{64}-\frac{1}{64}\sum_{n=0}^{\infty}(-1)^n\left(\frac{z-1}{4}\right)^{n+1}$

(ii) $\frac{1}{(z-1)^2}-\frac{1}{(z-1)^3}-\frac{1}{64}\sum_{n=0}^{\infty}(-1)^n\left(\frac{4}{z-1}\right)^{n+4}$

(49) $f(z)=e^2\left[\frac{1}{(z-1)^3}+\frac{2}{(z-1)^2}+\frac{2}{z-1}+\frac{4}{3}+\sum_{n=0}^{\infty}\frac{2^{n+4}(z-1)^{n+1}}{(n+4)!}\right]$ in $|z-1|>0$

EXERCISE 4.3

(16) $z=0$ is a double pole
(17) $z=0$ is a removable singularity
(18) $z=0$ is an isolated essentiel singularity
(19) $z=0$ is a simple pole
(20) $z=0$ is removable singularity and $z=2n\pi i$, $n\neq 0$ are simple poles
(21) $z=1$ is an isolated essential singularity
(22) $z=0$ is a pole of order 3
(23) $z=0$ is removable singularity and $z=2n\pi i$, $n\neq 0$ are simple poles
(24) $z=0$ is isolated essential singularity
(25) $z=0$ is a simple pole

(26) 1 (27) $\frac{3e}{2}$ (28) $2e^2$

(29) $-\frac{1}{2}$ (30) $\frac{1}{2\pi i}$ (31) $-4\pi i$

(32) $2\pi i$ (33) $-4\pi i$ (34) πi
(35) $-2\pi i$ (36) $4\pi i$ (37) $2\pi i$
(38) $2\pi i$ (39) 0 (40) 0
(41) πi (42) 0 (43) $4\pi i$
(44) $-4i$ (45) $\pi/\sqrt{3}$

(46) (i) Poles at 0, i -1, -i -1; residues $2, \frac{3i-1}{2}, \frac{-3i-1}{2}$

(ii) Poles at 0, 3i, -3i; residues $\frac{1}{9}, \frac{i\cos 3 - \sin 3}{54}, \frac{-i\cos 3 - \sin 3}{54}$

(47) (i) 1 (ii) $\frac{3}{10}$

(48) (i) 0 is a pole of order 3; residue $\frac{-4}{3}$

(ii) -4 is simple pole; residue $\frac{8}{25}$ 1 pole of order 2; residue $\frac{8}{25}$

(49) (i) $\frac{(-1)^{n+1}(2n)!}{(n-1)!(n+1)!}$ (ii) $\frac{-i(2n+2)!}{2^{2n-1}(n-1)!(n-1)!}$

(50) (i) $2e^{i\pi/4}, 2e^{i3\pi/4}, 2e^{i5\pi/4}, 2e^{i7\pi/4}$ are simple poles with reisdues

$\frac{1}{32\sqrt{2}}(-1-i), \frac{1}{32\sqrt{2}}(1-i), \frac{1}{32\sqrt{2}}(1+i), \frac{1}{32\sqrt{2}}(-1+i)$

(ii) $z = 1$ is a simple pole; residue $1/4$,

$z = -1$ is a double pole; residue $-1/4$

(51) (i) $ae^{i\pi/4}, ae^{i3\pi/4}, ae^{i5\pi/4}, ae^{i7\pi/4}$ are simple poles; residues are $\frac{-i}{4a^2}, \frac{i}{4a^2}, \frac{-i}{4a^2}, \frac{i}{4a^2}$

(ii) $z = -1$ is a double pole; residue $-14/25$;

$z = 2i$ is a simple pole; residue $\frac{1}{25}(7+i)$,

$z = -2i$ is a simple pole; residue $\frac{1}{25}(7-i)$

(52) (i) Pole of order 3 at $z = 0$; residue -1
Simple pole at $z = \pm i$; residue at $z = i$ is ½, residue at $z = -i$ is ½
(ii) Poles of order 2 at $z = \pm i$; residue at $z = i$ is $-i/4$, residue at $z = -i$ is $i/4$

(53) (i) -1 (ii) $\sin 1 - 1$ (54) (i) 0 (ii) $-1/24$

(55) $\frac{\pi i}{32}$ (56) $2\pi i(1 - 1/e)$

(57) $\frac{-\pi}{100}(4-3i)\sin 3i$ (58) $2\pi i$

(59) $\frac{1}{60}\pi^6 i$ (60) πi

(61) $2\pi i$ (62) πi

(63) $\frac{3\pi}{256}$ (64) $\pi e^i(-4+5i)$

(65) $16\pi i$

EXERCISE 4.4

(7) $2\pi/3$ (8) $2\pi/\sqrt{3}$ (9) $\pi/6$

(10) $2\pi/\sqrt{11}$ (11) $2\pi/\sqrt{1+a^2}$ (12) $\pi/6$

(13) $3\pi/8$ (14) $\pi/\sqrt{a^2-1}$ (15) 0

(16) $2a\pi/(a^2-1)^{3/2}$ (24) $\pi/2\sqrt{2}$ (25) $\pi/(a+b)$

(26) $5\pi/12$ (27) $\pi/3$ (28) $\frac{\pi\sqrt{3}}{6}$

(29) $\frac{\pi}{2\sqrt{2}}$ (30) $\frac{\pi}{16}$ (31) $\frac{\pi}{4}$

CHAPTER 5

LAPLACE TRANSFORMS

5.0 INTRODUCTION

The Laplace transform is a very powerful tool in the hands of engineers and scientists for finding solutions to initial-value problems. The subject "Laplace transforms" originated from the operational methods applied by the English Engineer Oliver Heaviside (1850-1925), to problems in electrical engineering. Heaviside's operational calculus is best introduced by means of a particular type of definite integrals (called Laplace transforms). Pierre de Laplace (1749-1827) used such transforms while developing the theory of probability.

The method of Laplace transforms is generally used to solve systems of differential equations, partial differential equations and integral equations under suitable initial and boundary conditions. Laplace transform converts a given initial-value problem to an algebraic equation or a system of algebraic equations. This method has the advantage of directly giving the solution of differential equations with given boundary values without the necessity of first finding the general solution and then evaluating from it the arbitrary constants, to obtain the particular solution. In this chapter we derive the Laplace transforms of elementary functions, their derivatives and integrals and Laplace transforms of periodic functions. Further we solve several initial value problems involving linear ordinary differential equations up to second order with constant coefficients.

5.1 LAPLACE TRANSFORMS OF ELEMENTARY FUNCTIONS

Definition 5.1.1 Let f(t) be a function defined for $t \geq 0$. Then the integral

$$F(s) = \int_0^\infty e^{-st} f(t).dt \qquad (1)$$

is called the **Laplace transform of f(t)**, provided the integral exists.

The Laplace transform of f(t) is usually denoted by L[f(t)] and the symbol L, which transforms f(t) into F(s) is called the **Laplace transform**.

$$\text{i.e., } L[f(t)] = F(s) \qquad (2)$$

The original function f(t) is called the **inverse Laplace transform** of F(s) and we write

$$L^{-1}[F(s)] = f(t) \qquad (3)$$

5.1.1 SUFFICIENT CONDITIONS FOR THE EXISTENCE OF LAPLACE TRANSFORM

Definition 5.1.2 A function f(t) is said to be of exponential order α, if there exists constants α and $M > 0$ such that $|f(t)| \le Me^{\alpha t}$, for $t \ge 0$.

For example, the functions $f_1(t) = t$ and $f_2(t) = t^2$ are of exponential order.
$|f_1(t)| = |t| \le e^t \Rightarrow f_1(t)$ is of exponential order 1.
and $|f_2(t)| = |t^2| \le e^{2t} \Rightarrow f_2(t)$ is of exponential order 2.

Theorem 5.1.1 (Sufficient conditions for the existence of Laplace transform)
If f(t) is a piecewise continuous function on $[0, \infty)$ and is of exponential order α for $t \ge 0$, then L[f(t)] exists for $s > \alpha$.

Proof:
If f(t) is piecewise continuous on [0, T] for $T > 0$, then $e^{-st}.f(t)$ is also piecewise continuous on [0, T].

Hence $\int_0^T e^{-st} f(t).dt$ exists.

$\therefore$ The existence of $L[F(t)] = \int_0^\infty e^{-st} f(t).dt$ depends on whether $\int_0^T e^{-st} f(t).dt$ converges or has a finite limit as $T \to \infty$.

$$\text{Now } \left|\int_0^T e^{-st} f(t).dt\right| \le \int_0^T e^{-st} |f(t)|.dt$$

$$\le M\int_0^T e^{-st}.e^{\alpha t}dt, \text{ since f(t) is of exponential order.}$$

$$\le M\int_0^T e^{(\alpha - s)t}dt$$

$$\le M\left[\frac{e^{(\alpha - s)t}}{\alpha - s}\right]_0^T$$

$$\le \frac{M}{\alpha - s}\left[e^{(\alpha - s)T} - 1\right]$$

$$\le \frac{M}{s - \alpha}\left[1 - e^{-(s-\alpha)T}\right]$$

$$\leq \frac{M}{s-\alpha} \quad \text{if } s > \alpha \text{ and } T \to \infty$$

Hence $|L[f(t)]| \leq \dfrac{M}{s-\alpha}$

Thus $\int_0^\infty e^{-st} f(t).dt$ exists for $s > \alpha$.

i.e., L[f(t)] exists for $s > \alpha$.

Note:

(i) We have $|L[f(t)]| \leq \dfrac{M}{s-\alpha} \quad \to 0 \text{ as } s \to \infty$

$\therefore \lim\limits_{s\to\infty} L[f(t)] = 0$

i.e., $\lim\limits_{s\to\infty} F(s) = 0$

(ii) The conditions given in Theorem 5.1.1 are only sufficient to guarantee the existence of Laplace transforms. However they are not necessary.

For example, consider the function $f(t) = \dfrac{1}{\sqrt{t}}$. When $t \to 0$, $f(t) \to \infty$.

Hence f(t) is not piecewise continuous on any interval [0, T].

However $L\left(1/\sqrt{t}\right) = \int_0^\infty e^{-st}.t^{-1/2} dt$

Putting $t = u^2$, we get, $dt = 2udu = 2.t^{1/2} du$ i.e., $t^{-1/2} dt = 2du$

$$\therefore \quad L\left(1/\sqrt{t}\right) = 2\int_0^\infty e^{-su^2} du$$

$$= \frac{2}{\sqrt{s}} \int_0^\infty e^{-z^2} dz \quad (\text{Putting } z = u\sqrt{s},\ dz = \sqrt{s}.du)$$

$$= \frac{2}{\sqrt{s}} \frac{\sqrt{\pi}}{2}$$

$$= \sqrt{\frac{\pi}{s}} \quad \text{for } s > 0.$$

Example 5.1.1 Find L[1]

Solution:

$$L[1] = \int_0^\infty e^{-st} dt = \left.\frac{e^{-st}}{-s}\right]_0^\infty = \frac{1}{s} \quad \text{for } s > 0.$$

Example 5.1.2 Find $L[t^n]$ for n = 0, 1, 2, 3, …

Solution:

$$L[t^n] = \int_0^\infty e^{-st}.t^n dt = \left[t^n \frac{e^{-st}}{-s}\right]_0^\infty - \int_0^\infty \frac{e^{-st}}{-s}.n.t^{n-1}dt$$

$$= \frac{n}{s}\int_0^\infty e^{-st}.t^{n-1}dt \text{ for } s > 0$$

i.e., $L\left[t^n\right] = \frac{n}{s}L\left[t^{n-1}\right]$

$$= \frac{n}{s}.\frac{n-1}{s}.\frac{n-2}{s}.\cdots.\frac{2}{s}.\frac{1}{s}L[1]$$

$$= \frac{n}{s}.\frac{n-1}{s}.\frac{n-2}{s}.\cdots.\frac{2}{s}.\frac{1}{s}.\frac{1}{s}$$

$$= \frac{n!}{s^{n+1}} \text{ for } s > 0$$

Example 5.1.3 Find $L\left[e^{at}\right]$

Solution:

$$L\left[e^{at}\right] = \int_0^\infty e^{-st}.e^{at}dt$$

$$= \int_0^\infty e^{-(s-a)t}dt$$

$$= \left[\frac{e^{-(s-a)t}}{-(s-a)}\right]_0^\infty = \frac{1}{s-a} \qquad \text{for } s > a$$

Example 5.1.4 If f(t) = sin at, find L[f(t)]

Solution:

$$L[\sin at] = \int_0^\infty e^{-st}\sin at\, dt$$

$$= \text{Imaginary part of } \int_0^\infty e^{-st}e^{iat}\, dt$$

$$= \text{Imaginary part of } \int_0^\infty e^{-(s-ia)t}\, dt$$

$$= \text{Imaginary part of } \frac{1}{s - ia}$$

$$= \text{Imaginary part of } \frac{s + ia}{s^2 + a^2} \quad = \frac{a}{s^2 + a^2}$$

Example 5.1.5 If f(t) = cos at, find L[f(t)].

Solution:

$$L[\cos at] = \int_0^\infty e^{-st} \cos at \, dt$$

$$= \text{Real part of } \int_0^\infty e^{-st} e^{iat} \, dt$$

$$= \text{Real part of } \int_0^\infty e^{-(s-ia)t} \, dt$$

$$= \text{Real part of } \frac{1}{s - ia}$$

$$= \text{Real part of } \frac{s + ia}{s^2 + a^2}$$

$$= \frac{s}{s^2 + a^2}$$

5.1.2 PROPERTIES OF LAPLACE TRANSFORMS

Property I (Linearity Property)

If c_1 and c_2 are any constants, then $L[c_1 f_1(t) + c_2 f_2(t)] = c_1 L[f_1(t)] + c_2 L[f_2(t)]$

Proof: $L[c_1 f_1(t) + c_2 f_2(t)] = \int_0^\infty e^{-st} (c_1 f_1(t) + c_2 f_2(t)) dt$

$$= c_1 \int_0^\infty e^{-st} f_1(t) \, dt + c_2 \int_0^\infty e^{-st} f_2(t) \, dt$$

$$= c_1 L[f_1(t)] + c_2 L[f_2(t)]$$

Property II (Shifting Property or First Shifting Theorem)

If L[f(t)] = F(s), then $L[e^{at} f(t)] = F(s - a)$, where a is a real constant.

Proof: $L[e^{at} f(t)] = \int_0^\infty e^{-st} . e^{at} f(t) \, dt$

$$= \int_0^\infty e^{-(s-a)t} f(t)dt$$

$$= F(s - a)$$

Property III (Change of Scale Property)

If L[f(t)] = F(s), then $L[f(at)] = \frac{1}{a} F\left(\frac{s}{a}\right)$

Proof: $L[f(at)] = \int_0^\infty e^{-st} f(at)\,dt$

$$= \int_0^\infty e^{-\frac{s}{a}u} f(u) \frac{1}{a} du \quad \text{, putting } at = u,\ adt = du$$

$$= \frac{1}{a} \int_0^\infty e^{-\frac{s}{a}u} f(u)\,du$$

$$= \frac{1}{a} F\left(\frac{s}{a}\right)$$

Property IV

If L[f(t)] = F(s), then $L[t^n f(t)] = (-1)^n \frac{d^n}{ds^n} F(s)$ where n = 1, 2, 3, …

Proof: $F(s) = \int_0^\infty e^{-st} f(t)\,dt$

$$\therefore \frac{d}{ds} F(s) = \frac{d}{ds} \int_0^\infty e^{-st} f(t)\,dt$$

$$= \int_0^\infty e^{-st} (-t) f(t)\,dt \qquad \text{(By Leibnitz's rule)}$$

$$= L[tf(t)]$$

$\therefore L[t\,f(t)] = -\frac{d}{ds} F(s)$, which proves the result for n = 1

Assume that result is true for n = m

i.e., $L[t^m f(t)] = (-1)^m \frac{d^m}{ds^m} (F(s))$

$$\int_0^\infty e^{-st} t^m f(t)\,dt = (-1)^m \frac{d^m}{ds^m} (F(s))$$

Then $\frac{d}{ds}\left[\int_0^\infty e^{-st} t^m f(t)\,dt\right] = (-1)^m \frac{d^{m+1}}{ds^{m+1}} (F(s))$

i.e., $\int_0^\infty e^{-st}(-t)t^m f(t)\,dt = (-1)^m \frac{d^{m+1}}{ds^{m+1}}(F(s))$ (By Leibnitz's rule)

i.e., $\int_0^\infty e^{-st} t^{m+1} f(t)\,dt = (-1)^{m+1} \frac{d^{m+1}}{ds^{m+1}}(F(s))$

i.e., $L\left[t^{m+1} f(t)\right] = (-1)^{m+!} \frac{d^{m+!}}{ds^{m+1}}(F(s))$

By induction principal, the result is true for all values of n.

i.e., $L\left[t^n f(t)\right] = (-1)^n \frac{d^n}{ds^n} F(s)$ where n = 1, 2, 3, …

Property V

If L[f(t)] = F(s), then $L\left[\frac{1}{t} f(t)\right] = \int_s^\infty F(s)\,ds$ provided the integral exists.

Proof: We have $F(s) = \int_0^\infty e^{-st} f(t)\,dt$

Integrating both sides w.r.t s from s to ∞,

$$\int_s^\infty F(s)\,ds = \int_s^\infty \left(\int_0^\infty e^{-st} f(t)\,dt \right) ds$$

$$= \int_0^\infty \int_s^\infty f(t)\, e^{-st}\, ds\, dt \quad \text{(By changing the order of integration)}$$

$$= \int_0^\infty f(t) \left(\int_s^\infty e^{-st} ds \right) dt$$

$$= \int_0^\infty f(t) \left(\left[\frac{e^{-st}}{-t} \right]_s^\infty \right) dt$$

$$= \int_0^\infty f(t) \frac{e^{-st}}{t}\, dt$$

$$= \int_0^\infty e^{-st} \frac{f(t)}{t}\, dt$$

$$= L\left[\frac{1}{t} f(t)\right]$$

Hence $L\left[\frac{1}{t} f(t)\right] = \int_s^\infty F(s)\,ds$

Example 5.1.6 Prove $L[\sinh at] = \frac{a}{s^2 - a^2}$ and $L[\cosh at] = \frac{s}{s^2 - a^2}$.

Solution:

$$L[\sinh at] = L\left[\frac{1}{2}\left(e^{at} - e^{-at}\right)\right]$$

$$= L\left[\frac{e^{at}}{2} - \frac{e^{-at}}{2}\right]$$

$$= \frac{1}{2}L\left[e^{at}\right] - \frac{1}{2}L\left[e^{-at}\right] \quad \text{(By property I)}$$

$$= \frac{1}{2}\frac{1}{s-a} - \frac{1}{2}\frac{1}{s+a}$$

$$= \frac{1}{2}\frac{(s+a)-(s-a)}{s^2 - a^2}$$

$$= \frac{a}{s^2 - a^2}$$

$$L[\cosh at] = L\left[\frac{e^{at} + e^{-at}}{2}\right]$$

$$= \frac{1}{2}L\left[e^{at}\right] + \frac{1}{2}L\left[e^{-at}\right]$$

$$= \frac{1}{2}\frac{1}{s-a} + \frac{1}{2}\frac{1}{s+a}$$

$$= \frac{s}{s^2 - a^2}$$

Example 5.1.7 Prove $L[e^{at}.t^n] = \frac{n!}{(s-a)^{n+1}}$, $n = 0,1,2,3,\cdots$

Solution:

We have $L[t^n] = \frac{n!}{s^{n+1}}$, $n = 0,1,2,3,\cdots$

By property II, $L[e^{at} f(t)] = F(s-a)$

Hence $L[e^{at}.t^n] = \frac{n!}{(s-a)^{n+1}}$

Note:

By applying property II, we can prove that,

(i) $L[e^{at}.\sin bt] = \frac{b}{(s-a)^2 + b^2}$

(ii) $L[e^{at}.\cos bt] = \frac{s-a}{(s-a)^2 + b^2}$

(iii) $L[e^{at}.\sinh bt] = \dfrac{b}{(s-a)^2 - b^2}$

(iv) $L[e^{at}.\cosh bt] = \dfrac{s-a}{(s-a)^2 - b^2}$

Example 5.1.8 Find $L[t^2 \cos 2t]$

Solution:

$$L[\cos 2t] = \frac{s}{s^2 + 2^2}$$

$$L\left[t^2 \cos 2t\right] = (-1)^2 \frac{d^2}{ds^2}\left(L\left[\cos 2t\right]\right) \qquad \text{(By property IV)}$$

$$= \frac{d^2}{ds^2}\left(\frac{s}{s^2+4}\right)$$

$$= \frac{d}{ds}\left(\frac{s^2+4-s.2s}{\left(s^2+4\right)^2}\right)$$

$$= \frac{d}{ds}\left(\frac{4-s^2}{\left(s^2+4\right)^2}\right)$$

$$= \frac{\left(s^2+4\right)^2(-2s) - \left(4-s^2\right)2\left(s^2+4\right)2s}{\left(s^2+4\right)^4}$$

$$= \frac{-2s^3 - 8s - 16s + 4s^3}{\left(s^2+4\right)^3}$$

$$= \frac{2s^3 - 24s}{\left(s^2+4\right)^3}$$

Example 5.1.9 Prove that $L[t^2 e^{-2t}.\cos t] = \dfrac{2\left(s^3 + 6s^2 + 9s + 2\right)}{\left(s^2 + 4s + 5\right)^3}$

Solution:

$$L\left[\cos t\right] = \frac{s}{s^2+1}$$

$$L\left[t^2 \cos t\right] = (-1)^2 \frac{d^2}{ds^2}\left(\frac{s}{s^2+1}\right) \qquad \text{(By property IV)}$$

$$= \frac{d}{ds}\left(\frac{s^2+1-s.2s}{\left(s^2+1\right)^2}\right)$$

$$= \frac{d}{ds}\left(\frac{1-s^2}{(s^2+1)^2}\right)$$

$$= \frac{(s^2+1)^2(-2s)-(1-s^2)2(s^2+1)2s}{(s^2+1)^4}$$

$$= \frac{-2s^3-2s-4s+4s^3}{(s^2+1)^3}$$

$$= \frac{2s^3-6s}{(s^2+1)^3}$$

$$\therefore\ L[e^{-2t}t^2.\cos t] = \frac{2(s+2)^3-6(s+2)}{((s+2)^2+1)^3} \qquad \text{(By property II)}$$

$$\text{i.e., } L[t^2e^{-2t}.\cos t] = \frac{2(s^3+6s^2+9s+2)}{(s^2+4s+5)^3}$$

Example 5.1.10 Prove $L\left[\frac{\sin at}{t}\right] = \cot^{-1}(s/a)$

Solution:

We have $L[\sin at] = \dfrac{a}{s^2+a^2}$

By property V, $L\left[\frac{1}{t}\sin at\right] = \int_s^\infty \frac{a}{s^2+a^2}\,ds$

$$\text{i.e., } L\left[\frac{\sin at}{t}\right] = a.\frac{1}{a}\tan^{-1}(s/a)\Big]_s^\infty$$

$$= \frac{\pi}{2} - \tan^{-1}(s/a) = \cot^{-1}(s/a)$$

$$\left[\text{Since } \pi/2 - \tan^{-1}(s/a) = x \Rightarrow \tan^{-1}(s/a) = \pi/2 - x\right.$$

$$\Rightarrow \frac{s}{a} = \tan(\pi/2 - x)$$

$$\left.\Rightarrow \frac{s}{a} = \cot x \Rightarrow x = \cot^{-1}(s/a)\right]$$

Example 5.1.11 Find the Laplace transform of $\dfrac{\cos at - \cos bt}{t}$

Solution:

$$L[\cos at - \cos bt] = L[\cos at] - L[\cos bt]$$

$$= \frac{s}{s^2+a^2} - \frac{s}{s^2+b^2}$$

$$L\left[\frac{\cos at - \cos bt}{t}\right] = \int_s^{\infty}\left(\frac{s}{s^2+a^2} - \frac{s}{s^2+b^2}\right)ds$$

$$= \frac{1}{2}\log(s^2+a^2) - \frac{1}{2}\log(s^2+b^2)\Big]_s^{\infty}$$

$$= \frac{1}{2}\log\frac{s^2+a^2}{s^2+b^2}\Big]_s^{\infty}$$

$$= \left(\lim_{s\to\infty}\frac{1}{2}\log\frac{s^2+a^2}{s^2+b^2}\right) - \frac{1}{2}\log\frac{s^2+a^2}{s^2+b^2}$$

$$= \lim_{s\to\infty}\frac{1}{2}\log\left(\frac{1+a^2/s^2}{1+b^2/s^2}\right) - \frac{1}{2}\log\frac{s^2+a^2}{s^2+b^2}$$

$$= 0 - \frac{1}{2}\log\frac{s^2+a^2}{s^2+b^2}$$

$$= \frac{1}{2}\log\frac{s^2+b^2}{s^2+a^2}$$

Example 5.1.12 Find the Laplace transform of $\frac{e^{-t}\sin t}{t}$

Solution:

$$L[\sin t] = \frac{1}{s^2+1}$$

$$\therefore L\left[\frac{\sin t}{t}\right] = \int_s^{\infty}\frac{1}{s^2+1}ds$$

$$= \tan^{-1}s\Big]_s^{\infty}$$

$$= \frac{\pi}{2} - \tan^{-1}s$$

$$= \cot^{-1}s$$

Now $L\left[\frac{e^{-t}\sin t}{t}\right] = \cot^{-1}(s+1)$.

Example 5.1.13 Find the Laplace transform of $\frac{1-\cos 2t}{t}$

Solution:

$$L[1\text{-}\cos 2t] = L[1] - L[\cos 2t]$$

$$= \frac{1}{s} - \frac{s}{s^2+4}$$

$$L\left[\frac{1-\cos 2t}{t}\right] = \int_s^{\infty}\left[\frac{1}{s} - \frac{s}{s^2+4}\right]ds$$

$$= \log s - \frac{1}{2}\log(s^2+4)\Big]_s^{\infty}$$

$$= \frac{1}{2}\log\left(s^2/(s^2+4)\right)\Big]_s^{\infty}$$

$$= \lim_{s\to\infty}\frac{1}{2}\log\left(\frac{s^2}{s^2+4}\right) - \frac{1}{2}\log\left(\frac{s^2}{s^2+4}\right)$$

$$= \lim_{s\to\infty}\frac{1}{2}\log\left(\frac{1}{1+4/s^2}\right) + \frac{1}{2}\log\frac{s^2+4}{s^2}$$

$$= 0 + \frac{1}{2}\log\left(\frac{s^2+4}{s^2}\right)$$

$$= \frac{1}{2}\log\left(\frac{s^2+4}{s^2}\right)$$

Example 5.1.14 Evaluate $\int_0^{\infty} t.e^{-2t}.\cos t.dt$

Solution:

Replacing 2 by s, the given integral becomes,

$$\int_0^{\infty} t.e^{-2t}.\cos t.dt = \int_0^{\infty} e^{-st}(t.\cos t)dt$$

$$= L[t\cos t]$$

$$= (-1)\frac{d}{ds}L[\cos t]$$

$$= (-1)\frac{d}{ds}\left(\frac{s}{s^2+1}\right)$$

$$= (-1)\frac{s^2+1-s.2s}{(s^2+1)^2}$$

$$= \frac{s^2-1}{(s^2+1)^2}$$

$$= \frac{2^2-1}{(2^2+1)^2} = \frac{3}{25} \quad \text{(Substituting } s = 2\text{)}$$

List of Properties:

If F(s) = L[f(t)], then

1. $L[e^{at}f(t)] = F(s - a)$
2. $L[f(at)] = \frac{1}{a}F(s/a)$
3. $L[t^n f(t)] = (-1)^n \frac{d^n}{ds^n}[F(s)]$ n = 1, 2, 3, …
4. $L\left[\frac{1}{t}f(t)\right] = \int_s^{\infty} F(s)\,ds$

In the following table we list the elementary functions and their Laplace transforms for ready reference. These functions and their Laplace transforms are frequently used while solving problems.

Table of Elementary Laplace transforms

No	f(t)	$F(s) = \int_0^{\infty} e^{-st} f(t)\,dt$
1	1	$\frac{1}{s}$
2	t^n $(n \geq 0)$	$\frac{n!}{s^{n+1}}$
3	e^{at}	$\frac{1}{s-a}$
4	$L[e^{at}.t^n]$	$\frac{n!}{(s-a)^{n+1}}$, $n = 0,1,2,3,\cdots$
5	sin at	$\frac{a}{s^2+a^2}$
6	cos at	$\frac{s}{s^2+a^2}$
7	sinh at	$\frac{a}{s^2-a^2}$
8	cosh at	$\frac{s}{s^2-a^2}$

5.1.3 UNIT STEP FUNCTION 0R HEAVISIDE'S FUNCTION

Definition 5.1.3 The unit step function is the function u(t) defined by

$$u(t) = \begin{cases} 0 & \text{if } t < 0 \\ 1 & \text{if } t \geq 0 \end{cases}$$

The unit step function u(t) is also known as Heaviside's function, named after the British Electrical Engineer Oliver Heaviside (1850-1925). The unit step function has a discontinuity at $t = 0$, with a jump of magnitude 1. If the jump discontinuity is at $t = a$,

$$\text{we define } u(t-a) = \begin{cases} 0 & \text{if } t < a \\ 1 & \text{if } t \geq a \end{cases}$$

u(t - a) is also denoted by $u_a(t)$. (Fig. 5.1)

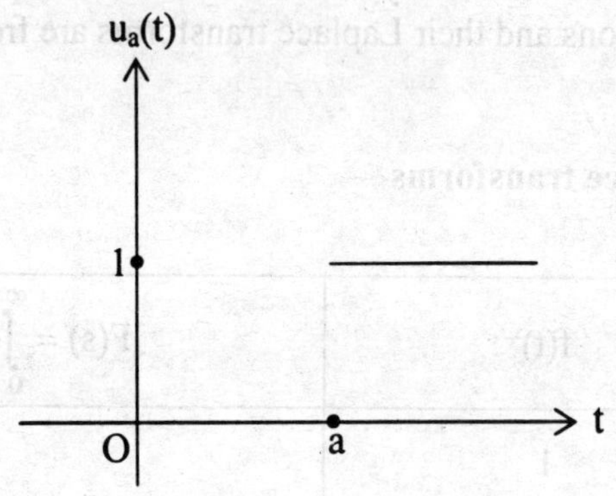

Fig. 5.1

Note:

(i) $L[u_a(t)] = \int_0^\infty e^{-st}.u_a(t)dt, \quad a \geq 0$

$$= \int_0^a e^{-st}.0.dt + \int_a^\infty e^{-st}.1.dt$$

$$= \left[\frac{e^{-st}}{-s}\right]_a^\infty = \frac{e^{-as}}{s}$$

(ii) If f(t) is any function, then $f(t).u_a(t) = \begin{cases} 0 & \text{if } t < a \\ f(t) & \text{if } t \geq a \end{cases}$

For example, if $f(t) = \begin{cases} 0 & \text{if } t < a \\ t^2 & \text{if } t \geq a \end{cases}$ then we can write $f(t) = t^2 u_a(t)$

(iii) If $a < b$, then $u_a(t) - u_b(t) = \begin{cases} 0 & \text{if } t < a \\ 1 & \text{if } a \le t < b \\ 0 & \text{if } t \ge b \end{cases}$

Theorem 5.1.2 (Second Shifting Property or Second Shifting Theorem)
If $L[f(t)] = F(s)$ and $a \ge 0$ then $L[f(t - a).u_a(t)] = e^{-as}. F(s)$

Proof: $L[f(t-a).u_a(t)] = \int_0^\infty e^{-st} f(t-a).u_a(t)dt$

$$= \int_0^a e^{-st} f(t-a).0.dt + \int_a^\infty e^{-st} f(t-a).1.dt$$

$$= \int_0^\infty e^{-s(u+a)} f(u)du \quad \text{(Putting t-a = u)}$$

$$= e^{-as} \int_0^\infty e^{-su} f(u)du$$

$$= e^{-as}. F(s)$$

(iv) We have $f(t).u_a(t) = f(t - a + a)\, u_a(t) = g(t - a).u_a(t)$, where $g(t) = f(t + a)$
Now by theorem 5.1.2, $L[g(t - a).u_a(t)] = e^{-as} L[g(t)] = e^{-as} L[f(t + a)]$.
i.e., $L[f(t).u_a(t)] = e^{-as} L[f(t + a)]$

Example 5.1.15 Find the Laplace transform of the function $f(t) = \begin{cases} 0 & 0 \le t \le 5 \\ (t-5)^2 & t > 5 \end{cases}$

Solution:
We can write $f(t) = (t - 5)^2.u_5(t)$

$\therefore\ L[f(t)] = L[(t - 5)^2.u_5(t)] = e^{-5s} L[t^2]$

$$= e^{-5s}.\frac{2}{s^3}$$

$$= \frac{2e^{-5s}}{s^3}$$

Example 5.1.16 Find the Laplace transforms of (i) $f(t) = \begin{cases} t^2 & 0 \le t \le 2 \\ 0 & t > 2 \end{cases}$

(ii) $f(t) = \begin{cases} 0 & 0 \le t \le 1 \\ t-1 & 1 \le t < 2 \\ 0 & t \ge 2 \end{cases}$ (iii) $f(t) = \begin{cases} 5 & 0 \le t < 2 \\ -5 & t \ge 2 \end{cases}$

Solution:

(i) f(t) can be written as $f(t) = t^2[1 - u_2(t)] = t^2 - t^2.\,u_2(t)$

$\therefore L[f(t)] = L[t^2] - L[t^2.u_2(t)]$

$$= \frac{2}{s^3} - e^{-2s}L[(t+2)^2]$$

$$= \frac{2}{s^3} - e^{-2s}L[t^2 + 4t + 4]$$

$$= \frac{2}{s^3} - e^{-2s}\left(\frac{2}{s^3} + 4.\frac{1}{s^2} + \frac{4}{s}\right)$$

$$= \frac{2}{s^3}\left(1 - e^{-2s}\right) - 4e^{-2s}\left(\frac{1}{s^2} + \frac{1}{s}\right)$$

(ii) f(t) can be written as $f(t) = (t - 1)(u_1(t) - u_2(t))$

$\therefore L[f(t)] = L[(t - 1).u_1(t)] - L[(t - 1).u_2(t)]$

$$= e^{-s}L[t] - e^{-2s}L[t - 1 + 2]$$

$$= e^{-s}L[t] - e^{-2s}.L[t] - e^{-2s}.\,L[1]$$

$$= e^{-s}.\frac{1}{s^2} - e^{-2s}\frac{1}{s^2} - e^{-2s}\frac{1}{s}$$

$$= \frac{1}{s^2}\left[e^{-s} - (1+s)e^{-2s}\right]$$

(iii) f(t) can be written as $f(t) = 10(1 - u_2(t)) - 5 = 5 - 10u_2(t)$

$\therefore\ L[f(t)] = L[5] - 10L[u_2(t)]$

$$= \frac{5}{s} - 10.\frac{e^{-2s}}{s}$$

$$= \frac{5}{s}(1 - 2e^{-2s})$$

EXERCISE 5.1

PART-A

Find the Laplace transform of each of the following functions. (Ex: 1-20)

1. $\cos^2 2t$
2. $t^3.e^{2t}$
3. $t \sin t$
4. $2t - 5$
5. $3t^2 + 2e^{-t}$
6. $5\cos 3t$
7. $\sinh 5t$
8. $(t^2 + 3)^2$

9. $e^{3t}\sin 2t$
10. $\cos(at + b)$
11. $e^{-2t}\sin 4t$
12. $t - \sinh 2t$
13. $e^{-at}\sinh bt$
14. $t \sinh at$
15. $t^2 e^t$
16. $(\cos t + \sin t)^2$
17. $6 - e^{3t}$
18. $3t^2 - 5e^{-2t} + 6$
19. $t \cos 2t$
20. e^{-2t+5}

PART-B

Find the Laplace transforms of the following functions. (Ex: 21-32)

21. $\sin 2t \cos 3t$
22. $\sin^3 2t$
23. $e^{4t}\sin 2t \cos t$
24. $e^{-t}\sin^2 t$
25. $\sin at \sin bt$
26. $e^{-t}(2\cos 5t - 3\sin 5t)$
27. $t^2\cos t$
28. $t^2\sin at$
29. $te^{-t}\sin 3t$
30. $\dfrac{e^{-t}\sin t}{t}$
31. $\dfrac{e^{at} - \cos bt}{t}$
32. $\dfrac{1-\cos t}{t^2}$

Answer the following questions. (Ex: 33-45)

33. Find the Laplace transform of $f(t) = \begin{cases} e^t & 0 < t < 1 \\ 0 & t > 1 \end{cases}$

34. Find the Laplace transform of $f(t) = \begin{cases} \sin t & 0 < t < \pi \\ 0 & t > \pi \end{cases}$

35. Find the Laplace transform of $f(t) = \begin{cases} 1 & 0 \le t < 1 \\ -1 & t \ge 1 \end{cases}$

36. Find the Laplace transform of $f(t) = \begin{cases} 2 & 0 \le t < 3 \\ 0 & t \ge 3 \end{cases}$

37. Find the Laplace transform of $f(t) = \begin{cases} 0 & 0 \le t < \pi \\ \sin t & t \ge \pi \end{cases}$

38. Prove $\int_0^\infty t.e^{-2t}.\sin t.dt = \frac{4}{25}$.

39. Prove $\int_0^\infty \frac{\sin mt}{t}.dt = \frac{\pi}{2}$ if $m > 0$

40. Prove $\int_0^\infty \frac{e^{-t}.\sin^2 t}{t}.dt = \frac{1}{4}\log 5$

41. Prove $\int_0^{\infty} \frac{e^{-t} - e^{-3t}}{t} .dt = \log 3$

42. Find the Laplace transform of $t^2 \int_0^t e^{-5u} .du$.

43. Find the Laplace transform of $t \int_0^t e^{-u} \sin 2u.du$.

44. Find the Laplace transforms of the following functions.
   (i) $(t^2-1)u_2(t)$
   (ii) $\sin t. u_{\pi}(t)$
   (iii) $e^{3-t}.u_3(t)$
   (iv) $(t-1)^2.u_1(t)$

45. Find the Laplace transforms of

(i) $f(t) = \begin{cases} 0 & 0 \le t < \pi/2 \\ \sin t & t \ge \pi/2 \end{cases}$ (ii) $f(t) = \begin{cases} k & 0 \le t < 2 \\ 0 & 2 \le t < 4 \\ k & t \ge 4 \end{cases}$

5.2 LAPLACE TRANSFORMS OF DERIVATIVES AND INTEGRALS

5.2.1 TRANSFORMS OF DERIVATIVES

Theorem 5.2.1 Let f(t), $t \ge 0$ be a continuous function and be of exponential order on $[0, \infty)$. If the Laplace transform of f′(t) exists, then $L[f'(t)] = s. L[f(t)] - f(0)$

i.e., $L[f'(t)] = s. F(s) - f(0)$ where $F(s) = L[f(t)]$

Proof:

$$L[f'(t)] = \int_0^{\infty} e^{-st} f'(t)\, dt \quad \text{(By definition)}$$

$$= e^{-st} f(t)\Big]_0^{\infty} - \int_0^{\infty} (-s) e^{-st} f(t)\, dt$$

$$= \lim_{t\to\infty} e^{-st} f(t) - f(0) + s. \int_0^{\infty} e^{-st} f(t) dt$$

$$= 0 - f(0) + s. L[f(t)] \quad \text{(Since f(t) is of exponential order)}$$

i.e., $L[f'(t)] = s. L[f(t)] - f(0)$

i.e., $L[f'(t)] = s. F(s) - f(0)$, where $F(s) = L[f(t)]$ (1)

Theorem 5.2.2 Let the function f(t) and its first (n -1) derivatives be continuous for $t \geq 0$ and $\lim_{t\to\infty} e^{-st} f^{(m)}(t) = 0$ for m = 0, 1, 2, …, n -1. Then if $L[f^{(n)}(t)]$ exists, we have

$$L[f^{(n)}(t)] = s^n L[f(t)] - s^{n-1}f(0) - s^{n-2}f'(0) - \dots - sf^{(n-2)}(0) - f^{(n-1)}(0)$$

i.e., $L[f^{(n)}(t)] = s^n F(s) - s^{n-1}f(0) - s^{n-2}f'(0) - \dots - s\, f^{(n-2)}(0) - f^{(n-1)}(0)$ where $F(s) = L[f(t)]$

Proof:

$$L[f^{(n)}(t)] = \int_0^\infty e^{-st} f^{(n)}(t)\,dt$$

$$= \left[e^{-st}f^{(n-1)}(t) - (-s)e^{-st}f^{(n-2)}(t) + (-s)^2 e^{-st}f^{(n-3)}(t) + \cdots + (-1)^{n-1}(-s)^{n-1}e^{-st}f(t)\right]_0^\infty$$

$$+ (-1)^n(-s)^n \int_0^\infty e^{-st} f(t)\,dt$$

(By the general rule of integration by parts)

$$= -f^{(n-1)}(0) - s\, f^{(n-2)}(0) - s^2 f^{(n-3)}(0) - \dots - s^{n-1}f(0) + s^n \int_0^\infty e^{-st} f(t)\,dt$$

$$= s^n L[f(t)] - s^{n-1}f(0) - s^{n-2}f'(0) - \dots - s\, f^{(n-2)}(0) - f^{(n-1)}(0)$$

i.e., $L[f^{(n)}(t)] = s^n F(s) - s^{n-1}f(0) - s^{n-2}f'(0) - \dots - s\, f^{(n-2)}(0) - f^{(n-1)}(0)$ (2)

5.2.2 TRANSFORMS OF INTEGRALS

Theorem 5.2.3 If $L[f(t)] = F(s)$, then $L\left[\int_0^t f(u)\,du\right] = \frac{1}{s}F(s)$.

Proof: Let $\varphi(t) = \int_0^t f(u)\,du$

Then $\varphi(0) = 0$ and $\varphi'(t) = f(t)$

$\therefore L[\varphi'(t)] = s.\, L[\varphi(t)] - \varphi(0)$

i.e., $L[f(t)] = s.\, L[\varphi(t)] - 0$

i.e., $F(s) = s.\, L[\varphi(t)]$

$\therefore L[\varphi(t)] = \frac{1}{s}F(s)$.

i.e., $L\left[\int_0^t f(u)\,du\right] = \frac{1}{s}F(s)$.

Note:

We have, from Theorem 5.2.3 $L\left[\int_0^t f(u)\,du\right] = \frac{1}{s}L[f(t)]$.

Example 5.2.1 If $f(t) = \sin^2(t)$, find $L[f(t)]$ using the formula for transform of derivatives.

Solution:

We have $L[f'(t)] = s. L[f(t)] - f(0)$ (1)

Now $f'(t) = 2 \sin t.\cos t = \sin 2t$

$$\therefore L[f'(t)] = L[\sin 2t] = \frac{2}{s^2+4}$$

$f(0) = \sin^2 0 = 0$

$$\therefore \text{From (1) we get, } \frac{2}{s^2+4} = s. L[\sin^2 t] - 0$$

$$\text{i.e., } L[\sin^2 t] = \frac{2}{s(s^2+4)}$$

Example 5.2.2 Find the Laplace transform of $f(t) = t^2$ using the formula for the transform of derivatives.

Solution:

We have $L[f''(t)] = s^2 L[f(t)] - sf(0) - f'(0)$ (1)

$f(t) = t^2$ $f(0) = 0$

$f'(t) = 2t$ $f'(0) = 0$ $f''(t) = 2$

$\therefore$ From (1) we get,

$L[2] = s^2 L[t^2] - s.0 - 0$

$$\therefore L[t^2] = \frac{1}{s^2}.L[2]$$

$$= \frac{1}{s^2}.\frac{2}{s} = \frac{2}{s^3}$$

Example 5.2.3 If $L[f(t)] = F(s)$, show that $L\left[\int_a^t f(u)\,du\right] = \frac{1}{s}F(s) - \frac{1}{s}\int_0^a f(u)\,du$, $a > 0$.

Solution:

$$L\left[\int_a^t f(u)\,du\right] = L\left[\int_0^t f(u)\,du - \int_0^a f(u)\,du\right] \quad (\text{as } a > 0)$$

$$= L\left[\int_0^t f(u)\,du\right] - L\left[\int_0^a f(u)\,du\right]$$

$$= \frac{1}{s}F(s) - \int_0^a f(u)\,du.L(1) \quad \left(\because \int_0^a f(u)\,du \text{ is a constant}\right)$$

$$= \frac{1}{s}F(s) - \int_0^a f(u)\,du.\frac{1}{s}$$

Hence $L\left[\int_a^t f(u)\,du\right] = \frac{1}{s}F(s) - \frac{1}{s}\int_0^a f(u)\,du$

Example 5.2.4 Prove that $L\left[\int_0^t \frac{e^t \sin t}{t}\,dt\right] = \frac{1}{s}\cot^{-1}(s-1).$

Solution:
By the formula for transform of integrals, we have,

$$L\left[\int_0^t \frac{e^t \sin t}{t}\,dt\right] = \frac{1}{s}L\left[\frac{e^t \sin t}{t}\right]. \qquad (1)$$

Now $L\left[\frac{\sin t}{t}\right] = \int_s^\infty L[\sin t]ds = \int_s^\infty \frac{1}{s^2+1}ds = \left[\tan^{-1}s\right]_s^\infty = \frac{\pi}{2} - \tan^{-1}s = \cot^{-1}s$

$\therefore L\left[\frac{e^t \sin t}{t}\right] = \cot^{-1}(s-1)$ (By shifting property)

$\therefore$ Equation (1) becomes $L\left[\int_0^t \frac{e^t \sin t}{t}\,dt\right] = \frac{1}{s}\cot^{-1}(s-1).$

Example 5.2.5 Find $L\left[\int_0^t x^3 e^{ax}\,dx\right].$

Solution:
By the formula for the transform of integrals, we have,

$$L\left[\int_0^t x^3 e^{ax}\,dx\right] = \frac{1}{s}L\left[t^3 e^{at}\right] \qquad (1)$$

Now $L[t^3] = \frac{3!}{s^4} = \frac{6}{s^4}$

$\therefore L[t^3 e^{at}] = \frac{6}{(s-a)^4}$

Substituting in (1), we get $L\left[\int_0^t x^3 e^{ax}\,dx\right]. = \frac{1}{s}.\frac{6}{(s-a)^4} = \frac{6}{s(s-a)^4}$

Example 5.2.6 Find the Laplace transform of $t\int_0^t e^{-u}\sin 2u\,du$

Solution:

$$L\left[t\int_0^t e^{-u}\sin 2u\,du\right] = -\frac{d}{ds}L\left[\int_0^t e^{-u}\sin 2u\,du\right] \qquad (1)$$

Now $L\left[\int_0^t e^{-u}\sin 2u\, du\right] = \frac{1}{s}L\left[e^{-t}\sin 2t\right]$

$L[\sin 2t] = \frac{2}{s^2+4}$

$\therefore L[e^{-t}\sin 2t] = \frac{2}{(s+1)^2+4} = \frac{2}{s^2+2s+5}$

$\therefore L\left[\int_0^t e^{-u}\sin 2u\, du\right] = \frac{1}{s}\cdot\frac{2}{(s^2+2s+5)} = \frac{2}{s^3+2s^2+5s}$

$\therefore L\left[t\int_0^t e^{-u}\sin 2u\, du\right] = -\frac{d}{ds}\left(\frac{2}{s^3+2s^2+5s}\right) = \frac{2(3s^2+4s+5)}{(s^3+2s^2+5s)^2}$

EXERCISE 5.2

PART-A

1. Write $L[f'(t)]$ in terms of $L[f(t)]$.
2. Write $L[f^{(n)}(t)]$ in terms of $L[f(t)]$.
3. If $L[f(t)] = F(s)$, write $L\left[\int_0^t f(u)\, du\right]$ in terms of $F(s)$.
4. If $L[f(t)] = F(s)$, express $L[f''(t)]$ in terms of $F(s)$, given that $f(0) = -2$, $f'(0) = 0$ and $f''(0) = 1$.
5. Find $L\left[\int_0^t x^2 e^x\, dx\right]$.

PART-B

6. Prove that $L[f'''(t)] = s^3L[f(t)] - s^2f(0) - sf'(0) - f''(0)$.
7. If $y' + 4y = 0$ and $y(0) = 7$, find $L[y(t)]$.
8. If $4f'' + f = 0$, $f(0) = 0$ and $f'(0) = 2$ show that $L[f(t)] = \frac{8}{4s^2+1}$.
9. Find $L\left[\int_0^t \frac{\sin t}{t}\, dt\right]$.
10. Find $L\left[\int_0^t e^{-t}\cos t\, dt\right]$.

11. Evaluate $L\left[\int_0^t x^4 e^x \, dx\right]$.

12. Find the Laplace transform of $\int_0^t \sin 2x \, dx$

13. Find the Laplace transform of $\int_0^t \cos^2 x \, dx$.

14. Find the Laplace transform of $t^2 \int_0^t e^{-5u} \, du$

15. Find the Laplace transform of $t \int_0^t e^{-2u} \cos 3u \, du$

5.3 INVERSE TRANSFORMS

In this section we consider the problem of finding f(t) for a given function F(s).

Definition 5.3.1 If $L[f(t)] = F(s) = \int_0^\infty e^{-st} f(t).dt$, then f(t) is called the **inverse Laplace transform** of F(s) and we write symbolically $L^{-1}[F(s)] = f(t)$ when L^{-1} is called the **inverse Laplace transformation operator.**

For example, we have $L[e^{at}] = \dfrac{1}{s-a}$ $\quad \therefore \; L^{-1}\left[\dfrac{1}{s-a}\right] = e^{at}$.

Note: (Lerch's Theorem)

(i) **If f(t) is piecewise continuous in every finite interval $0 \le t \le N$ and is of exponential order for $t > N$, then $L^{-1}[F(s)] = f(t)$ is unique.**

5.3.1 PROPERTIES OF INVERSE LAPLACE TRANSFORMS

Property I If $F_1(s)$ and $F_2(s)$ are Laplace transforms of $f_1(t)$ and $f_2(t)$ respectively, then
$$L^{-1}[c_1F_1(s) + c_2F_2(s)] = c_1L^{-1}[F_1(s)] + c_2L^{-1}[F_2(s)] = = c_1f_1(t) + c_2 f_2(t)$$

Property II If $L^{-1}[F(s)] = f(t)$, then $L^{-1}[F(s - a)] = e^{at} f(t$.

Property III If $L^{-1}[F(s)] = f(t)$, then $L^{-1}\left[\dfrac{d^n}{ds^n}F(s)\right] = (-1)^n t^n f(t) \quad n = 1, 2, 3, \ldots$

Property IV If $L^{-1}[F(s)] = f(t)$, then $L^{-1}\left[\int_s^\infty F(u)\,du\right] = \dfrac{1}{t} f(t)$

Property V If $L^{-1}[F(s)] = f(t)$, then $L^{-1}\left[e^{-as}F(s)\right] = \begin{cases} f(t-a), & t>a \\ 0 & t<a \end{cases}$

Table of Inverse Laplace transforms

No	F(s)	f(t)
1	$\frac{1}{s}$	1
2	$\frac{1}{s^{n+1}}$, n = 0, 1, 2, ...	$\frac{t^n}{n!}$
3	$\frac{1}{s-a}$	e^{at}
4	$\frac{1}{s^2+a^2}$	$\frac{1}{a}$ sin at
5	$\frac{s}{s^2+a^2}$	cos at
6	$\frac{1}{s^2-a^2}$	$\frac{1}{a}$ sinh at
7	$\frac{s}{s^2-a^2}$	cosh at

Example 5.3.1 Find $L^{-1}\left[\frac{5}{s+1}+\frac{3s}{s^2+9}-\frac{2}{s^2+16}\right]$

Solution:

$$L^{-1}\left[\frac{5}{s+1}+\frac{3s}{s^2+9}-\frac{2}{s^2+16}\right] = L^{-1}\left[\frac{5}{s+1}\right]+L^{-1}\left[\frac{3s}{s^2+9}\right]-L^{-1}\left[\frac{2}{s^2+16}\right]$$

$$= 5e^{-t} + 3.\cos 3t - 2.\frac{1}{4}\sin 4t$$

$$= 5e^{-t} + 3.\cos 3t - \frac{1}{2}\sin 4t$$

Example 5.3.2 Find the inverse transform of $\frac{s^2+3s-6}{s^3}$

Solution:

$$L^{-1}\left[\frac{s^2+3s-6}{s^3}\right] = L^{-1}\left[\frac{1}{s}\right]+3.L^{-1}\left[\frac{1}{s^2}\right]-6L^{-1}\left[\frac{1}{s^3}\right]$$

$$= 1 + 3t - 6.\frac{t^2}{2!}$$
$$= 1 + 3t - 3t^2.$$

Example 5.3.3 Find $L^{-1}\left[\dfrac{s+4}{s^2-4s+13}\right]$

Solution:

$$L^{-1}\left[\frac{s+4}{s^2-4s+13}\right] = L^{-1}\left[\frac{s+4}{(s-2)^2+9}\right]$$
$$= L^{-1}\left[\frac{s-2+6}{(s-2)^2+3^2}\right]$$
$$= L^{-1}\left[\frac{s-2}{(s-2)^2+3^2}\right] + 6.L^{-1}\left[\frac{1}{(s-2)^2+3^2}\right]$$
$$= e^{2t}L^{-1}\left[\frac{s}{s^2+3^2}\right] + 6.e^{2t}L^{-1}\left[\frac{1}{s^2+3^2}\right]$$
$$= e^{2t}\cos 3t + \frac{6}{3}.e^{2t}\sin 3t$$
$$= e^{2t}\cos 3t + 2.e^{2t}.\sin 3t$$

Example 5.3.4 Find $L^{-1}\left[\dfrac{s}{\left(s^2+a^2\right)^2}\right]$ and $L^{-1}\left[\dfrac{1}{\left(s^2+a^2\right)^2}\right]$

Solution:

$$L^{-1}\left[\frac{s}{\left(s^2+a^2\right)^2}\right] = -\frac{1}{2}L^{-1}\left[\frac{-2s}{\left(s^2+a^2\right)^2}\right]$$
$$= -\frac{1}{2}L^{-1}\left[\frac{d}{ds}\left(\frac{1}{s^2+a^2}\right)\right]$$
$$= -\frac{1}{2}(-1).t.L^{-1}\left[\frac{1}{s^2+a^2}\right] = \frac{t}{2}.\frac{1}{a}\sin at$$
$$= \frac{1}{2a}t.\sin at$$

Now $L^{-1}\left[\dfrac{s^2-a^2}{\left(s^2+a^2\right)^2}\right] = L^{-1}\left[\dfrac{s^2+a^2-2a^2}{\left(s^2+a^2\right)^2}\right]$

$$= L^{-1}\left[\frac{1}{s^2+a^2}\right] - 2a^2L^{-1}\left[\frac{1}{\left(s^2+a^2\right)^2}\right]$$

$$\therefore\ 2a^2L^{-1}\left[\frac{1}{(s^2+a^2)^2}\right]=L^{-1}\left[\frac{1}{s^2+a^2}\right]-L^{-1}\left[\frac{s^2-a^2}{(s^2+a^2)^2}\right]$$

$$=L^{-1}\left[\frac{1}{s^2+a^2}\right]-L^{-1}\left[(-1).\frac{d}{ds}\left(\frac{s}{s^2+a^2}\right)\right]$$

$$=L^{-1}\left[\frac{1}{s^2+a^2}\right]-t.L^{-1}\left[\frac{s}{s^2+a^2}\right] \qquad \text{(By Property III)}$$

$$=\frac{1}{a}\sin at - t\cos at$$

$$\therefore\ L^{-1}\left[\frac{1}{(s^2+a^2)^2}\right]=\frac{1}{2a^3}[\sin at - at\cos at]$$

Example 5.3.5 Find $L^{-1}\left[\frac{s^2-10s+13}{(s-7)(s^2-5s+6)}\right]$

Solution:

$$\frac{s^2-10s+13}{(s-7)(s^2-5s+6)}=\frac{s^2-10s+13}{(s-7)(s-2)(s-3)}$$

$$=\frac{A}{s-7}+\frac{B}{s-2}+\frac{C}{s-3} \qquad \text{(By partial fractions)}$$

$$\therefore\ s^2-10s+13=A(s-2)(s-3)+B(s-7)(s-3)+C(s-7)(s-2)$$

$s=7$, $49-70+13=A(5)(4)$ $\therefore\ A=-2/5$

$s=2$, $4-20+13=B(-5)(-1)$ $\therefore\ B=-3/5$

$s=3$, $9-30+13=C(-4)(1)$ $\therefore\ C=2$

$$\therefore\ L^{-1}\left[\frac{s^2-10s+13}{(s-7)(s^2-5s+6)}\right]=L^{-1}\left[\frac{-2/5}{s-7}\right]+L^{-1}\left[\frac{-3/5}{s-2}\right]+L^{-1}\left[\frac{2}{s-3}\right]$$

$$=-\tfrac{2}{5}.e^{7t}-\tfrac{3}{5}.e^{2t}+2.e^{3t}$$

Example 5.3.6 Find $L^{-1}\left[\frac{s^2+6}{(s^2+1)(s^2+4)}\right]$

Solution:

$$\frac{s^2+6}{(s^2+1)(s^2+4)}=\frac{As+B}{s^2+1}+\frac{Cs+D}{s^2+4} \qquad \text{(By partial fractions)}$$

$$\therefore\ s^2+6=(As+B)(s^2+4)+(Cs+D)(s^2+1)$$

Equating the coefficients of s^3, s^2, s and constants

$0=A+C$

$1=B+D$

$0 = 4A + C$
$6 = 4B + D$

Solving, $A = 0, C = 0;\ B = 5/3$ and $D = -2/3$

$$\frac{s^2+6}{(s^2+1)(s^2+4)} = \frac{5/3}{s^2+1} + \frac{(-2/3)}{s^2+4}$$

$$\therefore\quad L^{-1}\left[\frac{s^2+6}{(s^2+1)(s^2+4)}\right] = \frac{5}{3}L^{-1}\left[\frac{1}{s^2+1}\right] - \frac{2}{3}L^{-1}\left[\frac{1}{s^2+4}\right]$$

$$= -5/3 \sin t - 2/3 . 1/2 \sin 2t$$

$$= -(5/3 \sin t + 1/3 . \sin 2t)$$

Example 5.3.7 Find the inverse Laplace transform of $\dfrac{1+2s}{(s-1)^2(s+2)^2}$

Solution:

$$\frac{1+2s}{(s-1)^2(s+2)^2} = \frac{A}{s-1} + \frac{B}{(s-1)^2} + \frac{C}{s+2} + \frac{D}{(s+2)^2} \qquad \text{(By partial fractions)}$$

$\therefore\ 1 + 2s = A(s-1)(s+2)^2 + B(s+2)^2 + C(s-1)^2(s+2) + D(s-1)^2$

$s = 1,\quad 1 + 2 = B.9 \qquad \therefore\ B = 1/3$

$s = -2,\quad 1 - 4 = D.9 = \qquad \therefore\ D = -1/3$

Equating constants, $1 = -4a + 4B + 2C + D$

i.e., $0 = -4A + 2C$

Equating the coefficients of s^3, $0 = A + C$

Solving for A and C, we get, $A = 0, C = 0$

$$\therefore\quad \frac{1+2s}{(s-1)^2(s+2)^2} = \frac{1/3}{(s-1)^2} - \frac{1/3}{(s+2)^2}$$

$$L^{-1}\left[\frac{1+2s}{(s-1)^2(s+2)^2}\right] = \frac{1}{3}L^{-1}\left[\frac{1}{(s-1)^2}\right] - \frac{1}{3}L^{-1}\left[\frac{1}{(s+2)^2}\right]$$

$$= \frac{1}{3}e^t L^{-1}\left[\frac{1}{s^2}\right] - \frac{1}{3}e^{-2t}L^{-1}\left[\frac{1}{s^2}\right]$$

$$= \frac{1}{3}e^t \frac{t}{1!} - \frac{1}{3}e^{-2t}\frac{t}{1!}$$

$$= \frac{1}{3}t.e^t - \frac{1}{3}t.e^{-2t}$$

Example 5.3.8 Find $L^{-1}\left[\dfrac{s+2}{(s^2+4s+5)^2}\right]$

Solution:

$$L^{-1}\left[\frac{s+2}{(s^2+4s+5)^2}\right] = L^{-1}\left[\frac{s+2}{((s+2)^2+1)^2}\right]$$

$$= e^{-2t}L^{-1}\left[\frac{s}{(s^2+1)^2}\right]$$

$$= e^{-2t}\left(\frac{-1}{2}\right)L^{-1}\left[\frac{-2s}{(s^2+1)^2}\right]$$

$$= \frac{-1}{2}e^{-2t}L^{-1}\left[\frac{d}{ds}\left(\frac{1}{s^2+1}\right)\right]$$

$$= \frac{-1}{2}e^{-2t}(-1).t.L^{-1}\left[\frac{1}{s^2+1}\right] \quad \text{(By property III)}$$

$$= \frac{-1}{2}.t.e^{-2t}.\sin t$$

Example 5.3.9 Find $L^{-1}\left[\frac{e^{-\pi s}}{s^2+2s+2}\right]$

Solution:

$$L^{-1}\left[\frac{1}{s^2+2s+2}\right] = L^{-1}\left[\frac{1}{(s+1)^2+1}\right]$$

$$= e^{-t}L^{-1}\left[\frac{1}{s^2+1}\right] = e^{-t}.\sin t$$

$$\therefore L^{-1}\left[\frac{e^{-\pi s}}{s^2+2s+2}\right] = \begin{cases} e^{-(t-\pi)}.\sin(t-\pi) & t>\pi \\ 0 & t\le\pi \end{cases} \quad \text{(By property V)}$$

$$= \begin{cases} -e^{-(t-\pi)}.\sin t & t>\pi \\ 0 & t\le\pi \end{cases}$$

5.3.2 OTHER METHODS OF FINDING INVERSE LAPLACE TRANSFORMS

Theorem 5.3.1 If $L^{-1}[F(s)] = f(t)$ and $f(0) = 0$, then $L^{-1}[s.F(s)] = \frac{d}{dt}(f(t))$

Proof:

We have $L[f'(t)] = s.L[f(t)] - f(0)$

i.e., $L\left[\frac{d}{dt}(f(t))\right] = s.F(s)$

$$\therefore L^{-1}[s.F(s)] = \frac{d}{dt}(f(t))$$

Note:

In general $L^{-1}\left[s^n.F(s)\right]=\frac{d^n}{dt^n}(f(t))$ provided $f(0) = f'(0) = \ldots = f^{(n-1)} = 0$

Theorem 5.3.2 If $L^{-1}[F(s)] = f(t)$, then $L^{-1}\left[\frac{1}{s}F(s)\right] = \int_0^t f(t)dt$

Proof:

We have $L\left[\int_0^t f(t)dt\right] = \frac{1}{s}L[f(t)]$

i.e., $L\left[\int_0^t f(t)dt\right] = \frac{1}{s}F(s)$

$\therefore$ $L^{-1}\left[\frac{1}{s}F(s)\right] = \int_0^t f(t)dt$

Note:

In general, $L^{-1}\left[\frac{1}{s^n}F(s)\right] = \int_0^t\int_0^t\int_0^t\cdots\int_0^t\left(\int_0^t f(t)dt\right)dt\cdots dt\, dt$

Example 5.3.10 Find the inverse Laplace transforms of (i) $\frac{1}{s(s^2+4)}$ (ii) $\frac{1}{s(s+4)^3}$

(iii) $\frac{1}{s^2(s+2)}$

Solution:

(i) $L^{-1}\left[\frac{1}{s(s^2+4)}\right] = \int_0^t L^{-1}\left[\frac{1}{s^2+4}\right]dt$ (By Theorem 5.3.2)

$$= \int_0^t \frac{1}{2}\sin 2t\, dt$$

$$= \frac{1}{2}.\left[\frac{(-\cos 2t)}{2}\right]_0^t$$

$$= \frac{1}{4}[-\cos 2t + 1]$$

$$= \frac{1}{4}[1 - \cos 2t]$$

(ii) $L^{-1}\left[\frac{1}{(s+4)^3}\right] = e^{-4t}L^{-1}\left[\frac{1}{s^3}\right] = e^{-4t}.\frac{t^2}{2}$

$$L^{-1}\left[\frac{1}{s(s+4)^3}\right] = \int_0^t e^{-4t}\frac{t^2}{2}dt \qquad \text{(By Theorem 5.3.2)}$$

$$= \left.\frac{t^2}{2}\frac{e^{-4t}}{(-4)}\right]_0^t + \frac{1}{4}\int_0^t e^{-4t}.t\,dt$$

$$= -\frac{t^2}{8}e^{-4t} + \frac{1}{4}\left[t.\frac{e^{-4t}}{-4}\right]_0^t + \frac{1}{4}\int_0^t \frac{e^{-4t}}{4}dt$$

$$= -\frac{t^2}{8}e^{-4t} - \frac{t}{16}e^{-4t} + \frac{1}{16}\left[\frac{e^{-4t}}{-4}\right]_0^t$$

$$= -\frac{t^2}{8}e^{-4t} - \frac{t}{16}e^{-4t} - \frac{1}{64}(e^{-4t} - 1)$$

(iii) $L^{-1}\left[\frac{1}{(s+2)}\right] = e^{-2t}$

$$\therefore L^{-1}\left[\frac{1}{s(s+2)}\right] = \int_0^t e^{-2t}dt$$

$$= \left.\frac{e^{-2t}}{-2}\right]_0^t$$

$$= \frac{e^{-2t}}{-2} - \frac{1}{-2}$$

$$= \frac{1}{2}(1 - e^{-2t})$$

Then $L^{-1}\left[\frac{1}{s^2(s+2)}\right] = \int_0^t \frac{1}{2}(1 - e^{-2t})dt$

$$= \frac{1}{2}\left[t + \frac{e^{-2t}}{2}\right]_0^t$$

$$= \frac{1}{2}\left(t + \frac{e^{-2t}}{2}\right) - \frac{1}{2}\left(0 + \frac{1}{2}\right)$$

$$= \frac{1}{4}(e^{-2t} + 2t - 1)$$

Example 5.3.11 Find the inverse Laplace transforms of

(i) $\frac{s^2}{(s+a)^3}$ (ii) $\frac{s^2}{(s^2+a^2)^2}$ (iii) $\frac{1}{s^2+a^2}$

Solution:

$$L^{-1}\left[\frac{1}{(s+a)^3}\right] = e^{-at}L^{-1}\left[\frac{1}{s^3}\right]$$

$$= e^{-at}\frac{t^2}{2} = f(t)$$

Then f(0) = 0

$$\therefore\ L^{-1}\left[\frac{s}{(s+a)^3}\right] = \frac{d}{dt}(f(t)) \qquad \text{(By theorem 5.3.1)}$$

$$= \frac{d}{dt}\left(e^{-at}\frac{t^2}{2}\right)$$

$$= t.e^{-at} + \frac{t^2}{2}.e^{-at}(-a)$$

$$= t.e^{-at} - \frac{a}{2}t^2e^{-at}$$

Again applying Theorem 5.3.1

$$L^{-1}\left[\frac{s^2}{(s+a)^3}\right] = \frac{d}{dt}\left(t.e^{-at} - \frac{a}{2}t^2e^{-at}\right)$$

$$= 1.e^{-at} + t.e^{-at}(-a) - a.t.e^{-at} - \frac{at^2}{2}e^{-at}(-a)$$

$$= \frac{e^{-at}}{2}\left(a^2t^2 - 4at + 2\right)$$

(ii) $$L^{-1}\left[\frac{s}{(s^2+a^2)^2}\right] = \left(-\frac{1}{2}\right)L^{-1}\left[\frac{d}{dt}\left(\frac{1}{s^2+a^2}\right)\right]$$

$$= -\frac{1}{2}(-1)t.L^{-1}\left(\frac{1}{s^2+a^2}\right)$$

$$= \frac{1}{2}.t.\frac{1}{a}\sin at$$

$$= \frac{t}{2a}\sin at = f(t).$$

Then f(0) = 0

$$\therefore\ L^{-1}\left[\frac{s^2}{(s^2+a^2)^2}\right] = \frac{d}{dt}\left(\frac{t}{2a}\sin at\right) \qquad \text{(By Theorem 5.3.1)}$$

$$= \frac{1}{2a}\sin at + \frac{t}{2a}\cos at.a$$

$$= \frac{1}{2a}(\sin at + at\cos at)$$

(iii) $$L^{-1}\left[\frac{1}{(s^2+a^2)^2}\right] = L^{-1}\left[\frac{1}{s}.\frac{s}{(s^2+a^2)^2}\right] = \int_0^t L^{-1}\left[\frac{s}{(s^2+a^2)^2}\right]dt$$

$$= \int_0^t \frac{t}{2a}\sin at\,dt = \frac{1}{2a}.t.\frac{-\cos at}{a}\Bigg]_0^t - \frac{1}{2a}\int_0^t \frac{-\cos at}{a}\sin at\,dt$$

$$= -\frac{t}{2a^2}.\cos at + \frac{1}{2a^3}.\sin at = \frac{1}{2a^3}[\sin at - at\cos at]$$

Note:

It will be useful to remember the following inverse Laplace transforms

(i) $$L^{-1}\left[\frac{s^2}{(s^2+a^2)^2}\right] = \frac{1}{2a}(\sin at + at\cos at)$$

(ii) $$L^{-1}\left[\frac{s}{(s^2+a^2)^2}\right] = \frac{t}{2a}\sin at$$

(iii) $$L^{-1}\left[\frac{1}{(s^2+a^2)^2}\right] = \frac{1}{2a^3}[\sin at - at\cos at]$$

Example 5.3.12 Find the inverse Laplace transform of

(i) $\log\left(\frac{s+a}{s+b}\right)$ (ii) $\log\left(\frac{s+1}{(s+2)(s+3)}\right)$ (iii) $\tan^{-1}(2/s)$

Solution:

Let $f(t) = L^{-1}\log\left(\frac{s+a}{s+b}\right)$

Then $L^{-1}\left[\frac{d}{ds}\log\left(\frac{s+a}{s+b}\right)\right] = -t.f(t)$ (By Property III)

i.e., $-t.f(t) = L^{-1}\left[\frac{d}{ds}\log(s+a) - \frac{d}{ds}\log(s+b)\right]$

$$= L^{-1}\left[\frac{1}{s+a} - \frac{1}{s+b}\right]$$

$$= L^{-1}\left[\frac{1}{s+a}\right] - L^{-1}\left[\frac{1}{s+b}\right]$$

$$= e^{-at} - e^{-bt}$$

$$\therefore\ f(t) = -\frac{1}{t}\left(e^{-at} - e^{-bt}\right)$$

(ii) Let $f(t) = L^{-1}\log\left(\frac{s+1}{(s+2)(s+3)}\right)$

Then $L^{-1}\left[\frac{d}{ds}\log\left(\frac{s+1}{(s+2)(s+3)}\right)\right] = -t.f(t)$

$$\therefore\ -t.f(t) = L^{-1}\left[\frac{d}{ds}\log(s+1) - \frac{d}{ds}\log(s+2) - \frac{d}{ds}\log(s+3)\right]$$

$$= L^{-1}\left[\frac{1}{s+1} - \frac{1}{s+2} - \frac{1}{s+3}\right]$$

$$= L^{-1}\left[\frac{1}{s+1}\right] - L^{-1}\left[\frac{1}{s+2}\right] - L^{-1}\left[\frac{1}{s+3}\right] = e^{-t} - e^{-2t} - e^{-3t}$$

$$\therefore\ f(t) = -\frac{1}{t}\left(e^{-t} - e^{-2t} - e^{-3t}\right)$$

(iii) Let $f(t) = L^{-1}\tan^{-1}(2/s)$

Then $L^{-1}\left[\frac{d}{ds}\tan^{-1}(2/s)\right] = -t.f(t)$

$$\text{i.e., } -t.f(t) = L^{-1}\left[\frac{1}{1+(2/s)^2}\left(-2/s^2\right)\right]$$

$$= L^{-1}\left[\frac{-2}{s^2+4}\right]$$

$$= -2.L^{-1}\left[\frac{1}{s^2+4}\right]$$

$$= -2\frac{1}{2}.\sin 2t$$

$$= -\sin 2t$$

$$\therefore\ f(t) = \frac{1}{t}.\sin 2t$$

5.3.3 INVERSE LAPLACE TRANSFORMS OF UNIT STEP FUNCTIONS.

By the second shifting property of Laplace transforms (Theorem 5.1.2), we have $L\{f(t-a).u_a(t)] = e^{-as}.F(s)$ where $F(s) = L[f(t)]$.

Hence $L^{-1}[e^{-as} F(s)] = f(t-a).u_a(t)$ (1)

Using (1), the inverse Laplace transforms of a certain functions can be computed.

Example 5.3.13 Find the inverse Laplace transform of the following functions.

(i) $\dfrac{e^{-2s}}{s+3}$ (ii) $\dfrac{s.e^{-s/2}}{s^2+\pi^2}$

Solution:

$L^{-1}[e^{-as}F(s)] = f(t-a).u_a(t)$.

(i) Let $F(s) = \dfrac{1}{s+3}$ and a = 2

Then $f(t) = L^{-1}[F(s)] = L^{-1}\left[\dfrac{1}{s+3}\right] = e^{-3t}$

$\therefore L^{-1}\left[e^{-2s}\dfrac{1}{s+3}\right] = e^{-3(t-2)}.u_2(t)$

i.e., $L^{-1}\left[\dfrac{e^{-2s}}{s+3}\right] = e^{-3(t-2)}.u_2(t)$

(ii) Let $F(s) = \dfrac{s}{s^2+\pi^2}$ and $a = \frac{1}{2}$.

Then $f(t) = L^{-1}[F(s)] = L^{-1}\left[\dfrac{s}{s^2+\pi^2}\right] = \cos \pi t$

$\therefore L^{-1}\left[\dfrac{s.e^{-s/2}}{s^2+\pi^2}\right] = \cos \pi\left(t-\frac{1}{2}\right)u_{\frac{1}{2}}(t)$

$= \sin \pi t.u_{\frac{1}{2}}(t)$

Example 5.3.14 Find the inverse Laplace transforms of

(i) $\dfrac{e^{-s}}{(s-1)(s-2)}$ (ii) $\dfrac{(1+e^{-s\pi})^2}{s+5}$

Solution:

(i) $\dfrac{1}{(s-1)(s-2)} = \dfrac{1}{s-2} - \dfrac{1}{s-1}$

Let $F(s) = \dfrac{1}{s-2} - \dfrac{1}{s-1}$ and a = 1

Then $f(t) = L^{-1}[F(s)] = L^{-1}\left[\dfrac{1}{s-2}\right] - L^{-1}\left[\dfrac{1}{s-1}\right] = e^{2t} - e^{t}$

$\therefore L^{-1}\left[\dfrac{e^{-s}}{(s-1)(s-2)}\right] = \left[e^{2(t-1)} - e^{t-1}\right]u_1(t)$

(ii) $\dfrac{(1+e^{-s\pi})^2}{s+5} = \dfrac{1+2e^{-s\pi}+e^{-2s\pi}}{s+5} = \dfrac{1}{s+5}+\dfrac{2.e^{-s\pi}}{s+5}+\dfrac{e^{-2s\pi}}{s+5}$

$$L^{-1}\left[\frac{1}{s+5}\right] = e^{-5t}$$

$$L^{-1}\left[\frac{2}{s+5}\right] = 2e^{-5t}$$

$$\therefore\ L^{-1}\left[\frac{2.e^{-s\pi}}{s+5}\right] = 2\,e^{-5(t-\pi)}.u_\pi(t)$$

$$\therefore\ L^{-1}\left[\frac{e^{-2s\pi}}{s+5}\right] = e^{-5(t-2\pi)}.u_{2\pi}(t)$$

Hence $L^{-1}\left[\dfrac{(1+e^{-s\pi})^2}{s+5}\right] = L^{-1}\left[\dfrac{1}{s+5}\right]+L^{-1}\left[\dfrac{2.e^{-s\pi}}{s+5}\right]+L^{-1}\left[\dfrac{e^{-2s\pi}}{s+5}\right]$

$$= e^{-5t}+2\,e^{-5(t-\pi)}.u_\pi(t)+e^{-5(t-2\pi)}.u_{2\pi}(t)$$

EXERCISE 5.3

PART-A

1. Define inverse Laplace Transform.
2. If $L^{-1}[F(s)] = f(t)$, write $L^{-1}\left[\dfrac{d^n}{ds^n}F(s)\right]$ in terms of f(t).
3. If $L^{-1}[F(s)] = f(t)$, write $L^{-1}\left[\int_s^\infty F(u)\,du\right]$ in terms of f(t).

Find the inverse Laplace transforms of each of the following functions (Ex: 4-25)

4. $\dfrac{1}{s^n}$
5. $\dfrac{a}{s^2+a^2}$
6. $\dfrac{a}{s^2-a^2}$
7. $\dfrac{s}{s^2+16}$
8. $\dfrac{s}{s^2-9}$
9. $\dfrac{s+1}{s^2+4}$
10. $\dfrac{s}{(s+3)^2+4}$
11. $\dfrac{3s+7}{s^2-2s-3}$
12. $\dfrac{3s}{s^2+2s-8}$
13. $\dfrac{s^2}{(s-2)^2}$

14. $\frac{s+3}{s^2-4s+13}$

15. $\frac{s+2}{s^2+9}$

16. $\frac{1}{s(s+2)^2}$

17. $\frac{1}{s^2+4s+8}$

18. $\frac{1}{(s+5)^4}$

19. $\frac{3}{s^2+2s}$

20. $\frac{1}{4s^2+1}$

21. $\frac{1}{s(s^2+9)}$

22. $\frac{1}{s^2(s^2+4)}$

23. $\cot^{-1}(s/2)$

24. $\log\frac{1+s}{s}$

25. $\cot^{-1}(s+1)$

PART-B

Find the inverse Laplace transforms of the following functions (Ex: 26-33)

26. $\frac{4s+5}{(s-1)^2(s+2)}$

27. $\frac{2s^2-6s+5}{s^3-6s^2+11s-6}$

28. $\frac{2s-3}{s^2+4s+13}$

29. $\frac{s+3}{(s^2+6s+13)^2}$

30. $\frac{5s+3}{(s-1)(s^2+2s+5)}$

31. $\frac{3s-1}{(s-2)^2}$

32. $\frac{3s+2}{(s+3)^3}$

33. $\frac{3s+5}{s^2+6s+12}$

Find $f(t) = L^{-1}[F(s)]$ for the function F(s) given below (Ex: 34-41)

34. $\frac{1}{s(s+a)^3}$

35. $\frac{s+2}{s^2(s+1)(s-2)}$

36. $\frac{1}{s^3(s^2+1)}$

37. $\frac{1}{s^2(s^2+a^2)}$

38. $\frac{s^2+s-2}{s(s+3)(s-2)}$

39. $\frac{1}{s^4+3s^2}$

40. $\frac{s-a}{s^2(s^2+a^2)}$

41. $\frac{3}{s^2(s^2+4)(s^2+1)}$

Find the inverse Laplace transforms of the following functions. (Ex: 42-48)

42. $\dfrac{s^2}{(s+a)^3}$

43. $\dfrac{2as}{(s^2+a^2)^2}$

44. $\dfrac{s}{(s^2+1)(s^2+4)}$

45. $\dfrac{s}{(s-1)(s^2+1)}$

46. $\log\dfrac{s^2+1}{s(s+1)}$

47. $\tan^{-1}\left(2/s^2\right)$

48. $\dfrac{1}{2}\log\left(\dfrac{s^2+b^2}{s^2+a^2}\right)$

49. Find $L^{-1}\left[\dfrac{e^{-3s}}{(s-1)^4}\right]$

50. Find $L^{-1}\left[\log\dfrac{s(s+1)}{s^2+4}\right]$

51. Find $L^{-1}\left[s\log\dfrac{s-1}{s+1}\right]$

52. Find $L^{-1}\left[\dfrac{1}{(s+1)(s+2)}\right]$. Hence find $L^{-1}\left[\dfrac{s}{(s+1)(s+2)}\right]$

53. Find $L^{-1}\left[\dfrac{w}{s^2+w^2}\right]$. Hence find $L^{-1}\left[\dfrac{w}{s^2(s^2+w^2)}\right]$

54. Find $L^{-1}\left[\dfrac{4.e^{-s\pi/2}}{s^2+16}\right]$

55. Find $L^{-1}\left[\dfrac{4.e^{-2s}}{s^2}\right]$

56. Prove that $L^{-1}\left[\dfrac{s.e^{-s\pi}}{s^2+9}\right]=f(t)$ where $f(t)=\begin{cases}-\cos 3t & t>\pi\\ 0 & t<\pi\end{cases}$

57. Prove that $L^{-1}\left[\log\left(\dfrac{s^2+1}{s^2}\right)\right]=\dfrac{2}{t}(1-\cos t)$

58. Prove that $L^{-1}\left[\tanh^{-1}\left(\frac{1}{s}\right)\right] = \frac{1}{t}\sinh t$

59. Prove that $L^{-1}\left[\frac{1}{s}\sin\left(\frac{1}{s}\right)\right] = t - \frac{t^3}{(3!)^2} + \frac{t^5}{(5!)^2} - \frac{t^7}{(7!)^2} + \cdots$

60. Prove that $L^{-1}\left[\frac{1}{s}\cos\left(\frac{1}{s}\right)\right] = t - \frac{t^2}{(2!)^2} + \frac{t^4}{(4!)^2} - \frac{t^6}{(6!)^2} + \cdots$

Find the inverse Laplace transforms of the following functions. (Ex: 61-66)

61. $\frac{s.e^{-s\pi}}{s^2+9}$

62. $\frac{s(1+e^{-s\pi/2})}{s^2+4}$

63. $\frac{4.e^{-s\pi/2}}{s^2+16}$

64. $\frac{se^{-s/2}+\pi e^{-s}}{s^2+\pi^2}$

65. $\frac{e^{-cs}}{s^2(s+a)}$ $(c > 0)$

66. $\frac{e^{-s}}{(s+1)^3}$

5.4 CONVOLUTION THEOREM AND TRANSFORMS OF PERIODIC FUNCTIONS

In this section we prove the convolution theorem and use it to compute the inverse Laplace transforms of certain functions. Further we solve integral equations using convolution theorem. Computation of the Laplace transforms of periodic functions is also discussed in this section.

Definition 5.4.1 Let f(t) and g(t) be two functions defined in $[0, \infty)$. The **convolution** of f(t) and g(t), denoted by (f*g)(t) is defined by $(f * g)(t) = \int_0^t f(u)\, g(t-u)\, du, \quad t \geq 0.$

Note:

Convolution of two functions has the following properties.

(i) $f*g = g*f$ (Commutative law).

(ii) $(f*g)*h = f*(g*h)$ (Associative law).

(iii) $f*(g+h) = f*g + f*h$ (Distributive law).

(iv) $f*0 = 0*f = 0$ (where 0 is the constant function $0(t) = 0$ for $t > 0$).

5.4.1 CONVOLUTION THEOREM

Theorem 5.4.1(Convolution Theorem) Let f(t) and g(t) be two piecewise continuous functions on $[0, \infty)$ and be of exponential orders. Then $L[(f*g)(t)] = L[f(t)].L[g(t)]$.

Proof:

Let L[f(t)] = F(s) and L[g(t)] = G(s).

We have $(f*g)(t) = \int_0^t f(u).g(t-u)du$

$$\therefore\ L[(f*g)(t)] = \int_0^\infty e^{-st}.(f*g)(t)dt$$

$$= \int_0^\infty e^{-st}\int_0^t f(u).g(t-u).du.dt.$$

$$= \int_0^\infty\int_0^t e^{-st} f(u).g(t-u).du.dt. \qquad (1)$$

The region of integration in (1) is the area lying between the lines u = 0 and u = t. i.e., $0 \le t < \infty$, $0 \le u \le t$. (Fig. 5.2)

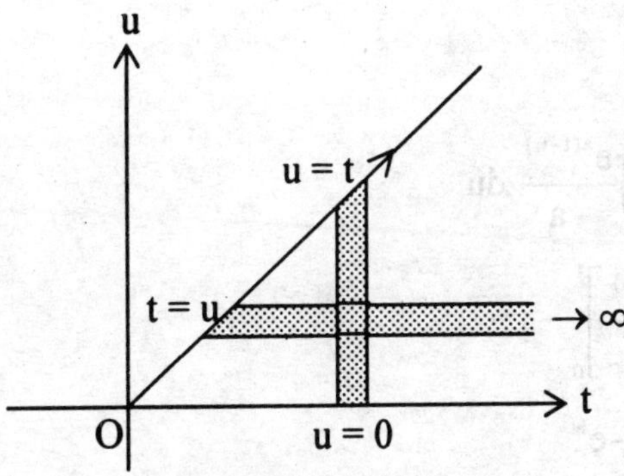

Fig. 5.2

Changing the order of integration we have $0 \le u < \infty$, $u \le t < \infty$

$$L[(f*g)(t)] = \int_0^\infty\int_u^\infty e^{-st} f(u).g(t-u).dt.du.$$

$$= \int_0^\infty e^{-su} f(u)\left(\int_u^\infty e^{-s(t-u)} g(t-u).dt\right)du$$

$$= \int_0^\infty e^{-su} f(u)\left(\int_0^\infty e^{-sv} g(v).dv\right)du \qquad \text{(By Putting t - u = v)}$$

$$= \int_0^\infty e^{-su} f(u).G(s)du$$

$$= \int_0^{\infty} e^{-su} f(u).du.\, G(s)$$

$$= F(s).\, G(s)$$

$$= L[f(t)].L[g(t)]$$ Hence the theorem.

Note:

By convolution theorem, we have $L[(f*g)(t)] = F(s).G(s)$

$\therefore\ L^{-1}[F(s).G(s)] = (f*g)(t)$

i.e., $L^{-1}[F(s).G(s)] = \int_0^t f(u).g(t-u)du$

Using this result we can compute the inverse Laplace transforms of certain functions.

Example 5.4.1 Find the convolutions (i) $t*e^{at}$ (ii) $e^{at} * e^{bt}$ $(a \neq b)$

Solution:

(i)
$$t*e^{at} = \int_0^t u.e^{a(t-u)}.du$$

$$= u.\frac{e^{a(t-u)}}{-a}\Big]_0^t - \int_0^t \frac{e^{a(t-u)}}{-a}.du$$

$$= -\frac{t}{a} + \frac{1}{a}.\frac{e^{a(t-u)}}{(-a)}\Big]_0^t$$

$$= -\frac{t}{a} - \frac{1}{a^2} + \frac{1}{a^2}e^{at}$$

$$= \frac{1}{a^2}\left[e^{at} - 1 - at\right]$$

(ii)
$$e^{at} * e^{bt} = \int_0^t e^{au}.e^{b(t-u)}.du$$

$$= \int_0^t e^{(a-b)u+bt}.du$$

$$= \frac{e^{(a-b)u+bt}}{a-b}\Big]_0^t$$

$$= \frac{e^{at}}{a-b} - \frac{e^{bt}}{a-b}$$

$$= \frac{1}{a-b}\left(e^{at} - e^{bt}\right)$$

Example 5.4.2 Find the inverse Laplace transforms of the following functions.

(i) $\frac{1}{s(s^2+4)}$ (ii) $\frac{1}{(s-2)(s-3)}$ (iii) $\frac{s}{(s^2+4)^2}$

Solution:

(i) Let $F(s) = 1/s$ and $G(s) = \frac{1}{s^2+4}$

Then $f(t) = L^{-1}[F(s)] = L^{-1}\left[1/s\right] = 1$

$$g(t) = L^{-1}[G(s)] = L^{-1}\left[\frac{1}{s^2+4}\right] = \frac{1}{2}\sin 2t$$

$$\therefore \; L^{-1}\left[\frac{1}{s(s^2+4)}\right] = \int_0^t f(u).g(t-u)du$$

$$= \int_0^t \frac{1}{2}.\sin 2(t-u)du$$

$$= -\frac{1}{2}.\left[\frac{\cos 2(t-u)}{(-2)}\right]_0^t = \frac{1}{4}(1-\cos 2t)$$

(ii) Let $F(s) = \frac{1}{s-2}$ and $G(s) = \frac{1}{s-3}$

Then $f(t) = L^{-1}[F(s)] = L^{-1}\left[\frac{1}{s-2}\right] = e^{2t}$

$$g(t) = L^{-1}[G(s)] = L^{-1}\left[\frac{1}{s-3}\right] = e^{3t}$$

$$\therefore \; L^{-1}\left[\frac{1}{(s-2)(s-3)}\right] = \int_0^t f(u).g(t-u)du$$

$$= \int_0^t e^{2u}.e^{3(t-u)}du$$

$$= \int_0^t e^{3t-u}du$$

$$= \left[\frac{e^{3t-u}}{(-1)}\right]_0^t$$

$$= \frac{e^{2t}}{-1} - \frac{e^{3t}}{-1}$$

$$= e^{3t} - e^{2t}$$

(iii) $L^{-1}\left[\frac{s}{(s^2+4)^2}\right]$

Let $F(s) = \frac{s}{s^2+4}$ and $G(s) = \frac{1}{s^2+4}$

Then $f(t) = L^{-1}\left[\frac{s}{s^2+4}\right] = \cos 2t$

$$g(t) = L^{-1}\left[\frac{1}{s^2+4}\right] = \frac{1}{2}\sin 2t$$

$$\therefore\ L^{-1}\left[\frac{1}{s(s^2+4)}\right] = \int_0^t f(u).g(t-u)du$$

$$= \int_0^t \cos 2u.\frac{1}{2}.\sin(2t-2u)du$$

$$= \frac{1}{4}\int_0^t [\sin 2t - \sin(4u-2t)]du$$

$$= \frac{1}{4}\left[u\sin 2t + \frac{\cos(4u-2t)}{4}\right]_0^t$$

$$= \frac{1}{4}\left[t\sin 2t + \frac{1}{4}\cos 2t - \frac{1}{4}\cos 2t\right]$$

$$= \frac{1}{4}t\sin 2t$$

Example 5.4.3 Using convolution theorem, evaluate $L^{-1}\left[\frac{1}{s^2(s+1)^2}\right]$

Solution:

Let $F(s) = \frac{1}{s^2}$ and $G(s) = \frac{1}{(s+1)^2}$

Then $f(t) = L^{-1}[F(s)] = L^{-1}\left[\frac{1}{s^2}\right] = t.$

$$g(t) = L^{-1}[G(s)] = L^{-1}\left[\frac{1}{(s+1)^2}\right]$$

$$= t.e^{-t}$$

$$\therefore\ L^{-1}\left[\frac{1}{s^2(s+1)^2}\right] = \int_0^t f(u).g(t-u)du$$

$$= \int_0^t u.(t-u)e^{-(t-u)}.du$$

$$= u.(t-u).\frac{e^{-(t-u)}}{1}\Big]_0^t - \int_0^t e^{-(t-u)}(t-2u).du$$

$$= \int_0^t (2u-t).e^{-(t-u)}.du$$

$$= (2u-t).e^{-(t-u)}\Big]_0^t - \int_0^t e^{-(t-u)}.2\,du$$

$$= t + te^{-t} - 2.e^{-(t-u)}\Big]_0^t$$

$$= t + te^{-t} - 2 + 2e^{-t}.$$

Example 5.4.4 Using Convolution, find $L^{-1}\left[\frac{8s}{(s^2+16)(s^2+1)^2}\right]$

Solution:

Let $F(s) = \frac{8}{s^2+16}$ and $G(s) = \frac{s}{(s^2+1)^2}$

Then $f(t) = L^{-1}\left[\frac{8}{s^2+16}\right] = 8.\frac{1}{4}\sin 4t = 2\sin 4t$

$$g(t) = L^{-1}\left[\frac{s}{(s^2+1)^2}\right] = \left(-\frac{1}{2}\right)L^{-1}\left[\frac{-2s}{(s^2+1)^2}\right]$$

$$= \left(-\frac{1}{2}\right)L^{-1}\left[\frac{d}{ds}\left(\frac{1}{s^2+1}\right)\right]$$

$$= \left(-\frac{1}{2}\right).(-1).t.\sin t$$

$$= \frac{1}{2}t.\sin t$$

Hence $L^{-1}\left[\frac{8s}{(s^2+16)(s^2+1)^2}\right] = \int_0^t f(u).g(t-u)du$

$$= \int_0^t 2\sin 4u.\frac{1}{2}.(t-u).\sin(t-u)du$$

$$= \int_0^t (t-u).\frac{1}{2}.[\cos(5u-t) - \cos(3u+t)]du$$

$$= \left(\frac{t-u}{2}\right)\left(\frac{\sin(5u-t)}{5} - \frac{\sin(3u+t)}{3}\right)_0^t -$$

$$\int_0^t \left(\frac{\sin(5u-t)}{5} - \frac{\sin(3u+t)}{3}\right)\left(-\frac{1}{2}\right).du$$

$$= \left(\frac{-t}{2}\right)\left(\frac{-\sin t}{5} - \frac{\sin t}{3}\right) + \frac{1}{2}\left(\frac{-\cos(5u-t)}{25} + \frac{\cos(3u+t)}{9}\right)_0^t$$

$$= \frac{4}{15}t\sin t + \frac{1}{2}\left(\frac{-\cos 4t}{25} + \frac{\cos 4t}{9}\right) - \frac{1}{2}\left(\frac{-\cos t}{25} + \frac{\cos t}{9}\right)$$

$$= \frac{4}{15}t\sin t + \frac{8}{225}\cos 4t - \frac{8}{225}\cos t$$

$$= \frac{1}{225}(60t\sin t + 8\cos 4t - 8\cos t)$$

Example 5.4.5 Solve the following integral equation using convolution theorem

$$f(t) = t + 6\int_0^t f(u).e^{(t-u)}du$$

Solution:

Let $(f*g)(t) = \int_0^t f(u).g(t-u)du$ where $g(t) = e^t$

Taking Laplace transform, we get

$L[(f*g)(t)] = F(s).G(s)$ where $G(s) = L[g(t)] = L[e^t] = \dfrac{1}{s-1}$

Given $f(t) = t + 6\int_0^t f(u).g(t-u).du$

Taking Laplace transform both sides, we get

$F(s) = L[t] + 6.F(s).G(s)$

i.e., $F(s) = \dfrac{1}{s^2} + 6.F(s).\dfrac{1}{s-1}$

$$F(s)\left[1 - \frac{6}{s-1}\right] = \frac{1}{s^2}$$

$$F(s) = \frac{s-1}{(s-7)s^2}$$

Then, $f(t) = L^{-1}\left[\dfrac{s-1}{(s-7)s^2}\right]$

i.e., $f(t) = L^{-1}\left[\frac{1}{s(s-7)}\right] - L^{-1}\left[\frac{1}{s^2(s-7)}\right]$ (1)

Now $L^{-1}\left[\frac{1}{s-7}\right] = e^{7t}$

$$\therefore\ L^{-1}\left[\frac{1}{s(s-7)}\right] = \int_0^t e^{7t}dt = \left[\frac{e^{7t}}{7}\right]_0^t = \frac{1}{7}\left(e^{7t}-1\right)$$

$$L^{-1}\left[\frac{1}{s^2(s-7)}\right] = \int_0^t \frac{1}{7}\left(e^{7t}-1\right)dt = \frac{1}{7}\left[\frac{e^{7t}}{7} - t\right]_0^t = \frac{1}{7}\left(\frac{e^{7t}}{7} - t - \frac{1}{7}\right)$$

$$\therefore\ \text{From (1), } f(t) = \frac{1}{7}\left(e^{7t}-1\right) - \frac{1}{7}\left(\frac{e^{7t}}{7} - t - \frac{1}{7}\right)$$

$$= \left(\frac{1}{7} - \frac{1}{49}\right)e^{7t} + \frac{t}{7} - \frac{1}{7} + \frac{1}{49}$$

$$= \frac{6}{49}e^{7t} + \frac{t}{7} - \frac{6}{49}$$

$$= \frac{1}{49}\left(6e^{7t} + 7t - 6\right)$$

Example 5.4.6. Use convolution theorem to solve the integral equations

$$f(t) = 1 + t + 2\int_0^t \sin u.f(t-u).du$$

Solution:

$$\text{Let } (f*g)(t) = \int_0^t \sin u.f(t-u)\,du$$

$$= \int_0^t g(u).f(t-u)\,du, \qquad \text{where } g(t) = \sin t$$

$$= \int_0^t f(u).g(t-u)\,du$$

Taking Laplace transform, we get,

$L[(f*g)(t)] = F(s).G(s)$ where $G(s) = L[g(t)] = L[\sin t] = \frac{1}{s^2+1}$

Given $f(t) = 1 + t + 2\int_0^t f(u).g(t-u).du$

Taking Laplace transform both sides, we get

$F(s) = L[1] + L[t] + 2.F(s).G(s)$

i.e., $F(s) = \frac{1}{s} + \frac{1}{s^2} + 2.F(s).\frac{1}{s^2+1}$

$\therefore \quad F(s)\left[1 - \frac{2}{s^2+1}\right] = \frac{s+1}{s^2}$

$F(s)\left[\frac{s^2-1}{s^2+1}\right] = \frac{s+1}{s^2}$

$F(s) = \frac{(s+1)(s^2+1)}{s^2(s^2-1)}$

$= \frac{s^2+1}{s^2(s-1)}$

$= \frac{1}{s-1} + \frac{1}{s^2(s-1)}$

$\therefore f(t) = L^{-1}\left[\frac{1}{s-1}\right] + L^{-1}\left[\frac{1}{s^2(s-1)}\right]$

$L^{-1}\left[\frac{1}{s-1}\right] = e^t$

$L^{-1}\left[\frac{1}{s(s-1)}\right] = \int_0^t e^t dt$

$= e^t - 1$

$L^{-1}\left[\frac{1}{s^2(s-1)}\right] = \int_0^t (e^t - 1)dt = [e^t - t]_0^t$

$= (e^t - t) - (1 - 0)$

$= e^t - t - 1.$

$\therefore \quad f(t) = e^t + e^t - t - 1.$

i.e., $f(t) = 2e^t - t - 1.$

5.4.2 LAPLACE TRANSFORMS OF PERIODIC FUNCTIONS

Definition 5.4.2 A function f(x) is said to be **periodic** if there exists a number $T > 0$ such that $f(x + T) = f(x)$ for all x belonging to the domain of definition of the function. The least number T possessing this property is called the **period** of the function f(x).

For example, the trigonometric functions sin x and cos x are periodic with period 2π. The Laplace transform of a periodic function f(x) with period T can be obtained by integrating over the interval $0 \le x \le T$. This is because the graph of the function $y = f(x)$ repeats itself after every interval of length T.

Theorem 5.4.2 Let f(t) be piecewise continuous on $[0, \infty)$, be of exponential order and periodic with period T. Then $L[f(t)] = \frac{1}{1-e^{-sT}}\int_0^T e^{-st}f(t).dt,\ s > 0$

Proof:

$$L[f(t)] = \int_0^\infty e^{-st}f(t).dt \qquad \text{(By definition)}$$

$$= \int_0^T e^{-st}f(t).dt + \int_T^{2T} e^{-st}f(t).dt + \int_{2T}^{3T} e^{-st}f(t).dt + \cdots$$

$$= I_1 + I_2 + I_3 + \ldots \qquad (1)$$

In the integral I_2, put $t = u + T$, so that $dt = du$ and

$$I_2 = \int_0^T e^{-s(u+T)}f(u+T).du = e^{-sT}\int_0^T e^{-su}f(u).du$$

$$= e^{-sT}.\, I_1$$

In the integral I_3, put $t = u + 2T$, so that $dt = du$ and

$$I_3 = \int_0^T e^{-s(u+2T)}f(u+2T).du = e^{-2sT}I_1$$

Similarly $I_4 = e^{-3sT}\ I_1$ e.t.c.

Substituting in (1), we get,

$$L[f(t)] = I_1 + e^{-sT}I_1 + e^{-2sT}.I_1 + e^{-3sT}.I_1 + \ldots$$

$$= (1 + e^{-sT} + e^{-2sT} + e^{-3sT} + \ldots)I_1$$

$$= (1 + x + x^2 + x^3 + \ldots)I_1 \quad \text{where } x = e^{-sT}$$

$$= (1 - x)^{-1}.I_1 \qquad (\text{Since } s > 0 \Rightarrow |x| = |e^{-sT}| < 1)$$

i.e., $L[f(t)] = (1 - e^{-sT})^{-1}.I_1$

$$= \frac{1}{1-e^{-sT}}\int_0^T e^{-st}f(t).dt$$

Example 5.4.7 Find the Laplace transform of the square wave function f(t) defined by

$$f(t) = \begin{cases} 1 & 0 \le t < T \\ -1 & T \le t < 2T \end{cases} \quad \text{and } f(t + 2T) = f(t).$$

Solution:

$$L[f(t)] = \frac{1}{1-e^{-s2T}}\int_0^{2T} e^{-st}f(t).dt \qquad \text{(Since f(t) is periodic with period 2T)}$$

$$= \frac{1}{1-e^{-2sT}}\left[\int_0^T e^{-st}.dt + \int_T^{2T} e^{-st}(-1).dt\right]$$

$$= \frac{1}{1-e^{-2sT}}\left[\frac{e^{-st}}{-s}\Big|_0^T - \frac{e^{-st}}{-s}\Big|_T^{2T}\right]$$

$$= \frac{1}{1-e^{-2sT}}\left[\frac{e^{-sT}}{-s}+\frac{1}{s}+\frac{e^{-2sT}}{s}-\frac{e^{-sT}}{s}\right]$$

$$= \frac{1}{s\left(1-e^{-2sT}\right)}\left[1-2e^{-sT}+e^{-2sT}\right]$$

$$= \frac{\left(1-e^{-sT}\right)^2}{s\left(1-e^{-sT}\right)\left(1+e^{-sT}\right)}$$

$$= \frac{1-e^{-sT}}{s\left(1+e^{-sT}\right)} = \frac{1}{s}\tan h\left(\frac{sT}{2}\right)$$

Example 5.4.8 Find the Laplace transform of sin at considering it as a periodic function with period $\frac{2\pi}{a}$

Solution:

Let f(t) = sin at; $T = \frac{2\pi}{a}$

Then $L(\sin at) = \frac{1}{1-e^{-s.2\pi/a}}\int_0^{2\pi/a} e^{-st}\sin at.dt = \frac{1}{\left(1-e^{-s.2\pi/a}\right)}[\,I\,]$ (1)

Now, $I = \int_0^{2\pi/a} e^{-st}\sin at.dt$

$$= \frac{e^{-st}}{-s}\sin at\Big]_0^{2\pi/a} - \int_0^{2\pi/a}\frac{e^{-st}}{-s}\cos at.a.dt$$

$$= \frac{a}{s}\frac{e^{-st}}{-s}\cos at\Big]_0^{2\pi/a} - \frac{a}{s}\int_0^{2\pi/a}\frac{e^{-st}}{-s}(-\sin at).a.dt$$

$$= -\frac{a}{s^2}e^{-2\pi s/a}+\frac{a}{s^2}-\frac{a^2}{s^2}.I$$

$$\therefore \left(1+\frac{a^2}{s^2}\right)I = -\frac{a}{s^2}e^{-2\pi s/a}+\frac{a}{s^2}$$

$$I = -\left(\frac{s^2}{s^2+a^2}\right)\left(\frac{a}{s^2}\right)\left(e^{-2\pi s/a}-1\right) = \left(\frac{a}{s^2+a^2}\right)\left(1-e^{-2\pi s/a}\right)$$

Substituting in (1), we get,

$$L(\sin at) = \frac{1}{\left(1-e^{-s.2\pi/a}\right)}\left(\frac{a}{s^2+a^2}\right)\left(1-e^{-2\pi s/a}\right)$$

$$= \frac{a}{s^2+a^2}$$

Example 5.4.9 Find the Laplace transform of the saw-tooth wave function f(t) defined by f(t) = t, $0 \le t \le a$, f(t + a) = f(t).

Solution:

f(t) is periodic with period a.(i.e., T = a)

$$\therefore \quad L[f(t)] = \frac{1}{1-e^{-sa}}\int_0^a e^{-st} f(t).dt$$

$$= \frac{1}{1-e^{-sa}}\int_0^a e^{-st}.t.dt$$

$$= \frac{1}{1-e^{-sa}}\left[t.\frac{e^{-st}}{-s}\bigg|_0^a + \int_0^a \frac{e^{-st}}{-s}dt\right]$$

$$= \frac{1}{1-e^{-sa}}\left[-\frac{a}{s}.e^{-sa} + \frac{1}{s}.\frac{e^{-st}}{-s}\bigg|_0^a\right]$$

$$= \frac{1}{1-e^{-sa}}\left[-\frac{a}{s}.e^{-sa} - \frac{1}{s^2}e^{-sa} + \frac{1}{s^2}\right]$$

$$= \frac{1}{1-e^{-sa}}\left[-\frac{a}{s}.e^{-sa} + \frac{1}{s^2}\left(1-e^{-sa}\right)\right]$$

$$= \frac{1}{s^2} - \frac{a.e^{-sa}}{s\left(1-e^{-sa}\right)}$$

Example 5.4.10 Find the Laplace transform of the periodic function

$$f(t) = \begin{cases} \cos t & 0 < t < \pi/2 \\ -\cos t & \pi/2 < t < 3\pi/2 \\ \cos t & 3\pi/2 < t < 2\pi \end{cases}$$

Solution:

f(t) is periodic with period 2π

$$\therefore L[f(t)] = \frac{1}{1-e^{-s.2\pi}}\int_0^{2\pi} e^{-st} f(t)\,dt$$

$$= \frac{1}{1-e^{-s.2\pi}}\left[\int_0^{\pi/2} e^{-st}\cos t\,dt + \int_{\pi/2}^{3\pi/2} e^{-st}(-\cos t)\,dt + \int_{3\pi/2}^{2\pi} e^{-st}\cos t\,dt\right]$$

$$= \frac{1}{1-e^{-s.2\pi}}\left[I_1 + I_2 + I_3\right] \qquad (1)$$

The integrals I_1, I_2, I_3, are evaluated by the method of integration by parts.

$$I_1 = \int_0^{\pi/2} e^{-st}\cos t\,dt$$

$$= e^{-st}\sin t\Big]_0^{\pi/2} - \int_0^{\pi/2}\sin t.e^{-st}(-s)\,dt$$

$$= e^{-s\pi/2} + .s\left[e^{-st}(-\cos t\}\Big|_0^{\pi/2} - \int_0^{\pi/2}(-\cos t).e^{-st}(-s)\,dt\right]$$

$$= e^{-s\pi/2} + s.1 - s^2\int_0^{\pi/2} e^{-st}.\cos t\,dt$$

$$\therefore\ (1+s^2)I_1 = e^{-s\pi/2} + s$$

$$I_1 = \frac{s + e^{-s\pi/2}}{1+s^2} \qquad (2)$$

$$I_2 = -\int_{\pi/2}^{3\pi/2} e^{-st}\cos t\,dt$$

$$= -e^{-st}\sin t\Big]_{\pi/2}^{3\pi/2} + \int_{\pi/2}^{3\pi/2}\sin t.e^{-st}(-s)\,dt$$

$$= -e^{-s3\pi/2}(-1) + e^{-s\pi/2} - s\left[e^{-st}(-\cos t\}\Big|_{\pi/2}^{3\pi/2} - \int_{\pi/2}^{3\pi/2}(-\cos t).e^{-st}(-s)\,dt\right]$$

$$= e^{-3s\pi/2} + e^{-s\pi/2} + s^2\int_{\pi/2}^{3\pi/2} e^{-st}.\cos t\,dt$$

$$\therefore\ (1+s^2)I_2 = e^{-3s\pi/2} + e^{-s\pi/2}$$

$$I_2 = \frac{e^{-3s\pi/2} + e^{-s\pi/2}}{1+s^2} \qquad (3)$$

$$I_3 = \int_{3\pi/2}^{2\pi} e^{-st}\cos t\,dt$$

$$= e^{-st}\sin t\Big]_{3\pi/2}^{2\pi} - \int_{3\pi/2}^{2\pi} \sin t.e^{-st}(-s)\,dt$$

$$= e^{-s3\pi/2} + s\left[e^{-st}(-\cos t\}\Big|_{3\pi/2}^{2\pi} - \int_{3\pi/2}^{2\pi} -\cos t.e^{-st}(-s)\,dt \right]$$

$$= e^{-3s\pi/2} + s.e^{-S2\pi}(-1) - s^2 \int_{3\pi/2}^{2\pi} e^{-st}.\cos t\,dt$$

$$(1+s^2)I_3 = e^{-3s\pi/2} - s.e^{-s2\pi}$$

$$I_3 = \frac{e^{-3s\pi/2} - s.e^{-s2\pi}}{1+s^2}$$

$$L[f(t)] = \frac{1}{1-e^{-s.2\pi}} \frac{\left[s + e^{-s\pi/2} + e^{-3s\pi/2} + e^{-s\pi/2} + e^{-3s\pi/2} - s.e^{-2s\pi} \right]}{1+s^2}$$

$$= \frac{\left[s + 2.e^{-s\pi/2} + 2e^{-3s\pi/2} - s.e^{-2s\pi} \right]}{(1+s^2)(1-e^{-.2s\pi})}$$

$$= \frac{s(1-e^{-2s\pi})}{(1+s^2)(1-e^{-2s\pi})} + \frac{2\,e^{-s\pi/2}(1+e^{-s\pi})}{(1+s^2)(1-e^{-2s\pi})}$$

$$= \frac{s}{(1+s^2)} + \frac{2\,e^{-s\pi/2}(1+e^{-s\pi})}{(1+s^2)(1-e^{-s\pi})(1+e^{-s\pi})}$$

$$= \left(\frac{s}{1+s^2}\right) + \left(\frac{2\,e^{-s\pi/2}}{(1+s^2)(1-e^{-s\pi})}\right)$$

EXERCISE 5.4

PART-A

1. Define convolution of two functions.
2. State any two properties of convolution.
3. State the convolution theorem.
4. L[f(t)] = F(s) and L[g(t)] = G(s). What is L[(f*g)(t)]?
5. $L[f(t)] = 1/s$ and $L[g(t)] = e^{as}$. What is L[(f*g)(t)] ?

6. If f(t) = 1, g(t) = 1 find (f*g)(t)
7. If f(t) = t, g(t) = e^t, find (f*g)(t).
8. If $L[f(t)] = \frac{1}{s}$ and $L[g(t)] = \frac{1}{s+2}$, find $L^{-1}\left[\frac{1}{s(s+2)}\right]$
9. If $L[f(t)] = \frac{1}{s^2}$ and $L[g(t)] = \frac{1}{s+2}$, find $L^{-1}\left[\frac{1}{s^2(s+2)}\right]$
10. Define periodic functions with an example.
11. If f(t) is periodic with period T, what is L[f(t)]?
12. If f(t) is periodic with period 2π, what is L[f(t)]?

PART-B

Using convolution theorem, evaluate the inverse Laplace transforms of the following functions (Ex: 13-20)

13. $\frac{1}{(s-a)(s-b)}$
14. $\frac{1}{(s^2+a^2)^2}$
15. $\frac{s}{(s^2+1(s^2+4)}$
16. $\frac{1}{(s^2+9)^2}$
17. $\frac{s}{(s^2+4)^2}$
18. $\frac{1}{(s+a)^2(s+b)^2}$ $a \neq b$
19. $\frac{1}{(s+1)(s+9)^2}$
20. $\frac{s^2}{(s^2+a^2)(s^2+b^2)}$

Solve the following integrals equations using convolution theorem. (Ex: 21-25)

21. $f(t) = t + e^{-2t} + \int_0^t f(u)e^{2(t-u)}du$

22. $f(t) = \cos t + e^{-t}\int_0^t f(u)e^{u}du$

23. $f(t) = e^t + \int_0^t u.f(t-u)\,du$

24. $f(t) = e^{-t} - \int_0^t f(u)\cos(t-u)\,du$

25. $4.f(t) = t - \int_0^t (e^u + e^{-u}).f(t-u)\,du$

26. Find the Laplace transform of the function $f(t)=\begin{cases}\sin wt & 0<t<\pi/w\\ 0 & \pi/w<t<2\pi/w\end{cases}$

27. Find the Laplace transform of the following saw-tooth wave function of period T given by $f(t)=t/T \quad 0<t<T$

28. Find the Laplace transform of the triangular wave function of period 2a given by
$$f(t)=\begin{cases}t & 0<t<a\\ 2a-t & a<t<2a\end{cases}$$

29. Find the Laplace transform of the full-wave rectifier f(t) = E sin wt, $0<t<\pi/w$.

30. Find the Laplace transform of the periodic function f(t) of period 2π defined by
$$f(t)=\begin{cases}0 & 0<t<\pi\\ t-\pi & \pi<t<2\pi\end{cases}$$

31. Find the Laplace transform of the square wave function f(t) given by
$$f(t)=\begin{cases}k & 0\le t<a\\ -k & a\le t<2a\end{cases}$$

5.5 APPLICATION TO DIFFERENTIAL EQUATIONS

In this section we discuss the Laplace transform method of solving differential equations. We consider here only linear differential equation with constant coefficients and simultaneous linear differential equations with constant coefficients. The general procedure for solving a linear differential equation in the variable y(t) with given initial conditions is as follows. First obtain the Laplace transforms of both sides of the differential equation. This results in an algebraic equation, which is a function of L[y(t)] = Y(s). The initial conditions are then substituted in the algebraic equation and then this equation is solved to obtain Y(s). The inverse Laplace transform of Y(s) is then computed and $L^{-1}[Y(s)] = y(t)$ is the solution of the given differential equation.

5.5.1 LINEAR DIFFERENTIAL EQUATIONS WITH CONSTANT COEFFICIENTS

Consider the linear differential equation of second order with constant coefficients,
$$a_0\frac{d^2y}{dt^2}+a_1\frac{dy}{dt}+a_2y=f(t) \qquad (1)$$
with initial conditions $y(0)=\alpha$, $y'(0)=\beta$ and a known function f(t).

The problem of finding a function y(t) satisfying the differential equation (1) and the given initial conditions is generally called an **initial value problem**.

Let L[f(t)] = F(s) and L[y(t)] = Y(s)

Taking Laplace transform of both sides of (1), we get,

$a_0\,L[y''(t)] + a_1L[y'(t)] + a_2L[y(t)] = L[f(t)]$

i.e., $a_0\,[s^2Y(s) - sy(s) - y'(0)] + a_1[sY(s) - y(0)] + a_2Y(s) = F(s)$.

i.e., $(a_0s^2 + a_1s + a_2)Y(s) - (a_0s + a_1)y(0) - a_0y'(0) = F(s)$

i.e., $Y(s) = \dfrac{F(s)}{a_0s^2 + a_1s + a_2} + \dfrac{(a_0s + a_1)Y(0) + a_0Y'(0)}{a_0s^2 + a_1s + a_2}$

Substituting for Y(0) and Y′(0), we get

$$Y(s) = \frac{F(s)}{a_0s^2 + a_1s + a_2} + \frac{(a_0s + a_1)\alpha + a_0\beta}{a_0s^2 + a_1s + a_2} \qquad (2)$$

Since F(s) is known, $L^{-1}[Y(s)] = y(t)$ can be calculated from (2).

Note:

The advantage of the Laplace transform method of solving the given linear differential equation over other methods is that the initial conditions $Y(0) = \alpha$ and $Y'(0) = \beta$ are automatically involved in the computation of the algebraic equation in Y(s). To obtain the particular solution Y(t) of the differential equation, one has to evaluate $L^{-1}[Y(s)]$ only and there is no need to find the general solution of the differential equation involving arbitrary constants.

Example 5.5.1 Solve by Laplace transform method, the differential equation $y'' + 4y' + 3y = e^{-t}$, $\quad y(0) = y'(0) = 1$.

Solution:

$y'' + 4y' + 3y = e^{-t}$

$\therefore L[y''] + 4L[y'] + 3L[y] = L[e^{-t}]$

i.e., $[s^2Y(s) - s.y(0) - y'(0)] + 4[sY(s) - y(0)] + 3Y(s) = \dfrac{1}{s+1}$ where $L[y(t)] = Y(s)$

i.e., $(s^2 + 4s + 3)Y(s) + (-s-1-4) = \dfrac{1}{s+1}$ (using initial conditions)

$$(s^2 + 4s + 3)Y(s) = \frac{1}{s+1} + (s+5)$$

$$Y(s) = \frac{1}{(s+1)(s^2+4s+3)} + \frac{s+5}{s^2+4s+3}$$

$$= \frac{1}{(s+1)(s+1)(s+3)} + \frac{s+5}{(s+1)(s+3)}$$

$$= \frac{1+(s+1)(s+5)}{(s+1)^2(s+3)}$$

$$= \frac{A}{s+1} + \frac{B}{(s+1)^2} + \frac{C}{s+3} \qquad \text{(By Partial fractions)}$$

$1 + (s + 1)(s + 5) = A(s + 1)(s + 3) + B(s + 3) + C(s + 1)^2$

$s = -1, \qquad 1 = 2B \qquad$ i.e., $B = ½$

s = -3, -3 = 4c i.e., C = -¾

Equating Coefficients of s^2, 1 = A + C

∴ A = 1-C = 1 + ¾ = 7⁄4

$$\text{Thus } Y(s) = \frac{7/4}{s+1} + \frac{1/2}{(s+1)^2} + \frac{-3/4}{s+3}$$

Taking inverse Laplace transform both sides, we get,

$$y(t) = \frac{7}{4}L^{-1}\left[\frac{1}{s+1}\right] + \frac{1}{2}L^{-1}\left[\frac{1}{(s+1)^2}\right] - \frac{3}{4}L^{-1}\left[\frac{1}{s+3}\right]$$

$$\text{i.e., } y(t) = \frac{7}{4}e^{-t} + \frac{1}{2}te^{-t} - \frac{3}{4}e^{-3t}$$

Example 5.5.2 Find the solution of the initial value problem 4y″-8y′ + 3y = sin t, y(0) = 0, y′(0) = 2.

Solution:

4y″ - 8y′ + 3y = sin t

∴ 4L[y″] - 8L[y′] + 3L[y] = L[sin t]

$$\text{i.e., } 4[s^2Y(s) - sy(0) - y'(0)] - 8[sY(s) - y(0)] + 3.Y(s) = \frac{1}{s^2+1} \quad \text{where } L[y(t)] = Y(s)$$

$$\text{i.e., } (4s^2-8s+3)Y(s) - 8 = \frac{1}{s^2+1} \qquad \text{(Using initial conditions)}$$

$$Y(s) = \frac{1}{(s^2+1)(4s^2-8s+3)} - \frac{8}{4s^2-8s+3}$$

$$= \frac{1}{(s^2+1)(2s-1)(2s-3)} - \frac{8}{(2s-1)(2s-3)}$$

$$= \frac{1-8(s^2+1)}{(s^2+1)(2s-1)(2s-3)}$$

$$= \frac{8s}{65(s^2+1)} - \frac{1}{65(s^2+1)} - \frac{22}{5(2s-1)} + \frac{54}{13(2s-3)} \qquad \text{(By Partial fractions)}$$

Taking inverse Laplace transform both sides, we get,

$$y(t) = \frac{8}{65}L^{-1}\left[\frac{s}{s^2+1}\right] - \frac{1}{65}L^{-1}\left[\frac{1}{s^2+1}\right] - \frac{11}{5}L^{-1}\left[\frac{1}{s-1/2}\right] + \frac{27}{13}L^{-1}\left[\frac{1}{s-3/2}\right]$$

$$\text{i.e., } y(t) = \frac{8}{65}\cos t - \frac{1}{65}\sin t - \frac{11}{5}e^{t/2} + \frac{27}{13}e^{3t/2}$$

Example 5.5.3 Solve the differential equation y‴-3y″ + 3y′-y = t^2e^t, given y(0) = 1, y′(0) = 0, y″(0) = -2.

Solution:

$y''' - 3y'' + 3y' - y = t^2e^t$.

Let $L[y(t)] = Y(s)$.

Taking Laplace transform on both sides, we get,

$[s^3 Y(s) - s^2y(0) - sy'(0) - y''(0)] - 3[s^2Y(s) - sy(0)] + 3[s.Y(s) - y(0)] - Y(s) = L[t^2e^t]$

$$(s^2 - 3s^2 + 3s - 1)\, Y(s) - s^2 + 2 + 3s - 3 = \frac{2}{(s-1)^3}$$

$$(s-1)^3 Y(s) = s^2 - 3s + 1 + \frac{2}{(s-1)^3}$$

$$Y(s) = \frac{s^2 - 3s + 1}{(s-1)^3} + \frac{2}{(s-1)^6}$$

$$Y(s) = \frac{(s-1)^2 - (s-1) - 1}{(s-1)^3} + \frac{2}{(s-1)^6}$$

$$= \frac{1}{s-1} - \frac{1}{(s-1)^2} - \frac{1}{(s-1)^3} + \frac{2}{(s-1)^6}$$

Taking inverse Laplace transform, we get,

$$y(t) = e^t - te^t - \frac{1}{2!}t^2e^t + \frac{2}{5!}t^5e^t$$

$$y(t) = e^t\left[1 - t - \frac{t^2}{2} - \frac{t^5}{60}\right]$$

Example 5.5.4 Use the Laplace transforms to solve the initial value problem $y'' + y = e^t \sin t$, $y(0) = 0$, $y'(0) = 0$.

Solution:

$y'' + y = e^t \sin t$

$\therefore L[y''] + L[y] = L[e^t \sin t]$

$$[s^2 Y(s) - sy(0) - y'(0)] + Y(s) = \frac{1}{(s-1)^2+1}, \text{ where } L[y(t)] = Y(s)$$

$$(s^2+1)Y(s) = \frac{1}{(s-1)^2+1} \qquad \text{(Using initial conditions)}$$

$$Y(s) = \frac{1}{(s^2+1)\left((s-1)^2+1\right)}$$

$$= \frac{As+B}{s^2+1} + \frac{Cs+D}{(s-1)^2+1} \qquad \text{(By Partial fractions)}$$

$$1 = (As+B)\left((s-1)^2+1\right) + (Cs+D)(s^2+1)$$

$s = 0, \qquad 1 = 2B + D$

s = 1, $1 = A + B + 2C + 2D$

Equating coefficients of s^3, $0 = A + C$

Equating coefficients of s^2, $0 = B - 2A + D$

Solving for A, B, C, D, we get $A = \frac{2}{5}$, $B = \frac{1}{5}$, $C = -\frac{2}{5}$ and $D = \frac{3}{5}$

$$\therefore\ Y(s) = \frac{\frac{2}{5}s + \frac{1}{5}}{s^2+1} + \frac{-\frac{2}{5}s + \frac{3}{5}}{(s-1)^2+1}$$

$$Y(s) = \frac{2}{5}.\frac{s}{s^2+1} + \frac{1}{5}.\frac{1}{s^2+1} - \frac{2}{5}.\frac{s-1}{(s-1)^2+1} + \frac{1}{5}\frac{1}{(s-1)^2+1}$$

Taking inverse Laplace transform, we get,

$$y(t) = \frac{2}{5}\cos t + \frac{1}{5}\sin t - \frac{2}{5}e^t\cos t + \frac{1}{5}e^t\sin t$$

$$= \frac{1}{5}(2\cos t + \sin t) + \frac{1}{5}e^t(\sin t - 2\cos t)$$

Example 5.5.5 Solve the differential equation $(D^2 + 9)y = \cos 2t$ with boundary values $y(0) = 1$, $y(\pi/2) = -1$

Solution:

$(D^2 + 9)y = \cos 2t$

$D^2y + 9y = \cos 2t$

$y'' + 9y = \cos 2t$

Taking Laplace transform both sides

$L(y'') + 9L(y) = L(\cos 2t)$

$$s^2Y(s) - sy(0) - y'(0) + 9Y(s) = \frac{s}{s^2+4}, \text{ where } L[y(t)] = Y(s)$$

$$(s^2 + 9)\,Y(s) - s - a = \frac{s}{s^2+4} \qquad \text{(Assuming } y'(0) = a)$$

$$Y(s) = \frac{s}{(s^2+4)(s^2+9)} + \frac{s}{s^2+9} + \frac{a}{s^2+9}$$

$$Y(s) = \frac{1}{5}.\frac{s}{s^2+4} - \frac{1}{5}.\frac{s}{s^2+9} + \frac{s}{s^2+9} + \frac{a}{s^2+9}$$

$$= \frac{1}{5}.\frac{s}{s^2+4} + \frac{4}{5}.\frac{s}{s^2+9} + \frac{a}{s^2+9}$$

Taking inverse Laplace transforms on both sides, we get

$$y(t) = \frac{1}{5}\cos 2t + \frac{4}{5}\cos 3t + \frac{a}{3}\sin 3t \qquad (1)$$

To compute a, use the boundary condition $y(\pi/2) = -1$

Substituting $y(\pi/2) = -1$ in (1), we get,

$$-1=\frac{1}{5}(-1)+\frac{4}{5}(0)+\frac{a}{3}(-1)$$

i.e., $a={}^{12}\!/_{5}$

Hence (1) becomes,

$$y(t)=\frac{1}{5}\cos 2t+\frac{4}{5}\cos 3t+\frac{4}{5}\sin 3t$$, which is the required solution.

Example 5.5.6 Solve the initial value problem $y'-4y+3\int_0^t y(u)du=t,\ y(0)=1$

Solution:

$$y'-4y+3\int_0^t y(u)du=t$$

Taking Laplace transforms both sides, we get

$$s.\,Y(s)-y(0)-4Y(s)+3\,L\left[\int_0^t y(u)du\right]=\frac{1}{s^2},\quad \text{where } L[y(t)]=Y(s)$$

i.e., $$s.\,Y(s)-y(0)-4Y(s)+3\frac{1}{s}L[y(t)]=\frac{1}{s^2}$$

$$s.\,Y(s)-y(0)-4Y(s)+\frac{3}{s}L[y(t)]=\frac{1}{s^2}$$

$$(s^2-4s+3)Y(s)=s+\frac{1}{s}\quad \text{since } y(0)=1$$

$$Y(s)=\frac{s^2+1}{s(s^2-4s+3)}$$

$$=\frac{s^2+1}{s(s-1)(s-3)}$$

$$=\frac{1}{3s}-\frac{1}{(s-1)}+\frac{5}{3(s-3)}\qquad \text{(By Partial fractions)}$$

Taking inverse Laplace transform, we get

$$y(t)=\frac{1}{3}-e^t+\frac{5}{3}e^{3t}$$

Example 5.5.7 Using convolution theorem, solve the initial value problem $y''+16y=\cos 4t,\ y(0)=0,\ y'(0)=0.$

Solution:

$y''+16y=\cos 4t$

$\therefore\quad L[y'']+16L[y]=L[\cos 4t]$

i.e., $s^2Y(s) - sy(0) - y'(0) + 16y(s) = \frac{s}{s^2+16}$, where $L[y(t)] = Y(s)$

i.e., $(s^2 + 16)Y(s) = \frac{s}{s^2+16}$ (Using initial conditions)

$$Y(s) = \frac{s}{(s^2+16)^2}$$

Taking inverse Laplace transform, we get,

$$y(t) = L^{-1}\left[\frac{s}{(s^2+16)^2}\right] \qquad (1)$$

Let $F(s) = \frac{s}{s^2+16}$ and $G(s) = \frac{1}{s^2+16}$

Then $f(t) = L^{-1}\left[\frac{s}{s^2+16}\right] = \cos 4t$

$$g(t) = L^{-1}\left[\frac{1}{s^2+16}\right] = \frac{1}{4}\sin 4t$$

Hence by convolution theorem,

$$L^{-1}\left[\frac{s}{(s^2+16)^2}\right] = \int_0^t f(u).g(t-u)du$$

$$= \int_0^t \cos 4u.\frac{1}{4}\sin(4t-4u)du$$

$$= \frac{1}{8}\int_0^t (\sin 4t - \sin(8u-4t)du$$

$$= \frac{1}{8}\left[u.\sin 4t + \frac{\cos(8u-4t)}{8}\right]_0^t$$

$$= \frac{1}{8}\left[t\sin 4t + \frac{1}{8}\cos 4t - \frac{1}{8}\cos 4t\right]$$

$$= \frac{1}{8}t\sin 4t$$

$\therefore$ Equation (1) becomes, $y(t) = \frac{1}{8}t\sin 4t$

Example 5.5.8 Solve the initial value problem $y'' + 4y = f(t)$, $y(0) = 0$, $y'(0) = 0$ and $f(t)$ is a periodic function defined by

$$f(t) = \sin t \quad 0 \le t < \pi$$

$$= 0 \quad \pi \le t < 2\pi$$

Solution:

Let L[y(t)] = Y(s) and L[f(t)] = F(s).

Given y″ + 4y = f(t).

Taking Laplace transform both sides, we get,

$$[s^2 Y(s) - sy(0) - y'(0)] + 4.Y(s) = F(s) \qquad (1)$$

Where $F(s) = \dfrac{1}{1-e^{-2\pi s}}\displaystyle\int_0^{2\pi} e^{-st} f(t)dt$

$$F(s) = \frac{1}{1-e^{-2\pi s}}\int_0^{\pi} e^{-st}\sin t\, dt \qquad (2)$$

Now $I = \displaystyle\int_0^{\pi} e^{-st}\sin t\, dt$

$$= \sin t.\frac{e^{-st}}{-s}\Bigg|_0^{\pi} - \int_0^{\pi}\frac{e^{-st}}{-s}\cos t\, dt$$

$$= \frac{1}{s}\left[\cos t.\frac{e^{-st}}{-s}\Bigg|_0^{\pi} - \int_0^{\pi}\frac{e^{-st}}{-s}(-\sin t)\, dt\right]$$

$$= -\frac{1}{s^2}\left(e^{-\pi s}-1\right) - \frac{1}{s^2}I$$

$\therefore \left(1+\dfrac{1}{s^2}\right)I = \dfrac{1+e^{-\pi s}}{s^2}$ i.e., $I = \dfrac{1+e^{-\pi s}}{1+s^2}$

$\therefore$ Equation (2) becomes $F(s) = \dfrac{1}{1-e^{-2\pi s}}.\dfrac{1+e^{-\pi s}}{1+s^2} = \dfrac{1}{(1+s^2)(1-e^{-\pi s})}$

Substituting for the initial values and F(s) in (1), we get,

$$(s^2+4)Y(s) = \frac{1}{(1+s^2)(1-e^{-\pi s})}$$

$$Y(s) = \frac{(1-e^{-\pi s})^{-1}}{(s^2+4)(1+s^2)}$$

$$= \frac{1}{(s^2+4)(1+s^2)}\left(1+e^{-\pi s}+e^{-2\pi s}+\cdots\right)$$

Now $L^{-1}\left[\dfrac{1}{(s^2+4)(1+s^2)}\right] = L^{-1}\left[\dfrac{1/3}{s^2+1}\right] - L^{-1}\left[\dfrac{1/3}{s^2+4}\right]$

$$= \frac{1}{3}\sin t - \frac{1}{6}.\sin 2t$$

$$= \frac{1}{6}(2\sin t - \sin 2t)$$

Let $g(t) = \frac{1}{6}(2\sin t - \sin 2t)$

Then $y(t) = L^{-1}[Y(s)] = g(t).u_0(t) + g(t-\pi).u_\pi(t) + g(t-2\pi).u_{2\pi}(t) + \ldots$ (3)

where $u_a(t)$ is the unit step function.

Substituting $g(t) = \frac{1}{6}(2\sin t - \sin 2t)$ in equation (3) we get the solution y(t) of the given initial value problem.

Example 5.5.9 Solve the initial value problem $\frac{d^4x}{dt^4} - x = 1$, $\dddot{x} = \ddot{x} = \dot{x} = x = 0$ at $t = 0$

Solution:

Let $L[x(t)] = X(s)$.

$$\frac{d^4x}{dt^4} - x = 1$$

$s^4.X(s) - s^3x(0) - s^2x'(0) - sx''(0) - x'''(0) - X(s) = 1/s$

i.e., $(s^4-1)X(s) = 1/s$

$$X(s) = \frac{1}{s(s^2-1)(s^2+1)}$$

$$X(s) = \frac{1}{s(s-1)(s+1)(s^2+1)}$$

$$= -\frac{1}{s} + \frac{\frac{1}{4}}{s-1} + \frac{\frac{1}{4}}{s+1} + \frac{\frac{1}{2}s}{s^2+1} \qquad \text{(By partial fractions)}$$

Taking inverse Laplace transform, we get,

$$x(t) = -1 + \frac{1}{4}e^t + \frac{1}{4}e^{-t} + \frac{1}{2}\cos t$$

5.5.2 SIMULTANEOUS LINEAR DIFFERENTIAL EQUATIONS WITH CONSTANT COEFFICIENTS

In this section we discuss the Laplace transform method of solving two or more simultaneous ordinary differential equations with given initial conditions. The procedure is similar to that of an initial value problem consisting of a single differential equation. Take Laplace transforms on both sides of the given equations and then solve these equations for the Laplace transforms of the required functions. It is illustrated in the following examples.

Example 5.5.10 Solve the following simultaneous equations by the Laplace transform method.

$\dot{x} - y = e^t$

$\dot{y} + x = \sin t$, given that $x(0) = 1$ and $y(0) = 0$

Solution:

Let L[x(t)] = X(s) and L[y(t)] = Y(s).

$$\dot{x} - y = e^t \tag{1}$$

Taking Laplace transforms, we get,

s. L[x(t)] - x(0) - L[y(t)] = L[e^t]

i.e., $sX(s) - 1 - Y(s) = \dfrac{1}{s-1}$

$$sX(s) - Y(s) = \frac{s}{s-1} \tag{2}$$

$$\dot{y} + x = \sin t \tag{3}$$

Taking Laplace transforms,

$s.Y(s) - y(0) + X(s) = \dfrac{1}{s^2+1}$

i.e., $$X(s) + s.Y(s) = \frac{1}{s^2+1} \tag{4}$$

Solving (2) and (4) for X(s) and Y(s), we get,

$$X(s) = \frac{\begin{vmatrix} \frac{s}{s-1} & -1 \\ \frac{1}{s^2+1} & s \end{vmatrix}}{\begin{vmatrix} s & -1 \\ 1 & s \end{vmatrix}} = \left(\frac{s^2}{s-1} + \frac{1}{s^2+1}\right) \cdot \frac{1}{(s^2+1)}$$

$$= \frac{s^2}{(s-1)(s^2+1)} + \frac{1}{(s^2+1)^2} = \frac{s/2}{s^2+1} + \frac{1/2}{s^2+1} + \frac{1/2}{s-1} + \frac{1}{(s^2+1)^2}$$

Taking inverse Laplace transforms,

$$x(t) = \frac{1}{2}L^{-1}\left[\frac{s}{s^2+1}\right] + \frac{1}{2}L^{-1}\left[\frac{1}{s^2+1}\right] + \frac{1}{2}L^{-1}\left[\frac{1}{s-1}\right] + L^{-1}\left[\frac{1}{(s^2+1)^2}\right]$$

$$= \frac{1}{2}\cos t + \frac{1}{2}\sin t + \frac{1}{2}e^t + \frac{1}{2}(\sin t - t\cos t)$$

i.e., $x(t) = \dfrac{1}{2}(\cos t + 2\sin t + e^t - t\cos t)$

From equation (1) we get,

$$y(t) = \dot{x} - e^t$$

$$= \frac{1}{2}(-\sin t + 2\cos t + e^t + t\sin t - \cos t) - e^t$$

$$= \frac{1}{2}(\cos t - \sin t - e^t + t\sin t)$$

x(t) and y(t) are the solutions of the given simultaneous equations.

Example 5.5.11 Solve the simultaneous equations

$\frac{dx}{dt} + y = \sin t, \frac{dy}{dt} + x = \cos t$; given that $x = 2$ and $y = 0$ when $t = 0$.

Solution:

Let L[x(t)] = X(s) and L[y(t)] = Y(s).

$$\frac{dx}{dt} + y = \sin t$$

Taking Laplace transforms, we get,

$$s.\ X(s) - x(0) + Y(s) = \frac{1}{s^2+1}$$

$$s.\ X(s) + Y(s) = 2 + \frac{1}{s^2+1} \qquad (1)$$

$$\frac{dy}{dt} + x = \cos t$$

Taking Laplace transforms, we get,

$$s.\ Y(s) - y(0) + X(s) = \frac{s}{s^2+1}$$

$$X(s) + s.\ Y(s) = \frac{s}{s^2+1} \qquad (2)$$

Solving equations (1) and (2) for X(s) and Y(s), we get,

$$X(s) = \frac{\begin{vmatrix} 2 + \frac{1}{s^2+1} & 1 \\ \frac{s}{s^2+1} & s \end{vmatrix}}{\begin{vmatrix} s & 1 \\ 1 & s \end{vmatrix}} = \left(2s + \frac{s}{s^2+1} - \frac{s}{s^2+1}\right)\frac{1}{s^2-1}$$

$$= \frac{2s}{s^2-1}$$

$$\therefore,\ X(s) = \frac{1}{s+1} + \frac{1}{s-1} \qquad \text{(By partial fractions)}$$

Taking inverse Laplace transforms, we get,

$x(t) = e^{-t} + e^{t}$

Given $\frac{dx}{dt} + y = \sin t$

$$\therefore y = \sin t - \frac{dx}{dt}$$

$$= \sin t - (-e^{-t} + e^{t})$$

$$\therefore y(t) = \sin t + e^{-t} - e^{t}$$

x(t) and y(t) are the solutions of the given simultaneous equations.

Example 5.5.12 Solve the following simultaneous equations using Laplace transforms $3\frac{dx}{dt}+\frac{dy}{dt}+2x=1,\ \frac{dx}{dt}+4\frac{dy}{dt}+3y=0$. Given x(0) = 0 and y(0) = 0.

Solution:

Let L[x(t)] = X(s) and L[y(t)] = Y(s).

$$3\frac{dx}{dt}+\frac{dy}{dt}+2x=1$$

Taking Laplace transform, we get,

$$3.[s.\,X(s)-x(0)]+s.\,Y(s)-y(0)\ +2.X(s)=\frac{1}{s}$$

$$\therefore\ (3s+2)X(s)+s.\,Y(s)=\frac{1}{s} \qquad (1)$$

$$\frac{dy}{dt}+4\frac{dy}{dt}+3y=0$$

Taking Laplace transform, we get,

$$sX(s)-x(0)+4[sY(s)-y(0)]+3.Y(s)=0$$

$$s.\,X(s)+(4s+3)Y(s)=0 \qquad (2)$$

Solving (1) and (2) for X(s) and Y(s), we get,

$$X(s)=\frac{\begin{vmatrix}\frac{1}{s} & s\\ 0 & 4s+3\end{vmatrix}}{\begin{vmatrix}3s+2 & s\\ s & 4s+3\end{vmatrix}}=\frac{1}{s}.(4s+3).\frac{1}{[(3s+2)(4s+3)-s^2]}$$

$$=\frac{4s+3}{s[11s^2+17s+6]}$$

$$=\frac{4s+3}{s(s+1)(11s+6)}$$

$$=\frac{A}{s}+\frac{B}{s-1}+\frac{11}{11s+6} \qquad \text{(By Partial fractions)}$$

$$4s+3=A(s+1)(11s+6)+Bs(11s+6)+Cs(s+1)$$

When s = 0, $A=\frac{1}{2}$

When s = -1, $B=-\frac{1}{5}$

Equating coefficients of s^2, $0=11A+11B+C$ i.e., $C=-\frac{33}{10}$

$$X(s)=\frac{\frac{1}{2}}{s}-\frac{\frac{1}{5}}{s+1}-\frac{\frac{33}{10}}{11s+6}$$

Taking inverse Laplace transforms, we get,

$$x(t) = \frac{1}{2} - \frac{1}{5}e^{-t} - \frac{3}{10}e^{-6/11\,t}$$

$$x(t) = \frac{1}{10}\left(5 - 2e^{-t} - 3e^{-6t/11}\right)$$

$$Y(s) = \frac{\begin{vmatrix} 3s+2 & 1/s \\ s & 0 \end{vmatrix}}{(s+1)(11s+6)} = \frac{-1}{(s+1)(11s+6)}$$

$$= \frac{A}{s+1} + \frac{B}{11s+6} \qquad \text{(By partial fractions)}$$

$-1 = A(11s + 6) + B(s + 1)$

When $s = -1$, $-1 = -5A$ i.e., $A = 1/5$
Equating constants, $-1 = 6A + B$

$$-1 = 6/5 + B$$

$$B = -11/5$$

Hence,

$$Y(s) = \frac{1/5}{s+1} - \frac{11/5}{11s+6}$$

$$= \frac{1}{5}.\frac{1}{s+1} - \frac{1}{5}.\frac{1}{s+6/11}$$

Taking inverse Laplace transform

$$y(t) = \frac{1}{5}e^{-t} - \frac{1}{5}e^{-\frac{6t}{11}}$$

$$y(t) = \frac{1}{5}\left(e^{-t} - e^{-\frac{6t}{11}}\right)$$

Example 5.5.13 Using Laplace transforms, solve the simultaneous differential equations. $m\ddot{x} = k\dot{y}$, $m\ddot{y} = -k\dot{x}$, given $x = y = \dot{y} = 0$ and $\dot{x} = c$ at $t = 0$.

Solution:

Let $L[x(t)] = X(s)$ and $L[y(t)] = Y(s)$

$m\ddot{x} = k\dot{y}$

Taking Laplace transforms

$$m[s^2.X(s) - s.X(0) - \dot{x}(0)] = k[s.Y(s) - y(0)]$$

$$\text{i.e., } s^2.X(s) - \frac{k}{m}s.Y(s) = c \qquad (1)$$

$m\ddot{y} = -k\dot{x}$

Taking Laplace transforms

$$m[s^2.Y(s) - s.y(0) - \dot{y}(0)] = k[s.X(s) - x(0)]$$

i.e., $\frac{k}{m}s.X(s) + s^2Y(s) = 0$ (2)

Solving (1) and (2) for X(s) and Y(s), we get

$$X(s) = \frac{\begin{vmatrix} c & -\frac{k}{m}s \\ 0 & s^2 \end{vmatrix}}{\begin{vmatrix} s^2 & -\frac{k}{m}s \\ \frac{k}{m}s & s^2 \end{vmatrix}} = \frac{c.s^2}{s^4 + \frac{k^2}{m^2}s^2} = \frac{c}{s^2 + \frac{k^2}{m^2}}$$

Taking inverse Laplace transform, we get

$$x(t) = \frac{c}{k/m}.\sin\frac{k}{m}t = \frac{mc}{k}\sin\frac{k}{m}t$$

$$\text{Also } Y(s) = \frac{\begin{vmatrix} s^2 & c \\ \frac{k}{m}s & 0 \end{vmatrix}}{s^4 + \frac{k^2}{m^2}s^2} = \frac{-c.\frac{k}{m}.s}{s^4 + \frac{k^2}{m^2}s^2}$$

$$= \frac{-c.\frac{k}{m}}{s\left(s^2 + \frac{k^2}{m^2}\right)}$$

$$= \frac{c}{k/m}\left(\frac{s}{s^2 + k^2/m^2} - \frac{1}{s}\right) \quad \text{(By partial fractions)}$$

Taking inverse Laplace transform, we get,

$$y(t) = \frac{mc}{k}\left(\cos\frac{k}{m}t - 1\right)$$

Example 5.5.14 Solve the simultaneous equations $\dot{x} - \dot{y} - 2x + 2y = 1 - 2t$, $\ddot{x} + 2\dot{y} + x = 0$, given $x = 0$, $y = 0$, $\dot{x} = 0$ when $t = 0$.

Solution:

Taking Laplace transforms on both sides of the given equations, we get

$$L[\dot{x}] - L[\dot{y}] - 2L[x] + 2L[y] = L[1] - 2L[t]$$

i.e., $sX(s) - x(0) - sY(s) + y(0) - 2X(s) + 2Y(s) = \frac{1}{s} - \frac{2}{s^2}$

i.e., $(s-2)X(s) - (s-2)Y(s) = \frac{s-2}{s^2}$ (1)

Again $L[\ddot{x}] - 2L[\dot{y}] + L[x] = 0$

$$s^2X(s) - s.x(0) - \dot{x}(0) + 2sY(s) - 2y(0) + X(s) = 0$$

$$(s^2+1)X(s) + 2s.Y(s) = 0 \qquad (2)$$

Solving (1) and (2) for X(s) and Y(s), we get,

$$X(s) = \frac{2}{s(s+1)^2} \text{ and } Y(s) = \frac{-(s^2+1)}{s^2(s+1)^2}$$

$$x(t) = L^{-1}\left[\frac{2}{s(s+1)^2}\right] = 2L^{-1}\left[\frac{1}{s} - \frac{1}{s+1} - \frac{1}{(s+1)^2}\right]$$

$$= 2.(1-e^{-t}-te^{-t})$$

$$y(t) = L^{-1}\left[\frac{-(s^2+1)}{s^2(s+1)^2}\right]$$

$$= -\int_0^t L^{-1}\left[\frac{s^2+1}{s(s+1)^2}\right]dt$$

$$= -\int_0^t L^{-1}\left[\frac{1}{s} - \frac{2}{(s+1)^2}\right]dt$$

$$= -\int_0^t (1-2.te^{-t})\,dt$$

$$= -t + \left[2t.\frac{e^{-t}}{-1}\right]_0^t - \int_0^t \frac{e^{-t}}{-1}.2dt$$

$$= -t - 2te^{-t} + \left[2.\frac{e^{-t}}{-1}\right]_0^t$$

$$= -t - 2te^{-t} - 2e^{-t} + 2$$

$$= 2 - t - 2(t+1)e^{-t}$$

EXERCISE 5.5

PART-A

1. Explain the Laplace transform method of solving linear differential equations.
2. Give the procedure for solving linear differential equations using Laplace transforms.
3. What is the advantage of solving initial value problems using Laplace transforms?

Solve the following initial value problems using Laplace transform (Ex: 4-12)

4. $y' = y,\ y(0) = 1.$

5. $\frac{dx}{dt} + x = \sin wt, \; x(0) = 2.$
6. $y' + 3y = 1, \; y(0) = 1.$
7. $y' + 2y = \sin t, y(0) = 1$
8. $y' + 3y = \cos t, y(0) = 0$
9. $x' - 4x = t, x(0) = -1.$
10. $x' - 2x = 1 + t, x(0) = 2.$
11. $x' + x = 1 + te^t, x(0) = 1.$
12. $y' + y = t \cos t, y(0) = 0.$

PART-B

Solve the following initial value problems using Laplace transforms (Ex: 13 -28)

13. $y'' + 4y = 0, y(0) = 1, y'(0) = 6.$
14. $y'' - 5y' + 4y = e^{2t}, \; y(0) = {}^{19}\!/_{12}, \; y'(0) = {}^{8}\!/_{3}.$
15. $2y''\text{-}y'\text{-}y = \cos t, y(0) = 1, y'(0) = 0.$
16. $y'' + 5y' + 4y = e^{3t}, y(0) = 0, y'(0) = 3.$
17. $y'' + 4y' + 13y = e^{-t}, y(0) = 0, y'(0) = 2.$
18. $y'' + 8y' + 16y = te^{-4t}, y(0) = 1, y'(0) = 2.$
19. $y'' + y = e^t \sin t, y(0) = 0, y'(0) = 0.$
20. $y'' + 6y' + 9y = 8t\,e^{2t}, y(0) = 0, y'(0) = -1.$
21. $y'' - 2y' + y = e^t, x(0) = 2, x'(0) = -1.$
22. $y'' - 3y' + 2y = 4t + e^{3t}, y(0) = 1, y'(0) = -1.$
23. $(D^2 + w^2)y = \cos wt, t > 0$, given that $y = 0$ and $Dy = 0$ at $t = 0$.
24. $(D^2 + 1)x = t \cos 2t, x = Dx = 0$ at $t = 0$.
25. $\frac{d^2y}{dt^2} + 2\frac{dy}{dt} + 5y = e^t \sin t$, where $y(0) = 0$ and $y'(0) = 1$.
26. $\frac{d^3y}{dt^3} + 2\frac{d^2y}{dt^2} - \frac{dy}{dt} - 2y = 0$, where $y = 1, \frac{dy}{dx} = 2, \; \frac{d^2y}{dx^2} = 2$ at $t = 0$.
27. $\frac{d^4y}{dt^4} - k^4y = 0$ where $y(0) = 1, y'(0) = y''(0) = y'''(0) = 0.$
28. $\frac{d^3y}{dt^3} - \frac{dy}{dt} = 2\cos t, \; y(0) = 3, y'(0) = 2, y''(0) = 1.$

Solve the following initial value problems using convolution theorem (Ex: 29 -33)

29. $y'' + y = \sin 3t, y(0) = y'(0) = 0.$
30. $y' - y = te^t \cos t, y(0) = 0.$
31. $y'' - w^2y = \cosh wt, y(0) = 1, y'(0) = 2.$

32. $y'' + 4y' + 4y = te^{-t}, y(0) = 0, y'(0) = 2.$
33. $y'' + 9y = \sin 3t, y(0) = 0, y'(0) = 0.$
34. Solve the initial value problem $y' + 3y + 2\int_0^t y(u)du = t, y(0) = 0.$
35. Solve the initial value problem $y' + 6y + 5\int_0^t y(u)du = 1 + t, y(0) = 1.$
36. Solve $y' - y - 6\int_0^t y(u)du = \sin t, y(0) = 2.$
37. Solve the initial value problem $y' + y = f(t), y(0) = 2$ where $f(t) = \begin{cases} 0 & 0 \le t \le \pi/2 \\ \cos t & t \ge \pi/2 \end{cases}$
38. Solve the initial value problem $y' + y = f(t), y(0) = 3,\ f(t) = \begin{cases} 1 & 0 \le t \le 1 \\ -1 & t \ge 1 \end{cases}$
39. Solve $y' + 2y = f(t), y(0) = 1,\ f(t) = \begin{cases} 1 & 0 \le t \le 2 \\ 0 & t \ge 2 \end{cases}$
40. Solve the initial value problem $y' + 7y + 12\int_0^t y(u)du = t.u_2(t), y(0) = 1.$
41. Solve the initial value problem $y' + 4\int_0^t y(u)du = t.u_1(t), y(0) = 2.$
42. Solve $y''-3y' + 2y = u_1(t), y(0) = 1, y'(0) = 1.$
43. Solve $y'' + 4y' + 3y = f(t), y(0) = 0, y'(0) = 1,\ f(t) = \begin{cases} -1 & 0 \le t \le 3 \\ 0 & t \ge 3 \end{cases}$

Solve the following initial value problems where f(t) is a periodic function. (Ex: 44-48)

44. $y'' + 8y' + 17y = f(t), y(0) = 0, y'(0) = 0,\ f(t) = \begin{cases} 1 & 0 < t < \pi \\ 0 & \pi < t < 2\pi \end{cases}$ and $f(t+2\pi) = f(t).$
45. $y'' + 9y = f(t), y(0) = 1, y'(0) = 0\ f(t) = \begin{cases} \cos t & 0 \le t < \pi \\ 0 & \pi \le t < 2\pi \end{cases}$ and $f(t + 2\pi) = f(t).$
46. $y'' + y = f(t), y(0) = 0, y'(0) = 0,\ f(t) = \begin{cases} 0 & 0 \le t < \pi \\ \sin t & \pi \le t < 2\pi \end{cases}$ and $f(t + 2\pi) = f(t).$
47. $y'' + 4y' + 5y = f(t), y(0) = 0, y'(0) = 0, f(t) = \begin{cases} 1 & 0 \le t < \pi \\ -1 & \pi \le t < 2\pi \end{cases}$ and $f(t + 2\pi) = f(t).$
48. $y'' + 3y' + 2y = f(t), y(0) = 0, y'(0) = 1$ and $f(t) = t, 0 \le t \le a$ and $f(t + a) = f(t).$

49. Solve the following simultaneous differential equations using Laplace transforms:
$\frac{dx}{dt}+3x+y=0, \frac{dy}{dx}-x+y=0$, given $x(0)=y(0)=1$.

50. Solve the simultaneous equations: $\frac{dx}{dt}+5x-2y=t, \frac{dy}{dt}+2x+y=0$, given $x=y=$ 0 when $t=0$.

51. Solve the simultaneous equations using Laplace transforms: $\ddot{x}+3x-2y=0,$
$\ddot{x}+\ddot{y}-3x+5y=0$, where it is given $x=0, y=0, \dot{x}=3, \dot{y}=2$ when $t=0$.

52. Solve the system of equations $\dot{y}+2z=1, 2y-\dot{z}=2t$, given that $y(0)=0$ and $z(0)=1$.

53. Solve the simultaneous equations using Laplace transform $\ddot{x}-2y=0, \ddot{y}+2x=0$ given that $x(0)=y(0)=1, \dot{x}(0)=\dot{y}(0)=0$.

54. A particle moves along a line so that its displacement x from a fixed point at any time t is given by $\ddot{x}+4\dot{x}+5x=80\sin 5t$. If the particle starts from rest initially, find its displacement at any time $t>0$.

55. An electron is projected into a uniform magnetic field which is perpendicular to its direction of motion. If x and y are coordinates of its position at any time t, the equations of motion are given by $m\ddot{x}=k\dot{y}$ and $m\ddot{y}=-k\dot{x}$, where the dots denote differentiations with respect to t, m and k are constants. The initial conditions are $x=y=\dot{y}=0$ and $\dot{x}=c$ at $t=0$. Use Laplace transforms method to find x and y at time t.

56. The currents i_1 and i_2 in a mesh are given by the differential equations
$\frac{di_1}{dt}-wi_2=a\cos pt, \frac{di_2}{dt}+wi_1=a\sin pt$. Find the currents i_1 and i_2 by Laplace transform method if $i_1=i_2=0$ at $t=0$.

57. A particle moves in the xy-plane so that its position (x, y) at any time t is given by $\ddot{x}+\alpha^2x=0, \ddot{y}+\beta^2y=0$. If the particle starts from rest at (a, b) find its position at time $t>0$ using Laplace transform.

58. Solve the initial value problem
$\frac{dx}{dt}+2x-3y=t, \frac{dy}{dt}-3x+2y=e^{2t}, x=0, y=0$ when $t=0$.

59. Solve the initial value problem
$\frac{dx}{dt}-3\frac{dy}{dt}=\frac{t}{3}, \frac{d^2y}{dt^2}+x-2y=0, x=0, y=0, \frac{dy}{dt}=0$ when $t=0$.

ANSWERS

EXERCISE 5.1

(1) $\frac{1}{2}\left(\frac{1}{s}+\frac{s}{s^2+16}\right)$ (2) $\frac{6}{(s-2)^4}$ (3) $\frac{2s}{(s^2+1)^2}$

(4) $\frac{2}{s^2}-\frac{5}{s}$ (5) $\frac{6}{s^3}+\frac{2}{s+1}$ (6) $\frac{5s}{s^2+9}$

(7) $\frac{5}{s^2-25}$ (8) $\frac{24}{s^5}+\frac{12}{s^3}+\frac{9}{s}$ (9) $\frac{2}{s^2-6s+13}$

(10) $\frac{s\cos b - a\sin b}{s^2+a^2}$ (11) $\frac{4}{s^2+4s+20}$ (12) $\frac{4+s^2}{s^2(4-s^2)}$

(13) $\frac{b}{(s+a)^2-b^2}$ (14) $\frac{2as}{(s^2-a^2)^2}$ (15) $\frac{2}{(s-1)^3}$

(16) $\frac{1}{s}+\frac{2}{s^2+4}$ (17) $\frac{5s-18}{s(s-3)}$ (18) $\frac{6}{s^3}-\frac{5}{s+2}+\frac{6}{s}$

(19) $\frac{s^2-4}{(s^2+4)^2}$ (20) $\frac{e^5}{s+2}$ (21) $\frac{12s}{(s^2+1)(s^2+25)}$

(22) $\frac{48}{(s^2+4)(s^2+36)}$ (23) $\frac{1}{2}\left[\frac{3}{(s-4)^2+9}+\frac{1}{(s-4)^2+1}\right]$

(24) $\frac{2}{(s+1)(s^2+2s+5)}$ (25) $\frac{2abs}{[s^2+(a+b)^2][s^2+(a-b)^2]}$

(26) $\frac{2s-13}{s^2+2s+26}$ (27) $\frac{2s^3-6s}{(s^2+1)^3}$ (28) $\frac{2a(3s^2-a^2)}{(s^2+a^2)^3}$

(29) $\frac{6(s+1)}{(s^2+2s+10)^2}$ (30) $\cot^{-1}(s+1)$ (31) $\frac{1}{2}\log\frac{s^2+b^2}{(s-a)^2}$

(32) $\cot^{-1}s-\frac{1}{2}s.\log\left(\frac{1+s^2}{s^2}\right)$ (33) $\frac{1}{1-s}\left[e^{1-s}-1\right]$ (34) $\frac{1+e^{-\pi s}}{1+s^2}$

(35) $\frac{1-2e^{-s}}{s}$ (36) $\frac{2(1-e^{-3s})}{s}$ (37) $\frac{-e^{-s\pi}}{s^2+1}$

(42) $\frac{2(3s^2+15s+25)}{s^3(s+5)^3}$ (43) $\frac{2(3s^2+4s+5)}{s^2(s^2+2s+5)^2}$

(44) (i) $\frac{e^{-2s}}{s^3}(2+4s+3s^2)$ (ii) $\frac{-e^{-\pi s}}{s^2+1}$ (iii) $\frac{1}{s+1}e^{-3s}$ (iv) $\frac{2}{s^3}e^{-s}$

(45) (i) $\frac{s}{s^2+1}.e^{-\pi s/2}$ (ii) $\frac{k}{s}\left(1-e^{-2s}+e^{-4s}\right)$

EXERCISE 5.2

(4) $s^3.F(s)+2s^2-1$ (5) $\frac{2}{s(s-1)^3}$ (7) $\frac{7}{s+4}$

(9) $\frac{1}{s}.\cot^{-1}s$ (10) $\frac{s+1}{s(s^2+2s+2)}$ (11) $\frac{4!}{s(s-1)^5}$

(12) $\frac{2}{s(s^2+4)}$ (13) $\frac{1}{2s}\left(\frac{1}{s}+\frac{s}{s^2+4}\right)$ (14) $\frac{2(3s^2+15s+25)}{s^3(s+5)^3}$

(15) $\frac{2(s^3+5s^2+8s+13)}{s^2(s^2+4s+13)^2}$

EXERCISE 5.3

(9) $\cos 2t + \frac{1}{2}\sin 2t$ (10) $e^{-3t}\left(\cos 2t-\frac{3}{2}\sin 2t\right)$ (11) $4e^{3t}-e^{-t}$

(12) $e^{2t}+2e^{-4t}$ (13) $e^{2t}(1+4t+2t^2)$ (14) $e^{2t}\left(\cos 3t+\frac{5}{3}\sin 3t\right)$

(15) $\cos 3t+\frac{2}{3}\sin 3t$ (16) $\frac{1}{4}-\frac{1}{4}e^{-2t}+\frac{1}{2}t.e^{-2t}$ (17) $\frac{e^{-2t}}{2}\sin 2t$

(18) $\frac{1}{6}t^3.e^{-5t}$ (19) $\frac{3}{2}(1-e^{-2t})$ (20) $\frac{1}{2}\sin \frac{t}{2}$

(21) $\frac{1}{9}(1-\cos 3t)$ (22) $\frac{1}{4}(t-\frac{1}{2}\sin 2t)$

(23) $\frac{1}{t}\sin 2t$ (24) $\frac{1}{t}(1-e^t)$

(25) $\frac{1}{t}e^t\sin t$ (26) $\frac{1}{3}e^t+3t\,e^t-\frac{1}{3}e^{-2t}$

(27) $\frac{1}{2}e^t-e^{2t}+\frac{5}{2}e^{3t}$ (28) $\frac{1}{3}e^{-2t}(6\cos 3t-7\sin 3t)$

(29) $3e^{-3t}t\sin t$ (30) $e^t - e^{-t}\cos 2t+\frac{3}{2}e^{-t}\sin 2t$

(31) $3e^{2t}+5t\,e^{2t}$ (32) $(6t-7t^2)\frac{e^{-3t}}{2}$

(33) $\frac{1}{\sqrt{3}}\left(3\sqrt{3}\cos\sqrt{3}t - 4\sin\sqrt{3}t\right)e^{-3t}$

(34) $\frac{1}{a^3}\left(1 - e^{-at} - ate^{-at} - \frac{a^2}{2}t^2e^{-at}\right)$

(35) $\frac{1}{3}(e^{2t} - e^{-t} - t)$

(36) $\frac{1}{2}t^2 + \cos t + 1$

(37) $\frac{1}{a^2}\left(t - \frac{1}{a}\sin at\right)$

(38) $\frac{1}{3} + \frac{4}{15}e^{-3t} + \frac{2}{5}e^{2t}$

(39) $\frac{1}{18}(3t^2 - 2t) + \frac{1}{27}(1 - e^{-3t})$

(40) $\frac{1}{a^2}(\sin at - \cos at + 1) - \frac{t}{a}$

(41) $\frac{3}{4}t - \sin t + \frac{1}{8}\sin 2t$

(42) $\frac{1}{2}(a^2t^2 - 4at + 2)e^{-at}$

(43) $t \sin at$

(44) $\frac{1}{3}(\cos t - \cos 2t)$

(45) $\frac{1}{2}(e^t + \sin t - \cos 2t)$

(46) $\frac{1}{t}(1 + e^{-t} - 2\cos t)$

(47) $2\sinh t.\ \sin t$

(48) $\frac{1}{t}(\cos at - \cos bt)$

(49) $\begin{cases} \frac{1}{6}(t-3)^3 e^{t-3} & t > 3 \\ 0 & t < 3 \end{cases}$

(50) $\frac{1}{t}(2\cos 2t - e^{-t} - 1)$

(51) $\frac{2}{t^2}(\sinh t - t\cosh t)$

(52) $e^{-t} - e^{-2t};\ 2e^{-2t} - e^{-t}$

(53) $\sin wt,\ \frac{1}{w^2}(wt - \sin wt)$

(54) $\begin{cases} \sin 4t & t \ge \pi/2 \\ 0 & t < \pi/2 \end{cases}$

(55) $\begin{cases} t-2 & t \ge 2 \\ 0 & t < 2 \end{cases}$

(61) $-\cos 3t.u_\pi(t)$

(62) $\cos 2t.\left[1 - u_{\pi/2}(t)\right]$

(63) $\sin 4t.u_{\pi/2}(t)$

(64) $\left[u_{1/2}(t) - u_1(t)\right]\sin \pi t$

(65) $\frac{1}{a^2}\left(a(t-c) - 1 + e^{-a(t-c)}u_c(t)\right)$

(66) $\frac{1}{2}e^{-(t-1)}(t-1)^2.u_1(t)$

EXERCISE 5.4

(6) t

(7) $e^t - t - 1$

(8) $\frac{1}{2}(1 - e^{-2t})$

(9) $\frac{1}{4}(e^{-2t} + 2t - 1)$

(13) $\dfrac{e^{at}-e^{bt}}{a-b}$ (14) $\dfrac{1}{2a^3}(\sin at - at\cos at)$

(15) $\dfrac{1}{3}(\cos t-\cos 2t)$ (16) $\dfrac{1}{54}(\sin 3t-3t\cos 3t)$

(17) $\dfrac{1}{4}t\sin 2t$ (18) $\dfrac{(2+(a-b)t)e^{-at}-(2-(a-b)t)e^{-bt}}{(a-b)^3}$

(19) $\dfrac{e^{-t}}{64}(1-e^{-8t}-8t.e^{-8t})$ (20) $\dfrac{1}{(b^2-a^2)}(b\sin bt - a\sin at)$

(21) $\dfrac{1}{45}(14e^{3t}-5+30t+36e^{-2t})$ (22) $f(t)=\cos t+\sin t$

(23) $f(t)=\dfrac{1}{4}(3e^t+2te^t+e^{-t})$ (24) $2e^t-e^{-t/2}\left[\cos\frac{\sqrt{3}}{2}t+\frac{1}{\sqrt{3}}\sin\frac{\sqrt{3}}{2}t\right]$

(25) $\dfrac{1}{8}\left[1+2t-\left\{\cosh\left(\frac{\sqrt{17}}{4}t\right)+\frac{1}{\sqrt{17}}\sinh\left(\frac{\sqrt{17}}{4}t\right)\right\}e^{-t/4}\right]$

(26) $\dfrac{w}{\left(1-e^{-2\pi s/w}\right)(s^2+w^2)}$ (27) $\dfrac{1}{s^2T}-\dfrac{e^{-sT}}{s(1-e^{-sT})}$

(28) $\dfrac{1}{s^2}\tanh(\frac{1}{2}as)$ (29) $\dfrac{Ew}{(s^2+w^2)}\coth(\pi s/2w)$

(30) $\dfrac{e^{-\pi s}-(1+\pi s)e^{-2\pi s}}{s^2(1-e^{-2\pi s})}$ (31) $\dfrac{k}{s}\tanh(\frac{as}{2})$

EXERCISE 5.5

(4) $y(t)=e^t$

(5) $x=\left(2+\dfrac{w}{w^2+1}\right)e^{-t}+\dfrac{1}{(w^2+1)}(\sin wt - w\cos wt)$

(6) $y(t)=\dfrac{1}{3}+\dfrac{2}{3}e^{-3t}$ (7) $y(t)=-\dfrac{1}{5}\cos t+\dfrac{2}{5}\sin t+\dfrac{6}{5}e^{-2t}$

(8) $y(t)=-\dfrac{3}{10}e^{-3t}+\dfrac{3}{10}\cos t+\dfrac{1}{10}\sin t$ (9) $x(t)=-\dfrac{1}{16}-\dfrac{1}{4}t-\dfrac{15}{16}e^{4t}$

(10) $x(t)=\dfrac{11}{4}e^{2t}-\dfrac{3}{4}-\dfrac{t}{2}$ (11) $x(t)=1+\dfrac{1}{4}(e^{-t}-e^t+2te^t)$

(12) $y(t)=\dfrac{1}{2}[t\cos t-(1-t)\sin t]$ (13) $y(t)=\cos 2t+3\sin 2t$

(14) $y(t)=-\dfrac{1}{2}e^{2t}+\dfrac{14}{9}e^t+\dfrac{19}{36}e^{4t}$

(15) $y(t) = -\frac{3}{10}\cos t - \frac{1}{10}\sin t + \frac{1}{2}e^{t} + \frac{4}{5}e^{-t/2}$

(16) $y(t) = \frac{11}{12}e^{-t} - \frac{20}{21}e^{-4t} + \frac{1}{28}e^{3t}$

(17) $y(t) = \frac{1}{10}\left(e^{-t} - e^{-2t}\cos 3t + \frac{19}{3}e^{-2t}\sin 3t\right)$

(18) $y(t) = \frac{1}{6}e^{-4t}(6 + 36t + t^3)$

(19) $\frac{1}{5}\left[(2\cos t + \sin t) + e^{t}(\sin t - 2\cos t)\right]$

(20) $\frac{1}{125}\left[(16 - 85t)e^{-3t} + (40t - 16)e^{2t}\right]$

(21) $y(t) = 2e^{t} - 3te^{t} + \frac{1}{2}t^2e^{t}$

(22) $y(t) = 2t + 3 + \frac{1}{2}(e^{3t} - e^{t}) - 2e^{2t}$

(23) $y(t) = \frac{1}{2w}\sin wt$

(24) $x(t) = \frac{4}{9}\sin 2t - \frac{5}{9}\sin t - \frac{1}{3}t\cos 2t$

(25) $y(t) = \frac{11}{3}e^{-t}(\sin t + \sin 2t)$

(26) $y(t) = \frac{1}{3}(5e^{t} + e^{-2t}) - e^{-t}$

(27) $y(t) = \frac{1}{2}(\cos kt + \cosh kt)$

(28) $y(t) = 2 + 2e^{t} - e^{-t} - \sin t.$

(29) $y(t) = \frac{3}{8}\sin t - \frac{1}{8}\sin 3t$

(30) $y(t) = (t\sin t + \cos t - 1)e^{t}$

(31) $y(t) = \frac{1}{2w}(2w\cosh wt + 4\sinh wt + t\sinh wt)$

(32) $y(t) = (3t + 2)e^{-2t} + (t - 2)e^{-t}$

(33) $y(t) = \frac{1}{18}\left[\sin 3t - 3t\cos 3t\right]$

(34) $y(t) = \frac{1}{2}e^{-2t} - e^{-t} + \frac{1}{2}.$

(35) $y(t) = \frac{1}{5} - \frac{1}{4}e^{-t} + \frac{21}{20}e^{-5t}$

(36) $y(t) = \frac{-7}{50}\cos t - \frac{1}{50}\sin t + \frac{63}{50}e^{3t} + \frac{22}{25}e^{-2t}$

(37) $2e^{-t}\left[e^{-(t-\pi/2)} - \sin t - \cos t\right]u_{\pi/2}(t)$

(38) $(1 + 2e^{-t})u_0(t) - 2[1 - e^{-(t-1)}]u_1(t).$

(39) $\frac{1}{2}\left[\left(1 - e^{-2t}\right)u_0(t) - \left(1 - e^{-2(t-2)}\right)u_2(t)\right]$

(40) $(4e^{-4t} - 3e^{-3t})u_0(t) + \frac{1}{12}\left(1 + 20e^{-3(t-2)} - 21.e^{-4(t-2)}\right)$

(41) $2\cos 2t.u_0(t) + \frac{1}{4}\left[1 - \cos 2(t-1) + 2\sin 2(t-1)\right]u_1(t)$

(42) $e^t u_0(t)+\frac{1}{2}\left[1-2e^{t-1}+e^{2(t-1)}\right]u_1(t)$

(43) $\frac{1}{3}\left(3e^{-t}-2e^{-3t}-1\right)u_0(t)+\frac{1}{6}\left[2+e^{-3(t-3)}-3e^{-(t-3)}\right]u_3(t)$

(44) $y(t)=\left[\frac{1}{17}-g(t)\right]u_0(t)-\left[\frac{1}{17}+e^{4\pi}g(t)\right]u_\pi(t)+\left[\frac{1}{17}-e^{8\pi}g(t)\right]u_{2\pi}(t)-\cdots$, where

$g(t)=\frac{e^{-4t}}{17}(4\sin t+\cos t)$

(45) $y(t)=\cos 3t.u_0(t)+\frac{1}{8}(\cos t-\cos 3t)[u_0(t)-u_\pi(t)+u_{2\pi}(t)-u_{3\pi}(t)+\cdots]$

(46) $y(t) = g(t-\pi)u_\pi(t) + g(t-2\pi)u_{2\pi}(t) + \ldots$, where $g(t) = ½ (t \cos t-\sin t)$

(47) $y(t)=\frac{1}{5}(1-g(t))u_0(t)-\frac{2}{5}(1-g(t-\pi))u_\pi(t)+\frac{2}{5}(1-g(t-2\pi))u_{2\pi}(t)+\cdots$,
where $g(t) = e^{-2t}(2 \sin t + \cos t)$

(48) $y(t)=\left(\frac{-3}{4}+\frac{t}{2}+2e^{-t}-\frac{5}{4}e^{-2t}\right)u_0(t)-a\left(\frac{1}{2}+g(t-a)\right)u_a(t)$

$-a\left(\frac{1}{2}+g(t-2a)\right)u_{2a}(t)+\cdots$, where $g(t)=\frac{1}{2}e^{-2t}-e^{-t}$

(49) $x(t) = (1-2t)e^{-2t}$, $y(t) = (1 + 2t)e^{-2t}$

(50) $x(t)=\frac{1}{27}+\frac{t}{9}-\frac{1}{27}e^{-3t}-\frac{2}{9}te^{-3t}$, $y(t)=\frac{4}{27}-\frac{2t}{9}-\frac{4}{27}e^{-3t}-\frac{2}{9}te^{-3t}$

(51) $x(t)=\frac{11}{4}\sin t+\frac{1}{12}\sin 3t$, $y(t)=\frac{11}{4}\sin t-\frac{1}{4}\sin 3t$

(52) $y(t) = t-\sin 2t$, $z(t) = \cos 2t$

(53) $x(t) = \cos t \cosh t + \sin t \sinh t$, $y(t) = \cos t \cosh t-\sin t \sinh t$

(54) $x(t) = -2(\cos 5t + \sin 5t) + 2e^{-2t}(\cos t + 7\sin t)$

(55) $x(t)=\frac{mc}{k}\sin\left(\frac{k}{m}t\right)$, $y(t)=\frac{mc}{k}\left(\cos\left(\frac{k}{m}t\right)-1\right)$

(56) $i_1=\frac{a}{p+w}(\sin wt+\sin pt)$, $i_2=\frac{a}{p+w}(\cos wt-\cos pt)$

(57) $x = a \cos \alpha t$, $y = b \cos \beta t$

(58) $x(t)=\frac{16}{175}e^{-5t}-\frac{13}{25}-\frac{2t}{5}+\frac{3}{7}e^{2t}$, $y(t)=-\frac{16}{175}e^{-5t}-\frac{12}{25}-\frac{3t}{5}+\frac{4}{7}e^{2t}$

(59) $x(t)=-\frac{1}{2}(3\cos t-t^2+3)$, $y(t)=\frac{1}{2}\left(-\cos t-\frac{t^2}{2}+1\right)$

INDEX

ABOUT THE AUTHORS

Dr. A Chandra Babu is the head of the post graduate department of Mathematics of The American College, Madurai. He has recvied Ph.D in Applied Mathematics from Madurai Kamaraj University. He has studied at the Indian Institute of Science, Bangalore and has worked in the department of Mathematics and Computer Science, at the Beloit College, Wisconsin, USA. He has thirty-five years of experience in teaching Mathematics and Computer Science at both undergraduate and postgraduate levels. His area of specialization is applications of Stochastic Processes in Stock Market Analysis and he has published many papers.

Dr. C R Seshan is the former head of the post graduate department of Mathematics of The American College, Madurai. He has received Ph.D in Operations Research from the Indian Institute of Science, Bangalore. He has thirty-five years of experience in teaching Pure and Applied Mathematics at both undergraduate and postgraduate levels. He is a member of the board of studies and Chairman of the board of examiners of several Universities and Autonomous Colleges. His area of specialization is Optimization Techniques and he has published several papers in national and international journals.